£85.00p

2317

KU-587-392

S
540
.A2
.ARN

Social Science Library
Oxford University Library Services
Manor Road
Oxford OX1 3UQ

WITHDRAWN

WITHDRAWN

300552164Q

UNIVERCITY OF OXFORD
INTERNATIONAL DEVELOPMENT
CENTRE LIBRARY
QEH

DATE CATALOGUED 16. 7. 90

CLASS-MARK L3
AR

WITHDRAWN

WITHDRAWN

AGRICULTURAL RESEARCH
AND TECHNOLOGY TRANSFER

AGRICULTURAL RESEARCH
AND
TECHNOLOGY TRANSFER

I. ARNON

Ing. Agr., Ph.D.

*Former Director, Volcani Institute of Agricultural Research;
Professor of Agronomy, Hebrew University of Jerusalem,
Israel*

WITHDRAWN

ELSEVIER APPLIED SCIENCE
LONDON and NEW YORK

ELSEVIER SCIENCE PUBLISHERS LTD
Crown House, Linton Road, Barking, Essex IG11 8JU, England

Sole Distributor in the USA and Canada
ELSEVIER SCIENCE PUBLISHING CO., INC.
655 Avenue of the Americas, New York, NY 10010, USA

WITH 10 TABLES AND 27 ILLUSTRATIONS

© 1989 ELSEVIER SCIENCE PUBLISHERS LTD

British Library Cataloguing in Publication Data
Agricultural research and technology transfer
1. Agriculture. Research
I. Arnon, I. (Isaac), *1909–*
630'.72

ISBN 1-85166-275-8

Library of Congress Cataloging-in-Publication Data
Arnon, Itzhak, 1909–
    Agricultural research and technology transfer/I. Arnon.
    p. 00 cm.
    Bibliography: p.
    Includes index.
    ISBN 1-85166-275-8
    1. Agriculture—Research. 2. Agriculture—Technology transfer.
I. Title.
S540.A2A755 1989
630'.72—dc 19

No responsibility is assumed by the Publisher for any injury and/or damage to persons or property as a matter of products liability, negligence or otherwise, or from any use or operation of any methods, products, instructions or ideas contained in the material herein.

Special regulations for readers in the USA

This publication has been registered with the Copyright Clearance Center Inc. (CCC), Salem, Massachusetts. Information can be obtained from the CCC about conditions under which photocopies of parts of this publication may be made in the USA. All other copyright questions, including photocopying outside of the USA, should be referred to the publisher.

All rights reserved. No parts of this publication may be reproduced, stored in a retrieval system, or transmitted in any form or by any means, electronic, mechanical, photocopying, recording, or otherwise, without the prior written permission of the publisher.

Photoset by Enset (Photosetting), Midsomer Norton, Bath, Avon
Printed in Great Britain at the University Press, Cambridge

To my children—Dan, his wife Naomi and Gideon; my grandchildren—Orna and her Yochanan, Tamar, Yoram and Amir; and to my great-granddaughter, Shai, for all of whom my greatest wish is peace in their time.

Contents

Part Three: Human Resources in Agricultural Research

Part Four: Organisational Structure at the Institutional Level

Part Five: Administration in the Service of Research

Part Six: The Transfer of Technology

Preface

Agricultural research was probably the first and is the most widespread form of organised research in the world, and one in which both the most developed and underdeveloped countries are engaged. Whilst most forms of research activity, such as in the field of medicine, have world-wide application, agricultural research, by its very nature, has to be regional; practically no research finding can be adopted without studying the results of its application under the infinite number of ecological situations with which the farmers of the world are faced.

The improvement of agricultural production is the essential first step whereby developing countries can hope to raise their standard of living. Research is therefore an activity in which no underdeveloped country can afford *not* to engage; nor can countries in which agriculture has reached a high level of development and sophistication afford to neglect agricultural research. It is not because of inertia or vested interests that highly industrialised countries maintain, mostly at public expense, a costly and complex infrastructure for agricultural research. Even when problems of overproduction weigh heavily on the economy, agricultural research is considered the essential key to further progress: the objectives and goals are simply changed and adapted to the needs of the economy.

All the more surprising the fact that the organisation and management of this vast and complex activity are so haphazard. In almost every country, the agricultural research organisation has 'grown up' from small beginnings, without this 'growth' having been planned or directed. This has resulted in innumerable organisational forms, different for each country. The stock explanation for this state of affairs is 'that each country has developed the agricultural research organisation adapted to its needs'. A dispassionate analysis of the situation will, however, usually indicate that the multitude of organisational forms is the result of *lack* of planning and not of planning, and that inter-departmental power politics, institutional prestige considerations and personality problems have had a greater hand in shaping the organisation, than has planning according to the specific needs of the country.

Most developing countries have either remnants of former research services, or are organising their agricultural research starting, from scratch.

During the first period after its inception, the organisation of agricultural research in Third World countries has generally been in a state of flux, changing with almost every change in the political constellation. It has suffered from a lack of understanding by the political leadership, hampered by excessive bureaucracy, hamstrung by lack of adequate facilities and by insufficient budgets. Linkages between research institutes and universities, liaison with extension services and contacts with farmers have generally been weak. Under these circumstances, research efforts have had little impact on the economy, reinforcing the negative image of research in the minds of the political leaders.

In recent years, the leadership of the agricultural research organisations in a number of developing countries are showing increasing concern with the poor image presented by their institutions, their limited contribution to the advancement of agriculture, and their limited effectiveness in achieving their declared goals. Re-evaluations and reorganisations of the national research systems follow each other, sometimes with counterproductive results. The choice they face is whether to let agricultural research grow haphazardly, in response to temporary situations and pressures, or whether to develop it according to a well thought-out organisational blueprint, planned according to the needs of the country and implemented gradually as manpower and resources become available.

Developed countries would probably be justified in taking a hard look at the elaborate and expensive infrastructure for agricultural research that has developed over the years. They might consider whether a planned reorganisation would not be justified, and whether public control of research policy is effective in assuring its orientation towards the needs of the community.

The management of the research organisations at all its levels is, in most cases, in the hands of veteran agricultural research workers who have risen from the ranks. This is as it should be. However, here we have people who, by training and inclination, have usually been conditioned to averseness to administration in all its manifestations. They are then made responsible for managerial activities in an extremely complex field, for which they have had little or no training whatsoever, and for which their only qualifications are their individual character traits and standing with their research colleagues. Administrative understanding is usually incidental, and rarely present.

The change-over from an activity in which the individual was highly competent to one for which he does not have even the rudiments of essential know-how may be a traumatic experience. A frequent solution is to place the effective management of the research organisation in the hands of a trained administrator, who rarely has experience in, or understanding of, research, whilst the veteran research worker remains a figure-head; this results in a situation fraught with difficulties and tension.

This was the kind of situation I faced when, in 1958, I was appointed Director of the Volcani Institute of Agricultural Research (later renamed the National Agricultural Research Service). For the first time in my professional life I found that I had undertaken an assignment for which I had no training, formal or otherwise.

My situation in this respect was not unique, as training for agricultural research management simply did not exist at the time.

My response to this dilemma, conditioned by long years as a research worker, was to search in the literature for guidance. In vain. For, with the exception of a few descriptive case histories of agricultural research in a limited number of developed countries, there was simply nothing to be found on the subject.

This dearth of literature on the management of agricultural research contrasted with the situation regarding management of industrial research. Though a much newer field, far less widespread, and infinitely more segmented, a considerable amount of literature on the organisation and administration, as well as planning of industrial research, was available to the scientist-administrator. This literature, in turn, had drawn largely on studies on general organisation and management problems, which had appeared since the beginning of the century, and in which the basic principles of management were discussed.

After an intensive study of this important source of information, much of which is relevant to the management and planning of agricultural research, and drawing on my own personal experience in the field, I prepared a text, published in 1968 by Elsevier Publishing Company Ltd, *Organisation and Administration of Agricultural Research*.

Since then the situation has changed beyond recognition. Numerous congresses, symposia, colloquia, workshops and missions have been devoted to all aspects of agricultural research management, organisation and planning. These have resulted in a veritable flood of publications, proceedings, reports, case studies, manuals, bulletins and a limited number of books. The spiralling of research costs, the increasing difficulties in mobilising research personnel for agriculture, the need to advise developing countries in organising national research systems and making the most efficient use of their limited human and material resources, have lent a special urgency to these activities.

Considerable effort has been invested in developing methods for the rational planning and programming of agricultural research. New concepts of the role of agricultural research have resulted in new or revised approaches such as pre-extension and Farming Systems research; training in agricultural research management has become available; the International Research Centres have become a major factor in their own right and in strengthening the national research systems.

The twenty years since the publication of the book mentioned above, was a period during which I had the opportunity of applying and testing in practice the ideas and proposals I had presented at the time. I also had the privilege of serving as a consultant on agricultural research and development for several international organisations and national research systems in a number of developing countries in Asia, Africa, and Latin America, which had given me an insight in the difficulties facing Third World countries and a feeling of personal involvement.

The time had come for a new review of the situation and the drawing of the necessary conclusions. This time, there was no need to search the literature on industrial research for guidelines on agricultural research management. Perusing the numerous texts relating directly to our subject and evaluating the viewpoints and experiences presented has required a complex, time-consuming but highly satisfying effort in preparing this text.

All the foregoing explains only the first part of the title of the present work. The reader may well ask: what justification does an agricultural research worker have to write on the complex subject of technology transfer, a multidisciplinary activity for which there are experts in extension methodology and communications, economists, sociologists and anthropologists, who are certainly more qualified? I therefore feel the need to justify this intrusion into an area in which I have not been professionally active, but to which I have devoted much thought and effort in promoting close links between the research and extension systems in my own country and elsewhere.

In the course of my work as a consultant, I could not help but observe how little impact agricultural research in most developing countries was having on the agriculture practised by the majority of farmers. To concern oneself with agricultural research whilst neglecting the reasons for its lack of impact, appeared to me to be an exercise in futility.

In every country to which I came, I would tell my local counterpart that before visiting agricultural experiment stations and the faculties of agriculture, I wished to see at first hand in how far the results of agricultural research were being adopted and by whom. This request usually caused surprise, and sometimes even consternation, and whilst it was never refused, there usually was considerable procrastination in compliance. One example will explain the relevance of this insistence on an unorthodox approach by a consultant.

On a mission to Mexico, I made the usual request, namely to visit typical farmsteads of various categories of farmers; I was told (after several mañanas) that I would have the opportunity to meet as many farmers as I wished, on a field day organised at one of the regional experiment stations. The results of highly relevant research shown at the station were really interesting. Many of the participating farmers arrived in their Cadillacs and Buicks, and a few even in their private planes. On the weekend, I asked a Mexican colleague to take me to some ejidos for a 'private' visit. At the first one we reached, I asked a group of farmers how often the extension officers visited the ejido. The answer was 'Jamas, somos la gente mas olvidada en México' (never, we are the most forgotten people in Mexico). It is encounters of this kind that influence one's perceptions and attitudes more than theories and arguments.

In the last two decades, extension methodology has been the subject of many seminars, workshops, colloquia, etc. which have resulted in a considerable volume of literature. However, to the best of my knowledge, rarely, if ever, have research workers participated in discussions on the one topic that has always been stressed as being of considerable importance: the need for effective linkage between research and extension, as a vital factor affecting the ability of both agencies to achieve their respective goals. Similarly, extension workers rarely participate in the discussions at meetings on research organisation and methodology organised by research workers.

I therefore felt that the time had come to treat the two topics, of research and extension, within the same framework. In a work that has no pretension to serve as a manual for extension workers, the lack of expertise of the author in extension methodology does not seem to be an important constraint. I am convinced that an understanding of the objectives, problems and viewpoints of both research and extension systems is essential for researchers and extension workers alike, if they are to achieve the fruitful relationship between the two systems so important to both. It is my sincere hope that this book will contribute to this mutual understanding.

The development of new technologies and their efficient transfer to the majority of the farmers is a complex process, dependent for its success on many political, social, economic and institutional factors. Therefore, the two vital activities of research and technology transfer cannot be effective and achieve their goals, if the political climate is biased in favour of urban development and industrialisation at the expense of agriculture; if policies are adopted that favour a small sector of the farming community and neglect the vast majority consisting of small farmers; if the infrastructure of markets, transport storage, etc. is deficient; if the essential services of health, education and other social amenities are not provided; if economic incentives, such as credit on reasonable terms, are not available to those who need them most; if bureaucracy and corruption are rampant; and if the vast majority of farmers are not helped to organise themselves, so as to be able to assert their just demands for equitable treatment.

The provision of these prerequisites for the modernisation and progress of agriculture is a vast subject, and though closely related to the twin topics of research and technology transfer, it is beyond the scope of this work, and has been treated in a companion volume.[†]

[†]Arnon I. (1987). *Modernisation of agriculture in developing countries: Resources, potentials, and problems* (Second Edition). John Wiley & Sons, Chichester.

Acknowledgements

This book has drawn largely on numerous studies and case histories of agricultural research and technology transfer worldwide that have been published in recent years. The sources of the material presented here are recognised in the usual way, in the lists of references at the end of each of the six parts of the book. However, I have drawn on the works of a number of specialists in various fields to an extent to which a simple reference cannot do adequate justice. I wish therefore to acknowledge my debt in particular to the following: H. Akino, E.R. de A. Alves, E. Ampuero, J. Casas, D. Elz, R.E. Evenson, Y. Hayami, H.R. Jain, R. Lecomte, B. Nestel, M. Piñeiro, P. Pinstrup-Andersen, N. Röling, V.W. Ruttan, I. Skorvgaard, B. Stavis, B. Swanson, E.J. Trigo, A. Weber, D.E. Van de Zaag, P.J. Zuurbier.

I also wish to express my appreciation to A. Bloom, G. Corley, D.W.E. Shannon and E.B. Knipling, who have read various chapters of the manuscript, for their valuable comments, and M. Beringer, G. Cooke, A. Dam Kofoed, Paul C. Ma, D. Malquori, A. Van Diest and A. Vez, for their personal communications on agricultural research and technology transfer in their respective countries.

I also wish to thank those authors and publishers who have generously allowed me to use copyright material from their publications; the courtesy is recognized in each case; as well as all those who have made their publications available to me. In particular, ISNAR has provided me all their publications, which have been a major source of information and documentation.

Finally, I wish to express my appreciation for the competent and painstaking work of the editorial staff of Elsevier Science Publishers Ltd.

I. ARNON

Part One

AGRICULTURAL RESEARCH SYSTEMS

Chapter 1

National Agricultural Research Systems: Selected Case Histories

This chapter presents a selected number of case histories representative of the various situations encountered world-wide: large and small countries; industrialised and agrarian countries; market economies and centrally directed economies; advanced and developing countries; former colonial powers and former colonies; mono-lingual and bi-lingual countries; countries in the temperate and the tropical regions, etc.

The lessons learnt from these case histories have contributed, to a large extent, to the discussions and proposals presented in the chapters that follow.

Great Britain

HISTORICAL DEVELOPMENT

Russell (1966) distinguishes five periods in the development of agricultural research in Great Britain, from its beginnings until the 1960s.

The first period: from the end of the 16th century to the end of the 18th century, research was pioneered by gifted amateurs, working on their own initiative, who established and recorded what were generally unrelated facts. The most important innovation of this period was the Norfolk rotation of clover, wheat and turnips. Arthur Young publicised the efforts of these innovators.

The second period: at the beginning of the 19th century, agricultural societies were being established with the object of defining and solving farmers' problems and protecting them against fraudulent practices.

3

These societies took the initiative in setting up laboratories for the analysis of soils and fertilisers and carrying out field experiments. During this period considerable progress was made in modern chemistry, which provided a basis for the subsequent development of agricultural science.

In 1843 the oldest agricultural research establishment in the world— the Rothamsted Experimental Station, was established by Lawes and Gilbert, funded at first by Lawes' phosphate fertiliser enterprise, and subsequently, after Lawes' death, by the Lawes Agricultural Trust.

The third period was characterised by the development of agricultural teaching at university level. The need for authoritative textbooks prompted systematic agricultural research, which though still voluntary and individual, started assuming an institutional character. During this time, the government provided funds for agricultural education but none for agricultural research.

The fourth period started shortly before the First World War. The Government was searching for means to overcome the depression in which agriculture found itself, and support for agricultural science was one of the means adopted for ensuring an increase in agricultural production. This support was, however, on a small scale, in the form of grants to universities and the privately sponsored institutes. It was decided that agricultural research should not be centralised, but divided among a number of institutions working in separate fields. Agricultural research became a full-time professional occupation. Contact with farmers during this period was very close, and the agricultural institutes also served as advisory centres.

Funding agricultural research with public money meant that some kind of official supervision became necessary. Starting from 1911, government grants to various institutes were made through a *Development Commission* and thus became known as State-aided Institutes. These grants were administered by the Departments of Agriculture of England and Wales, and the Department of Agriculture for Scotland (DAFS), respectively (Webster, 1970).

In 1931, the Agricultural Research Council (ARC) was established by Royal Charter to advise the Agricultural Departments on the programmes of the institutes and to act as scientific adviser to the Development Commission. A small budget was allocated to the council to initiate research in areas not covered by the research institutes.

The fifth period started after the Second World War, and ended in the early 1970s. In 1946 the Development Commission ceased to be concerned with agricultural research and the two Agricultural Departments assumed responsibility for the provision and the administration of grants for State-aided Institutes, acting on the advice of ARC. In 1956 the responsibility for the State-aided Institutes in England and Wales was transferred to ARC, whilst in Scotland it remained with the Department of Agriculture and Fisheries.

In 1959, the ARC was made responsible for food research, and it assumed control of three research institutes that had formerly been administered by the Department of Scientific and Industrial Research (Webster, 1970). Responsibility for the ARC's budget was transferred to the Department of Education and Science under advice of the Council for Scientific Policy.

These transfers, in addition to the establishment of the National Advisory Service in 1946, considerably weakened the link between researchers and farmers.

THE COMPONENTS OF THE NATIONAL AGRICULTURAL RESEARCH SYSTEM

In Great Britain, there are four agencies involved in agricultural research: the Agricultural and Food Research Council, the Ministry of Agriculture, Fisheries and Food, the Universities and the private sector (Fig. 1).

The Agricultural and Food Research Council

The Agricultural Research Council (ARC) formally changed its title in 1983 to The Agricultural and Food Research Council (AFRC) 'to reflect more clearly its aim and objectives, and in particular the increasing importance it attaches to research on food'.

The autonomy of the Agricultural Research Council is attributed to 'an historical accident, going back to the First World War and Lord Haldane's doctrine that government departments should not be allowed to control basic research because they would not understand it' (*Nature* (1981), **289**, 3).

Constitution
The AFRC consists of a Chairman, a Deputy Chairman and Secretary, and 18–21 members, appointed either by the Secretary of State for

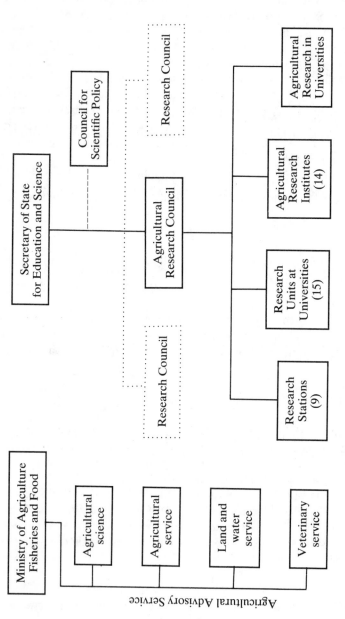

Fig. 1. The organisation of agricultural research in Great Britain. (From Robinson (1970), reproduced by permission of Ministry of Overseas Development, Great Britain.)

Education and Science, the Minister of Agriculture, Fisheries and Food, and the Secretary of State for Scotland.

Functions
The functions of the Council can be broadly indicated by stating that it is charged with:

—advancing scientific knowledge relevant to the agricultural, horticultural and food industries;
—applying this knowledge to increase the efficiency of these industries and those that support them, including the chemical and engineering industries;
—safeguarding and improving the quality of food for the community;
—protecting the environment and improving the welfare of animals.

These objectives are implemented by undertaking the following activities:

—organising and co-ordinating agricultural and food research in Great Britain;
—establishing and developing institutes for investigation and research relating to the advancement of agriculture, food production and processing;
—making grants for complementary investigations and research in academic institutions;
—supervising the research of the eight AFRC institutes located in England, Wales and Scotland and administering the funds provided by the Department of Education and Science, and the MAFF for research at these institutes;
—advising the Department of Agriculture and Fisheries for Scotland on the programmes of the five Scottish Agricultural Institutes.

No central direction of research is attempted by the Council, whose main function is to review research in progress and to encourage new research if deemed necessary. However, the Council sees it as its function to take positive measures to encourage work on problems of practical importance. A large measure of freedom is allowed scientists who receive grants from the Council to pursue research of their own choice (ARC, 1963).

Responsibility for the detailed research programme for each research establishment is vested in the senior staff actually engaged on the research, within limits of the financial allocation made available to them.

A review of the programme in each research establishment is made every five to six years by *ad hoc* 'visiting groups' of scientists appointed by the Research Council, who assess the programme of the establishment as a whole, and the work of the individual researchers. On the basis of their findings, the Council is in a position to evaluate the contribution made, in basic and applied research, in relation to its importance to agriculture, and continued financial allocation can then be based on the merits of the programme.

Scope of the Research Council's work

In the 1970s the Research Council administered approximately 75 per cent of the national funds available for agricultural research. As a rough indication of the scale of its work: its research institutes and units employed about 2500 scientific and experimental officers, plus the necessary supporting staff (Webster, 1970). In 1985/86, AFRC supervised research to the value of £112.3 million, including £7.1 million in support of agricultural research workers in universities.

The structure of the Research Council (Webster, 1970).

The Council's *Secretariat* carries out the administration of its affairs. It comprises the Secretary, the Deputy Secretary and Director of the central office, a team of seven scientific advisers to the Secretary, and a headquarters staff of about one hundred. The staff is organised in three divisions dealing with: establishment and finance, research on animals, and research on plants and soils, respectively. A Management Board was established in 1986.

The Council has set up five *Standing Committees*, concerned respectively with research on Plants and Soils, Animal Husbandry, Agricultural Engineering, Farm Buildings, and Farm Effluents. Each of these committees is composed of some members of the Council, and additional experts (mainly from the universities), representatives from the NAAS and DAFS, and some farmers. Each committee examines the research programmes and annual reports of the relevant institutes and considers the applications for research grants.

Each Standing Committee has under it a number of *Technical Committees* which meet at appropriate intervals, and are concerned with specific commodities. The composition of these committees is on similar lines as that of their respective Standing Committees; the members are, however, more narrowly specialised. They keep under review the progress in research in a particular subject and make recommendations to the appropriate Standing Committee, drawing attention to any research needs which are not being met and to areas where the research effort needs to be intensified or better co-ordinated. A committee may find it desirable to appoint a Working Party, on which a few members of the committee work with additional experts to make a more detailed examination of particular problems or fields of work.

The research establishment of the Research Council
The components of the Council's research establishment comprise the State-aided Institutes, the Council's Institutes, AFRC Units, and individual Research Grants.

State-aided Institutes are the older institutes, formerly funded by individuals or associations, but subsequently becoming dependent on government funding.

AFRC institutes were set up by the Council to deal with areas of research not covered by the older institutes. The first of these were established to deal with animal diseases, animal physiology and animal breeding. More recently, a Food Research Institute and a Meat Research Institute were established, reflecting the Council's increased concern with food problems.

AFRC units are operated at the universities on behalf of the Council. The object of setting up a unit is to enable a university scientist, who has established a reputation in a particular field of science of potential interest to the Council, to develop his research more quickly and extensively. He is given full freedom to plan his research programme, which is not subject to monitoring by the Council. The Units are usually not permanent, but are disbanded when the scientist around whom the Unit was built retires.

AFRC grants are mainly for scientists in universities, but occasionally in the institutions, to provide short-term assistance for projects which are of interest to the Council. These grants are usually for three years.

The Ministry of Agriculture, Fisheries and Food (MAFF)

MAFF carries out research in its own laboratories into veterinary problems, plant pathology, pest control, fisheries, and botanical matters. Within the Ministry functions the National Advisory Service, which in addition to its advisory role, carries out an extensive experimental programme on its own chain of experimental husbandry farms and horticultural stations.

Animal health. A Central Veterinary Laboratory is involved in a wide range of research into animal diseases. Another laboratory is mainly concerned with poultry diseases. The Veterinary Investigation Service operates 21 centres which carry out co-operative experiments with the central laboratories, including field trials and disease surveys.

Pest control. The Pest Infestation Control Laboratory is concerned with research on the control of harmful mammals, birds and insects and research on stored products infestation.

Plant pathology. A Plant Pathology Laboratory is responsible for pest and disease assessment, general plant pathology and entomology.

Plant physiology. The Royal Botanical Gardens at Kew undertakes research on plant physiology and the identification and classification of plants.

Agricultural Advisory Service Experimental Centres (Gardner, 1970)
With the establishment of the *National Agricultural Advisory Service* (NAAS) in 1946, a chain of 13 experimental husbandry farms and nine horticultural experiment stations were established throughout the country, to carry out experimental and development programmes designed to help in applying the results of research in a variety of ecological conditions and to investigate local problems.

All types of agricultural commodities produced in England and Wales are investigated; each centre specialises in those commodities which have the greatest economic importance in the region in which the centre is located.

Function. The major objective of the centres was originally to provide information for the NAAS field advisers. The functions of the centres have changed considerably from this original concept. In the early days it was assumed that the findings of research could be tested and evaluated individually, provided that the tests were undertaken on a sufficiently wide scale. Much of the work involved testing new varieties developed at the research institutes, new agrochemicals for plant protection, fertilisers, etc. However, farming systems are becoming more specialised and sophisticated and the centres are now requested not to limit themselves to testing and evaluating the findings of research, but to establish how these findings can be integrated into improved production systems

(Gardner, 1970). The need for the development of more economical production systems for use on a national scale might appear to limit the ability of the experimental centres to study local problems. However, with the centres located strategically throughout the country, they can fulfil a vital local role, while functioning as part of a national network (Webster, 1970). It is of interest to note that this concern with 'production systems' antedates and is identical in concept to 'Farming Systems Research' which is presently being widely promoted.

Direction. Each centre has an *Advisory Committee* appointed by the Minister of Agriculture, Fisheries and Food, to advise on the management of the centre, to identify local problems, and to comment on the programmes formulated by the NAAS Experiments Commodity Sub-committees.

The membership of the Advisory Committee includes the Director of the Centre, representatives of the local NAAS, nearby universities and AFRC institutes, and a number of progressive local farmers.

Research programmes. Centre directors and their staffs are encouraged to develop original ideas in collaboration with the NAAS Subject Matter Specialists (SMSs), and colleagues at the AFRC institutes.

Proposals are sometimes received at headquarters from farmers' organisations, while proposals from individual farmers generally arise in the course of discussions with the centre director or a member of his staff. These proposals are given sympathetic study, and are taken into the programme where this appears justified on economic grounds.

All research proposals are cleared by the NAAS directorate and not by reference to committees.

Some types of investigation are conducted on commercial farms rather than on an experimental centre; for example, where the problem needs to be studied under specific environmental conditions, or else when a new system of production or feasibility testing is best carried out under ordinary farming conditions.

This kind of work is carried out by the field advisers in close co-operation with their colleagues from the experimental centres (Gardner, 1970).

Agricultural research at the Universities

Darling (1970), writing of the attitude of the average English 'redbrick'

University in the past, states that 'they adopted an unrealistic "ivory tower" approach towards the rest of the national community', and 'considered that in the interests of academic freedom the University should not be involved in any way in local affairs'. Even the agricultural faculties were influenced by this approach, and in particular, the natural sciences departments of these faculties pursued scientific research for its own sake. These attitudes of course created a sense of alienation from the farming community, but this trend was reversed in recent years.

The involvement of AFRC in University research
Mention has already been made of the AFRC Units at the Universities. The council, however, has no part in the allocation of funds for the general financing of the Universities. This is the responsibility of the University Grants Committee.

A number of the AFRC institutes are administered by one or other University. In other cases, the research institutes provide lecturers to the Universities. There is generally close contact between research staff of the institutes and University researchers. Proximity to a University is an important criterion in siting a new Institute (Darling, 1970).

Private sector research

A number of chemical, machinery and feeding-stuffs firms carry out agricultural research, in addition to financing research projects at the Universities and other institutions.

In February 1986, the Government announced its intention of transferring responsibility for certain institutes to the private sector. This decision has already been implemented in a number of cases such as that of the Cambridge Plant Breeding Institute.

RESEARCH POLICY

Division of research responsibilities between AFRC and MAFF

In the 1960s, a situation had developed whereby a distinct division of research responsibilities between the Research Council and the MAFF was in evidence. Much of the research supported by the Council was of a fundamental nature and removed from the immediate problems of agricultural production. Decisions on choice of research topics were made by the researchers themselves.

The research carried out by MAFF in its various laboratories, and by NAAS experimental centres, was applied and adaptive research. To justify this division of activities, a distinction was made between 're-search' (a Council activity) and 'experimentation' (a Ministry responsibility) (Ruttan, 1982).

This distinction between Council research activities and those of MAFF gradually became blurred, and at present the kind of work undertaken by the two major components of the national research system is almost indistinguishable.

Research planning at AFRC

The Council is guided on policy issues by three Research Committees: Animals; Food; and Plants and Soils, which work with the corresponding Research Divisions of the Secretariat.

Research planning of MAFF and its NAAS

A national committee known as the National Agricultural Advisory Service Experiments and Development Committee is charged with the general monitoring of the experimental farms and horticultural stations and with the planning of the overall strategy for the centres. This committee is composed of senior advisers, who are specialists in different branches of agriculture and horticulture, science specialists, representative directors from the experimental husbandry farms and horticultural stations, and representatives of the finance and administrative divisions of MAFF. At this level there is no direct representation of the Council or of farmers.

The Experiments and Development Committee concerns itself only with the broad issues; for example, ensuring that priority is given to the most urgent and important problems, allocating resources effectively and fairly to each of the centres, etc.

Formulating policy in regards to each commodity is the task of a number of Commodity Subcommittees. These subcommittees comprise representatives of the relevant AFRC institutes, directors of the husbandry farms and horticultural stations concerned with the commodity encompassed by each subcommittee, and regional specialist advisers. Farmers or growers are not represented on the Commodity Subcommittees (Gardner, 1970).

Linkages

The Commodity Subcommittees provide the major formal framework for liaison between the AFRC research institutes and the NAAS experimental centres. NAAS has also posted liaison officers at the AFRC research stations to provide an improved two-way link (Fig. 2).

There is also informal contact at all levels: advisory officers visit research institutes and researchers visit experimental stations. Representatives of both organisations serve together on many committees and *ad hoc* meetings.

Liaison with farmers is mainly through the Agricultural and Horticultural Advisory Councils serving England and Wales, and through the Scottish Agricultural Improvement Council. These three councils are composed mainly of practising farmers and growers. They review the whole field of British agriculture, and suggest which are the most important problems which should be brought within the research programmes and advisory activities of the government institutions. These Councils provide a common meeting place for the Agricultural Departments, AFRC, and the farmers, where they can exchange views on matters of common interest (Gardner, 1970).

The Rothschild Report

The research policy of the Council led to a situation in which MAFF as well as the farming sector, felt that the needs of agriculture were not being adequately covered by research, and that much of the research effort of the Council was irrelevant to these needs. In the late 1960s, the Government commissioned a report from Lord Rothschild (Head of the Central Policy Review Staff on Government research and development) and sought advice, through the Council for Scientific Policy, on the most effective arrangements for organising and supporting basic and applied scientific research. As this situation is not exactly specific to the United Kingdom, it is felt that a somewhat detailed discussion of the report is justified.

The Report's recommendations

The report submitted to the Government in 1971 (Rothschild & Dainton, 1971) is based on the principle that 'applied R and D—that is R and D with a practical application as its objective—must be done on a customer–contractor basis. The customer says what he wants; the contractor does it (if he can); and the customer pays'. In the case of agricultural research,

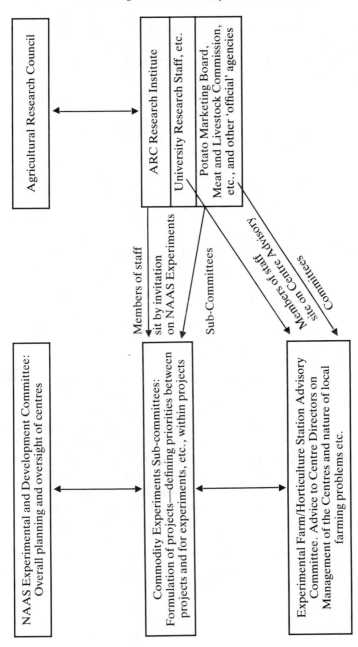

Fig. 2. Diagram of NAAS system for programming experiments and development work in co-operation with ARC and kindred organisations (From Gardner (1970), reproduced by permission of the Ministry of Overseas Development, UK).

the 'customer' is the Ministry of Agriculture, Fisheries and Food (MAFF) and the 'contractor' is the Agricultural Research Council (ARS).

The 'customer' should make, with advice or on his own initiative, the following decisions:

(a) that an R and D programme is needed to achieve a specific objective;

(b) how much can be spent on the programme;

(c) determine priorities between programmes.

The 'contractor' for agricultural research is the Research Council, which should not have the right to reject contracts for research requested by the MAFF without good reasons agreed with the latter.

Conversely, all contracts between the MAFF and the Council should, without exception, be paid for by the former. For this purpose, part of the funds provided by Government in the past directly to the Agricultural Research Council, has been transferred to the Ministry to help meet their needs for commissioned research.

An essential feature of this approach is that the Ministry should have its own Central Scientific Staff, consisting of a Chief Scientist, supported by a group of specialists, who are responsible for the scientific aspects of departmental policy, and make possible the work in partnership with their 'contractor', the Council.

In order to maintain some freedom to undertake research within the Council, not immediately related to a specific programme of work, it was proposed to allow a so-called 'general research surcharge' of about 10 per cent of the customers' programmes, which is wholly at the discretion of the Head of the Research Council. After considerable public debate, the Government accepted the basic principles outlined above.

In brief, the main change proposed was a redefinition of the respective responsibilities of the MAFF and the Council. In the past, the Agricultural Research Council was responsible both for research policy *and* its implementation and received its operating funds directly from Government. In the new situation, the Ministry defines its requirements, and therefore research policy, and has the means to pay for the execution of this policy; the ARS advises on the feasibility of meeting these requirements and undertakes the work.

Joint planning of agricultural research

In order to implement the Government decisions, the Council, the MAFF and the Department of Agriculture and Fisheries of Scotland

(DAFS) agreed to set up a *Joint Consultative Organisation* to advise them on programmes and priorities for government-financed research and development in food and agriculture. The organisation consists of Boards which advise the Council and the Departments, supported by more specialised committees.

Five R & D Boards were appointed:

(i) Commodity Boards
   Animals
   Arable Crops and Forage
   Horticulture

(ii) Special Subject Boards
   Food Science and Technology
   Engineering and Buildings.

The terms of reference of the Boards are as follows: 'To keep under review, with the help of specialist committees, the R & D needs of the British agricultural and food industries and of the consumer; to consider how far research findings have been successfully applied, in what fields existing knowledge could be used more effectively and what new knowledge should be sought; to assess and compare the Committee reports from the component fields; to arrange their recommendations in order of priority; to maintain liaison with other Boards; to report to the Agricultural Research Council and to the Agricultural Departments with recommendations regarding the initiation of new work and the use, continuation, modification or termination of existing work.'

Commodity Boards consider R & D requirements relating to the production, processing and utilisation of each commodity, including, where appropriate, their inputs.

The Special Subject Boards concern themselves with R & D requirements within their field for science and technologies generally relevant to the work on commodities but not normally specific to any single commodity. In addition, they may well be asked to provide advice to the Commodity Boards to help with determining requirements for specific commodities.

Each Board includes representatives of the several interests concerned with R & D, namely those of the farming and food industries, of science (including Directors of the research institutes), of policy, of the professional and technical services of the Departments, and of economics.

Each Board has set up a number of committees, whose membership is similar to that of the Boards, but with a stronger representation ot scientists, who constitute approximately half the membership. Each committee reviews its own special area and reports to its Board. Each Board then produces a composite annual report which is submitted to all three sponsors: the Council, the MAFF and DAFS.

The Joint Consultative Organisation has an advisory and not an executive function; it has no allocated budgets which it can disburse.

A Project System was introduced to provide the committees with basic information required for their work. In a computerised programme, all the research in progress is described, providing administrative information on the agricultural problem to which the project relates, and the action being taken to investigate the problem. Concurrently, a project-costing system was introduced which made it possible to evaluate the potential contribution of each project, in relation to its costs, to national needs. The programme also provides information on the current value of production of the various sectors of the industry and their economic prospects.

Subsequently, MAFF, DAFS and the Ministry of Agriculture in Northern Ireland all decided to adopt the same project system. Consequently almost all the agricultural research in the UK (except that at the Universities) is now classified in the same manner, and the information is available from one computer (Ulbricht, 1977).

The repercussions of the Rothschild reforms
When the 'Rothschild era' began, all the research councils affected were deeply offended. A heated debate followed the publication of what one eminent research worker called the 'infamous report'. In this debate three things were confused, according to Ulbricht (1977):

(a) whether scientists financed by government should be accountable for what they do and have a responsibility to meet the needs of the country;
(b) the application of the customer–contractor principle as a particular means of achieving the accountability;
(c) the wider issue of science policy: how can scientists through their research help to meet their country's needs when the government has no clearly defined long-term policy.

Very few scientists could reject the premise that they have the responsibility to meet the country's needs. The majority, however, rejected the principle of the customer–contractor relationship, and were unclear on the subject of science policy.

The basic premise of the Rothschild Report—namely, the customer–contractor principle—is now applied to most R & D funded by government departments. This principle was reaffirmed in a 1987 White Paper 'Civil Research and Development' in response to the first report of the House of Lords Select Committee on Science and Technology.

It should also be pointed out that in actual practice, the AFRC received 55 per cent of its budget in the form of contracts from the MAFF, and 45 per cent continued to come from the Department of Education and Science.

This was a significant departure from the recommendations of the Rothschild Report, thus allowing the Council to strike a balance between its support for applied research and long-term strategic and fundamental research (Ulbricht, 1977).

Evaluation of the reforms†
In the early years, the influence of the new policy initiated by the Rothschild Report was minimal. A note in *Nature* (1981) (beating a retreat from Rothschild, Vol. 289, p. 2), asks why there has been so little change in the patterns of AFRC's work since 1972 and assumes that the only reason is that MAFF's influence was too remote.

However, as time passed, the influence of the 'customers' has increased considerably, and reputedly, they are more satisfied than they were before 1972. The 'contractors' would probably still prefer to have absolute control over their funding, but a reasonable relationship between MAFF and AFRC appears to have developed, in which the 'contractors' have a significant influence on the research programme by indicating what science has to offer, and the 'clients' are satisfied that their problems are being dealt with. For example, following major cuts in the budget, the 'clients' had a very significant influence on how AFRC implemented the cuts it faced.

Presently, the main criticism voiced by the researchers regarding the 'new order' is that the customer–contractor principle has led to an increase in bureaucracy. The investment in additional administrative manpower was, however, not entirely unjustified. There were serious

†This section, and those that follow, are based (unless otherwise stated) on information kindly provided by the Chief Scientist of MAFF, Dr D.W.F. Shannon, and a former Chief Scientist, Dr G.W. Cooke. The opinions are those of the author.

weaknesses in the former administrative system which urgently required reform. The system could give precise information regarding the budget of a particular institute, but it would be in classical terms of staff costs, equipment, inputs, building overheads, etc. It could not, however, provide information vital for research planning and programming. For example, it was impossible to say what proportion of the total expenditure was devoted to various commodities or to major research areas, or to specific topics (Ulbricht, 1977). The Project System described above, was introduced in order to correct these deficiencies in the system; such a major reform could not be carried out without additional manpower.

Present trends and future perspectives

The Government of the UK is becoming increasingly concerned with the potential impact of R & D and whether it leads to a net benefit to the economy. Also, the Common Agricultural Policy of the EEC and the creation of surpluses of certain commodities have increased the need for government influence on the orientation of research and determination of priorities.

In an era of overproduction, increase in output is no longer the prime objective of agricultural research. There is an increasing concern with the efficiency of production, the quality of agricultural products and their use, and with the consequences of agricultural practices for the environment. The strengthening of strategic research to open up new possibilities and to meet contingencies became another important objective (AFRC, 1987).

These concerns, as well as the reductions in allocations for agricultural research (about 25 per cent in real terms since 1984/85), have resulted in a major review of agricultural research in the UK in recent years. They have already led to major redundancies and early retirement among research workers, and the closure of research facilities. The Council has already lost 20 per cent of its staff, and the former 27 English and Welsh Institutes have been reduced to eight, by closure, sale and consolidation of the remainder.

These constraints have also led to a reappraisal of the Council's research programme and the efficiency with which it is being carried out. In 1983, the Council published a Corporate Plan 1984/1988. The Plan indicated areas of research which were, in the Council's judgement, under-supported, were in balance, or were considered to be over-supported. It set out plans for the changes needed to redress the imbalance.

In particular, the Plan indicated that more resources should be allocated to food research. The change of name of the Council (Agriculture and Food Research Council) reflects the recognition that 'more attention needs to be given to the utilisation of agricultural output, and the role of the food processing industry in this respect' (AFRC, 1987).

The repercussions of budget cuts and policy changes are expected to increase over the coming years. Government is seeking greater involvement of the agricultural sector in R & D, both in the direction of research and its share of the financial burden, wherever the industry can obtain a direct benefit (Hayes, 1987).

This policy will result in further significant cuts in government funding of agricultural research, and a reorientation of government funds to strategic and basic research (the latter not necessarily to agriculture).

Naturally, this policy is not entirely to the liking of the farming sector, as exemplified by a letter from a prominent farmer and landowner to the *Farmers' Weekly* (20 November 1987), in which he concludes: 'There is a limit to the financial support available from outside sources. This subject must be well aired so that these reprehensible cuts can be reversed by objective persuasion' (Wilkinson, 1987).

In an editorial of the *Farmers' Weekly* (1987, **107**(19)), the Editor writes of further expected reductions in the agricultural R & D budget: 'This would be like bleeding an ailing body that has only just survived an acute haemorrhage. Exhortations for farmers and industry to take up the slack are now starting to wear thin. Diversification into novel enterprises, close and amicable liaison on environmental issues, means that hard-pressed farmers will need all the R & D help they can get in a changing world. The need and the challenge must not be ignored. Otherwise British consumers as well as producers will be the losers.'

The justification for two parallel research systems

An important question, not only in the context of the particular case history of agricultural research in Great Britain, but also of much wider significance, is whether, in view of the cuts in the funds available for agricultural research, there is objective justification to maintain the two separate research frameworks of AFRC and MAFF.

The original reason for the establishment of the Research Councils was to insulate research from political interference; apparently this was a valid concern at the time. It can hardly be still considered valid following the measures that have been taken to ensure that MAFF is able to have a major say on the way government funds allocated to research are used by the AFRC.

The original intention that AFRC would be concerned with basic research, whilst the MAFF research establishment would devote itself to applied research has long since been invalidated by developments in both institutions, and the kind of research carried out by AFRC, is, on the whole, indistinguishable from that of MAFF.

It is true that the existing research system in the United Kingdom has produced some of the best agricultural research in the world. The question raised above refers only to the cost-effectiveness of a dual system, with its inevitable overlapping and duplication of functions. This aspect is particularly relevant at a time when significant reductions in research funding are being made in many countries. This subject will be treated in more detail in Chapter 3.

The Netherlands

HISTORICAL BACKGROUND

The first university in the Netherlands was established in 1575, and ever since, the Dutch government has been committed to science (De Zeeuw, 1985).

The first agricultural experiment station in Holland was established in 1877 at the Agricultural School in Wageningen. Within a very short period, it became apparent that a far greater research effort was needed in order to ensure agricultural progress. A decision was therefore taken to increase agricultural research facilities. However, it was considered undesirable to create a strong central research institute by expanding the experiment station at Wageningen, and preference was given to establishing experiment stations in each of the agricultural regions of the country. Each of the experiment stations was subsequently split into two separate departments, for basic research and routine research respectively, each with its own personnel.

The next stage was a decision to establish, in Wageningen, specialised institutes such as those for farm mechanisation, phytopathology, plant breeding, etc.

In 1917, the agricultural school in Wageningen achieved college status. The college was vested with the threefold responsibility for teaching, research and extension in agriculture. This decision, however, was firmly opposed by the Senate and teaching staff of the faculty. Their objections were:

(a) Concern with practical problems of agriculture would adversely affect the scientific character of the college.
(b) The professors in charge of the research institutes would have to devote too much of their time to extension work.
(c) The professors would not have sufficient freedom in the choice of subjects for research.

And so, the first attempt in Europe to achieve the integration of education, research and extension died before birth.

Government was not alone in its awareness of the importance of agricultural research (and extension) for progress in agriculture. In the 1930s, the farmers' organisations began setting up their own applied research facilities, for example in sugar beet production and in animal nutrition (Zuurbier, 1983).

It was during the period 1945–60 that the research system in the Netherlands marked its major development; the establishment of numerous research institutions constituted a joint endeavour by the authorities and the agricultural sector. The system that was established during this period has remained largely unchanged since then (Zuurbier, 1984).

COMPONENTS OF THE NATIONAL RESEARCH SYSTEM

Agricultural research is carried out in the Netherlands under the aegis of the Ministry of Agriculture and Fisheries (MAF), the Agricultural Council TNO, and various other research councils, several universities, jointly by government and farmers' organisations and by the private sector.

The MAF runs 34 research institutes and experiment stations. In addition, MAF participates in the financing of Regional Research Centres.

The National Council for Applied Agricultural Research TNO maintains 32 research establishments in co-operation with MAF. The research of the TNO institutes is mostly oriented to topics such as processing of agricultural products; biological, biochemical and chemical products, nutrition and health.

The TNO institutes are partly dependent on contract research, for clients who pay for the services provided. Government contributes about 85 per cent of the budgets of these institutes (Zuurbier, 1983).

Joint government and crop associations' experiment stations

The crop associations are groups in which the farmers' organisations, the industrial and the commercial interests concerned with a specific crop or group of crops are represented. A member of the Extension Service, and occasionally an expert from the University, are co-opted into the Boards of each of these associations (Van Lennep, 1958).

The crop associations raise funds for research projects in which they are most directly interested. Farmers have to pay a special levy for this purpose. The government matches these funds on a fifty-fifty basis.

The stations are governed by Boards comprising representatives of the farming community and the Ministry; the farmers exercise a decisive influence on the stations' research programmes (De Zeeuw, 1985).

The Experiment Stations are usually situated in a main area of the crop production with which they are concerned. They are commodity or farming-type oriented, and deal with specific aspects of a certain branch or sector of agriculture (Van de Zaag, 1985).

The crop associations also fulfil an advisory role, by indicating the problems they consider require urgent attention. These proposals are submitted for consideration to the appropriate branch of the National Council for Applied Agricultural Research; they are then forwarded, with appropriate comments, to the National Council, which decides on the relative allocation to be made to each of its branches, within the framework of the budget approved by the MAF. The detailed proposals are also subject to approval by the Minister of Agriculture and Fisheries.

The regional research centres

A number of regional experiment farms have been established for location-specific, commodity-oriented research; 12 for crop production, 27 for horticulture, five for cattle, two for swine, and three for poultry management.

Other government institutions and services

These, such as the National Service for the Ijselmeer Polders and the National Institute for the Management of Natural Resources, carry out research relevant to their specialised activities.

The universities

The Agricultural University at Wageningen maintains 22 institutes and laboratories that are concerned mainly with fundamental research, but often staff members engage in applied research. These units are headed by professors of the University, who have complete freedom in the choice of their research projects. The Agricultural University is an autonomous body, formally responsible to the Minister of Agriculture and Fisheries, and fully financed by MAF.

Research results that need further work at farm level, are transmitted to the appropriate specialists working at the experiment farms of the research stations, at the regional experiment farms, or on commercial farms, depending on the nature of the problem. Research results relevant to merchants and processors, are usually introduced to these branches after validation by the Institute for Research on Storage and Processing of Agricultural Produce (Van de Zaag, 1985).

Other universities in the Netherlands, which are responsible to the Ministry of Education and Science, carry out research of relevance to agriculture. The most important contribution is made by the Faculty of Veterinary Science at the National University of Utrecht.

RESEARCH POLICY

'The Netherlands never has had an independent agricultural research policy, and it is my sincere hope that my country will stay in this blessed state as long as the sea level allows us to exist' (De Zeeuw, 1985). While this statement is fairly indicative of the allergy of Dutch scientists to anything like a dictated policy, nevertheless, national science policy is determined by a number of bodies, subject to government approval (OECD, 1965):

1. The Advisory Council for Science Policy, which advises the Council of Ministers on general science policy;
2. The Royal Academy of Sciences;
3. The Central Council for Fundamental Scientific Research (ZWO);
4. The Academic Council, which represents all the universities in the Netherlands;
5. The Council of Nuclear Energy; and
6. The Central Council for Applied Scientific Research (TNO).

The last mentioned consists of five branch organisations: The National Council for Food Research TNO; the National Council for Health Research TNO; the National Council for Defence Research TNO; the National Council for Industrial Research TNO; and the National Council for Applied Agricultural Research TNO. Any agricultural research carried out by the first four branch organisations is supervised by the delegate of the National Council for Applied Agricultural Research on the Board of the Central Council for Applied Scientific Research TNO.

The Advisory Council for Science Policy advises the Council of Ministers on the division of the science budget, pointing out its weaknesses, and proposing new developments. The Council is independent, and does not act as a spokesman for the scientific community (De Zeeuw, 1985).

The Minister of Agriculture is directly responsible for the research carried out by the National Council for Applied Agricultural Research; the departments of the Ministry of Agriculture; and the Agricultural University of Wageningen. He has no direct responsibility for research of relevance to agriculture as carried out by all other universities, the six councils mentioned previously, and their branch organisations, with the exception of the National Council for Applied Agricultural Research.

The Minister has to account for his policy to Parliament, which also influences research policy. Since the decrease in the number of farmers in the Netherlands, the agricultural vote has lost in importance. Nevertheless, the farm lobby is still very effective, and has been able to prevent disproportionate reductions in the agricultural budget (Van de Zaag, 1985).

In the MAF, a special department is concerned with research and extension, and one of its subdivisions is responsible for co-ordination.

THE NATIONAL COUNCIL FOR AGRICULTURAL RESEARCH (NRLO)

The Council is an independent advisory body, based on voluntary participation by MAF, the Veterinary Institute of the National University, TNO and farmers' organisations. NRLO serves as a meeting ground for government, science and the clients of science, namely: the farming community, agribusiness, and the bodies concerned with environmental protection (De Zeeuw, 1985).

The objective of this federation is to ensure the optimal use of available resources for agricultural research.

The structure of NRLO comprises (Zuurbier, 1983):

(a) *Departments* whose functions are to determine priorities for research and monitor co-ordination. There are departments for Natural Resources and Landscape Management, Plant Production, Animal Production, Processing and Marketing, Agriculture and Society.
(b) *Programme Advisory Committee:* Professional farmers' organisations have established commodity committees which collate and integrate the problems and requests of the farming community, and thereby influence the choice of research projects. Extension workers, agricultural entrepreneurs and researchers participate in this work.
(c) *Co-ordination Committees* consider research priorities and propose to whom the projects should be assigned.
(d) *'Contact' Committees* and *Working Groups* initiate meetings between research workers for the exchange of information and evaluation of research results.
(e) *Study Committees* undertake in-depth studies of specific aspects of agricultural research.

Every five years, the Council advises the MAF on the reprogramming of agricultural research in general.

The *Regional Research Centres* also have a procedure for research policy and programming. Annually, the researchers at the centres review the demands for research made by the staffs of the various experiment stations, the farmers of their respective regions, and the Board members of the Centre. From these requests, a list of research proposals is compiled and submitted to a special committee which determines priorities, after discussion with the interested parties: MAF, extension workers, researchers and farmers. A country-wide programme is then drawn up on a regional basis and submitted for approval to the governing boards of the regional centres.

FUNDAMENTAL AND APPLIED RESEARCH

The basic tenet of agricultural research in the Netherlands is that the Dutch agricultural research establishment limits itself in the choice of research subjects to those which cannot be left to others (De Zeeuw, 1985):

- It does not duplicate research and development carried out by the private sector.
- Only limited funds are earmarked for subjects that cannot be applied at the farm level within a period of 10–15 years.

Originally it was assumed that each component of the research system would undertake research of a specific nature, as follows (Zuurbier, 1983):

Fundamental Research—Universities and Colleges
Applied Research　　—The Research Institutes of MAF and TNO
Management Research—The Regional Experiment Centres and Experiment Stations.

The private sector forms a category apart, engaging in fundamental, applied and development research.

In the course of time, these distinctions between the kind of research carried out by the universities, institutes and regional centres have become increasingly blurred.

RESEARCH PROGRAMMING AND PLANNING

Every institute has a Board of Supervisors, consisting of representatives of the universities, of other research institutes, of the MAF, and of the farmers' organisations. The Boards are supposed to approve and supervise research programmes; this role is, however, increasingly taken over by the MAF, which holds the purse-strings, reducing the function of the Boards to a purely advisory role.

At the Agricultural Experiment Stations, half the members of the Boards of Supervisors represent the farmers' organisations, and influence of the latter on the research programmes is significant, in contrast to their limited impact on the research programmes of the institutes. They have practically no influence on university research (Van Diest, A., pers. comm. 1988).

A Five-Year Agricultural Research Plan (Meerjarenvisie Landbouwkundig Onderzoek)

This Plan has been adopted in the Netherlands, for preparing an outline of the orientation of agricultural research for five years ahead, against a backdrop of social requirements and problems, as well as scientific developments and available research capacity. The Research Project Administration and Information System provides an important source of data and other information for preparing the Plan. The Programme Advisory Committees, together with professional associations and research institutions are involved in providing information and indicating overall orientation and priorities; the Co-ordination Committees provide information on the research potentials.

The departments of NRLO then prepare a development plan, based on the information, ideas, and proposals of the Programme Advisory Committees, the Co-ordination Committees and the research management participants. A comprehensive research plan for a five-year period is then finalised by the NRLO (Zuurbier, 1983).

CO-ORDINATION OF RESEARCH

Co-ordination between agricultural research and other scientific activities is the responsibility of the Minister of Science and Education, who is responsible for the quality of the Dutch scientific effort in general.

The co-ordination task has proven to be very difficult, if not impossible. De Zeeuw (1985) ascribes this difficulty to 'the Dutch character being what it is, means that everyone does not like integrated or overall approaches'. A more convincing explanation for the difficulties of co-ordination, at all levels, is another statement by the same author: 'our research institutions, our agricultural experiment stations, our agricultural university, are all incorporated in an extensive network of formal and informal connections. This network is so closely knit, and the flow of information through it is so intricate, that it really defies description'.

With a complex and fragmented research system as the one that has evolved in the Netherlands, one could hardly expect co-ordination to be other than difficult. This is compounded by the lack of formal linkages between agricultural research institutes which function under the aegis of the Directorate for Agricultural Research and the University Institutes. In many cases, staff members co-operate with the former, but only on a mutually voluntary basis (Van Diest, A., 1988 pers. comm.). One formal linkage between University and MAF institutes is that postgraduate students can be trained at the institutes, but only in conducting research. Staff members of the institutes can be appointed as part-time professors at the University (Van Diest, A., 1988 pers. comm.).

Small wonder that much thought and effort have been devoted in the Netherlands to the development of an effective method for the co-ordination and direction of the national research effort. Various solutions were tried and subsequently abandoned.

De Zeeuw (1985) comes to the conclusion that basically, policy decisions in agricultural research originate in informal linkages at the personal and small-group level. The decision in *statu nescendi*, as you might call it, then spreads around, gains support, and in the end is formalised by those that are formally responsible.

RESEARCH OBJECTIVES

Far more emphasis is now being given to environmental problems and the quality of agricultural produce. The trend is towards a technology based on a relatively low input of agro-chemicals to avoid further rises in overproduction in the EC, but high enough to guarantee a reasonable income for the farmer. Of particular importance is a reduction in the present level of pollution of soil, water, and the atmosphere (Van Diest, A., 1988, pers. comm.).

RESEARCH FUNDING

Since the 1970s, the Dutch science budget has been in a steady decline. 'Once ranked among the big spenders in the science league, the Dutch science budget is now among the lowest in the industrial world' (De Zeeuw, 1985). Naturally, the budget for agricultural research has also been affected, and is expected to diminish by 10 per cent in the late 1980s.

The total amount allocated to 100 agricultural institutes, experiment stations and agricultural university departments in 1986 was 375 million guilders, representing 0·04 per cent of the gross value of agricultural produce (Van Diest, A., 1988, pers. comm.).

The Directorate of Agricultural Research (DAR) receives its funds from MAF, and distributes them among the research institutes under its aegis. The Agricultural University is also funded by MAF, but directly, and not through DAR.

A process of 'privatisation' of agricultural research is being implemented gradually, whereby the farmers' organisations and industries will have to pay more for the research they request (Van Diest, A., 1988, pers. comm.).

FACTORS FOR THE SUCCESS OF THE DUTCH NARS

The farmers

The economic success of Dutch agriculture is largely due to the following characteristics of the Dutch farmer (De Zeeuw, 1985):

- a strong desire to be independent, to produce more, to lower costs of production, and to improve the quality of his produce;
- a keen sense for consumer demand, and the ability to react quickly to changes in the market;
- a readiness to adopt innovations: 'the farmers almost literally pull preliminary results out of the scientists' hands';
- a marked will and ability to co-operate. Farmers' organisations are powerful, and are able to bring considerable pressure on government;
- skill in farm management, resulting from excellent professional training in many kinds of agricultural schools.

The research and extension workers

The impact of these workers on agriculture is due to the following factors:

- the full participation of research workers, extension workers and the farmers, at all levels, in deciding on the problems to be investigated;
- the high degree of decentralisation in the decision-making process for research and extension, thereby avoiding top-down instructions on what should be undertaken;
- a social climate, in which research workers are interested in assuring that the results of their work are applied by the farmers, and extension workers are motivated to transfer relevant results of research to their 'clients'.

With such a farming sector, a committed and highly competent research establishment and extension service, and strong government support, agricultural research has been highly effective.

RECAPITULATION

The major characteristics of agricultural research in the Netherlands are: a very complex and diffuse national system of research; the need to invest considerable efforts in co-ordination; informal linkages between researchers at the personal level; the practical orientation of much of the research and its validation before diffusion; the considerable influence of the farming sector on the orientation of research.

Belgium (Lecomte, R., 1988, pers. comm.)

STRUCTURE OF THE NARS

In Belgium, the entire responsibility for public-funded agricultural research is vested in the Ministry of Agriculture. A special Department in the Ministry—Administration of Agronomic Research—headed by a Director General is in charge of overall administration of the agricultural research establishment.

Research institutions

The Ministry maintains seven scientific institutions:

The Agronomic Research Centre of Gembloux (for French-speaking Wallonia)
The Agronomic Research Centre of Ghent (for Flemish-speaking Flanders)
The National Institute for Veterinary Research
The Institute for Agricultural Economy
The Forestry Research Station
The Institute for Chemical Research
The National Botanical Gardens.

Each institution has its own *Scientific Council* whose main functions are the preparation of the annual research programme and the appointment of the leadership of the institute. The scientific administration of each institute is completely independent in the implementation of its research programme, subject to its prior approval by the Ministry.

Staff

The total of research workers of the seven institutions is 235 permanent staff, and another 90 scientists are employed under contract for limited-time projects.

Openings in the research service are announced in the Official Journal; candidates are interviewed in each institution by a jury of six persons of whom three are Professors representing the Faculties of Agronomy and of Veterinary Science. The candidates approved by the jury are appointed for a probationary period of two years, which can be renewed twice, subject to approval by the jury. After a maximum of six years they are appointed as Assistants. They can be promoted to the next grade (First Assistant) if they present a PhD thesis. After 10 years of service, they can be promoted to a higher grade, subject to the approval of the Minister.

Three special grades are reserved for scientific management functions: Head of Section, Director of Station, and Head of Institute, respectively.

Salaries of research personnel are more favourable than those of other public servants, and are comparable to those of the scientific and teaching personnel of the Universities.

Funding

The major source of funds for research is provided from the budget of the Ministry of Agriculture. In 1987 the allocation for agricultural research amounted to a little over 1·5 million B.Fr. Since 1977, research allocations have shown a steady decline in real terms, amounting to about 10 per cent for the decade, in line with the austerity policy of government.

Additional sources of funding for research are provided by regional or provincial authorities, the private sector, the OECD, the World Bank, and the International Agency for Atomic Energy. These funds vary from year to year and are assigned to specific projects.

RESEARCH POLICY AND OBJECTIVES

Consultative Committee for Scientific Research in Agriculture

This committee was reactivated in 1984, and comprises representatives of the farmers' associations, the Faculties of Agronomy and of Veterinary Science, the Ministry of Agriculture, the Administration of Scientific Policy, the Institute for the Encouragement of Research, the National Fund for Research, the Scientific Institutions of the Ministry of Agriculture, and the Federation of Industries. The chairman is the Secretary-General of the Ministry of Agriculture.

The Consultative Committee advises the Minister of Agriculture on the overall objectives of agricultural research and its priorities.

Research objectives and priorities

The major objectives of agricultural research in Belgium have evolved considerably in the course of the last decade, under the influence of economic, social and scientific developments.

Among these factors, are the decisions taken by the Common Market in relation to prices of agricultural commodities, and the financing of the Common Agricultural Policy. These decisions have resulted in a reorientation of research objectives from increasing yields, to optimisation of economic production in farming systems.

The major social factor influencing research orientation is the steady decline in the number of farms, and the concomitant increase in the size of the remaining enterprises.

Major developments in biological and chemical sciences have enabled important new applications in agricultural research: in-vitro tissue culture, genetic engineering, cloning, etc., are being used in numerous research projects.

The principal *priorities* in agricultural research, as determined by the Consultative Committee for 1986 are:

- Research on plant physiology and genetics, and animal breeding based on new techniques of biotechnology and other advanced techniques.
- Research aimed at obtaining new, high-value non-food products, using products, by-products or residues of agricultural production and fisheries.
- Research aimed at promoting the production of commodities that are not in oversupply.
- Research on reducing the costs of production in all sectors, by limiting the use of agrochemicals whilst maintaining soil fertility, improving the agriculture–environment relationship, and reducing toxic residues on products.

- Research on the protection of soils, crops and forests against acid rain.
- Socio-economic and biological research in maritime fisheries.
- Study of hydrological resources.

RELATIONS WITH FACULTIES OF AGRICULTURE

There is no formal relationship between the research institutions and the Faculties of Agriculture. Informal relations are very strong in some cases, and less so in others, depending mainly on the personal relationships between the leaders of the respective institutions.

The only formal link between faculties and research institutes, is the possibility for graduate and postgraduate students to carry out their research theses at one or other research institution.

In a few cases, faculty and institution staff co-operate on certain research projects; these linkages are entirely spontaneous and based on personal relations.

LINKAGES WITH THE EXTENSION SERVICE

Research results are regularly diffused by the Extension Service. Meetings between research and extension workers are organised on a regular basis by the research institutions; conversely, the Extension Service carries out field experiments in collaboration with research workers in planning, implementation and evaluation.

In each research institution, certain senior research workers assume responsibility for liaison with the extension service.

INSTITUTE FOR ENCOURAGEMENT OF SCIENTIFIC RESEARCH IN INDUSTRY AND AGRICULTURE (IRSIA)

IRSIA was established immediately after World War II, in order to promote research in important areas by subsidising limited-term contract research. The major beneficiaries of these subsidies, in the field of agricultural research, are the Faculties of Agronomy, the Universities, private-sector research institutions and certain provincial and local research-development institutions.

PRIVATE SECTOR RESEARCH

Independently of the Ministry of Agriculture and IRSIA, a significant amount of research is carried out by industrial sectors concerned with plant protection, agricultural machinery and food processing.

Switzerland

THE FEDERAL STRUCTURE OF AGRICULTURAL RESEARCH

The basic structure of agricultural research in Switzerland has not changed markedly in the course of the last half-century; however, the roles and competences of the several federal agronomic research institutes have been defined by an ordinance promulgated in 1975; the information that follows is based on this ordinance (Conseil Fédéral Suisse, 1975).

There are seven federal agronomic research stations, all under the jurisdiction of the Division of Agriculture of the Federal Department of Public Economy. Two of these stations are multi-disciplinary, and have regional vocations. Their respective locations are: *Changins* for French-speaking regions and Haut-Valois; and *Zurich-Reckenholz* for the German (excepting Haut-Valois), Italian and Romanche speaking regions.

The other federal stations are: (a) commodity-oriented—Animal Husbandry, Dairy Products, Horticulture; (b) cross-commodity— Agricultural Economy and Engineering; Agricultural Chemistry and Environmental Hygiene.

The federal stations have been established in different locations; all have responsibility for research, extension and certain regulatory services.

Consultative committees

The Division of Agriculture has established two consultative committees:

The Committee for Agronomic Research comprising representatives from the Universities and the Polytechnic Schools. The Directors of the stations participate in the deliberations as experts.

The Committee for Agronomic Research Stations comprising farmers, representatives of agricultural associations and of the food industry. They advise on the research programmes in progress, which are approved for four years.

Conference of Station Directors

This conference comprises all the Directors of the Federal Research Stations and is chaired by the Director of the Federal Division of Agriculture. The conference meets at regular intervals to discuss the problems and the tasks of their stations and possible avenues of collaboration.

It is of interest to note that there is no common framework for deliberations between academia, the clients of research, and those responsible for research implementation, each group meeting separately.

RESEARCH OBJECTIVES

After World War II, agricultural policy in Switzerland was addressed to increasing agricultural production in all commodities. This policy was translated into the research objectives of breeding high-yielding varieties, achieving effective control of pests and diseases and the development of efficient management practices. Agriculture became fully mechanised, and farm management highly efficient.

The increased production resulting from these developments was at first easily absorbed by a rapidly increasing population, with considerable purchasing power. However, problems of excess production eventually appeared, and became progressively more acute in the 1970s. It became necessary to limit production in certain commodities, in particular animal products.

Concurrently, public concern has increased regarding ecology in general, and protection of the environment in particular. These concerns have influenced government policy and have resulted in the adaptation of the objectives of agricultural research to the new public policies.

The major thrust of research policy is now being increasingly directed to the following objectives (Vez, 1987):

- Instead of attempting to achieve high yields, to develop less costly management practices, and breeding less-demanding varieties.

- Diversification of production, for better adaptation to demand. Priority to be given to protein-rich oil crops, triticale, medicinal and aromatic herbs, and new kinds of vegetables and fruits.
- Improvement of quality: improved nutritive and organoleptic characteristics of varieties; limitation or absence of toxic substances.
- Maintenance of soil fertility: the development of techniques for the long-term maintenance and improvement of soil fertility.
- Integrated control of plant diseases and pests.
- Energy and nitrogen conservation.

REGULATORY FUNCTIONS

In contrast to agricultural research services in other developed countries, the Federal Research Stations are also involved in certain regulatory and control functions: quality control of fertilisers, pesticides, animal fodder, dairy products, seed testing, plant and seed certification, testing of machinery and materials, phytosanitary inspection, routine analyses, etc.

These non-research activities absorb about 50 per cent of the resources of the research stations (Vez, 1987).

LINKAGES

Linkages between the Federal Research Stations and the single Faculty of Agronomy (Zürich) are satisfactory. Research workers are given teaching assignments at the Faculty, or direct seminars. The students can do research at the stations for graduate and postgraduate work, under the supervision of a professor. A similar arrangement is in force with the Faculties of Science of several Swiss Universities (Vez, A., 1988, pers. comm.).

Linkages with farmers occur in four ways (Vez, A., 1988, pers. comm.):

- Research workers participate in training courses for farmers.
- On-farm trials and pilot farms for integrated techniques.
- Research workers provide expert consultations when the extension workers cannot solve a specific problem.

• Farmers can influence the research programmes through their representatives on the Committee for Agronomic Research Stations.

IMPACT OF RESEARCH ON FARMING PRACTICE
(Vez, A., 1988, pers. comm.)

Crop varieties have become more resistant to diseases; new crops have been introduced, such as triticale; the production of pulses, such as horse-beans, peas and soya, has been expanded; the insecticides used are more specific and less persistent; bio-technological control of plant pests and diseases is in wide use; the quantity of insecticides in viticulture and fruit production has been reduced by half; preventive control is no longer practised. Nitrogen fertiliser levels have been significantly reduced, whilst the use of phosphoric and potassic fertilisers has stabilised.

RESEARCH STAFF

The conditions of work of the research workers are those defined by statute for workers in the public service. Promotion is limited by administrative hierarchy, and not primarily by merit and scientific achievements (Vez, A., 1988, pers. comm.).

A decision taken by the Federal Parliament in 1973, has resulted in a 5 per cent reduction in the number of scientists employed by the Stations (Vez, A., 1988, pers. comm.).

PRIVATE SECTOR

Several professional organisations (Sugar Beet Centre, Association for the Development of Forage Crops, Swiss Federation of Plant Breeders, etc.) are involved in developing new technology. These associations maintain excellent relations with the Federal Research Stations.

The main thrust of the private sector is in the development of new agro-chemicals (plant insecticides, fungicides, growth substances), improved varieties, seeds, machinery and equipment.

France†

THE BEGINNINGS

The pioneers of agricultural research in France were Lavoisier, who concerned himself with experimentation on his own estate, and Boussingault, who in 1834 established what can be considered as the first experiment station. This initiative was shortly followed by the establishment of a number of private laboratories for the analysis of fertilisers and soils. Farmers not only submitted samples for testing to these laboratories, but also started requesting advice, so that the laboratories gradually developed into experimental stations, investigating the problems raised by the farmers.

In 1921, Roux, then Director of the Ministry of Agriculture, made the first attempt to organise these nascent experiment stations into a single administrative framework, linked to the Ministry of Agriculture, but maintaining a large degree of autonomy. The first agricultural research centre was established by him in 1926. Research workers were recruited by open competition and those accepted received preliminary training in appropriate laboratories.

In 1934, Laval, then Minister of Agriculture, decided to arrest the development of the research centre; its semi-autonomous status was revoked and it became a sub-unit of the Direction of Agricultural Education in the Ministry.

THE NATIONAL AGRONOMIC RESEARCH INSTITUTE

It was only after World War II, when the critical situation of French agriculture made itself felt, that the need for a dynamic agricultural research organisation was fully realised, and the decision to establish the Institut National de Recherches Agronomique (INRA) was taken. Prolonged discussions ensued as to whether the organisational form to be adopted should be vertical, based on branches of agricultural production, or horizontal, based on disciplines. The decision was finally taken to adopt the horizontal approach for the Research Centre, which would

†The information on the history and organisation of agricultural research in France is based on information provided verbally by M. Drouineaux, Inspecteur Général de la Recherche Agronomique, and in writing by Mr. Rebischung (1964).

be strengthened by regional stations oriented towards specific commodities (Fig. 3).

The responsibilities of the INRA first encompassed research in crop production only: in 1951 its responsibility was extended to include animal husbandry; in 1958, agricultural economy was added; and in 1964, forestry research. Research on agricultural engineering is still carried out outside the framework of the INRA.

INRA is therefore responsible for the organisation and implementation of all governmental scientific research pertaining to agriculture, the publication of research results, and the maintenance of liaison with other public services concerned with agriculture. The fields covered by INRA include the improvement and development of crop production and animal husbandry; the conservation and transformation of agricultural products, including forestry products and sweet-water fish; and economic and sociological research. It is also responsible for ascertaining the practical implications of the research carried out under its auspices and for its diffusion. All government funds earmarked for agricultural research are under the control of INRA, whether for financing its own research efforts or for work subsidised by INRA and carried out under its supervision in institutions which are not under its direct control.

The research programme adopted by the institute is determined by the Conseil Supérieur de la Recherche Agronomique, which functions within the framework of the Ministry of Agriculture. However, the membership of the Council, its mode of operation and its budget are under the joint jurisdiction of the Minister of State for Scientific Research and the Minister of Agriculture. A permanent Scientific Committee is appointed by the Council.

Whilst INRA is a publicly supported institution, it has the special status of 'personnalité civile', which endows it with a large measure of financial autonomy and flexibility in the administration of its research units.

The research work in all the fields for which INRA is responsible is carried out by 17 departments, as follows: Agronomy, Bioclimatology, Plant Physiology, Genetics and Plant Breeding, Phytopathology, Agricultural Zoology, Phytopharmacy, Animal Nutrition, Animal Physiology, Animal Genetics, Animal Husbandry, Veterinary Research, Food Technology (crops), Food Technology (animal products), Economic and Social Research, Forestry, Hydrobiology.

Each department has a Central Station. These Central Stations are grouped in two national research centres: one for crop production at

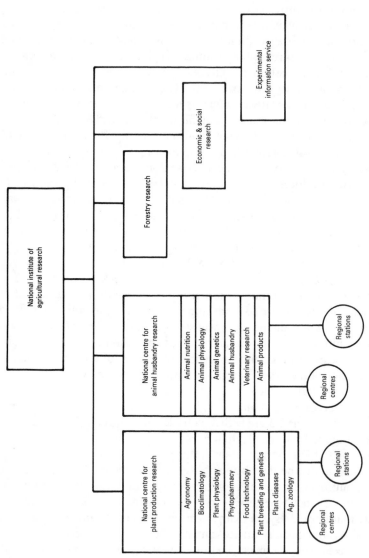

Fig. 3. The organisation of agricultural research in France.

Versailles, with eleven stations; and one for animal husbandry at Jouy-en-Joras, with seven stations. Each department also maintains a number of stations in the provinces. There are 68 such regional stations, which more or less specialise in one type of production (such as horticulture, maize production, bee-keeping, etc.). Most of these stations are grouped in nine regional centres, though some are isolated.

In principle, the national centres were to be concerned mainly with basic research, whilst the work of the regional centres was to be of a more applied nature. However, it has been found difficult to maintain, in practice, this rather artificial classification, and the differences in the nature of the research carried out in the national and the regional centres are becoming progressively less apparent.

A special unit, called the 'Service d'Expérimentation et d'Information', carries out field experimentation and is responsible for the liaison between research and the extension services, both governmental and private.

UNIVERSITY RESEARCH

The faculties of agriculture also carry out research. They do not have any independent sources of funds; the resources for research are derived from INRA, which finances the work carried out at 33 stations attached to the faculties of agriculture and directed by the senior teaching staff. This arrangement makes it possible for INRA to integrate the research work of faculty members into its own programme.

PRIVATE SECTOR RESEARCH

In addition to the government institutions mentioned above, agricultural research work in the field of plant breeding, pesticides and fertilisers is carried out in a number of laboratories and experimental stations maintained and financed by private commercial interests. A few examples are the plant breeding work carried out by the famous firm of Vilmorin, the Cooperative Society for Research and Experimentation of the Eastern Pyréneés, the Technical Institute for the Sugar Beet Industry, the Technical Institute for Oenology, etc. If these institutions accept state subsidies, their work is carried out under the scientific supervision of the INRA.

Germany

EARLY HISTORY

At the beginning of the 19th century W. Von Humboldt conceived the idea of a new type of university, which would combine teaching with research. This concept contrasted with the traditional European concept of universities concerned almost exclusively with teaching, modelled on the medieval schools of theology.

As agricultural science was not incorporated in the new university system, Albrecht Thaer founded Academies of Agriculture, separate from the universities, whose vocation was mainly teaching (Weber, 1979).

In Saxony, the publication of Liebig's treatise on plant nutrition had provoked considerable interest in developing methods for applying science to agriculture. After learning of the Scottish attempts to establish the 'Chemistry Association', a group of landowners petitioned the Government of Saxony to use tax moneys to hire a number of scientists to carry out agricultural experimentation. According to this proposal, agricultural research would be a state-sponsored undertaking and the researchers employees of the State. The Saxon Government was sceptical of the ability of farmers to benefit from scientific research, and denied the petition. However, several years later, a wealthy landowner made available part of his estate for an experimental farm, to be equipped with a scientific laboratory. A group of farmers' societies took the initiative to draft a charter for this undertaking, which the Saxon government subsequently legalised by statute, and an annual state appropriation was approved for it. Thus, in 1852, the first government agricultural experiment station, as it was named, was established at Moeckern, in Saxony. In its charter, it was specified that 'the Station would be devoted to the advancement of agriculture by means of scientific investigation carried out in close connection with practical experimentation', and stressed the importance of co-operation between scientists and farmers in order to achieve this aim. The programme of investigations was to include the following:

1. Conditions affecting the growth of plants;
2. composition of plants and their feeding value;
3. meteorological conditions;
4. introduction and testing of new crops; and
5. testing of farm machinery and implements.

The Saxon approach was characterised by:

1. The establishment of a specialised institution, operating as a state-chartered station;
2. detailed research directives provided by a Board on which sat representatives of several agricultural societies;
3. concentration on research exclusively, with no concern for the diffusion of research results;
4. situation on one location, with the necessary facilities for research in laboratory and field.

The experiment station at Moeckern served, for many years, as a model for agricultural experiment stations that were subsequently established in Germany and various parts of the world. It made a major contribution to the development of techniques for agricultural field experimentation.

In the following half-century, a network of 76 experiment stations was established in Germany, and it was at this time and at these stations that the scientific basis for the teaching of agricultural sciences was laid for the Faculties of Agriculture which were established in the second half of the 19th century (Weber, 1986).

Of this period Ruttan (1982) writes:

By the end of the century, leadership in the application of science to the problems of agriculture had passed from England to Germany. The German model was adopted in Austria, Italy, USA and Japan. During the latter half of the 19th century, it was almost obligatory for anyone with serious interest in agricultural science to study in Germany, and German experience in the organisation of agricultural research and education provided the models for institution-building in other countries.

The flourishing of the faculties of agriculture, integrated into the university system, led to a decline of the Academies, which lost their important role in the German agricultural education system (Weber, 1979).

However, a time came when the universities could no longer fill the need for applied research; in response to this situation, the various German states established specialised research institutes, without a teaching vocation (Elz, 1984). They were strongly supported by local farmers' organisations and by the Chambers of Agriculture.

During the entire period during which an agricultural research system was being progressively established in Germany, there was no Ministry of Agriculture to direct or co-ordinate the process. It was only after World War I, that the Weimar Republic established a Ministry for Food, Agriculture and Forestry.

The National Socialist regime led to a steep decline in agricultural research following the isolation of the German research community, the migration of eminent scientists, the preoccupation with military research, and the tremendous losses inflicted by the war.

After World War II, the West German agricultural research system had to be re-established. A large research budget was allocated by the Federal Ministry of Food, Agriculture and Forestry, and new Federal research institutes were established (Weber, 1979).

PRESENT STRUCTURE OF THE AGRICULTURAL RESEARCH SYSTEM IN THE FEDERAL REPUBLIC

Agricultural research in the Federal Republic of Germany is conducted by: (a) the Federal and Provincial Governments; (b) the Universities and the Max-Planck Gesellschaft; and (c) Private sector institutions (Beringer, 1988).

The Federal Research Institutes

There are seven institutes, mostly commodity oriented: Agriculture, Cereal and Potato Technology, Dairy Science, Meat, Nutrition, Lipids, and Forestry, respectively. They carry out basic and applied research in technology, ecology and toxicology. The Biological Federal Institute for Agriculture and Forestry, founded in 1905 for plant protection, has established a network of stations throughout Germany.

Provincial Agricultural Research Stations

There are 11 stations, mainly concerned with soil, feed and seed-testing as well as plant protection.

Universities and Max-Planck Institutes (MPIs)

There are seven Faculties of Agriculture, three Faculties of Horticulture, three Faculties of Forestry, and four Faculties of Veterinary Science.

The MPI for Plant Breeding was originally founded to support research in the basic sciences, but progressively incorporated genetics, plant breeding, agricultural engineering and others.

AGRICULTURAL RESEARCH IN DEVELOPING COUNTRIES

Agricultural Research in the Less Developed Countries (LDCs) is supported by the Federal Government. The work is carried out by the German universities, and co-ordinated by the Study Group on Tropical and Subtropical Agricultural Research (ATSAF).

PRIVATE SECTOR RESEARCH

Industry maintains a number of specialised research institutes: such as for pesticides and fertilisers (BASF, Bayer, Kali+Salz, Höchst); agricultural engineering (Deutz, Mercedes, etc.); irrigation companies, and animal feed industry.

These institutes engage in product development, toxicological and ecological testing, marketing research and basic research in their respective fields.

NUMBER OF AGRICULTURAL RESEARCH WORKERS

There are about 2300 scientists engaged in the public institutions and over 1000 in private sector research (Weber, 1979).

AGRICULTURAL RESEARCH COUNCIL

An Agricultural Research Council (Landwirtschaftlicher Forschungsrat) advises Government on questions of organisation, financing and co-ordination of agricultural research. The Council comprises scientists and agricultural practitioners. It has no research institutions under its direct control.

The Nordic Countries (Skovgaard, 1984, 1987)
(Sweden, Norway, Denmark, Finland and Iceland)

INTRODUCTION

Agriculture in the Nordic countries is under very heavy pressure. Despite the political desire to maintain a large number of relatively small holdings, the difficulties in achieving this objective are very great. Agricultural prices have been fixed at levels that encourage production in excess of the home markets' needs. Production is also increasing more rapidly than consumption. The surplus has to be sold on the world market at prices that do not cover costs. The losses entailed make it difficult to ensure a reasonable level of income to farmers and the continued existence of small holdings.

The objectives and problems of agricultural policy are identical in all the Nordic countries and affect research policy in the same ways. The most important problem is no longer to grow enough food, and political efforts are therefore addressed to a growing list of secondary objectives, many of which involved conflicting goals.

These secondary objectives relate to production, income, efficiency, quality, environmental protection, and resource conservation.

Agriculture has undergone an extensive process of adjustment to new realities. The major tendencies in the restructuring of agriculture have been more specialised production planning and increased use of technology. The major constraints encountered are the increasing proportion of fixed costs, a reduced capacity for adjusting production, and increased financial risk.

AGENCIES INVOLVED IN AGRICULTURAL RESEARCH

In all the Nordic countries, Universities, Research Councils, Ministries of Agriculture and the private sector are involved in agricultural research. And yet, notwithstanding their cultural affinities, identical objectives and problems, they have evolved national research systems which differ in the roles assigned to the different components of the systems, and the relative financial support provided from public funds.

The Universities

In all the Nordic countries there is an extensive university system which carries out basic research in relation to its educational objectives.

In *Sweden*, the University is the major institution engaged in agricultural research; it is the only Nordic country in which the University is responsible for teaching, research, and advisory work.

In *Norway* the Universities are mainly restricted to fundamental research.

In *Denmark*, in addition to fundamental research, the Universities conduct a comprehensive agricultural research programme.

In *Finland* the Universities devote themselves to basic research and to applied food research.

In *Iceland*, the University's role in research is very restricted.

Research Councils

Each of the Nordic countries has Research Councils for Farming and Forestry, and for Veterinary Medicine (amongst others). There are, however, differences between the countries in the composition of the councils and their administrative affiliation. In *Denmark* and *Finland*, the councils are mainly concerned with basic research and belong to the Ministry of Education; in *Norway* and *Sweden* the councils are more application-oriented, and are attached to the Ministry of Agriculture.

Ministries of agriculture

In *Iceland, Norway,* and *Sweden* the Ministry of Agriculture is responsible for education and research in agriculture and forestry. In *Denmark* and *Finland*, the Ministry of Agriculture maintains specialised institutes for applied research; the work of these institutes is user-controlled.

THE NATIONAL AGRICULTURAL RESEARCH SYSTEMS

Sweden

Swedish University of Agricultural Sciences
The research departments of the University are responsible for most of the public-funded applied agricultural research in the country. About

70 per cent of the University's budget is spent on research and experiments, including basic biological research relating to agriculture, forestry and veterinary medicine.

In 1983, the University, in part co-operation with the Research Council, prepared a development programme, with the following priority areas: biotechnology, natural resources, forestry production, livestock diseases, food research, energy and industrial products, horticulture, and the production-marketing (processing) chain.

In contrast to many other European countries, the Swedish government decided in the early 1980s to intensify agricultural research and development and increased the budget for 1984 by over 8 per cent, with special emphasis for research in the following areas: alternative production systems in agriculture and horticulture; breeding of fish and shellfish; forestry research; acid rain; and biotechnology.

Norway

Most of the funding for agricultural, forestry and veterinary research comes from the budget of the Agricultural Department.

Universities
The Agricultural University of Norway and the Norwegian College of Veterinary Medicine receive over 40 per cent of the research budget of the Agricultural Department. They are mainly concerned with basic research in their respective fields.

Ministry of Agriculture
The Ministry maintains a number of institutes and experimental farms for applied research in agriculture, forestry and veterinary science. Problems that are specific to small agro-ecological areas are also dealt with by these experimental farms.

The Norwegian Agricultural Research Council
The Council functions under the aegis of the Agricultural Department; it has its own funds, of which two-thirds come from football pools. It also finances the training of research workers.

The Council works in close co-operation with eight advisory agents appointed by the Agricultural Department, covering research on commodities (agricultural, horticultural and animal) and disciplines (technological, environmental, agronomic and social).

Research projects are initiated by the Council in certain priority areas and promising new technologies: e.g., automation technology, energy saving, alternative farming systems, nutrition policy.

Denmark

Ministry of Education
Public-funded research is carried out to a large extent by the institutions of higher education, under the administration of the Ministry of Education, which is also administratively responsible for the conduct and central planning of research.

The *Royal Veterinary and Agricultural University* conducts a comprehensive research programme in connection with its teaching activities. This research is funded from the University's budget, and accounts for 23 per cent of all public R & D spending in the agricultural sector. There is extensive co-operation between University researchers and those working at the experiment stations financed by the Ministry of Agriculture.

The Danish Agricultural and Veterinary Research Council
This has the following functions:

- to advise public authorities and institutions on official support for research, the application of the results of research projects, and the education of research personnel;
- to keep abreast of developments in Danish and foreign research and to participate in international research work;
- to train research staff;
- to take the initiative in supporting research projects of potential interest;
- to co-ordinate Danish research activities.

The Council disposes of about 7 per cent of the total public funds available for agricultural research.

Ministry of Agriculture
The Ministry maintains a number of institutions for agricultural research, and disposes of about 65 per cent of the total public funds earmarked for agricultural research. These institutions comprise: The Veterinary Institute for Virus Research, the Institute for Poultry Diseases, the Institute of Animal Science, the Institute of Farm Management and

Agricultural Economics, the Dairy Research Institute, the Research Service for Soil and Plant Sciences, the Government Forestry Research Services.

In recent years, the Ministry of Agriculture has strengthened the activities of the Research Institutes by providing special, time-limited funds, for priority areas, such as reducing energy use, utilisation of waste products, irrigation, environmentally safe plant protection, improvement of quality of horticultural products, and farming systems.

Joint Committee for Agricultural Research and Experimentation
This was established in 1961, in order to advise the Ministry of Agriculture on the distribution of R & D appropriations and on the co-ordination of the work of its institutes, and co-operation between them. The Committee also prepares proposals for research programmes and their distribution between the institutes.

The Committee has 24 members representing the users of research through their organisations, and the research institutions.

Other ministries involved in agricultural research
These include *Atomic Energy* (Atomic Research Station), *Environment* (Government Food Institute, Institute of Soil Ecology), *Social Affairs* (National Institute of Social Research), *Housing* (National Institute of Building Research).

Private Sector
R & D are conducted by a number of private organisations, such as the Academy of Technical Science, the Poultry Research Station, and the Danish Meat Research Institute.

A number of private organisations also provide financial support to research programmes carried out by government institutions.

Finland

The publicly financed agricultural research in Finland is organised along the same lines as in Denmark.

The *Faculty of Agronomy and Forestry* at the University of Helsinki, under the Ministry of Education, conducts basic research. In 1923, applied agricultural research was transferred from the Faculty to the Ministry of Agriculture (Phillips *et al.*, 1953).

The *National Research Council for Agriculture and Forestry*, under the Finnish Academy, is one of seven research councils which belong administratively to the Ministry of Education. The Council participates in research planning, finances research projects, and helps to finance the training of research workers.

The *Ministry of Agriculture* maintains a number of research institutes and experiment stations throughout the country. The stations do not have common targets or priorities.

An *Agricultural Research Delegation*, comprising representatives from various public authorities, non-agricultural disciplines, agricultural organisations, etc., advises the Ministry on planning, co-ordination, and financing of the work of the research institutes. The Delegation also has its own budget, to promote urgent research projects, to be implemented at the Ministry's institutes which are at the University.

Iceland

All agricultural research is publicly funded, through the Ministry of Agriculture, which is also responsible for the higher education in agriculture provided by the Agricultural College.

Most of the agricultural research is carried out at the *Agricultural Research Institute*, which has five experiment stations at different parts of the country. The staff of 50 research workers is divided among nine departments: Soil Investigations, Crop Production, Land Utilisation, Biostatistics and Mathematics, Livestock Production, Theory of Feeding, Agricultural Technology, Food Technology, Analyses and Control.

Research in veterinary medicine, freshwater fish and forestry is also carried out under the Ministry of Agriculture.

REGIONAL CO-OPERATION IN AGRICULTURE RESEARCH

Jointly financed R & D in the Nordic countries takes place (a) under the Nordic Council of Ministers; (b) by co-operation between the National Research Councils; and (c) by co-operation between Universities and other Institutes of Higher Education.

(a) The *Nordic Council of Ministers* mainly promotes co-operative basic research initiated by the Nordic Ministers of Education and research projects emanating from various sector ministries, including the Ministries of Agriculture. Support takes the form of provision of resources for the operation of inter-Nordic research institutes, and funding of specific research projects to be implemented in the national research institutions.

(b) The *National Research Councils* have been particularly active in fostering co-operative research projects.

(c) The *Agricultural Universities* have as a special activity the inter-Nordic training of agricultural research workers.

Ad hoc co-operation in agricultural research

A number of *Collegiate Bodies* have provided an important framework for inter-Nordic co-operation in agricultural research. The oldest of these bodies is the *Nordic Association of Agricultural Researchers*, established in 1918, and serves as a framework for co-operative activities of about 2500 scientists, instructors, and advisers.

The Association organises a large number of symposia, seminars, and workshops. A congress is held every four years, at which joint research projects are discussed. The Association has also been very active in the organisation of courses for researchers.

The *Nordic Forestry Union* organises a congress every fourth year, as well as seminars, and deals with problems of professional importance to forestry in the Nordic countries. The *Nordic Committee for Veterinary Research Co-operation* was established to strengthen inter-Nordic veterinary and interdisciplinary co-operation.

Formal co-operation in agricultural research

The *Nordic Contact Agency for Agricultural Research* was appointed by the national agricultural research councils to initiate and support inter-Nordic research in agriculture, forestry, and veterinary science. It helps to secure grants for research projects and training courses for researchers, initiates the planning and implementation of joint research projects, and undertakes an overview of Nordic research activities.

The Agency has no funds of its own, but recommends national financing projects in which two or more countries co-operate.

FUNDING OF RESEARCH

A comparison of government financed agricultural research in the Nordic countries is shown in Table 1.

TABLE 1

Agricultural research as percentage of publicly financed R & D activities in the Nordic countries in 1979

	Agricultural R & D as percentage of:	
	Publicly financed R & D	*Value added in the agricultural sector*
Denmark	8·7	1·08
Finland	12·3	0·87
Norway	9·3	1·50
Sweden	7·4	2·75

Source: The report: Dansk Jordbrugsforskning, Statens jordbrugs- og veterinærvidenskabelige Forskningsråd, 1984 (Skovgaard, 1987).

The United States of America

HISTORY OF THE DEVELOPMENT OF RESEARCH IN THE UNITED STATES

A very detailed and interesting account of the development of agricultural research in the United States has been given by Knoblauch *et al.* (1962). American scientists, inspired by the European experience in applying research to agriculture, started propagandising the need for establishing experiment stations in America. Their persistence bore fruit, and the first agricultural experiment station in the United States, the Connecticut State Agricultural Experiment Station, was officially established in 1877.

The State charter defined the duty of the station unequivocally as 'the conduct of scientific investigation and experiments'. It was also decided that the station would have no organic connection with a university. On the assumption that only small plots were needed for experimentation, and that the important facilities required were laboratories, glasshouses and library, which could be concentrated in a limited area, an experimental farm was declared to be unnecessary and a suburban site was therefore chosen. It was further decided that the top administrative post, of Director, was to be filled by a highly competent scientist; that the station was to be governed by a Board of Control appointed by the State, and that it would have complete autonomy in the allocation of its financial resources, from whatever source they were derived.

Subsequently, more state experiment stations, entirely separate from the Agricultural Colleges, were established.

In the meanwhile, the establishment of a Department of Agriculture was being debated, one bloc proposing that the Department should be scientific and non-political, responsible for scientific experimentation in agriculture, the other bloc favouring a Bureau of Agriculture and Statistics within the Department of the Interior, concerned with fact-gathering but without authority to conduct research. It was finally decided to establish a Department of Agriculture that 'would conduct practical and scientific experiments but would also have duties of a non-scientific nature'.

The third milestone in the history of agriculture research in the United States was the establishment of the Land Grant Colleges, so called because the Congress of the United States, in 1862, decided to donate public lands to the States, the sale of which would provide funds for the establishment of colleges, at least one in each state, in which the leading subject would be 'branches of learning related to agriculture and the mechanic arts' (Eddy, 1957).

The Morrill Act of 1862, establishing Land Grant Colleges, did not explicitly state that the colleges were obliged to carry out agricultural research but it did permit them to carry out simple experimentation and even to purchase sites for experimental farms. Many college administrators felt that the requirement to experiment was viewed only as an aid in student training.

However, the colleges were very sensitive to, and dependent on, public support. Farmers were bringing to the colleges problems for which the answers were not available, and the colleges were becoming increasingly aware that to engage in experimentation was practically unavoidable, if they were to maintain favourable relations with the farming community. The governing board of the Michigan Agricultural College, for instance, in 1862 made its academic professors responsible for research, decided on a laboratory and field approach to research, and appointed a faculty member to supervise the departmental programmes. It took many years, however, before an appropriate administrative setting for agricultural research evolved at the Land Grant Colleges. In a few of the colleges, groups of professors, with or without the help of the trustees, mobilised private funds for setting up college experimental stations. These funds were, however, usually inadequate for effective work on a reasonable scale.

An interstate movement of research-minded professors started, therefore, to agitate for State support for establishing agricultural stations at the land grant colleges. The reasons given for joining the stations with the colleges were:

1. The college could most economically supply the resources needed for operating a station: abundant farm lands, laboratories, the service of specialists.
2. A college-connected centre for research would be of great benefit to the students, as a means of practical education.

In legislation passed in 1884, it was decided to establish federally subsidised stations as 'departments' of the Land Grant Colleges. The stations would be placed under the direction of the college trustees who were empowered to appoint a director and staff. It also made the stations entirely independent of the control of the Department of Agriculture, who retained only advisory powers.

There followed a long period of groping for an overall philosophy of research applicable to the Land Grant Colleges. The principal controversial problems that were debated and required solution are discussed below.

1. Would there be a basic incompatibility of interests between professional education and professional researchers in the integrated college station system?

One group was of the opinion that research, in an agricultural college, was auxiliary to the basic purpose of the college teaching. It was argued that the staff, doing station work on a part-time basis, would elect to specialise in research and devote to it their undivided attention, at the expense of their teaching responsibilities. A further concern was that the research people would become so involved in fundamental research that they would neglect the immediate problems of the farmer, and thereby forfeit the goodwill of the farming community with catastrophic consequences for the college, dependent as it was on good public relations. They therefore demanded that the college president should have full authority over the station.

The opposing group saw no conflict of interests between college teaching and agricultural research, but considered both activities as complementary and vital to the interests of agriculture. They were opposed to a combination system in which the college president would do double duty as station director. It was also thought that professors, already overworked with teaching responsibilities, would be unable to assume full responsibility for research. They considered that the widespread belief that a person must be a teacher to be a good researcher was a fallacy, and that it was absurd to pretend that the daily grind of teaching was an inspiration to research. It was argued that as a result of the 'combination system' the professors carried the teaching load, whilst assistants and students did the station work. They therefore demanded a full-time director for the station, and a clear demarcation between the duties of teacher and researcher. The teaching of the research worker should be confined as far as possible to the training of advanced students in his special field.

2. *Was a system of agricultural stations, separate and independent of the college, preferable to the college–station combination?*

Many of the station officers considered the independent station as ideal and the connection with the colleges as a concession to circumstances. Those who favoured the college–station relationship felt that it was essential for ensuring the scientific level of the station. Without the help that only the colleges could supply, the 'unattached' stations would only be 'helping the farmer to raise more corn to feed more hogs to buy more land to raise more corn to feed more hogs'[†], in other words, they would be concerned solely with what the scientists called 'lower' problems, and the farmers called 'practical' problems.

3. *Was the main emphasis of the agricultural stations to be placed on 'original research' or on 'spreading practical information'?*

For many years the stations were very sensitive to pressure from farmers' groups for 'practical' experiments and rapid results. People untrained in scientific disciplines dominated the policy-making of the station. Some station directors contended that the tax-supported stations, vulnerable as they were to local pressure, were surrendering to non-scientific standards and carrying out only a small amount of original investigation. They protested that the viewpoint of the scientist, rather than of the farmer should dictate the policies of the stations, without negating the duty of the station to serve the farmers' welfare and the public interest. They argued that the true interests of the farmer would best be served if the stations were instrumental in improving and even revolutionising agricultural technology. This aim could not be achieved unless the stations were enabled to create a scientific atmosphere, work

†From the presidential address of H.P. Armsby, to the Association of American Colleges and Experiment Stations, in 1899 (Knoblauch *et al.*, 1962).

according to scientific procedures and concentrate on problems the importance of which only the scientist, and not the farmer, could evaluate correctly. They argued that discoveries of permanent value could come only from 'the most abstract and profound research'.

The opposing view, held by most station directors in the late 1890s, was that there already existed a huge fund of technological knowledge and that the stations should function primarily as diffusers of improved methods, until the farming community had fully exploited the knowledge already available. Only then would devotion of the major efforts of the stations to 'original' research be justified.

4. *What should be the relationship between federal research and the work of state stations?*

The period from 1900 onwards was characterised by the efforts of the US Department of Agriculture (USDA) to control the work of the college stations and the efforts of the stations to preserve their independence from federal intervention. The Department had become an effective scientific agency in its own right, with a large number of promising scientists, considerable funds for research and increasing support from Congress, and was soon conducting a full-scale research programme of its own. The view was held that the experiment stations constituted a national system and that integration of their work was essential to ensure economy and efficiency. This could be achieved only by a centralised administration.

The State Agriculture Experiment Stations (SAESs) held that their autonomy should be upheld—that the role of the USDA should be that of adviser and contributor, rather than of administrator.

The formula that was finally evolved was that the Agricultural Research Service (ARS) of the USDA was to be primarily responsible for research on problems of national and regional concern to agriculture, whilst the SAESs would be primarily responsible for research on problems within the borders of their respective states, and work in co-operation with ARS to contribute to national and broad regional research efforts (USDA, 1980).

The following statement 'that basic research is a responsibility of both the ARS and of the SAES, and should be advanced in each institution and in each area as far as is feasible', made any distinction between the work of the federal and state institutions practically unenforceable. To agree that both were to engage in 'discovering, verifying, and establishing the practical bearings of scientific facts and principles of agriculture', and at the same time attempt to distinguish between subjects of national interest and those of state interest only, was completely illusory. Work on streptomycin at the New Jersey experiment station, on the development of hybrid corn at the Connecticut station, on male sterility of sorghum at the Texas station, on rust research at the Minnesota station, to mention only a few examples, was of an importance which not only transcended the borders of the states concerned, but was of immeasurable benefit to most, if not all, nations of the world.

The 1890 institutions

Because of racial separation in the South, black people were not permitted to attend the new Land Grant Colleges. In 1890, Congress passed the second Morrill Act, specifically to support black Land Grant colleges, since called '1890 institutions'.

THE NEW CENTURY

By the beginning of the 20th century, the following Bureaux of the USDA were operating: Plant Industry, Entomology, Soils, Biological Survey and Weather, and in 1922 a Bureau of Agricultural Economics was added. An Office of Experiment Stations handled the transmission of the federal funds to the state experiment stations.

However, it took a half-century to organise an effective and stable system of agricultural research and extension in the USA, and 'it was not until the early 1920s that it was possible to claim with some confidence that a national agricultural research and extension system had been effectively institutionalised at both the federal and state level' (Ruttan, 1982).

Federal-level policy making and co-ordination

In 1945, the first attempt was made to co-ordinate federal and state research capacity, when the first of the co-ordinated federal–state commodity research programmes was initiated: a national maize improvement programme. Subsequently, similar programmes were initiated for other crops.

The situation in both federal and state research organisations remained unchanged until 1953, when the Agricultural Research Service was created. In the course of the years, the bureau chiefs had each built up ties with the particular farm industry served by their respective bureaux and with the congressmen who shared their constituency. 'The congressmen liked to place new facilities where the voters could see them, and each bureau chief tended his own satrapy without caring what the others did' (Wade, 1973*e*).

Advisory committees

Wishing to democratise agricultural research, Congress provided in 1946 for a non-scientific advisory body: *The National Agricultural Research Advisory Committee* which was to advise the Administration on broad policy programmes, ensure coverage of all important areas in agriculture and review the objectives of agricultural research as changing economic or technological factors made this necessary. A more specific advisory function was exercised by the *Commodity and Functional Advisory Committees*—which provided a direct link between the USDA and the various professional groups (farmers, processors, distributors) for whom the research involved was of vital interest.

During the 1950s and until the early 1970s, the use of advisory committees was quite effective in bringing to the attention of the USDA the problems which agricultural interests felt required a special research effort; they also made it possible for the farming community to learn at first hand the programme proposals for research by the ARS.

However, since the 1970s, there has been a growing disenchantment with this type of collaborative effort. Although there was full appreciation of the importance of having outside reviews of USDA research, the combined committees required 2000 days of travel, considerable time on the part of those who were members of the committees and, more important, a substantial drain on the time of USDA scientists who were required to prepare comprehensive background papers describing research of the Department and identifying additional research needs (Moseman *et al.*, 1981). These authors write of 'the beginning of a committee bureaucracy syndrome, created by legislative action, that had little authority and did little good'. The advisory committee structures created by successive Acts, are excellent examples of 'uncontrollable costs of social governance that contribute little, if anything, of value to the agricultural research establishment'. Recent (1980) executive orders were issued to discontinue the practice of using advisory committees. The issue of administration by, and through, committees is a continuing one. There has been an attempt to bridge this gap by informal discussions with certain types of industry representation, such as representatives of associations who were invited to consider specific questions at specific times. This has not been done on a continuing basis (Moseman *et al*, 1981).

The 1953 reorganisation

The first major restructuring of USDA research in the 20th century was undertaken in 1953. The Bureaux were renamed Divisions, but the power structure remained unchanged. The Administrator remained effectively powerless to co-ordinate, or to shift resources from one Division to another, without the consent of the Division Chief and of the Chairman of the House Agricultural Appropriations Subcommittee (Wade, 1973e).

One rationale for the 1953 reorganisation was to respond to the Hoover Commission Report of 1949, which described the USDA as 'a loose confederation of independent bureaux and agencies' (Moore, 1967).

The reorganisation of 1953 abolished the Office of Experiment Stations. The work of the Office was taken over by two new ARS divisions: the State Experiment Station Division and the Territorial Experiment Station Division. This issue remained a discordant factor in USDA–SAES relationships, until 1961, when a new independent agency, the Cooperative State Experiment Station Service, was created, with an Administrator reporting directly to the Secretary of Agriculture (the name of the agency was later changed to Cooperative State Research Service).

The reorganisation of 1953 also had a continuing deleterious effect on the stature and respect for agricultural research in succeeding years.

The success of the Eisenhower–Benson administration in bringing about the drastic reorganisation of agricultural research in 1953, had the effect of subjecting the research structure of USDA to pressures for drastic reorganisations following changes in political administrations, thereby effectively ending the stability and immunity from political interference that the Department had enjoyed for 40 years (Moseman *et al.*, 1981).

A period of rapid change, 1960–80

The period from 1960 to 1980 was marked 'by much volatility concerning the prestige and support for research'.

The President's Science Advisory Committee
The unsettled situation in agricultural research caused the President's Science Advisory Committee (PSAC) to appoint a Panel on Science and Agriculture, focused mainly on the USDA. Among the Panel's recommendations (PSAC, 1962) are the following:

• The present commodity-oriented agricultural structure of the ARS should be modified toward a scientific discipline structure (similar to that of the universities).
• Many of the USDA field stations should be closed, or transferred to the SAES. Existing or new research facilities should be associated with a university.
• The organisation of regional research, with its cumbersome administration, needs to be improved.
• The Agricultural Research Centre at Beltsville should be developed into a recognized national and international centre of basic science.
• A new grant programme, designed for the individual research worker interested in basic research, should be initiated by USDA.
• Opportunities for scientific growth and achievement of agricultural scientists should be improved, and less specific direction by their administrators as to objectives and procedures of research.
• Basic research in sciences related to agriculture should be substantially expanded relative to applied research.
• A committee on agricultural science, composed of scientists representing the disciplines important to agricultural research.

The PSAC review 'was conducted within an environment that was generally critical and hostile to the USDA. The dominant university representation on the Committee on Agricultural Science (appointed in 1962), and the Agriculture Panel of PSAC, and the attitude of staff members of the White House Office of Science and Technology, combined to reflect the low esteem of research in the USDA, and also the viewpoint that university personnel should have a dominant role in the planning and direction of USDA research' (Moseman *et al.*, 1981).

The recommendations of the Agriculture panel, with their bias to basic research and university procedures, completely ignored the mission-oriented role of the USDA.

The National Academy of Sciences (NAS) Committee on Research Advisory to the USDA (1969–72)
The NAS Committee on Research Advisory to the USDA was appointed

in 1969, at the request of the Secretary of Agriculture to: (a) examine and evaluate the quality of science in agricultural research; (b) ascertain gaps in agricultural research and make such recommendations as might be appropriate; and (c) ascertain the extent to which scientists in the basic disciplines relate their research to agriculture and the extent to which agricultural scientists contribute to basic science.

The report was published in 1972 (Report of the Committee on Research Advisory to the USDA, 1972). The findings of this committee were analysed by Wade (1973*a–e*).

The major criticisms levelled against the agricultural research system were briefly as follows:

- *Conservatism*: 'Patterns of growth have been determined by history rather than the needs of time, and when its growth was curtailed as at present, it finds adaptation painful or impossible.'
- *Bureaucracy*: 'Its productive workers are its scientists, but it is ruled by a higher caste that consists chiefly of administrators.'
- *Politicisation*: The agricultural research system is said to be 'governed with the rational dictates of planning, priority setting, and coordination'. In fact, it is politicised from crown to grass roots.
- *Duplication*: 'The federal and the state research systems are doing essentially the same kind of work.' There is an excessive number of field laboratories with a low level of co-ordination and integration of USDA/SAES efforts.
- *Fragmentation of effort*: 'Federal research stations were distributed piecemeal over the country on a strictly pork-barrel basis.'
- *Poor quality of research*: 'Much of the agricultural research is outmoded, pedestrian and inefficient. Support given to basic research is grossly inadequate.'

The 1972 Committee has not been sparing in its criticisms of the SAES system. The research is described as 'fragmentary, duplicative, non-additive, and usually without the critical mass necessary to achieve significant results'. The report further states:

The principal reef on which agricultural research founders is the jealously guarded autonomy of the 53 State stations. The SAES directors supposedly plan research through the *Experiment Station Committee on Organization and Policy* (ESCOP). In practice, ESCOP is mainly a lobbying organization with little effective influence on the policies of individual state experiment stations.

A similar degree of impotence characterises another body charged with coordinating state research: the *Cooperative State Research Service* (CSRS). The CSRS is an agency of the USDA, charged with disbursing federal funds to the State stations and with reviewing the projects the States propose to undertake with the funds. According to the NAS committee, the CSRS review system is 'inadequate to screen effectively for duplication and does not judge scientific merit. If it was really effective as a coordinating mechanism, it would then be unacceptable to SAES directors'.

Like the PSAC report, the NAS Committee Report reflected more concern about the freedom of action and independence of the individual research worker than with the design and conduct of mission-oriented research directed to the solution of national agricultural constraints and problems (Moseman *et al.*, 1981). The strong academic bias of the Committee and its panels is reflected in many of the comments and proposals of the report, such as: continued strong emphasis on basic research in such areas as photosynthesis, nitrogen fixation, membrane phenomena, etc.; the evaluation of productivity of research in terms of numbers of publications, without consideration of the fact that 'problem solving research is measured in terms of more productive plants and animals, new pest and disease control practices, and similar contributions, which require more time in the field and less time at the desk' (Moseman *et al.*, 1981); the consistently negative tone with respect to problem-solving, mission-oriented research and the strong endorsement of individual initiative in the choice of research subjects.

Whilst most of the attention has been focused on the scathing criticisms of the NAS Committee report, the report did also make a number of constructive proposals such as: that changes be made in the USDA research administration to ensure a greater participation in the decision-making process by active researchers and the lower echelon administrators who are closest to the programmes; that mechanisms be developed for easy replacement of unsatisfactory administrators; that USDA should recruit the best scientific leadership available; that unproductive branch stations should be closed and resources transferred to areas of greater scientific productivity; that SAES should make greater use of peer judgment in evaluating scientific personnel; that the USDA provide increased opportunity for active researchers to advance to the top salary classifications without having to become administrators; that there should be broad subject matter programme reviews at least every five years, etc.

Wade (1973*e*) concludes his review of the scathing criticisms of the US agricultural research system with the bottom line: 'yet if the state of agricultural research is as debilitated as the NAS committee believes, how has US agricultural production come to be the marvel of the world?'.

The 1972 reorganisation of ARS

The ARS was again reorganised in 1972 (quite independently of the NAS Committee report), in order to simplify administration. Wade (1973*e*) states that the real purpose of the reorganisation was to reduce the power of the Division chiefs and to channel more authority through the ARS administrator's office. The centralised Divisions established in 1953 were abolished.

The major thrust of the reorganisation was to assign the line operating authority to the field, with four Regional Deputy Administrators, to be located within each of the four SAES Regions. Each Region would have an administrative services staff, together with programme planning, development and evaluation staff, and information and biometrical service support. Each of the Regions was to be subdivided into a number of Research Area Centres.

The reorganisation of 1972 did not appear to have improved the image of USDA research. Almost a decade after its inception, the Office of Technology Assessment (OTA, 1981) writes:

AR is not organised to manage, conduct, and be responsive to broad regional, national and international needs and interests of the United States. When the 1972 reorganisation of USDA's Agricultural Research Service transferred line responsibility to four regional administrators, the National Programme Staff, USDA (NPS) was left without direct line responsibility for programme development, staff selection, resource allocation, etc. This caused AR to lose much, if not all, of its ability to plan, manage, and conduct research on broad regional and national problems. AR's research has become more oriented to local and state issues. Not only does this provide opportunities for duplication, but it increases the likelihood that: (1) broad regional and national interest will not receive adequate attention and (2) Federal funds appropriated for these purposes will be diverted or used inefficiently.

The Science and Education Administration (SEA)

The SEA was formed in 1978, integrating for the first time, at the federal level, the USDA's Agricultural Research Service, the Cooperative State Research Service, the Extension Service, and the National Library, with the possibility of centralised planning and coordination. However, the desired melding of these four disparate entities was apparently never quite realised, and the concept was abandoned in 1982, when the four agencies resumed their autonomy.

The response of ARS to criticism

Ever since its inception, but in particular since the 1970s, ARS has undergone 'a barrage of criticism from Congress, Executive agencies of the Government and others. According to its critics, ARS had inadequate control over its appropriate funds, lacked central management and planning, had allowed its research projects to become too localised, and had failed to address significant national problems' (Carter, 1987). The most scathing criticism on these matters and others, was made by the NAS Committee (1972) whose findings were discussed above. Again quoting Carter (1987): 'These were serious charges. Clearly something had to be done if we were to survive as a mission-oriented agency. This led to a thorough self-analysis, which showed that although some of the criticism was frivolous, some was certainly well-founded. Needed changes were made.'

Structural changes
The old ARS organisation of four administrative regions, described by some as 'four separate agencies', was abolished and a new system of 11 geographic areas was adopted, with each area director directly accountable to the Administrator.

One layer of programme management and administrative support was eliminated, certain key functions were centralised in lean Headquarters staff and others were delegated to Area Offices. The current 11 Area Offices are a significant reduction from the former 25 field-reporting relationships in the previous organisation.

The 1983 streamlining and other initiatives over a five-year period allowed ARS to redirect approximately $12 million from overhead to bench research efforts. In addition, the simplified structure enabled an improvement in the management of ARS's management of its national research programme and its capability to provide administrative support.

Operational changes
ARS felt the need to undertake the following operational changes (Carter, 1987):

• to improve co-operation between the Agency and industry, to make research findings more accessible;
• to improve the reporting to the public, through the news media, of the latest findings of the research;
• redirecting the research effort.

The Program Plan

Carter (1987) stresses that of all these changes, the reorientation of the research effort is the most significant. It has led to the elaboration of a Program Plan, which calls for research that will develop the means to reach the Agency's six major objectives:

• managing and conserving natural resources;
• increasing the productivity and quality of plant crops;
• increasing the productivity and quality of animal production;
• making maximum use of farm products for domestic use and export;
• improving human health through improved nutrition;
• developing scientific information systems.

The simple redefinition of objectives in the ARS Program Plan was not the essence or the only constructive response to prior criticisms.

The Plan provided a basic framework for program direction and priority setting. It set targets for accelerated programs. It established a commitment to allocate base resources from lesser priority areas to higher priority areas. It established policies and a centralised decision-making process for program implementation. It streamlined program and management structure. Compared to the 10-year period before 1983, these collective changes were revolutionary (Knipling, E.B., 1988, pers. comm.).

The Program Plan comprises three major parts (USDA, 1983): a Strategic Plan, an Implementation Plan, and an Operational Plan (Fig. 4).

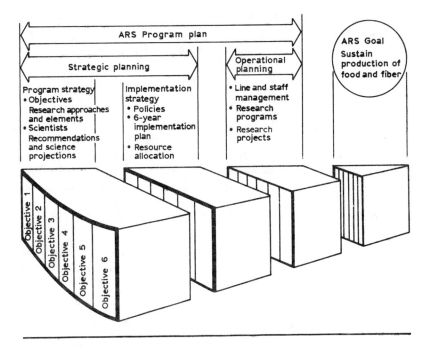

Fig. 4. The relationship of strategic and operational planning to the ARS goal and objectives. (From USDA (1983), reproduced by permission of USDA.)

The Strategic Plan

This identifies all the research that *should* be done, without regard to cost or the resources needed. Over 500 ARS scientists were consulted, as well as users of research, co-operators and supporters, including industry and commodity associations. The Strategic Plan also outlines the principles and strategies of ARS.

Three broad strategies, represented at different points in each of the six objectives, emerged as necessary for ensuring national security in terms of the quantity and quality of future supplies of food and fibre. They are (USDA, 1983):

- develop knowledge of, and technology for conserving and protecting the country's endowment of basic resources, such as germplasm, soil, and water;
- develop knowledge of, and technology for exploiting known production potentials by closing the gap between average and recorded

yields, reducing losses, and improving the efficiency of current production and marketing systems;
- develop totally new technologies for producing, processing, and
  marketing the vast quantities of food, fibre, and other agricultural
  products that will be needed in the next century and beyond.

The Implementation Plan

From the list of research needs enumerated in the Strategic Plan, projects
have to be selected that are appropriate for ARS in terms of need and
available resources. The development and maintenance of the Six-Year
Implementation Plan are responsibilities of the Deputy Administrator
of ARS and the National Program Staff (NPS).

Periodically, the NPS adjusts the Six-Year Plan based on budgetary
decisions and national priorities.

'No totally objective formula is possible for setting priorities and
allocating resources for research. Priority setting for research is the
exercise of informed judgment' (USDA, 1983).

The main criteria used by the NPS for setting priorities and allocating
resources are:

- consistency with the objectives and goals of Congress, the USDA,
  and the Agency;
- need for the research, as expressed by ARS scientists, user groups,
  federal agencies, and the general scientific community;
- potential benefits expected from achieving the stated objectives;
- research capabilities and capacity of scientists, laboratory and
  programme;
- probability of success;
- cost of conducting the research;
- amount and kind of research effort conducted by other research
  organisations.

The Operational Plan

The plan includes the day-to-day management of some 3000 ARS research projects, of which about 2000 are conducted by ARS researchers
and the rest are implemented under contract by researchers in universities and industry. The Operational Plan also encompasses the process
of evaluating the projects and making their results known. Scientists are
required to report annually on the progress of their research.

Workshops

Workshops are problem-solving or research-planning meetings on specific topics. They may be functional or commodity oriented, and may be national, regional or local in scope. Workshops review the state of the art in a field or discipline, identify problem areas, co-ordinate plans, and act as open forums which encourage the creative exchange of ideas.

CURRENT STATUS

The Agricultural Research Service (ARS)

The ARS is the USDA's principal research agency. It has long-standing working relationships with the other research agencies within the Department, the State Agricultural Experiment Stations, and the private research sector. The ARS also works closely with the action agencies in the Department and serves as the research arm for many of them.

The mission of ARS is: 'To plan, develop, and implement research that is designed to produce the new technologies required to assure the continuing vitality of the Nation's food and agricultural enterprise' (USDA, 1983).

As a federal agency, ARS concentrates on research problems that are of national or regional scope. Such research requires unified planning, continuity of effort, and the ability and flexibility to commit the necessary resources to high-priority needs that only a centrally funded national organisation can address.

The ARS research facilities

The ARS has developed over the years unique facilities and laboratories for all the disciplines required for agricultural research. These are located strategically across the major farm and rangeland ecosystems and climatic zones of the USA. They include among others:

The Agricultural Research Centre at Beltsville. Most of the work of the Centre is concerned with basic research that can be applied on a nation-wide scale; many of the projects are carried out in co-operation with the State Agricultural Experiment Stations (SAESs).

The Beltsville staff has played an important leadership role 'in continually assessing geographic potentials and requirements for agricultural research and development throughout the United States, and in coordinating and servicing the many national and regional research programmes cooperative with the SAESs' (Moseman *et al.*, 1981).

The regional research laboratories. The USDA established four regional research laboratories to engage in finding new industrial uses and outlets for surplus agricultural commodities. The laboratory in Pennsylvania is concerned with allergens, animal fats, meats, dairy and hides; the one in Illinois, with cereal and forage crops, oil seeds and fermentation; the one at Louisiana, with industrial crops, and the one in California, with enzymes, fruits, poultry, and wool.

Regional research stations. There are several regional research stations, located throughout the USA, which work in co-operation with the experiment stations in the neighbouring states.

The Research Management Information System

The National Program staff maintain a computer system—the *Research Management Information System* (RMIS)—which provides programme managers with information on precisely where, how and for what projects research funds are being spent and how much the projects cost (Carter, 1987). This makes it possible for a programme manager to co-ordinate the work at different sites and eliminate duplication. It also links ARS with a broader system of the USDA: the *Current Research Information System*, which documents all USDA implemented or sponsored research.

The State Agricultural Experiment Stations

These are independent agencies, usually an integral part of the land-grant college or state university. Each state station discharges its responsibility for agricultural improvement within its state boundaries, but at the same time, it recognises inter-state and federal responsibility for agricultural improvement in the nation as a whole.

The agricultural college usually has three areas of responsibility: agricultural education (the college proper), agricultural research (experiment station) and extension (extension service). Each of the three areas has a qualified scientist as administrator.

A fairly uniform system has evolved in all the land-grant colleges whereby teaching, research and extension have been merged into individual departments; the head of each department is responsible for the co-ordination of these three fields of activity in his subject. He is directly responsible to the Dean of the College in all matters pertaining to teaching, to the Director of the Experiment Station regarding research, and to the Director of Extension for extension matters. A Vice-President for Agriculture (or other appropriate title) has overall responsibility for the three services. Most of the scientists, in addition to their research work at the station, spend part of their time teaching in the appropriate departments of the College of Agriculture. The experiment station provides the facilities and funds for equipment, supplies, labour, etc., and pays a proportion of the salaries of the scientists according to the time spent on research; the University proper covers the remainder of their salaries. The system, on the whole, functions very well; however, whilst research and teaching are very well integrated, the relation between research and extension still appears to leave much to be desired.

Over the years, the structure of SAESs has changed little. Stations typically include a central station and headquarters, which is generally located on the campus of the state's land-grant university, and a number of branch stations located throughout the state.

Stations are organised by departments according to the various scientific disciplines represented on their staffs, such as departments of animal science, entomology, plant pathology, etc. These departments are usually the same as those of the academic unit and, in most cases, also include extension (OTA, 1981).

Although the SAESs retain their traditional focus in serving farmers and the agricultural sector of the state, their role is being substantially modified (Moseman *et al.*, 1981):

(a) Societal needs transcend the needs of the agricultural sector:

- Environmental concerns, ranging from chemicals to water and air quality to aesthetics and the regulatory and legal activities related thereto.
- The spectrum of equity and civil rights concerns, ranging from equal employment opportunity to ethnic equality and small and part-time farmers.
- The concerns of consumers, which have brought impetus to research on food quality, safety, and nutrition research.
- The resurgence of university participation in international technical assistance programmes.

(b) Limitations in funds from state and federal sources lead researchers increasingly to seek grant funds that may be available for non-agricultural or nonpriority topics. SAES directors have lost much of their control over programmes because of outside funding.

(c) Many new staff members do not have a background in farming and see more opportunity for recognition and rewards in research that is not closely identifiable with the needs of farmers.

Overall, despite these influences, SAESs work hard to maintain their identity with the farming community and to work with farm organisations, who continue to be their strongest supporters.

Private sector research

The private sector has made very significant contributions to the productivity and efficiency of American agriculture.

Major fields of interest in industry research have been in agrochemicals, feeds, pesticides, fertilisers, machinery and forest products. The emphasis in agricultural chemicals has been on the protection of plants and animals. Food research has been concentrated mainly on product development and processing.

In general, the work conducted by the private sector tended to complement the work done by USDA and SAES.

RECAPITULATION

There are few countries in the world in which the structure and policies of the agricultural research system have been debated continuously since its inception as in the case of the USA; in very few countries has agricultural research enjoyed such a wide measure of sustained political and grass-roots support and at the same time has been submitted to so much searching and sometimes devastating criticisms.

'Agricultural research in the USA has been criticised in the last decade for a lack of focus or direction, a lack of commitment to the solving of problems or removing constraints in agriculture, and the lack of a sense of urgency. There is also the criticism of a lack of leadership and direction, levelled equally at the USDA and SAES' (Wade, 1973).

The American system of agricultural research has been credited, and rightly so, with a major share in the extraordinary productivity of American agriculture. Over the years it has introduced innovations in every aspect of agriculture, not only in the USA, but in modern agriculture the world over.

How can one accommodate the scathing criticisms outlined above, of which Carter (1987) writes: 'These were serious charges . . . although some of the criticism was frivolous, some was certainly well-founded,' with the indisputable achievements of the system?

It is possible to attribute the extraordinary and indisputable achievements of the US research system, notwithstanding its organisational and administrative defects, to: the goodwill and unfailing support of national and state leaders and of the farming community; the dedication of well-trained research workers to the solution of the farmers' problems, which constituted for them an interesting and worthwhile challenge; the very size of the system, which could ensure complete coverage of practically all problems encountered; a country rich enough to afford a significant amount of duplication and overlapping, but also sufficiently aware of the importance of research for the development of agriculture and the high rates of return to be expected from this investment; and last, but not least, to a clientele willing and even eager to apply the results of research.

Japan

HISTORICAL BACKGROUND

In 1868, the Meiji restoration established a strong central government, replacing the loose federation of over 200 baronies which previously constituted feudal Japan.

Data indicate that the Meiji regime inherited a man/land ratio which was even more unfavourable than that in present-day South and Southeast Asia. However, land productivity was higher, and therefore labour productivity in agriculture in Meiji Japan was at the same level as in Asia today (Hayami & Akino, 1977).

When Japan opened its doors to foreign countries, it was exposed to the same dangers of colonisation as the countries in mainland Asia. The Meiji government realised that the only way to avoid this danger was by developing industry and building a strong army. Funding for this purpose, in a predominantly agrarian country, could only come from agriculture, either by heavy taxation, or by improving productivity (Hayami & Akino, 1977).

The Meiji government chose the second alternative, and undertook, for this purpose, systematic 'technology borrowing' from more advanced countries. Japanese officials combed all countries with an advanced agriculture, such as the USA, England and Germany, for scientific advances applicable in Japanese agriculture (Brown, 1965).

Attempts to transfer these findings indiscriminately inevitably resulted in many costly failures, and the need for establishing local research facilities, even for only evaluating practices developed elsewhere, was soon realised.

In Japan, it was the central government, rather than the prefectural governments that took the initiative in establishing the agricultural research infrastructure (Ruttan, 1982). According to Chang (1964), Japan was the first country in Asia to establish a government experiment station; the Tree Experiment Station was set up as early as 1878. It was followed by an Experimental Farm for Staple Cereals and Vegetables in 1886. In 1893, a National Agricultural Experiment Station was established with several branch stations (Ruttan, 1982).

The initial work consisted mainly of field tests, comparing different varieties and the effects of different management practices. These experiments did, however, result in improved practices and greater productivity. However, the need was felt for a more intensive research effort.

The prefectural agricultural research system developed somewhat slower than the national system; things began to change at the end of the 19th century, and since 1910 more than two-thirds of the investments for agricultural research in Japan are made by the prefectural governments (Ruttan, 1982).

PRESENT STRUCTURE OF THE NATIONAL RESEARCH SYSTEM

Up to World War II, each institute or research station was managed by different administrative branches of the Ministry of Agriculture, Forestry and Fisheries: agricultural experiment stations by the Division of Agricultural Production, animal husbandry experiment stations by the Division of Animal Industry, etc.

After World War II, a thorough reorganisation of agricultural research was undertaken, and all national research institutes and stations were placed under a central body, the *Agriculture, Forestry, and Fisheries Research Council*, responsible for planning and co-ordination of the work of all the agricultural research institutes and stations of the Ministry. The duties of the Council include: advising the Ministry on the operation of all its research institutions; providing basic guidelines for research programmes aimed at benefiting the farmer in the fields of agriculture, livestock, fisheries and forestry; co-ordinating the work of the institutes and stations; raising the standard of qualifications of the research workers; making financial allocations to local research bodies; and maintaining an index of all research projects undertaken by the Ministry's research units. The Council has 16 project leaders, each responsible for a specific field such as rice culture, forage crops, livestock, horticulture, soils, fertilisers, etc.

Research units were classified into three categories according to the type of research undertaken: the national agricultural research institutes, concerned with basic research; the regional experiment stations, doing applied research; and the prefectural experiment stations, doing adaptive research.

The National Research Institutes

The *National Institute of Agricultural Science* was established in Tokyo in 1950. It undertakes fundamental research in agricultural sciences (such as genetics and plant physiology), is responsible for research on problems that are of importance to the country as a whole or to a number of regions, and research that requires expensive or specialised equipment.

In addition to the National Institute of Agricultural Science, there are also a number of other national research institutes, each concerned with specific commodities or disciplines such as horticulture, sericulture, tea production, forestry, food, livestock, agricultural engineering, and agricultural economics.

Regional Experiment Stations

Japan is divided into eight agricultural regions, each of which has a Regional Agricultural Experiment Station which studies the specific problems of the region it serves. One of these stations, situated in the Saitama Prefecture, has been raised to the status of Central Agricultural Experiment Station, and carries out experiments both of national significance and of importance to the region in which it is sited.

Prefectural Experiment Stations

Japan is also divided into 46 prefectures or provinces, each having one or more experiment stations, in which the research results of the national and regional institutes are tested under local conditions.

Staff

In 1984, the number of researchers engaged in the national institutes was about 2500, and in the regional experiment stations about 800. The Research Council had about 250 employees (RAPA/FAO, 1986).

In addition to the above, in 1982 there were approximately 7000 researchers at 379 prefectural stations.

The number of agricultural research workers in Japan is about equal that of the US Federal–State system; the 'research-intensity per unit of area or production is much higher in Japan than in the USA' (Ruttan, 1982).

Exchange of researchers

Researchers from the universities, from other ministries or agencies are invited to carry out research in the MAFF institutes, as visiting researchers. Conversely, researchers from the MAFF institutes are sent to the universities and other agencies. There is also an active exchange of researchers between MAFF institutes (RAPA/FAO, 1986). These assignments are, however, of very short duration (one to three months) and are apparently more in the nature of visits than research assignments.

Training

In-service training in specialised fields, within the country and abroad, is provided for administrators and researchers at national and prefectural level.

UNIVERSITIES

Japanese universities are under the jurisdiction of the Ministry of Education and, traditionally, they were not concerned with agricultural research. The system of national and prefectural research developed separately from the institutions of higher education. Hayami & Akino (1977) ascribe this to the Government's desire to achieve rapid practical results for the farmers.

The involvement of the universities in agricultural research has increased in recent years, and fundamental research related to agriculture is conducted, much of it as a product of graduate work (Ruttan, 1982). The Ministry of Agriculture and Fisheries is also supporting fundamental research at the universities which cannot be carried out readily at the national and prefectural institutes, but which is relevant to their work (RAPA/FAO, 1986).

PRIVATE SECTOR RESEARCH

A unique feature in Japan is the enormous number (over 400!) of private and semi-private research organisations, which are concerned with such varied fields as agricultural policies and economics, basic agricultural sciences, sericulture, food processing, livestock, fisheries and fish products, forestry and forest products, etc.

The MAFF supports private sector research financially, and the national institutes co-operate with the private institutions and exchange information with them. Both sides apply for patents for results obtained from joint research.

Measuring both research expenditures and numbers of personnel, Hayami & Akino (1977) found that about 59 per cent of agricultural research in Japan is conducted at government institutions, nearly 38 per cent at universities, and only 3 per cent in the private sector.

Israel

BEFORE THE ESTABLISHMENT OF THE STATE OF ISRAEL

The history of agricultural research in Israel (or rather Palestine as it was in the days of the British Mandate) is unique in the sense that research had to serve two different communities of farmers—Arab and Jewish. The Arab farmers, in common with all the farmers in the neighbouring countries, practised the typical Mediterranean agriculture, which, to all intents and purposes, had remained practically unchanged since biblical days. There were no Jewish farmers in the accepted sense of the term. The villages were peopled by men and women who had chosen farming as a vocation from a sense of national duty, who felt that a return to the land was the first essential step to the redemption of the country—people with a high intellectual level, a great measure of devotion and a dedication to hard work, but with absolutely no knowledge whatsoever of agriculture. The only agricultural tradition that the settlers could fall back on was that of their Arab neighbours.

The Agricultural Research Station, which was founded in 1921 by the Zionist Organisation, was therefore faced with the unique situation of having to devise ways and means of enabling the new settlers to make a decent living from a soil depleted by centuries of abuse, under difficult climatic conditions. They had to manage without reliable data on climate, vegetation, soil types or any other essential information, in a region characterised by its backward agriculture and the complete absence of research institutions. Any improved practices would have to be carried out by people completely lacking in professional training or tradition. In the words of I. Volcani: 'We placed unprepared people on unprepared soil under unprepared conditions.' It was therefore essential that the station undertake the dual task of research and education.

The first problem to be tackled was to devise a completely new approach to agricultural development and planning. The 'Organic Farm', the brainchild of the first Director of the Station, I. Volcani, became the prototype of all farms established by the Jewish Agency. The new conception was based on an integration of animal and field husbandry— two occupations that had been traditionally separate in the Mediterranean region since the days of the Bible. This approach made it possible to replace the traditional soil-exhausting crop rotations by cropping systems aimed at building up soil fertility, so that modern tillage methods, fertiliser use and improved varieties could in their turn progressively increase yields.

The results of the research work were disseminated by extension specialists, who were members of the appropriate departments in the Research Station.

In 1942, following the initiative and promptings of the Director of the Agricultural Research Station, the Hebrew University decided to establish a Faculty of Agriculture, not on the University campus in Jerusalem but in close proximity to the Research Station. An agreement was signed between the University and the Jewish Agency, whereby the senior research workers of the Station would serve as teachers at the Faculty, which would therefore not have a separate academic staff, but only a separate administrative establishment.

The Mandatory Government established a centre for agricultural experimentation at Acre in 1927. Typically, the choice of the site was determined by the fact that state land, previously owned by the Sultan, was available in this area. The fact that the site was geographically remote and the land unsuitable was not taken into account. A number of separate experimental stations for agriculture, horticulture, animal husbandry, poultry and forestry were established, each dependent on the appropriate department of the Ministry of Agriculture.

A manager was appointed to handle the administrative aspects of the centre, with direct responsibility for the animal husbandry station, each of the other stations being under the direct control of the respective Chief Officers of the Ministry of Agriculture in Jerusalem. These officers were also responsible for a number of other, smaller stations that were established in various parts of the country, as well as for advisory and regulatory services.

The stations at Acre were headed by men who had no academic training (only in 1937 was a Station Superintendent with a University degree in agriculture appointed) but had practical experience in farming, unlimited devotion to their work, and made the utmost efforts to overcome the limitations resulting from the lack of formal education by keeping abreast of developments, each in his field. The labourers were mainly recruited from the neighbouring Arab villages and consisted of mostly illiterate workers, but with the skills associated with primitive agriculture.

No equipment or laboratories for scientific work were available—the only laboratory, situated in a small wooden hut, was that of a lone entomologist.

Notwithstanding all these limitations, the Centre had a considerable impact on farming practices and made important contributions to raising the productivity of crops and animals.

As a result of relatively simple experimentations, improved varieties of cereals replaced the traditional land-varieties, appropriate fertiliser practices were evolved, improved breeds of animals and poultry made available, new crops investigated and new branches of horticulture developed. Certified seed was produced and distributed to farmers, and advice as to insect control, based on experimentation, was provided.

Evidently, these successes were possible, notwithstanding the primitive means available, because of the generally low level of agriculture at the time. However, many developing countries are today in the same situation, and the case history of Palestine indicates what can be done under great limitations, provided there is enthusiasm and devotion to the tasks involved.

AFTER THE ESTABLISHMENT OF THE STATE OF ISRAEL

The early days

After independence, the State of Israel inherited the dual frameworks of agricultural research and experimentation of the Jewish Agency and of the Mandatory Government. The declared government policy was to vest responsibility for all agricultural research, education (excluding education at university level) and advisory services in a single department of the Ministry of Agriculture (The Department of Agricultural Education, Research and Extension). Formally, the Department took over authority for the Agricultural Research Station at Rehovot and the Experimental Stations established by the Mandatory Government in the areas which had become part of Israel. In practice, the following situation developed: the Agricultural Research Station at Rehovot maintained, to all intents and purposes, its autonomy; the Agricultural Experiment Station (formerly part of the Acre Centre) was incorporated in the Department of Agricultural Education, Research and Extension; all the other experimental stations (horticultural, animal husbandry, forestry, fisheries) remained dependent on the respective departments of the Ministry; and a number of new research units were established by the Director of Agricultural Education, Research and Extension.

For many years, all attempts to integrate all these elements into a single organisation structure, according to the declared government policy, failed in the face of the difficulties in solving personal problems, overcoming vested interests, etc. Only many years after the establishment of the State was it possible, progressively and piecemeal, to achieve a reorganisation of agricultural research and, with small anachronistic exceptions, to establish a rational structure. The special Department for Agricultural Education, Research and Extension was dissolved, education was transferred from the Ministry of Agriculture to the Ministry of Education, extension became a unified service, and the responsibility for all agricultural research was vested in a central research institute— The Volcani Institute of Agricultural Research, to which all experimental stations and research units were gradually transferred.

Relations between research, education and extension

It was stated above that early in the history of the Research Station extension was part of its responsibility. The senior staff of the Research

Station also served as teachers at the Faculty of Agriculture, situated on the same campus. There was therefore full integration of research, education and extension.

This situation was not maintained for long. Responsibility for extension was first transferred to the respective departments in the Ministry concerned with commodity production, and subsequently consolidated into a single service in the Ministry. Research Station and Faculty started growing apart, though no valid objective reasons for this development could be discerned. Separate staff was appointed to teaching posts in the Faculty, separate research was initiated and the chasm between the two bodies deepened. The proportion of joint Research–Faculty staff became smaller and smaller, as time went by. This was felt to be an untenable situation, by the responsible authorities in the Ministry and the University.

In 1960 the Minister of Agriculture and the President of the Hebrew University signed an agreement whereby the Agricultural Research Station and the Faculty of Agriculture were merged into one organisation: The National and University Institute of Agriculture. In it were integrated all research and teaching in agriculture at university level. A board of trustees was appointed, the Minister of Agriculture and the President of the Hebrew University serving alternately as chairman, and the responsibility for activating the policies decided on by the Board was vested in a Directorate of five members, of which the Dean of the Faculty and the Director of Research were ex-officio members. The first steps undertaken were to merge all parallel departments of the two original bodies and establish a single, unified administration. Notwithstanding the high hopes placed in the merger and the inherent rationality of the concept itself, its implementation was not successful and was eventually abandoned after several years of tension and frustrations. The reasons for the failure were mainly subjective: personal antagonisms and vested interests. There were, however, also objective shortcomings and difficulties:

1. It was a mistake to set up a single administration for such inherently diverse functions as teaching and research. The means for coping with the individual and collective problems of the student body are entirely different from those required for running experimental farms and stations, and separate administrative set-ups are therefore needed, even when there is one common source of directives.
2. It is extremely difficult to unite in a single framework, researchers

who have the privilege of academic freedom and others who are committed to oriented research. The former are not prepared to forgo their privilege, whilst the Ministry cannot forgo its right to direct research into those channels which it considers essential.

A new attempt was made to prevent a complete divorce between the Research Institute and the Faculty. An agreement was signed, of which the main provisions were:

- Faculty members who were prepared to assume the necessary responsibilities and obligations would be assigned research functions within the Research Institute;
- Members of the research staff of the Institute who had the necessary qualifications, would be entitled to serve as instructors for graduate and post-graduate students in the Faculty; they would also have priority in any teaching assignments that would become vacant in the Faculty;
- Joint services, such as libraries, publications, etc., would be maintained as far as possible.

The Directorate continued to function, mainly as a co-ordinating body, entrusted with the implementation of the new agreement.

The US–Israel Binational Agricultural Research and Development Fund (BARD)

BARD was established in 1977 by the governments of USA and Israel for the purpose of promoting and supporting research and development in agriculture on projects of mutual interest for both countries.

In the agreement between the two countries, the following principles were adopted (Halstead, 1988):

- Research may be initiated by either country and must be co-operative between US and Israeli scientists.
- Research will be of mutual benefit to both countries.
- All proposals will be reviewed independently in each country.
- Recommendations for awards will be made by a joint US–Israeli Technical Advisory Committee, which will base its judgement on the results of the detailed reviews of each country.
- These recommendations will be evaluated by the Board of Directors, made up of three Americans and three Israelis appointed by their respective government departments.

• The research budget will be based on the income from the endowment contributed by both governments.

A decade of co-operation has proven that the fund has been very successful, through a 'lean and flexible organisation, with low overhead and minimal procedures, while addressing the agricultural priorities of both countries' (Halstead, 1988).

Morocco†

THE COLONIAL PERIOD

Agricultural research was initiated in Morocco in 1919, when the French established an Agricultural Experiment Garden in Rabat. A few years later, the Garden developed into the Agricultural Experiment Service.

In the course of the years, research departments and experiment stations were added. With each expansion, the names and the status of what was becoming a research system were changed; what remained unchanged throughout this period was the orientation of the research efforts exclusively to the needs of the colonial power and the 'colons'.

Just prior to independence, the Service de la Recherche Agronomique et de l'Expérimentation Agricole (SRAE) consisted of six large Departments based on disciplines: Soils, Botany and Plant Genetics, Agricultural Chemistry, Technology, and a General Service Department. The system possessed 10 experiment stations and farms. All these research units were concentrated in the most favoured regions of the country, largely settled by the French colons.

The quality of the research carried out by SRAE was of a very high standard, practically equal to that of research in metropolitan France. The research staff consisted entirely of French nationals; actually the French research institutions regularly sent young research workers to Morocco for in-service training at SRAE.

As a result, the agriculture practised by the colons was completely up to date in its application of modern technology, and a typical dual economy developed in the country, with the native farmers continuing to practise traditional, mainly subsistence, farming.

---

†This case history is based largely on the report of an ISNAR mission (ISNAR, 1984*a*).

AFTER INDEPENDENCE

After independence, a Moroccan national was appointed Director of SRAE; however, the staff remained practically unchanged, with only 8 per cent of the 127 researchers being Moroccans and the remainder French expatriates. The large majority remained concentrated in Rabat, notwithstanding the increasing number of experiment stations.

The period 1956 to 1966 is characterised by expansion of the research system and continual changes in its organisation and status, a process that we will not detail.

In 1962, a French–Moroccan accord for co-operation in agricultural research was signed and the network of experiment stations continued to expand. In 1966, the SRAE lost its independent status to become a department in the Ministry of Agriculture and Agrarian Reform (MARA).

The number of expatriates in the research service diminished gradually, so that in three years (1963–66), their proportion had dropped from 92 per cent to 67 per cent of the total; however, the numbers of research workers had also been reduced during the same period, from 127 to 86. The main reason for this reduction was that many newly trained researchers left the research organisation for better remunerated posts in government administration and elsewhere.

The departure of the expatriates, and the migration of the nationals caused a marked imbalance between the number of scientists in relation to the significant increase in research facilities, which by 1973 had expanded to cover the country in its entirety, including the more marginal zones.

After 1976, the number of researchers increased rapidly, to reach 228 in 1984; the research network also continued to expand, and increased to 48 experiment stations by 1982.

This expansion did not affect the overall research procedures and objectives. These were still oriented towards the most advanced agricultural sector. The research work itself remained concentrated in the laboratories and experiment stations of the Service. The quality of the research deteriorated rapidly and its creditability diminished. The major reasons for this process were: (a) the lack of balance between the expansion of the infrastructure and the available human resources, and (b) the fact that the research had not adapted its methods and objectives to the completely changed national and social context within which it was operating.

THE INSTITUT NATIONAL DE RECHERCHES AGRONOMIQUES

In 1981, the National Institute of Agronomic Research (INRA) was established by a special Law, with financial autonomy, and took over the staff and facilities of the research department of MARA.

INRA is managed by a Administrative Council, of which the Prime Minister (or his delegate) is Chairman, and comprises as members the Ministers of Agriculture, Finance, Planning, Higher Education and Scientific Research, Interior, and Health, the President of the Chambers of Agriculture, and two representatives of the Agricultural Associations. The Administrative Council delegates part of its functions to an Executive Council, under the chairmanship of the Director of Education and Research of MARA. The Director of INRA assists in the deliberations of both Councils, without having voting rights.

Structure of INRA

Divisions
INRA comprises four major Divisions: research and experimentation; regional services; administration; and supporting services. The *Division of Research and Experimentation* is the most important, with about 75 per cent of the research personnel of INRA. The Division consists of three *Services—Plant Production*, with 10 Commodity Stations and 3 'Discipline' Stations; *Ecology*, with 10 very small Stations, all situated in Rabat; *Animal Husbandry*, described as an 'embryonic service' with 5 agronomists and 6 Stations, established with foreign help but practically inactive.

The Regional Services consist of several *Regional Centres*, each linked directly to the Directorate of INRA. Their function is to co-ordinate and administer the *Regional Experiment Stations*. These are of three major types: *general*, which undertake experimentation on behalf of any of the Centres, *single-commodity* and *seed-production* stations.

Fermes d'Application (pilot farms), 22 in number, are small stations with one or two workers, in which research findings are tested under conditions which are as near as possible to those under which the farmer operates.

Financial resources
The major source of funds is from the government budget; additional

sources are derived from the sale of farm products and from contractual services. Foreign aid has also made significant contributions, mainly in equipment, training and the provision of consultants.

The total investment in 1984 for INRA amounted to 0·32 per cent of the agricultural GDP.

By law, INRA enjoys 'financial autonomy', and each Director has a 'budget'; in practice, all expenditures have to be authorised by the Financial Controller in Rabat, involving long and complicated procedures which effectively hamstring most field operations. This results in the paradoxical situation that a budget, inadequate in itself, still leaves 'unused' funds at the end of the year!

Human resources

In 1984, the total number of INRA employees was 2225, of whom 10 per cent were Ingénieurs Agronomes, engaged in research and services; 21 per cent were technical assistants; and 69 per cent administrative personnel. Only a few expatriates were still serving in INRA.

A major weakness is the constant migration of research workers from INRA after a relatively short period of service. During the 12 years between 1963 and 1975, 112 research personnel were newly recruited and 53 left the service. The average service period for the research personnel does not exceed five years! All this is a certain indicator of unsatisfactory working conditions leading to lack of continuity in the research projects.

Foreign Aid

Bilateral agreements. France no longer has the monopoly of assistance to agricultural research, and is presently reduced to providing training for researchers and aid for a few research projects. Other bi-lateral agreements are in force with Germany (mainly forage crop research), Spain and the US (mainly dryland farming). Previous aid agreements with Eastern Bloc countries (Bulgaria, Rumania, Poland) have lapsed.

International aid. INRA is co-operating with a number of international bodies on research projects, with ISNAR for advice on strengthening the national research system, and with the World Bank for the development of experiment stations.

OTHER INSTITUTIONS INVOLVED IN AGRICULTURAL RESEARCH

INRA has no recognised monopoly for agricultural research in Morocco. Besides the *Universities*, other *Departments of MARA* are active in research in their respective fields; the number of researchers involved is at least equal to that serving in INRA. *Provincial Directorates*, which are responsible for integrated development projects, also undertake research activities. The *Societés d'Etat* (Statutory Boards) and various *Commodity Associations* are also involved in research.

REVIEW AND CONCLUSIONS

In the course of the history of agricultural research in Morocco, there has been a constant increase in the number of experiment stations, constituting a large and complex network of 78 stations and 64 experiment farms, linked to 14 Regional Centres. This number may not be excessive in relation to the size of the country and its numerous ecological zones, but it is completely disproportionate to the human and financial resources available to INRA, especially in view of the excessive concentration of researchers in Rabat, leaving an average of fewer than four workers for each of the regional research stations. The situation is further complicated by difficulties of communication and transport, leaving very small groups of researchers in practical isolation from their peers.

A further acknowledged weakness of the system is that the location of the experiment stations has not always been well chosen. Also, the cost of maintenance of such a complex infrastructure accounts for a considerable proportion of the funds available to INRA, leaving too little for research proper.

INRA has come belatedly to the conclusion that the number of experiment stations needs to be curtailed, by closure of some, and consolidation of others. Since 1981, an effort has been made to accelerate the decentralisation of research personnel.

The fragmentation of research activities among a large number of bodies is a further drawback. Whilst the research activities of the universities is a legitimate and integral part of their functions, the undertaking of research activities, funded from public sources, by semi-public and other bodies is bound to weaken INRA (cf. p. 225–6). The major justification for the present situation is that INRA does not have the means to respond to the needs of the different bodies engaging in their 'own' research, thereby creating a vicious circle which perpetuates an unsatisfactory situation.

A more serious allegation is that there is a lack of confidence in the quality of the answers to be expected from INRA, and the unduly long delays until such answers are forthcoming. In this connection Faraj (1986) writes:

> Among the structural factors, we find the analysis and constant criticism that users of research—farmers, professional organisations, services of the Ministry of Agriculture—voice every time the role of agricultural research is discussed: 'research is searching for itself', 'research is not useful to farmers', 'research is done in theory,' etc. Overall, the significance of research is neither perceived or felt.

How far these criticisms are justified, or are allegations raised to justify the fragmentation of research, is difficult for an outsider to judge, but they do indicate a very serious situation and present a poor image of the major institution responsible for agricultural research in Morocco.

There is no gainsaying that the national research system as a whole, consisting of a number of disparate bodies without co-operation or co-ordination between them, leads to competition for scarce resources and unnecessary duplication. As to INRA itself, its human resources are inadequate in both number and professional formation, in relation to the size of the country, the diversity of its regions, the importance of agriculture in the economy and the many problems it faces. The problem is aggravated by the inability of INRA to attract competent staff and to keep those it has recruited, for reasons beyond its control.

The constraints in human resources contrasts with the extensive and expensive infrastructure, much of it financed by ill-advised foreign aid, which in the short run is a financial and physical burden, causing a shortage of funds for research activities and an excessive dispersal of the research force.

A further factor which hamstrings the research effort of INRA, is a centralised, complex and rigid administration, dependent on a bureaucracy outside of INRA.

Finally, it should be stressed that the overall national investment in agricultural research is insufficient, amounting to less than 0·4 per cent of the agricultural GDP.

We have described in detail the weaknesses of the NARS in Morocco, not because it is characteristic of this particular country, but on the contrary, because these shortcomings are common to many of the LDCs in general, and of Africa in particular. What is special in this case, is simply the frank exposé of the actual situation in the literature, something that is generally lacking elsewhere.

Malawi

AGRICULTURAL BACKGROUND

After independence, Malawi concentrated its efforts in development by fostering production by estates, whilst smallholder farming was relatively neglected. In recent years, however, more attention has been paid to small farmers, and the physical infrastructure (access roads, health facilities, water supply) has been improved.

The population is 6·3 million people, of whom 85 per cent are directly in agricultural production, and 90 per cent of all arable land is already under cultivation.

Malawi has a better than average economic performance in the region; agriculture provides only 40 per cent of the GDP, of which 84 per cent is contributed by smallholders. However, no more than 25 per cent of their production enters the monetary economy. Industry contributes about 20 per cent, and services almost as much as agriculture (Malawi Government, 1982).

Food production has generally kept pace with population increase, so that the country is substantially self-sufficient in food supply, excepting in years of drought. Three agricultural commodities, tobacco, sugar and tea, are the major source of foreign exchange. Maize is the major food crop (Maida, 1982).

A National Rural Development Programme (NRDP) is responsible for all agricultural development projects; since the end of the 1970s, NRDP has provided an extensive range of farm services and agricultural inputs to improve the productivity of small farmers.

The country is divided into eight agricultural development divisions (ADDs), subdivided into 40 development areas.

STRUCTURE OF THE NARS

Department of agricultural research (DAR)

The Department, as part of the Ministry of Agriculture (MOA), is responsible for research to serve smallholders and for plantation crops, excepting tobacco and tea.

Research activities of DAR are carried out in three major and eight minor research stations (Gadabu, 1988, pers. comm.), and about 220 district trial sites.

Major research stations
Bvumbwe Research Station serves the Southern Region; it has a wide range of studies on soil fertility and moisture regime; variety and management trials on vegetables, fruit trees, maize, wheat, cassava; plant protection; crop storage. *Chitedze Research Station* serves the Central Region; its activities include both crop and livestock research; farm machinery and pasture improvement. *Lunyangwa Research Station* serves the Northern Region, and is also the main centre for cassava and sweet potato research (Gadabu, 1988, pers. comm.).

Only three of the stations have an adequate infrastructure. The distribution of the stations over the country is uneven, and some stations may be poorly located (ISNAR, 1982*a*).

Animal health and industry department

Also part of the MOA, the Department shares responsibility for livestock and poultry research with DAR. However, the respective areas for research on animal nutrition, cattle and poultry management, and animal breeding programmes between the two departments are not well defined (ISNAR, 1982).

Other agencies involved in agricultural research

The *Tobacco Research Authority* is a statutory body which conducts research on tobacco breeding, agronomy and curing programmes, at two stations. The *Tea Research Foundation* took over responsibility for tea research from government and has achieved a world reputation in its field. The research is largely funded from a cess derived from the sale of tea.

The *University of Mali*'s two Agricultural Colleges (Bunda and Chancellor) do research to support agricultural development. In consultation with MOA, Bunda College leads the national programmes for beans and pigs.

Research at the colleges is funded by the University Research and Publications Committee, but the sums made available are small. Some college staff receive support for their research from foreign agencies, such as USAID (bean research) and UNESCO (study of structural changes in Malawian economy). The Bunda College staff also participate in trials on farmers' fields and at the MOA experiment stations, with support from MOA.

Research personnel

Effective management in DAR has been hampered by lack of staff continuity, both at Headquarters and in the experiment stations, but the situation is improving following the return of increasing numbers of researchers from postgraduate training abroad.

Most researchers are recruited after receiving a BSc degree at the Agricultural Colleges. Some received training whilst working for DAR, and upon graduation were reappointed as professional officers.

MOA research staff serve within the Malawi Civil Service. Once recruited, most researchers remain in the same grade, and the same salary scale, indefinitely. Promotion posts are limited in number, and promotion is only possible when a post becomes vacant. Selection for promotion posts is by the Public Service Commission, on the basis of written or oral examination, with emphasis on knowledge of administrative procedures.

At the time of independence, the Malawi research system had 23 researchers, of whom 21 were expatriates (Maida, 1982). Since then, most of the expatriates have been replaced by Malawians.

DAR has a total establishment of about 500 professional and technical posts, of whom about 20 per cent are professional research officers. About 80 per cent of the professional officers work at the three main research stations.

Research planning

The Planning Division of MOA helps to plan development projects, which involves some related socio-economic analysis. The Economic Planning Division of the Office of the President evaluates project proposals for economic feasibility, budgetary implications, and conformance with government policy. However, no senior research personnel seem to be involved regularly in national level meetings to set agricultural development policy. There is apparently no overall planning of the national research programme and co-ordination of the research efforts of the various units is weak.

Funding

Malawi is investing less than 0·5 per cent of its agricultural GDP in agricultural research.

Services

DAR provides a number of technical services, including soil and plant analyses, soils survey, crop protection, plant quarantine, and seed certification.

Farmers' participation

Farmers do have an influence in the determination of the research programme. They draw the attention of researchers and extension workers to their problems; a few commodities farmers' associations are represented at meetings with research personnel to discuss research programmes (Gadabu, 1988, pers. comm.).

RECOMMENDATIONS FOR IMPROVING
THE RESEARCH SYSTEM

An ISNAR team has made a detailed study of the Malawi NARS, and has listed a number of important recommendations, which have also served as the basis for a NARS development programme to be implemented with the assistance of the World Bank.

Among these recommendations, which incidently reflect the weaknesses of the present system:

- To increase the investment in agricultural research, as rapidly as the system can effectively absorb it, until it reaches at least 1 per cent of agricultural GDP. As the technology adopted by farmers becomes more complex, it should move towards 2 per cent.
- A network of experiment stations should be planned after careful analysis of the needs of the major agro-ecological regions of the country and the agricultural problems that need to be addressed. Where appropriate, the existing stations should be strengthened, and those that are poorly located, closed down.
- National co-ordinated commodity research programmes should be

developed for the major smallholder enterprises, and closely linked with extension.

- Research management procedures should be adopted that relate budgets to specific research programmes and services, and that facilitate evaluation of progress and decisions on necessary changes.
- Every effort should be made to involve college staff and post-graduate students in the national research programme.
- Socio-economic research should be carried out to analyse the effects of agricultural policy decisions on various groups of producers.
- A special extended salary scale should be adopted for researchers.
- A plan of the scientific manpower needs of the NARS should be developed, and the necessary steps taken for gradual implementation. The possibility of training researchers to the MSc level at Bunda College should be examined (ISNAR, 1982).

Reorganisation (Sibale, 1986)

Following the review carried out by the ISNAR team, DAR appointed a Local Preparation Team (LPT) to make a more detailed analysis of the existing agricultural research system. A team of consultants from the International Agricultural Development Service (IADS) was engaged to work with LPT to prepare an Agricultural Research Strategy Plan and a National Agricultural Research Project Document, and an Agricultural Research Master Plan. The three documents were completed by 1983. The Government of Malawi subsequently appointed a Steering Committee to oversee the implementation of the reorganisation.

Senegal

THE EVOLUTION OF AGRICULTURAL RESEARCH

Research on crop production, animal husbandry, and oceanography have developed independently, under different organisational structures. In this study of agricultural research in Senegal we will confine ourselves to crop production.

The period 1921–38

Agricultural research was initiated in Senegal when an experiment station for groundnuts was established at Bambey, in 1921. Research concentrated on varietal screening of groundnuts, and the improvement of simple tools for production.

In 1933, work was expanded to include research on millet, sorghum and cowpeas. In 1936, a chemical laboratory was established, and the first pedological studies were undertaken.

The application of the results of this research programme resulted in a slight increase in the yields of groundnuts and cereals but the yield levels achieved (0·8 ton/ha for groundnuts and 0·4 ton/ha for the cereals) were still characteristic of a relatively primitive agriculture.

The period 1938–50

In 1938, France decided to establish a *Federal Centre for Research* at Bambey, responsible for research relevant to the entire dry tropical zone under French rule, from Senegal to Niger. About 20 French researchers undertook investigations on the intensification of agriculture, with priority for groundnuts. Breeding of this crop was expanded, more knowledge on the soils of the region was obtained, recommendations on fertiliser use were established, and tillage problems on small farms investigated. Improved tools were made available to farmers and an improved breed of draught bullocks was developed. A 'package' of improved practices for groundnut and cereal production was established and recommended to farmers, resulting in significant increases in yields to 1·5 ton/ha, on the experiment station's fields.

After Independence

The period of French guardianship
In 1960, with Independence, the Federal Research Services of the colonial period were dissolved. The Governments of Senegal and of France concluded an agreement whereby the management of agricultural research was entrusted to several French institutes.

Bambey became the National Centre for Agricultural Research, and its management entrusted to the French Institut de Recherches Agronomiques Tropicales (IRAT). Research in various fields was to be undertaken by the relevant French institutes: for groundnuts, the Research Institute for Oils and Oil Crops (IRHO); for animal husbandry and veterinary science, the Tropical Institute for Medical and Veterinary Research (IEMUT); for forestry research, the Technical Centre for

Tropical Forestry; and for fishery research, the Office for Overseas and Technical Research (ORSTOM).

During this period, agricultural research was supervised by a Committee of Research Co-ordination, chaired by the Minister of Rural Development, which defined the research objectives and entrusted their implementation to the research centres managed by the French institutes mentioned above.

THE SENEGALESE INSTITUTE OF AGRICULTURAL RESEARCH (ISRA)

In 1974, ISRA was established with the mandate of 'undertaking and developing research on crop production, animal husbandry and fisheries of relevance to the economic and social development of Senegal', and the Institute was placed under the control of the Ministry of Scientific Research.

Management and direction
Overall management and direction are provided by the following organs: an *Administration Council* responsible for defining the annual research programmes, establishing a training programme for researchers, deciding on development programmes based on the results of research, and resolving all administrative problems, a *Scientific and Technical Committee* which examines all research and development programmes, reports annually to the Administrative Council on the progress achieved; recommends new research orientations, and improvements in research structures, a *Consultative Commission for Agricultural Research* makes recommendations to the Interministerial Committee for Research. The chairman of this Commission also serves as chairman of the Scientific and Technical Council, which includes amongst its members the Director-General of ISRA, Heads of Departments, a representative of the Ministry of Scientific Research, the Dean of the Faculty of Science, and one or more foreign specialists.

Departments

ISRA comprises seven Departments: Agronomy and Bioclimatology; Rural Socio-economy; Forestry and Hydrobiology: Animal Husbandry and Veterinary Science; Fisheries; Rural Engineering; and Soils and Hydrology.

The National Centre for Agricultural Research, the other research centres and stations, with their personnel and assets, were transferred to ISRA.

Problems resulting from the evolution of the research structures
(Arnon, 1976)

ISRA, after its establishment, could not start with a clean slate, each period in the evolution of agricultural research in the country having left its imprint on research structure and content.

Lack of balance in the research programmes
   Imbalances in the research programmes are:

   (a) between commodities—a major part of the available resources have been devoted to groundnuts and to a lesser extent to cereals, whilst little effort has been invested in industrial and forage crops, and practically none in vegetable and fruit production.
   (b) between cropping systems—the major effort has been invested in rain-fed agriculture, whilst irrigated cropping has been relatively neglected.
   (c) between regions—a major research effort has been concentrated on the Central Region, which is the ecologically most favoured part of the country and also the most densely populated. As a result, the disparities between regions have increased considerably.

Siting of the research centres, regional stations and substations
The siting of research structures has not evolved in line with the structural changes that have taken place in the research system. Some of these stations were established to cope with specific problems that have become obsolete, and have therefore lost their *raison d'être*; others are not well located to serve their present vocation.

Lack of balance between sources of funding
The high proportion of foreign funds for research, in relation to national funding, is a serious constraint to the adoption of a research programme based strictly on national priorities. Overall shortage of funds tends to favour existing projects and discourages new lines of research.

Expatriates and national research workers
There is a marked delay in the training of Senegalese research personnel in relation to requirements, and the means to accelerate the process are still lacking. Consequently, expatriates still constituted, in the early 1980s, more than half the research staff.

Proposals for improvement

Structure of ISRA
Whilst there were commodity-oriented research departments for animal husbandry, forests and fisheries, there were none for crops; research on crops being confined to discipline departments. It was therefore proposed to add departments for research on field and industrial crops, and horticulture, respectively.

Regionalisation
A reassignment of human and material resources to ensure a more equitable distribution of the research effort between the different regions was deemed necessary. The number of Regional Centres to be reduced to the absolute minimum needed was essential in order to ensure a minimum 'mass' of researchers for each one. Each Centre should also be multi-commodity, in order to facilitate diversification of production, and integration of crop and animal husbandry.

Planning
ISRA had established organs for the orientation and co-ordination of research, but programming of research was weak and the project evaluation process was cumbersome and ineffective. Planning was carried out without consideration for budgetary constraints, so that almost all proposals are approved by the evaluating committee, resulting in a programme proposal whose cost was generally double the budget available. The result, arbitrary elimination of projects (generally the innovative ones) by the financial authorities. An improved and streamlined procedure for research programming was therefore urgently needed. Hitherto neglected or discriminated production areas needed to be included in the research programme. Allocations for research from national budgets needed to be increased to about 1 per cent of the agricultural GDP.

Training
A training programme for the next decade to reduce, and finally elimi-

nate, the overdependence on expatriate research workers was a major requirement.

Pre-extension research

Senegal has pioneered a new concept of agricultural research: Pre-Extension Research, which was the precursor of, and model for, Farming Systems Research (cf. p.343–7). It was a major contribution to the adoption of new technology by the small farmers of Senegal, because new findings of research were first tested under the conditions under which the farmers operated, the constraints to their adoption were identified, and when necessary, the farming systems adapted to the new technology.

Ivory Coast†

AGRICULTURE IN THE IVORY COAST—A MODEL OF DEVELOPMENT

Achievements

Since Independence, economic development in Ivory Coast has been exceptionally rapid, in relation to other LDCs. The average GDP growth rate of 7·7 per cent per annum between 1960 and 1980 has been surpassed by only a few oil-producing countries. Prior to 1960, the average per capita income was among the lowest in Africa south of the Sahara, by 1980 it was the second highest (US $1250).

These achievements are all the more remarkable because they were mainly derived from the agricultural sector. The national average agricultural growth rate between 1960 and 1980 was 3·9 per cent, unsurpassed elsewhere in Africa. Agriculture employed 79 per cent of the economically active population, and generated about 90 per cent of the export earnings.

Ivory Coast has become the world's largest producer and exporter of cacao, the largest African producer of palm oil, the largest producer of cotton and sugar in West Africa, and the third largest exporter of coffee in the world.

†This case history is based on the report of an ISNAR mission of which the author was a member (ISNAR, 1984*b*).

Drawbacks

This spectacular growth has also had its price:

- Little or no attention was paid to the conservation of natural resources. Land has been depleted by exhaustive practices and loss of forest cover.
- Increased social and regional disparities between the forest and the savanna zones. Average differences between the two regions are 1 to 4 in monetary terms, and 1 to 2 in global revenue terms, when home consumption is included.
- Excessive reliance on foreign markets, prone to considerable fluctuations.
- Growth rates have dropped, from an annual 4·2 per cent in the decade 1960 to 1970, to 3·4 per cent in the decade 1970 to 1980. Production levels for the major crops have also stagnated.

NEW DEVELOPMENT STRATEGIES

The former main objectives of the agricultural development strategy were to achieve a high agricultural growth rate so as to increase rural incomes, job opportunities and improve the foreign trade balance. These objectives have been expanded to include better coverage of food needs, reduction of the inter- and intra-regional disparities, and improvement of living and working conditions. The measures adopted include the development of co-operatives, to serve as partners to the development agencies; the extension of credit and reduction of subsidies and grants; and settlement aid for young, beginning farmers.

ORGANISATIONAL STRUCTURE

The report of the ISNAR mission to Ivory Coast states: 'The conglomeration of agricultural research institutions in Ivory Coast. . . form an intricate puzzle in which not all the pieces fit together' (ISNAR, 1984*b*).

Agricultural research, together with research in other areas, is under the Ministry of Scientific Research (MRS). The technical ministries responsible for various aspects of agricultural development (agriculture, livestock production, water and forests, marketing), maintain links with

MRS through interministerial committees, but their influence on research policy is mainly advisory.

The agricultural research is carried out by some 20 independent institutions, with varied statutes, supervisory structures, and modes of financing. Until 1971, when MRS was created, there was a multiplicity of individual research programmes that were resistant to control and coordination. Ever since its inception, MRS has strived to integrate these institutions into a national system.

Institutions under joint MRS–French control .

The most important research institutions in which 80 per cent of Ivorian and expatriate scientists work have been inherited from the colonial period, and French influence and aid were evidenced in all apects of their work.

Office of Overseas Scientific and Technical Research (ORSTOM)
ORSTOM is largely supported by France. It is divided into scientific sections, most of which are involved in research. It serves the whole country, and most of its work is carried out with the GERDAT group and development agencies.

Group for Study and Research for the Development of Tropical Agronomy (GERDAT)
The GERDAT group comprises eight research institutes, of which five, known as 'Forest Institutes', work on specific industrial or export crops: timber (CTFT), rubber (IRCA), coffee and cocoa (IRCC), citrus, pineapples, bananas and other fruits (IFRA), oil and coconut palms (IRHO). Each institute has its own central station, secondary station, substations, etc., its own legal status (different from each other), and its own accounts. There are practically no relations between the different institutes.

Three other institutes, known as the Savanna Institutes, have been formally grouped under the Savanna Institute (IDESSA), and converted into departments for food crops, fibre plants, sugar crops and animal products.

ORSTOM and the eight GERDAT institutes have their headquarters in France. Each has a network of research laboratories and centres in France and overseas. All nine have independent research strategies, which are defined by their respective executive boards. The units in the Ivory Coast are the biggest outside France.

Institutions controlled by MRS

Several specialised national research institutes have been established, and are mainly staffed by Ivorians. The two most important are the Ivorian Centre for Technological Research (CIRT), and the Ivorian Centre for Economic and Social Research (CIRES).

Over half the CIRT staff work on agro-food technology, and over half the CIRES staff work on rural economics and sociology. Since 1979, structural ties of CIRES with MRS have been replaced by links with the Ministry of National Education.

Institutions not under MRS jurisdiction

Institutions of Higher Learning
The National School of Agriculture and the Abidjan Faculty of Sciences have over 20 faculty members who take part or supervise research programmes sponsored by MRS.

Special Institutes
These, such as the Institute for Tropical Ecology (IET), and the National Floristic Centre (CNF), have relatively small laboratories. There is also a joint service for electronic microscopy (GERME).

Non-institutional agricultural research

Research activities conducted by various ministries, development agencies, special development projects funded through foreign aid, and other public entities, such as the Centre for Agricultural Machinery (CIMA), employ close to 100 senior specialists, mostly expatriates, recruited directly or through bilateral agreements from France, Germany, etc., or through international agreements with FAO, IBRD, etc. The funds involved in this research may equal that allocated to 'official' research.

There is a certain amount of overlapping, and no co-ordination, between the MRS-sponsored research and that of the various agencies. For example, IDESSA's Centre of Zootechnical Research (CRZ) and the Society for the Development of Animal Production (SODEPRA) both carry out research on cattle, sheep, pigs, poultry, animal pathology and nutrition, but have no co-ordination or official contacts between them, because they belong to different ministries. When the Ministry of Animal Production and SODEPRA organised a national seminar on animal production, CRZ, belonging to MRS, was not invited to attend, attesting to the lack of relations between institutes engaged in the same field of research.

RESEARCH PROGRAMMING

There are very few countries that have made as great an effort as Ivory Coast to develop a programming system for research activities and financing.

A trial was first made to adopt the French 'Graphe d'Evaluation' system, itself based on the Planning, Programming, Budgeting System (PPBS) developed by the Rand Corporation in the USA.

This particular method was given up, but since 1971, MRS has continued its efforts to develop a system for programming research that could facilitate, at national level, control of the disparate research system described above, and thereby replace institutional financing by programme-specific financing, congruent with national development policy.

The MRS programming system

The programming system uses three complementary steps: (a) standardised identification of research activities, (b) evaluation of the costs of the research, and (c) screening of research projects, and ranking according to priorities.

In the beginning, each research institution established its own programme, which was reviewed by a special committee, the Commission de Synthese.

By 1976, the institutions' individual programming was replaced by seven major, theme-specific, multi-disciplinary and multi-organisational committees, which dealt with problems directly related to the priorities of the National Development Plan, whose objectives were also theme-specific.

The priority objectives of the 1981–86 Plan were: improvement of food production, greater attention to the hitherto neglected savanna zone, and small farmer production.

Shortcomings of the system
Despite the numerous changes, the programming procedure still had a number of shortcomings:

- MRS programming endeavours continued to be hampered by the fact that certain institutions had their own very substantial resources, that national funds and manpower were insufficient, and that many research and development activities were outside the control of MRS.
- Committee meetings were too numerous and the treatment of programming was too superficial. Representatives of the Development Agencies could not devote much time to a thorough analysis of the proposals and current projects, as their responsibility for development projects was a full-time job. They attended the meetings mainly to remain informed on research activities, rather than participate actively in actual decision-making. Most decisions were therefore left to MRS and the individual research institutes.
- *All* projects were reviewed annually, resulting in considerable waste of time for all the participants.
- Budget ceilings were not set beforehand, leaving programme committees free to make an excessive number of proposals.
- Calculations were based on averages, rather than on real costs; as a result, some projects assigned to MRS co-financed projects were underfunded.

These deficiencies resulted in cumbersome administration of the programming process, a waste of time for institution executives and scientists, and the need for retroactive financial adjustments.

The impact of the programming

Though some improvement in co-operation and co-ordination between the institutions engaged in agricultural research, which previously was non-existent, has been noted, this major objective of the programming was far from being achieved after 10 years of effort.

Relations between the GERDAT institutions had changed very little. There was no structured co-operation between the Departments of IDESSA and their counterparts in the Forest Institutes. Nor had there been any marked effect on relations between similar disciplines in different institutions, whose people did not even know each other. Most of the scientists were expatriates who did not feel the need for such contacts in the Ivory Coast, as they had adequate scientific support from their home institutes.

Notwithstanding its shortcomings, the programming system has been-able to achieve some control over research conducted in the country. The foreign institutions working in the Ivory Coast, even though they were respected for the quality of their work, and appreciated for their financial support, were made aware of the need to conform their activities to national priorities.

The Ivorian programming system has even had an influence in other countries, through foreign institutions operating in the Ivory Coast, or through visiting scientists.

Research programmes

Research on *industrial and export crops* occupies more than 60 per cent of the researchers, and absorbs 75 per cent of the funds allocated to commodity-oriented programmes. It has had considerable impact on the production of large plantations of oil-palms, rubber, bananas and pine-apples as well as on the small-scale production of cotton. It has had little effect on small coffee or cocoa plantations.

Research on forests (production and ecosystems), animal production, and on *food crops* (which account for close to 40 per cent of the agricul-tural GDP) occupies only 21 per cent of the researchers, and accounts for only 17 per cent of the funds allocated for commodity research. The most important food crops (plantain, yam, cassava, etc.) receive only 25 per cent of the allocation to food crops; the remainder is used by a large rainfed rice production programme.

Reporting the results of research

Provisional and final results of research are scattered throughout various publications, such as scientific reports, annual progress reports, specialised journals, etc., which are not always available in Ivory Coast. The only scientific publication that reports on agricultural research in the Ivory Coast is *Cahiers*, edited by CIRES. Documentation is kept within each institute and is difficult to find.

RESEARCH STAFF

The total number of scientists engaged in agricultural research in 1981 was 212, of whom only 28 per cent were Ivorians. In the institutes co-managed with France the proportion was only 15 per cent. This is a major problem of the agricultural research system, and though more nationals were being trained, this was done mainly to round out existing teams, and not to prepare Ivorian staff for a take-over, through training in key disciplines and in research management.

FUNDING OF RESEARCH

The total outlay for agricultural research from public funds amounts to only 0·52 per cent of the agricultural GDP, which is very low. Foreign assistance to agricultural research is dominated by France, who is providing an annual budget of over US$ 1 million, mainly for paying the salaries for about 90 French scientists.

SUGGESTED IMPROVEMENTS FOR THE NARS

The ISNAR mission to the Ivory Coast made numerous proposals relating to a large variety of structural and functional aspects of the agricultural research system.

The first main proposal was the creation of a *National Agricultural Research Institute*, to integrate progressively all the disparate elements engaged in agricultural research into a single framework. This predicates a considerable increase in the national share of human and financial resources, without which it will be impossible to achieve national control over the system.

The second major proposal was the creation of a *Research Training Centre*, managed by the research institutions, to provide third-cycle training for future research workers, high-level ministry officials, executive-level staff from public and private companies concerned with agricultural development, and students and senior staff from neighbouring countries.

The creation of a *National Documentation, Information and Scientific Publications Service* was proposed, to serve the needs of research, higher education, ministries, development, etc.

Proposals were also made for strengthening the *Research Programming* and simplifying the procedures, along the lines described in Chapter 7.

Kenya

AGRICULTURAL BACKGROUND

Agriculture accounts for 30 per cent of the country's GDP, and 70 per cent of its exports; it provides employment for 78 per cent of the country's labour force. The bulk of the farming sector are subsistence farmers. Large farms in the high potential area are being divided into smaller commercial farms (Wapakala, 1986).

Until the 1980s, agriculture has depended largely on very productive farming on 18 per cent of the land resources with high or medium potential. The remaining 72 per cent is semi-arid to arid, and has only been used by nomadic pastoralists. With increasing population pressure, the semi-arid areas are being increasingly settled.

The evolution of agricultural research—before Independence

A number of research institutions were established early in the century: the Agricultural Laboratories, Coffee Research Services, Veterinary Research Laboratories and a Plant Breeding Station. A significant development of experiment stations took place between 1940 and 1950, when 16 national and regional experiment stations were established under the Department of Agriculture (Muturi, 1981).

Several agricultural research stations were established by the East African High Commission to concentrate on problems common to the three East African countries: the East African Agricultural and Forestry Research Organisation (EAAFRO), the East African Veterinary Research Organisation (EAVRO), the East African Trypanosomiasis Research Organisation (EATRO), and the Tropical Pesticides Research Institute (TPRI). Research in marine and freshwater fisheries was also institutionalised at the East African Community level (Muturi, 1981).

AGENCIES INVOLVED IN AGRICULTURAL RESEARCH

Agricultural research in Kenya is carried out by four government ministries, the Universities, International Institutions, and the private sector.

Government ministries

The *Ministry of Agriculture* has a Scientific Research Division (SRD) managed by a Director of Research. There are three categories of research stations under SRD. The *National Agricultural Research Stations* (NARS) have country-wide responsibility for major commodities, such as maize, wheat, sugarcane, horticultural crops; integrated research, as for dryland farming; disciplines, such as plant protection; or services, such as soil analysis. *Regional Research Stations* (RRS) engage in research specific to the region in which they are located, and co-operate with the NARS to facilitate country-wide testing of new varieties or management practices. Small *Sub-stations* or *Testing Sites* are operated by NARS or RRS, and provide facilities for controlled experiments in ecological niches, and for the implementation of pre-extension trials on farmers' fields.

The Ministry of Agriculture is also responsible for the Research Stations for coffee and for tea, which operate under the management of the respective Statutory Boards. A similar arrangement is in operation with the National Irrigation Board.

The *Ministry of Livestock Development* has a Research Division, responsible for a network of livestock research stations and discipline-oriented national research stations for animal production, veterinary sciences and for range management, respectively.

Other ministries are the *Ministry of Environment and Natural Resources*, which has research programmes in forest resources management, wildlife and range management, and fisheries and the *Ministry of Water Development* which supports research in an area of interest to agriculture.

Other institutions and agencies

These are involved in various aspects of agricultural research, such as the Kenya Seed Company, the National Seed Control Service and several agricultural development corporations. Others in the private sector, such as the Wellcome Laboratories and Kenya Canners, also undertake applied research.

The universities

Operating under the aegis of the Ministry of Higher Education, they comprise the Faculty of Agriculture of the University of Nairobi, the Kenyatta University College, and Egerton College.

Many members of the faculty are personally involved in the research activities of various ministries. About 40 per cent of the staff's time is devoted to research, and many serve on research advisory boards.

The International Research Centres

These are active in Kenya. The International Centre for Research on Agroforestry Systems (ICRAF), the International Laboratory for Research on Animal Diseases (ILRAD) and the International Centre of Insect Physiology and Ecology (ICIPE) are based in Kenya; in addition, the International Crops Research Institute for the Semi-arid Tropics (ICRISAT), the International Potato Institute (CIP), the International Livestock Centre for Africa (ILCA), the International Centre for the Improvement of Maize and Wheat (CIMMYT) have outreach programmes in the country.

Bilateral programmes

During the recent past, Kenya has had no fewer than 15 bilateral agencies involved in agricultural research contributing approximately US$ 5 million per year (FAO, 1984). There is little or no contact between the various projects, even when they are complementary.

PLANNING AGRICULTURAL RESEARCH

An ongoing debate on a desirable institutional arrangement for formulating science policy started in 1965 (Muturi, 1981).

When Kenya became independent in 1963, there was no centralised responsibility for defining policy for science, technology and research in all sectors of the economy, including agriculture. Responsibility for research policy lay with individual ministries and there was no co-ordination between them.

The establishment of the National Council for Science and Technology (NCST) was to be a major milestone in Kenya's aspirations to base its future development strategies on modern scientific and technological processes (Wapakala, 1986).

In addition to NCST—which was to establish policies and strategies for science and technology, to assess the financial resources required, and their allocation to different agencies—several other organs were established through the Science and Technology Act:

The Sectoral *Advisory Research Committees* (ARCs), for agriculture, industry, and natural sciences, for the promotion and co-ordination of research, are composed mainly of research administrators, and advise on details of the research programmes required to implement priority projects arising out of policy formulated by NCST and to review progress in implementation.

The *Agricultural Sciences Advisory Committee* (ASARC) is the ARC for the agricultural sector, and is concerned with research on agriculture, animal husbandry, forestry, wildlife and water development. This committee is required to work closely with the management committees of the research institutes, the technical services of government and the private sector (Muturi, 1981).

The *Statutory Research Institutes*: representatives of the NCST and of ASARC are entitled, by law, to participate in the management boards of the research institutes, thereby providing a link between the implementing organs and the advisory and supervisory functions.

Specialist Research Advisory Committees (SRAC) have been established to deal with commodities (maize, sugar cane, wheat, pyrethrum, etc.) or with specific disciplines (soil science, plant pathology, entomology, etc.) of country-wide concern.

Provincial Research Advisory Committees chaired by Provincial Directors of Agriculture, and comprising researchers, extension workers and farmers' representatives, translate national policies into regional research projects.

The complex decision-making system described above is in stark contrast with the way agricultural research programmes are determined in practice. In 1986, the research programmes were still being determined at the institutional level.

The Government policies for research and development, established through its Development Plan and policy papers, provide only brief and general guidelines. The ministries, their research institutions and scientists are expected to translate these policies into feasible projects. The initiative for the initiation of research programmes lies mainly with the research station staff. This is as it should be. However, many of the staff are young and inexperienced, and often lack guidelines from a well-informed, organised farming sector or extension service. What emerges as a research programme 'may be a collation of individual interpretations of (i) the mandate of the station, and (ii) the problems of the farmers in the particular locality' (ISNAR, 1985).

At the higher level of the ministry, an ISNAR team saw no clear evidence of a sustained attempt to establish programme priorities or to relate priorities to the resources available (ISNAR, 1981).

Wapakala (1986) summarises the situation as follows:

The present organisation is not geared to meet the manifold current and projected needs of Kenya's growing agricultural industry; within the research programme, some projects are well conceived and executed, while others are virtually worthless. All too often there is wasteful duplication and diversification of effort with a total disregard of priorities in the light of economic justification; there is a lack of coordination and assessment of research priorities by both foreign assistance donors and the Kenya Government, and there appeared to be no clear-cut concept of donors' role in Kenya's development process.

Available resources

Infrastructure
The agricultural research system's assets were built up during a long period by the former colonial power, the Ministry of Agriculture, the government parastatals, and the former East African Community. These comprise research buildings in Nairobi, Muguga, and Kabete, and about 40 national and regional research stations and substations located throughout the country. Additional facilities exist at the Tea Research Foundation, Agricultural Development Corporation and the Government Boards; as well as at the Universities.

These physical facilities for agricultural research are adequate in scope but not in quality. Many buildings and structures in the major research stations are not functional because of lack of facilities and equipment (ISNAR, 1981).

Funding
Agricultural research in the 1980s accounted for 70 per cent of the gross national expenditure for research.

Funding for maintenance and operation is inadequate, mainly because of spreading limited resources over too large a number of stations.

Human resources

An ISNAR mission in 1985 noted 'that lack of manpower for agricultural research continues to be one of the most important constraints to the planning, organisation and management of research in Kenya' (ISNAR, 1985). Out of a total of 566 research staff, only one-third had research-oriented training at the postgraduate level (MSc, equivalent diploma, PhD). At many of the research stations, the staff lack 'critical mass'.

THE NEED FOR REORGANISATION

The phenomenal increase in the number of research stations over the years has not always been justified by agro-ecological considerations or other needs. In some cases the stations have evolved from testing sites or have been established as a means of retaining assigned land, or of establishing a government presence in the area.

This lack of policy and planning has led to a 'a proliferation resulting in the extension of agricultural research resources over a large number of stations, some of which are hardly viable' (ISNAR, 1985).

It is clear that this system needs to be rationalised, if an efficient agricultural research service is to be established.

Under the provisions of the 1979 Science and Technology Act, the Kenya Agricultural Research Institute (KARI) was created to operate as a semi-autonomous parastatal research institute, under the Ministry of Agriculture. KARI was intended to combine the research services of the Ministry of Agriculture and Livestock Development and the Ministry of Environment and Natural Resources with those of the former East African Community to form a comprehensive research organisation, capable of providing national co-ordination, management, and execution of research in agriculture, livestock and forestry. These roles have not been realised, and KARI has remained limited to the functions formerly under the East African Community at Muguga (Wapakala, 1986).

In furtherance of the objective to strengthen Kenya's agricultural research system, ISNAR was invited to undertake a review of the system in 1981. This was followed by two other ISNAR missions in 1982 and 1984. Other missions and committees, before the ISNAR involvement, including an FAO mission headed by the author in 1978, have been requested by the Government of Kenya to advise on strengthening and rationalising the organisation of agricultural research in Kenya.

ISNAR (1985) in its report to the Government of Kenya made the following recommendations:

- develop a more unified, comprehensive organisation for the planning, promotion and execution of research;
- improve the mechanisms for the determination of research priorities and allocation of resources;
- strengthen communications and information flow within the system, and between the system and other arms of government and non-government agencies;
- improve procedures for the effective transfer of research results to the extension services and the farmers;
- increase the numbers of well-trained research workers;
- improve the research environment;
- more socio-economic research;
- adopt a small-farmer oriented research strategy.

Similar, if not identical recommendations had been made by other missions, and the most obvious way to implement these recommendations would have been to operate KARI as envisaged by the 1979 Science and Technology Act.

Unfortunately, the numerous efforts to strengthen agricultural research have not resulted in any substantial changes. Indeed, 'agricultural research organisation, management and infrastructure have deteriorated and may soon reach an all-time low that will make it difficult for research to meet the needs of the nation for agricultural growth and development' (ISNAR, 1985).

India

PRE-INDEPENDENCE

India was a major producer of agricultural raw materials for Britain as well as a major market for her industrial products. These interests were reflected in Britain's approach to agricultural research and development in what was her most important colony.

In order to improve agricultural production, a number of model farms, agricultural schools and colleges, as well as agricultural research institutes were established, beginning from the mid-19th century (Menon, 1971).

In 1905, the first agricultural research institute was established at Pusa, with the help of a donation by an American philanthropist. An experiment farm for training students, and a cattle farm were attached to the institute. It was named the Imperial (now Indian) Agricultural Research Institute, and it was intended to serve as the apex of a unified research organisation with ramifications in all the provinces; in each province colleges of agriculture were to be established which would also engage in research (Ranhawa, 1958).

The Agricultural Research Council

In 1919, it was decided to transfer the responsibility for agriculture from the Government of India to provincial governments. This meant that the central government had no further control over research, teaching and policy-making in agriculture, and that there was no longer an agency with authority to coordinate the research and policy of the individual provinces. However, the provinces did not have the financial resources necessary to initiate effective research programmes.

In 1928, a Royal Commission was appointed to enquire into the state of agriculture and rural economy in India. The Commission found that the periodic conferences of Ministers and Directors of Agriculture were not sufficient to ensure effective co-ordination of the work of the provinces, and proposed the establishment of a Central Council of Agricultural Research, to be entrusted with the responsibility for the promotion, guidance and coordination of agricultural research throughout India (Ranhawa, 1958).

Though entrusted with wide responsibilities, the Council never had the opportunity to play a prominent role in co-ordinating and promoting agricultural research throughout India. It had no control over the Research Institutes which were administered by the Ministry of Food and Agriculture. It did not operate Institutes of its own. It gave *ad hoc* grants to various research institutes, universities and other research organisations (Menon, 1971).

THE AGRICULTURAL RESEARCH SYSTEM
AFTER INDEPENDENCE

The Constitution of India, which came into effect in 1950, provided for the Union Government to be responsible for the promotion of special studies on research; for the co-ordination and determination of standards in institutes for higher education and for research, and scientific and technical institutions.

The States were to be responsible for agriculture and animal husbandry, including education and research; for plant protection and the prevention of animal diseases. In practice, agricultural research was badly neglected in many of the States. Support for agricultural research was provided largely by the Union Ministry of Food and Agriculture, through a number of central research institutes, commodity committees, and directorates, which had been established before independence.

Research institutes of the Ministry of Food and Agriculture

The Indian Agricultural Research Institute (ICAR)
This has developed into the major agricultural research complex in India, covering basic and applied research in a wide range of subjects. It also maintains a post-graduate training centre. In 1984, ICAR supervised directly about 30 Central Research Institutes (Venkatesan, 1985).

The Indian Veterinary Research Institute
In addition to research facilities, this institute also has a framework for post-graduate studies.

Other research institutes of the MFA
In addition to the two major research institutes mentioned above, MFA has a number of additional central institutes such as for rice research, potato research, sugarcane breeding, fisheries research and dairy research.

All these institutes were treated as subordinate, separate units of MFA, and individually responsible to different sections or administrative heads in the Ministry. Not even the research supported directly by the Ministry was subject to adequate internal co-ordination (Menon, 1971). They were also generally ineffective in producing scientific or administrative leadership for the large number of central and state research efforts (Ruttan, 1982).

National institutes and laboratories of ministries other than MFA

A number of institutions not associated with MFA have contributed to the advancement of agricultural technology, such as the National Physical

Laboratory, the National Chemical Laboratóry, the Central Laboratory for Scientific and Industrial Research, the School of Tropical Medicine, etc.

Central Commodity Committees

These were *not* advisory bodies, but semi-autonomous bodies with their own research institutes, financed by government grants and by the proceeds of a special commodity tax. Individual Commodity Committees were made responsible for research in cotton, jute, sugar cane, tobacco, coconuts, oilseeds and areca nuts. For each crop, specialist work on breeding, physiology, agronomy, plant protection, economics and technology was carried out. The field of activity of the Indian Council of Agricultural Research became limited to food crops, grasses, forage crops, spices and horticulture, as well as to general problems such as plant protection, fertilisers, dry farming, control of animal diseases, animal nutrition and breeding, etc. Various attempts were made to co-ordinate the work of the Commodity Committees:

1. The Vice-President of the Indian Council of Agricultural Research was made President of all the Commodity Committees, in order to serve as a co-ordinating link among them.
2. A special framework was created, in which all the Directors of the Research Institutes of the Commodity Committees and the Senators of the Committees on the Board of Research and Extension of the Indian Council of Agricultural Research were brought together to discuss matters of common interest in the fields of research and administration (Ranhawa, 1958).

The Commodity Committees made significant contributions to agricultural progress in their respective fields. The fragmentation of research on a commodity basis, precluded attention to research on cross-commodity problems, such as soil management, farming systems, etc. Nor were the efforts at co-ordination and avoidance of duplication very successful (Menon, 1971).

Regional co-ordinated research stations

These stations were also set up, in which research was organised on a cross-commodity basis, by including a major crop, such as cotton, and all the other crops with which cotton is grown in rotation. Co-ordination of these 'composite' stations with the work of the state stations treating the problems of the same commodities was entrusted to a committee headed by the Agricultural Commissioner with the Government of India.

Ad hoc research schemes
Financed by ICAR out of the cess funds, these schemes are operated at selected centres for short periods of 3–5 years. They are concerned with basic and applied research.

Colleges of agriculture

The numerous colleges of agriculture in India made only minor contributions to agricultural research in the past. Their main function was teaching, and they had no responsibility for research or extension in agriculture (Moseman & Hill, 1964). Their aloofness from the problems of the farming community is indicated by the fact they were not involved in any way in the community development programme of the Government of India, which was started in 1952.

Agricultural universities

In 1948, a University Education Commission was appointed, which recommended a complete change of policy in regard to the functions of the agricultural colleges, and proposed the establishment of 'rural universities', patterned on the US land grant colleges. Impetus to implement these proposals was given by the establishment, in 1954 and 1959, of two joint Indian–American teams on Agricultural Research and Education.

The Government of India decided on the establishment of an autonomous agricultural university in each State and hoped to achieve this aim by 1971. The first two agricultural universities were established in Uttar Pradesh and in the Punjab. All the agricultural research programmes and experiment stations of these states have been made the responsibility of the universities.

By the 1970s, all major states had decided to establish agricultural universities. By 1984, 23 of these were functioning in different states (Jain, 1985). These universities have dispensed with staff selection through the State Service Commission, and recruit their people through their own selection committees consisting of scientists. The procedures are the same as those used by the Council and the Central Institutes, thereby facilitating interchange of research personnel (Menon, 1971).

The State Agricultural Universities (SAUs) derive their statutory powers from the University Grants Commission of the Ministry of Education and the respective state governments which exercise supervisory jurisdiction over them and monitor progress in teaching and research (Jain, 1985). However, ICAR also exercises considerable influence over the SAUs, partly because it funds a major proportion of the research and education activities, and partly because the SAUs accept the general leadership of ICAR (Venkatesan, 1985).

The problems arising from the establishment of Agricultural Universities in India are described in detail in Chapter 3. The controversies on the subject of the SAUs can be illustrated by two citations: they have been called 'one of the most significant landmarks in the history of agricultural education in India' (Review Committee on Agricultural Universities, 1978) as well as 'an example of the thoughtless transfer of inapplicable experience from one civilisation to another' by Hunter (1978), a long-time observer of Indian rural development.

RESTRUCTURING AGRICULTURAL RESEARCH

The system described above, which was largely a legacy of the colonial period, had many shortcomings. There was no effective co-ordination of the work carried out by the Agricultural Research Institutes under the MFA, the Central Commodity Committees, the State Governments, and the more recently established institutes of the Central Council itself. Procedures required for the approval of research projects were complicated and protracted.

A number of expert teams were nominated to examine the set-up, and to make recommendations for improvement, in 1955, 1959, 1962 and in 1963. The recommendations of these teams tended to overlap, primarily because of the lack or slowness in implementation of the recommendations of previous teams. The last team, consisting of eminent scientists from India, the UK, and the USA, emphasised the need for radical change in order to achieve a more efficient and economic use of rare resources in scientific manpower, money and equipment.

Unlike previous other reports, that of the 1963 Research Review team, submitted in 1964, received prompt attention from the Government of India, increasingly concerned at the time with India's inability to meet its food requirements. Government decided to adopt the team's recommendations and to reorganise India's national research system.

The reform itself was strongly influenced by USAID programmes (Ruttan, 1982).

The focal point of the reorganisation was the strengthening of ICAR, and assuring its status as an autonomous research organisation. The Council was made responsible for all aspects of agricultural research and education, and co-ordination of research work on an all-Indian basis. It was also to have a strong extension wing to provide a link between ICAR and its 'clients'.

All the Central Research institutes were to be transferred from MFA to ICAR. Menon (1971) stresses that the Institutes became *'constituent units'* of ICAR, and not *organisationally subordinated* to the Council, in accordance with ICAR's policy in regards to scientific institutions. In practice, this signified that the institutes functioned as the executive arm of ICAR, and all research programmes of the Council were to be implemented by the Institutes.

It was also decided to abolish all Commodity Committees, and the Commodity Research Institutes were also to become constituent units of ICAR. Attempts were made to align the large number of *ad hoc* research schemes sanctioned by the Commodity Committees with the All-India Co-ordinated Research Projects.

Whilst the broad policies and priorities in agricultural research are laid down by high officials of Government, ICAR was given considerable freedom to develop priorities and programmes, subject to existing resources and financial constraints (Menon, 1971).

The All-India Co-ordinated Research Projects (AICRP)

The first co-ordinated, nation-wide crop improvement project, the Coordinated Maize Improvement Scheme, was initiated by ICAR in 1957, with active co-operation of the Rockefeller Foundation. Through screening of introduced maize germ-plasm and locally available strains and varieties, in multi-location trails, four excellent hybrids were released within four years for widescale cultivation.

Following on the unqualified success of the first AICRP, in achieving the full possible benefits from available human, financial and institutional resources, irrespective of whether they were in a national or state framework, ICAR decided to accept the AICRP as a model for national programmes for other major crops, and subsequently for livestock research too (Menon, 1971).

The AICRPs have gradually developed into a three-tier organisation: (a) crop-oriented schemes responsible for developing suitable varieties and broad management practices for individual crops; (b) adapting the research results to the various ecoagricultural zones of the country; and (c) consists of national demonstrations in which the results achieved by the AICRPs are demonstrated on the farmers' fields by qualified scientists (Mahapatra, 1971).

Organisational structure of the AICRPs
Typically, an AICRP consists of a National Co-ordinating Centre and a number of Co-operating Centres, located in different agricultural universities and central institutes, depending on the distribution of the particular crop throughout the country. The Co-ordinating Centre is headed by a Project Co-ordinator, and assisted by a small group of scientists from different disciplines; the Centre is located in one of the central institutes or agricultural universities.

The Project Co-ordinator, an eminent scientist, not only monitors the work of the different co-operating centres, but also organises a research programme of his own, which is designed to strengthen the work of the centres; he also acts as a link between the Co-operating Centres of the project and International Research Centres, for exchange of genetic materials (Jain, H.R. 1985).

Planning an AICRP
The Project Co-ordinator, aided by a team of scientists, prepares the outline of the project and identifies the Centres which should participate in it. Their proposal is considered by ICAR, and is then submitted to the Planning Commission and the Ministry of Finance. After it is approved, it is implemented by ICAR (Kanwar, 1971).

At an annual workshop, the researchers engage in the implementation of an AICRP, review the programme and progress of the project, and plan the work programme for the next year.

Participation of the agricultural universities in the AICRPs
Most of the funds for the SAUs come from ICAR, as the States are generally very limited in the funds they can allocate to them.

Under such a regime, when most of the funds, and hence research directives, come from one and the same source, 'there is little, if anything, to distinguish between the research carried out by the SAUs and that of the Institutes of ICAR' (Venkatesan, 1985). As a result of this situation, the SAUs do not have an appreciable research programme outside the AICRPs.

The National Agricultural Project (NAP)

NAP was established in 1974 with the assistance of the World Bank. The Project, administered by ICAR, was intended to meet the needs for interdisciplinary research of the 116 agroecological zones identified in India.

More specifically, the zones were to be the units for the management of research in each state, with location-specific research to be carried out in the zonal stations. All the research resources in a zone were to be integrated under the supervision of an Associate Director of Research at each Zonal Station, who was also charged with initiating rural development. Strong linkages were to be established between the research staff of the SAUs, officials of the MFA, and farmers in the area (Jain T.C., 1985).

For these purposes the following steps were undertaken (Venkatesan 1985; Jain T.C., 1985):

- Zonal Agricultural Research Stations were established or existing stations strengthened in each of the Zones. Each station was to have a number of supporting stations, in accordance with the size and heterogeneity of the Zone.
- In each Zone, a Zonal Research Advisory Committee would define research priorities. Research, extension and the farmers would be represented on the Committee.
- The NAUs would be encouraged to engage in location-specific research within the framework of NAP, and funded by the latter.
- Seasonal Zonal workshops were to be held twice yearly, before the rainy season and the winter season respectively. These workshops were to be attended by researchers, senior officials of the State Department of Agriculture, and farmers' representatives. Immediately after problems are identified, the whole research programme is reviewed, recommendations for field testing are made and recommendations to farmers are finalised.

By 1985, NAP was being implemented in most Indian states (Jain H.K. 1985).

Indonesia

INTRODUCTION

The Indonesian Archipelago stretches over 5000 km, from the Western tip of Sumatra to the Eastern border of Irian Jaya. It consists of 13 000 islands, of which 1000 are inhabited, with many different cultures and agroecological conditions. The country also has many different farming sectors and farming systems, requiring differentiated services from research and extension institutions (Baharsjah, 1985). The extent of the country, the varied ecological conditions, the many commodities grown, and the lack of adequate communications make it very difficult to establish and maintain efficient agricultural research institutions and to co-ordinate between them. Because of these conditions, and the limited resources available, such efficiency and co-ordination are doubly important.

HISTORICAL BACKGROUND

Agricultural research began with the establishment of the State Botanical Garden in 1817. In 1876, an Economic Garden was specifically entrusted with the introduction and testing of new crops and varieties. The most successful of these were then taken over by estate corporations which established their own experiment stations for sugar cane, coffee, tea, cocoa, tobacco, and other crops.

In 1918, the Economic Garden was taken over by the Ministry of Agriculture, Trade and Industry and renamed 'The General Agricultural Research Station'. This station was concerned only with crops. Subsequently the Ministry established institutes for veterinary research, forestry, and fisheries research. Laboratories for agroecology, pedology and chemistry followed. These institutes worked separately from each other.

During World War II, research activities were intensified to meet wartime needs, and reached their highest peak. Five state and eight private sector agricultural research stations were operating. The latter, sponsored by big plantation companies, did excellent work and achieved worldwide scientific reputations. 'The pride in their past achievements has, however, created continuing difficulties in cooperation and coordination' (Mangundojo, 1971).

At the time of independence, Indonesia inherited an extensive network of agricultural research institutions, but had no trained staff to man them. After diplomatic relations were re-established with the Netherlands, many Dutch research workers returned to the Indonesian Institutes. A decade of calm, from 1950 to 1960, allowed many Indonesian research workers to obtain academic degrees, both at home and abroad.

However, a deterioration in the relationship with the Netherlands led to the departure of the Dutch researchers, before sufficient qualified Indonesians were available to replace them. The situation further deteriorated because of internal political conflicts.

Agricultural affairs were handled by four different ministries. The Ministry of Agriculture alone had 15 research institutes distributed among five different directorates. The research institutes were not only under different ministries, but also had different legal status.

During the period 1960–68, a number of reorganisations were made, but these were motivated more by political pressures than considerations of efficiency (Karama, 1986). The results do not appear to have been very satisfactory, for in 1969, when the first National Development Plan was adopted, the need was felt for a more effective research system. Government, with the assistance of USAID, appointed an Agricultural Research Survey Team, which subsequently recommended the establishment of a strong agricultural research organisation.

The Team's proposal was not implemented; apparently it was felt that an additional reorganisation would only cause another disruption of the work.

Another proposal made by the Team was more successful: the setting up of a National Co-ordinated Research Project, along the lines of the successful Indian experience (cf. p. 120). As this proposal did not involve any organisation upheavals, it did not engender the usual distrust and antagonism following proposals that affected the status of the Institutes. In 1970, a National Co-ordinated Rice Project was initiated, with the objective of integrating all research efforts on rice in a single national programme. Participation was, however, to be voluntary.

THE 1974 REORGANISATION

In 1974, the Ministry of Agriculture was reorganised, with financial assistance from the World Bank. An agricultural Research Board was established with responsibility for research policy in agriculture, and the funding and co-ordination of research programmes. Within the Ministry, two Agencies were established: The *Agency for Agricultural Research and Development* (AARD), responsible for all agricultural research at the national level; and the *Agency for Agricultural Education, Training and Extension* (AAETE).

Since its establishment in 1975, the organisation structure of AARD underwent several changes. Its first steps were to integrate the Research Institutes of the Ministry of Agriculture. However, the control span of AARD became excessive, and the research system had become too large to be planned, programmed and evaluated by AARD alone (Karama, 1986).

Five Research Co-ordinating Centres in five commodity areas were established: Food Crops, Fisheries, Livestock, Forestry, and Industrial Crops (ISNAR, 1986) which took over planning, programming and evaluation functions.

AARD's institutions and national mandates

AARD comprises the following units (RAPA/FAO, 1986): A Secretariat; five Research Co-ordinating Centres in five commodity areas; two Research Centres for Soils and Agro-Economy, respectively; two Service Centres: Statistics & Data Processing, the National Library of Agricultural Sciences; 23 Research Institutes in different agroecological regions; the majority based on commodities, and a few cross-commodity; 42 Research Stations; 151 Experiment Farms and Ponds; and a Board of Estate Crops Management.

In 1984, further changes were made at the Research Institute, Research Station, and Experiment Farm levels, involving a rationalisation of the organisation and research missions of the Research Institutes grouped under the five Research Co-ordinating Centres.

Each Research Institute received more authority to establish its own programme and more autonomy in its budgetary and supervisory activities. Each Institute was given a mandate to become a centre of excellence for a particular commodity and/or agroclimatic area; other institutes, experiment stations and laboratories were to help to support the total research programme for that mandate (Baharsjah, 1985).

Research priorities

Since the late 1970s, research has been focused on diversifying food production, the development of rainfed agriculture, supporting agricultural development in transmigration areas, and above all, improving the productivity of small farms (FAO, 1984).

Most of AARD's Research Centres and Institutes are assigned to do Farming Systems Research, but the number of actual on-farm research projects is restricted because of budgetary constraints (Baharsjah, 1985).

Research complexes

In order to provide for location-specific adaptive research, ten Research Complexes have been set up throughout the archipelago. Within each complex are germ-plasm centres, experiment stations and experiment farms, and laboratories. Though the complexes are not part of the AARD structure, they make their special contribution to the national agricultural research effort (RAPA/FAO, 1986).

Research staff

During the period 1975–83, US$ 500 million in government and donor funds were invested in establishing a strong research staff and adequate facilities for their work. During this period the number of researchers doubled, and the facilities kept pace. In 1985, the number of research staff was 1350, of whom 8 per cent were PhDs, 17 per cent MScs, and the remainder BScs or Ingénieurs. The objective is for AARD to have a research staff of over 3400 college trained scientists and technicians by 1995, and a total staff of 10 000 (RAPA/FAO, 1986).

Korea

INTRODUCTION

Korea is one of the world's most densely populated countries, with 385 people per km^2. The country is dependent on small-scale farming, with an average of 1 ha per farmstead. Of the total land area, only 22 per cent is cultivable, and of this area, 59 per cent is used for rice production.

Until the 1960s, Korea was one of the poorest developing countries, with heavy dependence on agriculture and foreign aid. Since then, agriculture's relative contribution to the GNP is declining, as the result of the rapid industrialisation process, but it still plays a significant role in the national economy.

Korea is one of the few LDCs that continued to invest in agriculture whilst it was fostering industrial development. The industrial sector has attracted manpower from agriculture, resulting in the need to adopt a policy of selective mechanisation for small farms. Government has constructed large fertiliser plants.

Illiteracy has been reduced to 8 per cent, and land reform has changed many tenants into landowners (Chung & Dong, 1984).

In 1971, the results of an election indicated that support for the Government had eroded in the rural areas. The Korean Government responded with a rice-pricing policy favourable to farmers, the strengthening of the extension service, the formulation of the New Village Movement, and a rapid increase in rural infrastructure. The overriding strategic, political and economic objective was to achieve rice self-sufficiency (RAPA/FAO, 1986).

HISTORICAL BACKGROUND

Organised agricultural research started in 1905, when an agricultural demonstration farm was established by the Ministry of Agriculture and Commerce. In 1929, the farm was renamed the Agricultural Experiment Station and several branch stations were established in different localities. In 1946, the station was renamed the Central Agricultural Experiment Station. The Central Livestock and Horticulture Institutes were established in 1952–53 (Kim, 1987).

In 1957, the Institute of Agriculture was founded, combining a total of 29 organisations. The Institute's first task was to rehabilitate badly damaged installations and build new infrastructures, with US aid (Kim, 1987).

The Institute made rapid progress in many fields of research.

In 1962, the Office of Rural Development Administration was founded, replacing the former Institute of Agriculture, to develop new varieties, improved management practices and the dissemination of agricultural technology. In the 1970s, the Koreans launched their own 'Green Revolution'; farmers were encouraged to use more fertilisers, plant protection chemicals, machinery and capital inputs. In a short time, Korea was transformed from a rice importer to a rice exporter (Elz, 1984).

The food consumption pattern has changed, with a greater demand for high quality foods, such as meat, dairy products, vegetables and fruits (RAPA/FAO, 1986).

STRUCTURE OF THE OFFICE OF RURAL DEVELOPMENT ADMINISTRATION (RDA)

The RDA comprises six Bureaux at Headquarters, 14 Research Institutes, and nine Provincial RDA offices.

The RDA Headquarters co-ordinate the work carried out by the Research Institutes, and the Provincial RDAs carry out adaptive research and extension (RAPA/FAO, 1986).

Research programming

Research programming is carried out as follows (Kim, 1987):

- Information on the problems faced by agriculture is obtained from the Ministry of Agriculture, the extension services, the farmers, and the researchers.
- Research proposals are prepared by the researchers, and submitted to the Institutional Cooperation Committee, comprising researchers, university professors, extension workers, and administrative personnel.
- The Committee evaluates the validity of the research objectives, methodology, scientific relevance and economic feasibility.
- The projects are approved on an annual basis, but evaluated twice yearly.

Research funding

The investment in agricultural research has increased four-fold in the decade 1975–85, to reach US$ 32 million, of which 88 per cent is funded by the central Government.

Food and cash crops research accounts for 60 per cent of the research budget. Since 1986, emphasis has been transferred from research on food crops to cash crops, in order to improve the farmers' income.

In order to strengthen and modernise research capacity, a foreign loan of over US$ 100 million has been negotiated, and earmarked for the purchase of research equipment for genetic engineering, plant physiology, mechanisation and new computer systems (Kim, 1987).

Research staff

The number of research workers of RDA has increased almost three fold during the period 1962–85, and numbers about 1000 scientists; the number of extension workers during the same period increased 2·5 times and has reached 8000 workers in 1985. These figures represent one research worker per 2000 farm households, and one extension worker per 250 households (Kim, 1987).

Training

Training abroad includes graduate studies, and short-term mission-oriented training of several months to one year. In the decade 1975–85, the number of researchers trained abroad increased steeply, funded mostly by the host countries or the International Institutes.

A serious problem has been that most of the trainees left RDA for positions at the Universities and the private sector, mainly because of the low salaries, the early retirement age, and the low status. In order to counteract this tendency, Government changed the system of grading, and improved remuneration. The turnover rate was significantly reduced as a result of these measures (Kim, 1987).

Many of the researchers now enrol at the Korean Universities for advanced degrees. An institutional co-operative scheme has been established between the Universities and RDA to conduct joint research projects and joint training schemes for the RDA researchers, as well as to enable appointments of the latter to lecture at the Universities.

By 1985, 45 per cent of the research workers of RDA held MSc degrees, and 15 per cent were PhDs (Kim, 1987).

OTHER MINISTRIES AND INSTITUTIONS INVOLVED IN
AGRICULTURAL RESEARCH (RAPA/FAO, 1986)

Ministries, and the institutions they administer, also involved in agricultural research are:

● the Ministry of Finance: Tobacco & Ginseng Research Institute;

- the Ministry of Home Affairs: Forest Resources Research Centre, Forestry Research Institute, Forest Genetics Research Institute; and
- the Ministry of Education: Universities and Colleges.

Sri Lanka

STRUCTURE OF THE NARS

In 1986, agricultural research was being carried out by 20 separate research institutes, under 10 ministries and the Office of the President (RAPA/FAO, 1986). All research institutions come under the auspices of their respective ministries; however, those controlled by Boards are semi-autonomous and have greater flexibility of operation than those coming under line ministries.

The ministries and their departments

Ministry of Agricultural Development and Research is responsible for research on all crops, excepting rubber, tea, coconuts and cashew, and has four separate research units:

- *The Department of Agriculture* is responsible for research in food crops. The Department maintains nine Regional Centres and two Commodity Institutes. There are lead stations for particular commodities and nationally co-ordinated programmes for the major crops. The lead stations and the programme co-ordinators maintain linkages with the appropriate International Centres. There are also 24 units for adaptive research, for on-farm testing, one in each district. At each Regional Centre, there is a Regional *Technical Working Group*, comprising research and extension workers.
- *The Department of Minor Export Crops* conducts research on cocoa, coffee and spices. This Department has a main research station, and substations for cocoa, cinnamon, and cocoa and cinnamon, respectively. A multi-commodity station is under the control of the Agricultural Diversification Authority.
- *The Agrarian Research and Training Institute* is an autonomous unit, organised in four research divisions: Agricultural Planning and Evaluation, Production Economics and Extension Irrigation,

Water Management and Agrarian Relations, and Market and Food Policies.

- *The Sugarcane Research Institute* carries out mainly research of an adaptive character, because the basic agronomic requirements of cane are already known.

Ministry of Lands and Land Development has two Departments concerned with agricultural research:

- *The Irrigation Department*, which carries out research on the engineering aspects of water management and other small research programmes.
- *The Forest Department* carries out research on species grown in pure stands and forest management studies.

Ministry of Rural Industrial Development has a *Department of Animal Production and Health*, which has seven divisions, including:

- a *Research Division* responsible for the Veterinary Regional Investigation Centres, for a Vaccine Laboratory, and an Animal Virus Laboratory; and
- a *Veterinary Research Institute* engaged in research on pasture and forage crops, animal nutrition, animal breeding, bacteriology, parasitology, and reproductive disorders.

There is no formal mechanism for co-ordination of the research programmes of the two ministries responsible for crops and livestock respectively (Gunawardena, 1986).

Ministry of Plantation Industries is responsible for plantation crops, excepting coconut and sugar cane. The two main export crops, rubber and tea, are each served by a specialised Research Institute. Funding is through cesses on exports.

Ministry of Coconut Industry's Coconut Board is responsible for research on production and processing. The *Coconut Research Institute* has nine out-station units.

Ministry of Fisheries carries out its research in the *Fisheries Research Branch*, the *Institute of Fish Technology* and the *National Aquatic Resources Agency*.

Ministry of Higher Education maintains three Faculties of Agriculture, a Faculty of Veterinary Science, and a Post-graduate Institute of Agriculture. These institutions are primarily engaged in teaching; funding for research is inadequate and depends almost entirely on funds from the Office of the President or from external sources.

The National Resources, Energy and Science Authority

This is the only unit in the research complex which reports directly to the President, whom it advises on policies concerning science, natural resource development and energy. Other functions of the Authority are to initiate research in these areas and to disseminate information.

Co-ordination

No co-ordinating agency such as a Research Council is responsible for formulating national policies and priorities, nor is there any formal mechanism for interministerial co-ordination of research (Panabokka, 1985).

FUNDING

The multi-commodity research institutes receive their funding from annual appropriations, fluctuating according to the state of the national economy. The single-commodity institutes are funded through the income from cesses and from commercial activities.

The overall expenditure for agricultural research was 0·36 per cent of the agricultural GDP in 1975, and 0·77 per cent in 1982 (RAPA/FAO, 1986).

RESEARCH STAFF

There are three categories of research workers: (a) research officers, generally holding a first or second class degree in science or agriculture; these generally undergo a course of post-graduate training overseas; (b) experimental officers, who are generally pass-degree holders in science or agriculture; and (c) agricultural instructors, who hold a diploma from the Schools of Agriculture (Jogaratnam, 1971).

In 1985, there were 500 research workers employed in all the units described above (Panabokka, 1985), of whom almost half have either an MSc or PhD degree. There are a further 100 well-trained staff at the Faculties of Agriculture, Veterinary Medicine, and Animal Science (RAPA/FAO, 1986).

WEAKNESSES OF THE NARS

The NARS in Sri Lanka is extremely fragmented: agricultural research is carried out in 15 separate research institutes and departments, which function under seven ministries and the Office of the President, without any serious effort to co-ordinate these activities. Gunawardena (1986) lists the following shortcomings of the NARS in Sri Lanka:

- weak linkages between policy makers, producers and researchers;
- inadequate channels of communication within the line ministries, between policy makers and researcher institutions;
- absence of a forum to develop national research priorities;
- inadequate level of operational funds;
- low ratio of support staff per scientist;
- lack of a systematic training programme;
- restricted use of programme budgeting among research institutions;
- insufficient emphasis on research areas which require participation by more than one ministry or research institution.

PRIVATE SECTOR RESEARCH

Private sector research plays a very limited role in Sri Lanka. (Panabokka, 1985).

The Philippines

AGRICULTURAL RESEARCH UNTIL THE 1970s

Agricultural research in the Philippines was intensified in the 1950s; it did not, however, make a substantial impact on the agricultural economy in the two decades that followed, despite large sums of government funds spent annually on agricultural research, because of the absence of planning and co-ordination at the national level. Efforts to provide direction for research planning were fragmented and ineffective (RAPA/FAO, 1986).

In 1971, a Research System Technical Panel was appointed, to review the situation and which identified (among others) the following shortcomings of the existing system (RAPA/FAO, 1986):

- research projects are often proposed as a device to increase budget allocations; the funds are eventually diverted to other purposes, because of the low priority accorded to research by the agencies involved;
- inefficient training of research personnel;
- lack of operational budgets;
- fragmentation of responsibility and resource allocation;
- insufficient planning of research projects;
- excessive number of research stations and facilities;
- vague and loosely implemented procedures for preparation, screening and implementation of research projects;
- cumbersome procedures for spending research funds;
- inadequate and ineffective planning of research.

THE PHILIPPINE COUNCIL FOR AGRICULTURE AND RESOURCES RESEARCH (PCARR)

In order to develop an effective mechanism for the use of the scarce resources of manpower, funds and infrastructure, the Review Panel recommended the formation of a central authority 'with a formidable budgetary clout' over all government-funded research in agriculture and natural resources. (RAPA/FAO, 1986).

Following these recommendations, the Philippine Council for Agriculture Research was established, to provide planning, co-ordination and direction for a national research programme in agriculture, forestry and fisheries. Later, mines research was added to the functions of the Council, which was renamed the Philippine Council for Agriculture and Resources Research (PCARR).

The Council was given the mandate to review all research proposals; only those approved by PCARR were eligible for government funding. The inclusion of the Budget Minister and a representative of the National Economic and Development Authority as members of the Governing Board of PCARR ensured relevance of the research programme to the national development goals.

Present structure of the NARS

By 1978, progress had been made in bringing about a more systematic agricultural research system out of the once fragmented and inefficient research operations, comprising a network of research centres and co-operating stations as follows (Drilon & Librero, 1981):

- *National Research Stations*, single or multi-commodity, conduct basic and applied research. The single-commodity stations are funded by tax levies.
- *Regional Research Stations* conduct applied research for the major commodities of their respective regions. Specified research agencies in a region may co-ordinate their work and share their resources in the framework of a Consortium.
- *Co-operating Field Stations* provide facilities or sites for adaptive trials. To date, there are 130 Field Stations in the system.

Each of the research stations may be a main station for one or more commodities, and a co-operating station for several other commodities (RAPA/FAO, 1986).

National Commodity Research Teams
Twenty-nine Commodity Research Teams were established for crops, livestock, forestry, fisheries, soil and water resources, socio-economics, and mines in order to make the national commodity programmes more responsive to, and representative of Philippine agriculture and resources.

Human resources
In 1981, the research staff in the country comprised slightly over 3000 scientists, of whom 12 per cent were PhDs, 35 per cent MScs and the remainder BScs (RAPA/FAO, 1986).

The universities and colleges of agriculture
These institutions are carrying out a significant number of government-funded, development-oriented research projects.

Achievements of PCARR

An International External Review Team was requested to review eleven years of activities of PCARR and drew the following conclusions (RAPA/FAO, 1986):

During the period under review, PCARR had made significant progress towards achieving its objectives. Planning and management of research have been improved; leadership in strengthening the capability of the country to implement a research programme has been provided; a nation-wide system of national and regional research has been established; research priorities related to development goals have been defined; linkages with extension, the farming community, among research agencies within the country and with the International Research Centres have been strengthened.

Support for research has grown, research results have been disseminated and yields and incomes of farmers have increased.

Thailand

Thailand is one of the few countries in Asia in which all agriculture-related departments are under one and the same Ministry of Agriculture and Co-operatives, which is responsible for agriculture, fisheries, forestry and rural development. Nevertheless, agricultural research in Thailand is carried out under five ministries and the Bureau of Universities (Isarangkura, 1981).

AGENCIES INVOLVED IN AGRICULTURAL RESEARCH

The *Ministry of Agriculture and Co-operatives* (MOAC): research is carried out in seven departments: Agriculture (85 experiment stations); Land Development (60 Land Development Centres); Livestock (35 experiment stations); Fisheries; Forests; Agricultural Economics; the Office of the Undersecretary (Four Regional Offices).

The *Ministry of Science, Technology and Power* has an Applied Scientific Research Corporation, which concentrates on agro-industrial research.

The *Ministry of Industry* has a Sugar and Sugarcane Institute, which carries out simple varietal tests and research on cultural practices for sugarcane (whereas more advanced research on sugar-cane is carried out by the Kasetsart University and MOAC).

The *Public Welfare Department* of the *Ministry of Interior* is involved in agricultural development of the highland areas of northern Thailand and carries out research in these areas.

The *Ministry of Education*: research at the 40 agricultural colleges of the ministry is not significant. The four Universities under the *Bureau of Universities* are actively involved in agricultural research.

Most of the research stations are ill-equipped and poorly staffed (RAPA/FAO, 1986).

THE NATIONAL RESEARCH COUNCIL (NRC)

In 1982, NRC formulated an elaborate National Agricultural Research Plan for the period 1982–86, but it lacks authority so that its recommendations are not given due attention by the Budget Bureau (RAPA/FAO, 1986).

The NRC receives a lump sum budget to support programmes which it considers to be of high priority. There is, however, no effective mechanism for co-ordinating the various research programmes, nor specific national guidelines for resource allocation (Isarangkura, 1981).

Agricultural research programmes originate mainly from the departments located in Bangkok; the plans formulated far from the field are frequently unrealistic and generally show little connection with local needs (RAPA/FAO, 1986).

There are no clear linkages between the research and extension services (Isarangkura, 1981).

MANPOWER

Most of the senior research workers (especially those that hold PhD and MSc degrees) work in Bangkok, while the centres and regional stations have only a few BSc level researchers. The Australian Government is providing over 400 fellowships reserved for candidates who undertake to work after the completion of their training at one of the regional research centres.

Taiwan

INTRODUCTION

Taiwan is one of the few developing countries in which industry and commerce have boomed in the course of the decade 1975 to 1985, without this development having interfered with the long-standing commitment of the government to the fostering of agricultural progress.

Industrial development has, however, had a marked impact on agriculture in the country in general, and on the orientation of research in particular. This is reflected in the contemporary national research objectives, which are more similar to those of a developed country than of a LDC: the development of labour-saving production methods, germplasm bank, biotechnology, biological insect control, horticultural production in a protected environment, agricultural machinery, food science and food processing techniques, agricultural pollution control, farming systems for hill agriculture, erosion control, flood control, aquaculture systems, etc. (COA, 1987).

The data in Table 2 illustrate the dramatic changes that have taken place in Taiwan, simultaneously in agriculture and industry, during the period 1952 to 1985.

TABLE 2

	1952	*1985*
Percentage of NDP		
Agriculture	35·9	6·9
Industry	18·0	45·2
Percentage of total employment.		
Agriculture	56·1	17·5
Industry	16·9	41·4
Value of agricultural production (US$ millions)	707	6777
Value of agricultural exports (US$ millions)	114	2108

The annual percentage growth of agricultural production has consistently been higher than the annual percentage growth of population, and as a result, average calorie intake per capita has increased by 35 per cent (from 2078 in 1952, to 2815 in 1985) and of protein by 70 per cent (from 49 g to 83·3 g).

STRUCTURE OF THE NARS

The agricultural progress described above has led to 'a mushrooming' of new agricultural organisations: both administrative agencies and institutions carrying out basic and applied research.

Administrative agencies

The administrative agencies comprise: the *National Science Council* (NSC), the *Science and Technology Advisory Group*, the *Research, Development and Evaluation Committee* (RDEC), and the *Council of Agriculture* (COA).

These are involved in planning, co-ordinating, controlling, supervising, and evaluating the implementation of agricultural research and development plans.

Since 1979, agricultural R & D has been incorporated into the National Science and Technology Development Plan. In the implementation of the Plan, major research projects are placed under the direct control of the RDEC of the Executive Yuan, whilst secondary projects are under the control of COA.

Research institutions

There are about 40 research institutions, at or above provincial level, most of them government-run. They comprise:
  (a) *Central government institutions*: Institute of Botany and Institute of Zoology of the Academia Sinica.
  (b) *Provincial institutions*: Institutes and Improvement Stations for basic and applied research on crop production, livestock, fisheries, forestry, wines, tobacco, sericulture, chemicals and toxic substances. Each institute has from two to nine branches.
  (c) *Universities or Colleges*, seven in all, engage in agricultural research in addition to their teaching functions, in co-operation with national or provincial research institutes.
  (d) *Institutions supported by enterprises or foundations*: these include Research Institutes for sugar, animal industry, food industry, engineering, bananas and pigs, respectively.
The Asian Vegetable Research and Development Centre and the Food and Fertiliser Technology Centre for the Asian and Pacific Region are also located in Taiwan.

Human resources

The total of agricultural scientists and technicians engaged in agricultural research in 1987 amounts to 1957. Of these, 20 per cent are PhDs and 30 per cent hold MS degrees.

As a result of the economic development in Taiwan, increasing numbers of technical personnel have been drawn from agriculture to other sectors of the economy. This has made it essential to recruit and train additional personnel for agricultural research management and implementation. The supply of personnel at the BS level is considered sufficient to meet demand, and manpower planning for the next decade will therefore emphasise the requirement for additional PhD and MS degree-holders. It is estimated that over 1200 additional research personnel will be required over 10 years.

Planning procedures

Planning and evaluating procedures for mission-oriented research that is under the responsibility of COA are as follows:

(1) Funds are budgeted annually for an agricultural research programme as approved by the Executive Yuan.
(2) Direction and content of research projects in accordance with national priorities are resolved separately at the National Science and Technology Conference and the Science and Technology Advisory Group and at the National Science Conference.
(3) A *Sponsoring Agency* is designated to effect co-ordination with co-sponsoring and executing agencies, with whom the Sponsor consults on the division of tasks, allocation of budgets and scheduling of work. The Sponsor then submits an implementation plan to the Agricultural Development Group for screening and approval, and requests funds from the Co-ordination Group or other sources.

Allocation of government subsidies for agricultural research

In order to promote agricultural research, government subsidies are allocated on a yearly basis through the COA to various institutions for short- and medium-term research. Any agricultural research worker can request such a subsidy through the institution at which he works. The COA also initiates projects, and then seeks out institutions to implement the projects, either separately or co-operatively.

Brazil

INTRODUCTION

Brazil is a vast country, and its enormous size poses problems for the organisation of agricultural research which are intrinsically different to those of most other countries of Southern America. It has technologically advanced regions, and problem regions in the first stages of development. Until the 1960s, increases in agricultural production were mainly due to the expansion of land areas and increases in farm employment, a reflection of the abundance of land and rural labour in Brazil.

Since the late 1960s, increasing productivity of land and labour became a declared goal of official policy. The role of research in achieving this goal was however downplayed, on the assumption that sufficient technological know-how was available to ensure an increase in productivity, on condition that a sufficient effort in extension was invested, credit for the purchase of inputs was provided, and minimum prices were established to provide suitable incentives for production.

The structure of agricultural research underwent a number of changes, but none of these 'succeeded in providing Brazil with a research system capable of handling agricultural problems' (Pastore & Alves, 1984).

With the exception of two states, Sao Paulo and Rio Gran de Sul, financial and human resources allocated in the past to agricultural research were extremely limited. Even so, they were allocated to a wide variety of topics; the researchers tended to choose research subjects according to their own predilections, and, not rarely, 'were more eager to duplicate an investigation recently published abroad, than to solve the farmer's problem' (Pastore & Alves, 1984). Most research personnel had little contact with farmers and with each other, and very little knowledge of the basic national needs of agriculture.

A formal system for defining priorities in research was nonexistent and the 'directing of science and technology toward the solution of the entrepreneur's problems was considered heretic thinking. Research teams were rarely used' (Pastore & Alves, 1984). The almost complete lack of co-ordination led to considerable duplication. Only about 10 per cent of the researchers had adequate training for their work. All the available research facilities were underutilised (Pastore & Alves, 1984). The salary policy made it impossible to compete for competent scientists with other sectors, or to promote personnel in accordance with their productivity.

In view of the abundance of land, and the absence of strong farmers' associations, there was little pressure to improve the NARS.

PRE-1973 STRUCTURE OF NARS

Research was carried out by institutions linked to the Ministry of Agriculture, the various State Secretaries of Agriculture, some federal agencies, universities and private agencies (Lopes, 1985). Liaison between these institutions was weak or non-existent. The proliferation of these institutions created a diffuse, fragmented, and weak national system of agricultural research.

Each State had created its own system of agricultural research, but because of the great differences in economic and social development, nation-wide agricultural research was extremely unbalanced, both geographically and by commodities.

Research in coffee in San Paulo, and in cacao in S. Bahia had been very intensive. The States of San Paulo and Rio Gran de Sul had the oldest and most developed research systems in Brazil; in contrast, the Northern and Central parts of Brazil agricultural research was totally inadequate (Lopes, 1985).

EMPRESA BRASILEIRA DE PESQUISA AGROPECUARIA
(EMBRAPA)

In the 1970s, increased domestic demand for food, and in particular the political imperitive to feed an exploding urban population, as well as the need to earn foreign exchange, created pressures to increase agricultural production. Attempts to increase productivity through the supply of subsidised credit, and a strengthening of the extension effort, were only marginally successful (Ruttan, 1982). Only then was it recognized that a 'systematic effort to organise and expand agricultural research would be necessary to produce self-sustained growth of agricultural productivity' (Elz, 1984).

A committee appointed by the Minister of Agriculture in 1972, recommended the establishment of a federal public corporation for the planning, co-ordination and administration of agricultural research as best suited to remedy the situation.

As a result, EMBRAPA started operating in 1973, replacing the National Department of Agricultural Research of the Ministry of Agriculture. Though EMBRAPA remained under the Ministry of Agriculture, it is not subject to civil service hiring conditions, and is free to hire qualified personnel at national and international market rates and to augment its public-funded resources by contract work and other means.

At first, a major effort was made to replace the existing fragmentation of the research system by a highly centralised system. However, Brazil is too vast a country for the Federal Government to monopolise all agricultural research. Because of the strong opposition of the States, EMBRAPA had to give up its original intention of bringing the State agricultural research systems under its control (Ruttan, 1982). Instead, a three-level system was initiated:

- the establishment and implementation of a federal research system;
- strengthening existing State research systems, and establishing new ones where they are lacking; and
- ensuring co-operation between and within the state research systems.

Structure (Rivaldo, 1987)

The principal components of the EMBRAPA research system are the National Research Centres, Resource Centres, State Systems and State or Territorial Research Units (Fig. 5).

National Research Centres are characterised by a high concentration of financial and human resources on a limited number of commodities (not more than three per centre). The research teams are multidisciplinary. Special attention is given to research that transcends state frontiers and the limits of geopolitical and ecological regions.

Resource Centres are dedicated to the development of natural resources, and concentrate on plant–soil–environment and animal–soil–environment problems. The basic objective of these Centres is to develop production systems suitable for the ecological system under study.

State Research Systems: where these are adequate the State systems will be responsible for the direct conduct of research in their areas; where the research system is inadequate, co-ordination of research will be at first the responsibility of EMBRAPA, to be gradually transferred to the State system as it becomes established or strengthened.

State or Territorial Research Units are devoted to adapting technologies to the specific conditions of homogeneous agricultural zones of the

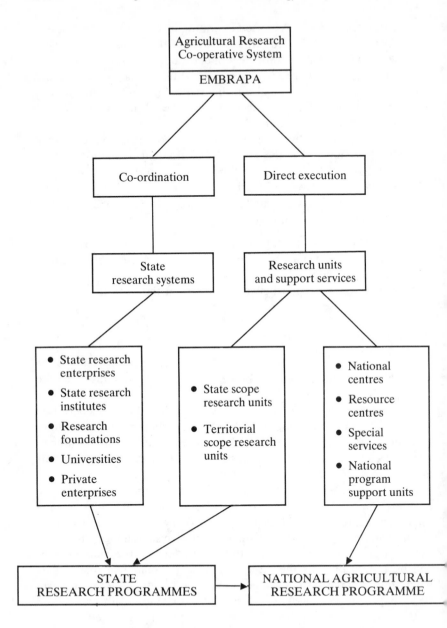

Fig. 5. Institutional model of EMBRAPA research. (From Rivaldo (1987), reproduced by permission of ISNAR, The Hague.)

States in which the Units are located. These Units may be linked to EMBRAPA, state organisations, universities, or private sector initiatives.

Policy

EMBRAPA's overall approach is that fundamental research should be left to the universities, applied, short-term research to the State research institutes, whilst EMBRAPA should concentrate on:

(a) strengthening the research systems of the 25 Brazilian states;
(b) establish a limited number of commodity-oriented national research centres; and
(c) define priorities and adapt these continuously to changing circumstances.

This policy has resulted, for example, in a significant increase in emphasis on export commodities; an increase, followed by a decline, in emphasis on livestock research; a steady and marked decline in biological research (Evenson, 1984).

Funding

In principle, EMBRAPA was to be the sole channel for allocation of funds for all research programmes, and thereby be in a strong position to influence their operations (Lopes, 1985). Actually, EMBRAPA controls about 60 per cent of all resources for agricultural research in Brazil, representing about 0·5 per cent of agricultural GDP.

By 1981, enormous investments to develop EMBRAPA were made by the Federal Government. Allocations for agricultural research were increased 40-fold: from US$ 5 million in 1973, to US$ 200 million in 1981 (Alves, 1983).

National research programmes

A major tool for co-ordinating the efforts of all the institutions involved in agricultural research is a new programming model adopted by EMBRAPA in 1980: the *Circular Model of Research Programming* (Rivaldo, 1987).

The Circular Model consists of two distinct but complementary programmatic concepts: the National Research Programme, and the Research Project.

The *National Research Programme* (PNP) is always related to commodities, a resource, or a major problem. It is defined by a group comprising EMBRAPA researchers, technologists of regional development bodies, co-operatives, credit agencies, extension services, and producers.

The PNP includes guidelines and political and research directives; short, medium and long-term objectives; and research priorities. The PNPs are periodically revised and updated. Their co-ordination is assigned by the Executive Directorate to an appropriate research unit or institute.

The *Research Project* relates to a specific problem within the PNP, and serves as the basic programming unit. The problem to be resolved is clearly defined, as is the methodological strategy aimed at its solution. The research project is reviewed annually.

National research programmes have been established for the major agricultural commodities: wheat, sugarcane, rice, maize, beans, soybeans, coffee, rubber, beef and dairy livestock. Three programmes were established to deal with major resource problems: the problem soils of the Cerrado, the semi-arid Nordeste, and the tropical conditions of Amazonia.

The infrastructure of each programme consists of a major central station and a network of regional stations located in the major agro-ecological regions in which the respective commodities are produced on a significant scale. The universities participate in these programmes, doing the basic work required.

Research orientation

The National Integrated Programmes are oriented mainly towards commodities of major commercial importance, in particular those produced for export. The technology that was developed was mainly useful to the bigger commercial enterprises. Mechanised agriculture has been given considerable weight, as well as the adoption of modern inputs. A system of subsidies and oriented credit was adopted to promote these objectives.

The special interests of the mass of small farmers was deliberately neglected as evidenced by the following statement made by the President of EMBRAPA (Alves, 1984):

If urban growth of industrial development in the course of the next 30 years (the time of maturation of many research projects) will have absorbed a significant proportion of the rural population, it is not worthwhile ('vale la pena') to dedicate a significant part of the researchers' time to topics which are relevant to family-sized or subsistence farms, such as multiple cropping, intermediary technologies, etc. It is obvious that research should be oriented towards commercial agriculture, whether already in place or which will be developed.

The policy expressed by Alves, is in marked contrast to that adopted in recent years in most developing countries, in which concern for the improvement of the productivity of the small farmer has become a major objective of agricultural research (cf. p. 338–40).

Staff

The staff that EMBRAPA inherited from the Ministry of Agriculture was inadequate, in number and training, for the great challenge confronting the new institution.

Brazil had about 3000 full- or part-time agricultural researchers of whom about one-third were employed by the Ministry of Agriculture. Of these, only 10 per cent had completed graduate courses, and only three held PhDs (Alves, 1984).

One of the first acts of EMBRAPA was to dismiss over half the bureaucratic personnel inherited from the Department of Agricultural Research. A major training programme for research workers was undertaken, so as to ensure that 80 per cent of the research personnel should hold Master's or Doctor's degrees. A massive programme of recruitment was also undertaken.

The strategy adopted was to closely evaluate the existing research staff, and to choose the appropriate candidates for training in Brazil and abroad. A recruitment campaign was launched to attract capable young people to EMBRAPA. Most were sent directly into graduate programmes, others underwent a period of adaptation of one year in the research units, before entering the graduate programme (Alves, 1984).

The cost of the training programme undertaken by EMBRAPA was 'probably the largest such investment to be made in a single programme by any nation ever' (Evenson, 1984). In the years 1974 to 1982, US$ 33 million were spent on the training programme in which 2300 individuals participated. The average annual per person cost of training increased from US$ 17 000 in 1974 to US$ 83 000 in 1982, as a result of shifting emphasis from MSc to PhD.

To ensure the success of the research effort, a policy of promotion based on merit was created to reward talent and work. The cost per research worker increased from US$ 7800 to US$ 37000, due to the shift to higher pay scales based on graduate education and merit of the research workers (Alves, 1984).

Venezuela

HISTORICAL BACKGROUND

In 1937, a Centre for Agronomic Research was established and was followed in 1941 by the Centre for Veterinary Research. In the course of the years that followed, many other bodies, such as the Universities, various Ministries, and private sector bodies became involved in agricultural research, which remained largely unstructured until 1960.

In 1961, the 'Directorate of Research' was set up as a unit within the Ministry of Agriculture and Animal Husbandry (MAC) and was entrusted with the following functions (Morillo, 1971): orientation, supervision, co-ordination, analysis, evaluation and diffusion of results of agricultural research. The Headquarters of the Directorate of Research were located in Caracas, near the seat of Government; the implementation of research took place in two research centres located at Maracay, and was supported by a number of Experiment Stations and Experiment Fields situated in various ecological regions of the country.

In 1969, following the reorganisation of MAC, several new units were initiated; among these were two concerned with agricultural research and were therefore incorporated into the Directorate of Research: (a) a Project Office, responsible for the design and analysis of experiments, economic evaluation of research proposals and their results, and the creation and maintenance of a central archive of projects; and (b) an Agricultural Communications Office, to integrate the libraries and technical secretariats of the Research Centres into a national system.

Also in 1969, were initiated the National Research Programmes, each headed by a Co-ordinator (Morillo, 1971).

The functions of planning and initiating the scientific and technological development of the country became the responsibility of the National Council for Scientific and Technological Research (CONICIT), an autonomous institution directly responsible to the President of the Republic. CONICIT appointed a commission for agricultural sciences to advise on all matters pertaining to agriculture.

More specifically concerned with agricultural research is the National Council for Agricultural Research (CONIA), founded in 1959 as an advisory body within the MAC, comprising senior officials of the ministry, and representatives of public and private bodies concerned with the development of agriculture.

In 1972, following the 'Decree of Regionalisation', the Directorate of Research of MAC was again restructured; it was to comprise a National Centre of Agronomic Research, and three Regional Research Centres, including all the experiment stations in their respective regions.

In the course of some 14 years, a constant reappraisal of the organisation of agricultural research had actually resulted in increasing the fragmentation of the national research effort and a proliferation of separate research units.

THE NATIONAL FOUNDATION FOR AGRICULTURAL RESEARCH (FONAIAP)

In 1975, the entire situation of agricultural research in Venezuela was again scrutinised, and far-reaching conclusions were drawn and subsequently implemented. The Directorate of Research of MAC was abolished and all its functions and assets were transferred to the National Foundation for Agricultural Research (FONAIAP).

FONAIAP had actually been established in 1961, as an autonomous service within MAC. Its functions at the time were defined as: (a) promoting agricultural research; (b) collaborating with other relevant organisations, and (c) propagation, storage and sale of improved seeds.

The decision to transfer to FONAIAP the responsibilities previously held by the Directorate of Research, completely changed the nature of FONAIAP, from being an inter-organisational institution as originally intended, to become an autonomous operational organisation with considerable responsibilities.

The following functions were assigned to FONAIAP: to co-ordinate the research programmes of MAC with those of other institutions; to collaborate in the training of research personnel; to plan research, and facilitate the diffusion of research results.

Structure

Headquarters

An *Administrative Board* is responsible for planning the activities of FONAIAP, the appointment of personnel, the determination of its budget, and the monitoring of its operations. The Board is chaired by the President of CONIA; the General Manager of FONAIAP is the Secretary of the Board, but has no vote.

The General Manager (Gerente General) is responsible for the routine administration of FONAIAP. A *Management Council*, comprising the Directors of the Regional Centres, and the Managers of the Administrative, Technical and Development Departments, respectively, serves in an advisory capacity to the General Manager.

Research units

The *National Centre of Agronomic and Zootechnical Research* (CENIAP) is the sole institution with research responsibility on a national scale. Besides management and supporting services, CENIAP consists of four research institutes: Agronomic Research (IIA); General Agricultural Research (IIAG); Zootechnical Research; and Veterinary Research. The Centre also maintains an experiment station and three experiment farms.

The *Technical Council* of the Centre consists of the heads of all its research units, including experiment stations, and acts in an advisory capacity to the Director of CENIAP.

Regional Centres are responsible for the planning, implementation and supervision of public-funded research in their respective regions. Each Regional Centre has a Technical Council to advise the Director. In order to ensure effective links with other institutions active in the region, each Centre has a *Regional Consultative Council* comprising the Regional Director of MAC, Directors of Agricultural schools, representatives of Farmers' Associations, the Director of the Regional Centre, and the Heads of the Experiment Stations.

Evaluation of the structure and functions of FONAIAP

The fragmentation of agricultural research in Venezuela remained practically unchanged, even after the complete reorganisation of the MAC system and the establishment of FONAIAP, as evidenced by the following extract from a FONAIAP report (FONAIAP, 1981):

There is no national plan for agricultural research. Involved in agricultural research besides FONAIAP are: the National Universities, through their faculties and Agronomic, Zootecnic and Veterinary Research Institutes; various autonomous public bodies, such as the National Council for Scientific and Technological Research, The Venezuelian Institute of Scientific Research; various ministerial departments, such as Irrigation and Fisheries (MAC), Environmental Research (Ministry of Environment and Renewable Natural Resources); various Foundations, such as Service for the Farmer.

Each of these institutions has its own research programme; there is no common strategy, no clear definition of objectives that relate to the National Plan; no concern with priorities; no definition of the area of responsibility of each research institute.

Nor was the situation within FONAIAP satisfactory. Small wonder that in the years after FONAIAP had assumed its new role as the national agricultural research institution, a general feeling of dissatisfaction and malaise pervaded the organisation. It should be remembered that, actually, FONAIAP was but a continuation of the Directorate of Research of MAC; it had inherited not only its role, assets and infrastructure, but also the weaknesses and constraints which had led to the need for reorganisation. Apparently, simply detaching the research unit from within the framework of MAC and providing it with a semiautonomous status, was not sufficient to eliminate the shortcomings that had characterised the national research in its previous setting.

Whilst some of the complaints voiced by the FONAIAP leadership referred to the lack of funds and limited possibilities for the training of research personnel, dissatisfaction with the state of affairs was mainly concerned with shortcomings in the research programme and its implementation. In his report to the Junta Administradora, the Gerente General (1978) states: 'many of the research projects that are included in the programme are not initiated, and many of those that are initiated, are not completed'.

This self-criticism is in itself admirable, but the felt need to change the situation led the leadership of FONAIAP in 1979, to request an outside body, the Instituto de Estudios Superiores de Administración (IESA) to undertake a study of the situation, and to make proposals for improvement. The findings of the group of experts from IESA were presented to FONAIAP in 1980, in two volumes: the first devoted to the 'diagnostic of the structural organisation and its functioning'; the second, presented the concrete proposals for improvement (IESA, 1980). It is worth noting, that this was the first time that the author, who has had the privilege of serving as a consultant for agricultural research in many developing countries, has encountered a study in depth of this kind, commissioned by a research organisation, and carried out by a local institution.

It should also be stressed that the conclusions formulated by the IESA group were based not only on their own expertise, but also on the opinions of the leadership and research staff of FONAIAP.

Very briefly, the major shortcomings of FONAIAP were ascribed to overcentralisation, lack of effective intra-organisational communication, and overextension.

The 'Prescriptions' proposed by IESA are only marginally concerned with the organisational structure of FONAIAP; they are targeted mainly on the role assigned to the various units of the system and the way they communicate with each other.

The organogram proposed by IESA has the same overall structure as the one adopted when FONAIAP was established, and it retains the four basic levels of functioning described above, with certain important modifications.

The main changes proposed are at the management and implementation levels, whose functions are reformulated. The Junta Administradora had been charged with general, instrumental, and operative functions, so that not only were certain functions overcentralised, but its real and essential function—to formulate policy and supervise its implementation—had become obscured. Therefore, the group proposed to replace the Administrative Council (Junta Administradora) by an Executive Committee (*Comité Ejecutivo*), the change in name implying a change in focus and direction; the Committee becoming responsible for overall direction, control and evaluation of the activities of FONAIAP, definition of priorities, and the establishment, control, and regulation of external relationships.

An effective system of information and control, which can provide data and general up-to-date information on the activities of the various units of FONAIAP (research, administration, personnel, etc.), is considered to be a precondition for rational administration under conditions of increased autonomy by units responsible for the implementation of FONAIAP's research programme. An important change proposed by the IESA group is the appointment of committees for areas of activity and for 'working groups', to serve as instruments of co-ordination and channels of internal communication–information.

It was also postulated that the units responsible for implementing the research programmes approved by the Executive Committee, and the Regional Research Centres, should have full operational autonomy. Management of the human and material resources involved in the research projects carried out at the respective Centres needed to be decentralised at this level.

Finally there was the problem of overextension of the area of activities. At the time of the evaluation, FONAIAP, in addition to its research functions, was also responsible for the Centro de Fomento de la Producción which engaged in the production and certification of seeds, vaccines and other zootherapeutic products. The IESA group therefore recommended the transfer of this Centro to MAC.

The major lesson to be learnt from the FONAIAP evaluation, is that a logical organisational structure, an autonomous status and a dedicated leadership, are not, in themselves, sufficient to ensure the efficient functioning of an agricultural research organisation.

Planning the research programme

The emphasis given in recent years by the leadership of FONAIAP to improving the planning of the research programme of the Institute, was motivated by the desire to ensure that the research carried out should be consistent with: (a) the Five-year Development Plan of the State of Venezuela; (b) the Operative Plan of MAC, which was a more detailed version of the national plan, in regards to agriculture; and (c) the conclusions drawn by FONAIAP from its own studies of the problems of agriculture in the country.

The General Management therefore decided to revise the whole system of research planning, in order to achieve the following declared objectives:

(a) to establish a programme based on priorities congruent with the National Plan;
(b) to achieve better interdisciplinary action;
(c) to enable an adequate evaluation of the results of research; and
(d) to make more efficient use of available resources.

The first step was taken in 1979, with the elaboration of a short-term programme: the 'Plan Indicativo de 1980', which was to serve as a prelude to medium- and long-term planning. The next step was the preparation of the Five-Year National Agricultural Research Programme (FONAIAP, 1981).

Project selection

FONAIAP has designed a method for assigning priorities to commodities and for project selection based on four criteria; all commodities are ranked, in decreasing order of importance (FONAIAP, 1983):

(a) according to their contribution to the agricultural GNP;
(b) according to the area harvested;
(c) according to the amount of official credit (Bs/commodity) invested;
(d) according to annual rates of growth.

A fifth criterion can be added, namely the strategic value of a commodity in relation to its local or regional importance, its prospects for future development, its social importance, and its role as a substitute for imports.

Human resources

FONAIAP is fortunate in that a significant proportion of its research personnel has had the benefit of good academic training, and gained much experience on the special problems of agricultural production in a tropical environment. This is considered to be the most valuable asset of FONAIAP, with an inherently considerable potential, if properly motivated and if the essential conditions for productive work are provided.

The leadership of FONAIAP aims to:

- strengthen the motivation of the research personnel by creating an environment conducive to efficient and creative action;
- improving remuneration and social conditions;
- enabling professional advancement by adequate recognition of achievements;
- ensuring stability of employment;
- clearly defined institutional objectives, and adequate financing and supporting services to achieve these objectives.

Transfer of research results

As in many other countries, the transfer of new research findings to the farmer through the extension service of MAC was far from satisfactory.

FONAIAP has a special unit for Technology Transfer, the Gerencia de Transferencia de Resultados, whose basic objective is to improve and strengthen the relations between FONAIAP, the extension workers and the farmers. The functions of this unit have been defined as follows (Gerente General, 1980):

- Establish firm relationships with the farmers and their institutions;
- Strengthen the Offices of Agricultural Communication of CENIAP, and of the Regional Centres;
- Make better use of modern methods of social communication; and
- Publish research results in a popular language.

In addition to the above, FONAIAP has initiated a programme of Farming Systems Research, in which the Extension Service is also involved.

Argentina

THE NATIONAL INSTITUTE OF AGRICULTURAL TECHNOLOGY (INTA)

INTA was established in response to the stagnation of agricultural production in Argentina in the 1950s, and can be considered as the recognition by government of the role that an improved technology could fulfil in the revitalisation of agriculture in the country (Trigo *et al.*, 1982).

The strengthening of agricultural research was part of an overall development strategy which included other policies, such as regulation of prices, and the provision of credit, which greatly facilitated the adoption of new practices by many farmers.

Until the establishment of INTA, agricultural research, extension and development had been the direct responsibility of the Ministry of Agriculture, which was incorporated in 1962 in the Ministry of Economy with the status of Secretariat de Agricultura. The Ministry of Economy assumed control of all areas related to the national economy, including agriculture.

The first decade of INTA's activities was devoted to a restructuring of the agricultural research and extension services inherited from the Secretariat de Agricultura, to establishing additional research centres and experiment stations, and to embarking on an extensive programme of staff training.

INTA has three characteristics particular to this agricultural research institute:

- INTA combines responsibility for research and extension, with jurisdiction over the entire national territory;
- it has a source of funding that is independent of Government budgets;
- it has enjoyed relative stability and continuity of work under changing political regimes.

The situation in the 1970s—a period of crisis

During the ten years following its establishment, INTA enjoyed considerable public support and appreciation. However, in the 1970s this initial support weakened progressively, due to a number of factors, leading to a destabilising effect on the Institute. In 1973, this process culminated in an institutional crisis, marked by a deficiency of resources, inner conflicts, paralysis of activities and out-migration of staff (Trigo *et al.*, 1982).

This situation caused a gradual slowdown and disorientation of the research efforts, leading to a further loss of public support. Training programmes supported by foreign aid came to a virtual end, and the efforts to establish an independent infrastructure for postgraduate training ground to a halt.

In 1976, INTA went into a period of recovery (Ardila *et al.*, 1981). However, in 1984, Nogueira stated that: 'objectives, strategy, priorities and communication within INTA have become complex and unclear'. INTA is no longer the only agricultural research institute in the country. There are many new public and private institutions working in the field, some collaborating, some competing with the national institute (Nogueira, 1984).

Legal status

INTA is a non-profit, legal entity, with nation-wide jurisdiction; its activities are regionalised and in accordance with the policies and directives of the national Government. The relations of INTA with Government are through the Secretariat of Agriculture and Animal Husbandry (SAG).

INTA's mission has been defined as follows:

- Research on all problems relating to natural resources and agricultural production technology.
- Research on the conservation and processing of agricultural products.
- Agricultural extension and assistance to the farmer and his family in education, health and cultural services; community improvement.
- All development activities essential to enable the diffusion and adoption of the results of INTA's research.

All functions involving control or inspection of agricultural production are expressly excluded from INTA's jurisdiction.

The legal status of INTA has remained practically unchanged since its inception. The relative insulation from political change enjoyed by INTA is to a large degree due to the nature of its sources of funding, which ensures its relative independence in regards to direct government funding. However, this autonomy is tempered by the fact that the President and Vice-President of the Executive Council are nominated by the Secretary of Agriculture and Livestock. Political changes result, at the most, in the replacement of the President and Vice-President; the remaining members of the Council remain unchanged and thereby ensure the continuity of the work of INTA.

Conceptual framework

The concept underlying INTA's philosophy is to maximise regionalisation and to minimise centralisation. As a result of this concept INTA is organised at three levels: (a) *formulation of policy*—which is the prerogative of the Consejo Directivo (see below); (b) overall *direction and administration*—which is the responsibility of the Dirección Nacional; and (c) *implementation* carried out by the National Research Centre for Agriculture and Livestock and a network of experiment stations established in the major agro-ecological regions of the country.

Overall direction and management

General policy of the Institute is determined by a board of directors consisting of a president and vice-president, representing the Secretary of State of Agriculture; three members representing the agricultural producers and co-operative associations, and a representative of the faculties of agriculture and veterinary medicine. The director general, the four assistant directors general, and the director for administration are also members of the board, but without a vote. The board of directors has the authority to formulate regulations; to appoint the director general and assistant directors of the Institute, as well as the directors of the experiment stations and research centres; to appoint, promote and discharge personnel; to contract for services of foreign personnel; to grant scholarships; to administer the National Agricultural Technology Fund.

The executive organ of INTA is its Dirección General which comprises a Director General, aided by four assistant directors general (extension;

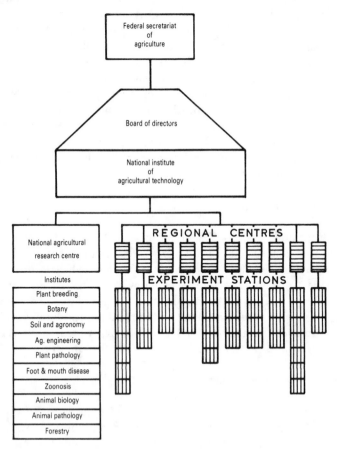

Fig. 6. The organisation of agricultural research in the Argentine.

research; special projects; planning and evaluation) and a director of administration. The Director General is responsible for the proper implementation of the policies determined by the board of directors, co-ordinating the technical and administrative activities of the Institute, administering the experiment stations and research centres, and keeping the board informed of all its activities.

Organisation structure

The Institute maintains a National Agricultural Research Centre and a number of Regional Centres, each of these comprising a number of Experiment Stations (Fig. 6). The National Centre is concerned with basic research and the development of research programmes on problems of national significance. It consists of three Research Centres, for Natural Resources, Agronomic Sciences, and Veterinary Sciences (including meat technology), respectively.

The 13 Regional Research Centres are based on 32 experiment stations working on problems of regional significance, and are also responsible for the extension work and agricultural development in their respective regions. Each Regional Centre also has a leadership role for one or more National Programmes.

The Directors of the Regional Centres have a considerable degree of autonomy and responsibility in the implementation of their work.

Because the Regional Centres are the major operational units, the National Centre is, in effect, mainly concerned with overall management and administration and the provision of services to its three constituent Research Centres.

Research policy and programming

Research policy at the *national level* is determined by the Board of Directors of INTA. At the *provincial level*, each province nominates an Agricultural Technology Council composed of representatives of INTA, the local Government and the agricultural producers. Each experiment station also has a local advisory board composed of technicians of the station, farmers and representatives of local organisations.

A staff group is responsible to the Director General of INTA for the overall co-ordination of research and extension between the Institute, regional centres and experiment stations.

The entire research and extension work of the Institute is broken down into *'programmes'*, mostly centred on an agricultural commodity (such as corn, wheat, milk production, etc.) or on a group of commodities (such as oil crops, stone fruits, etc.). The research work in the disciplines (economics, genetics, plant pathology, etc.) is considered as auxiliary to the commodity programmes.

Each programme is headed by a *co-ordinator* responsible for the overall orientation of the programme, working in co-operation with other co-ordinators. Each programme in turn is broken down into *'work plans'* with specific goals and a well-defined budget; the work plans are the smallest units of work at the experiment stations, the Institute or the extension agency.

Financing

The financing of the work of INTA is based on the *National Agricultural Technology Fund*, which derives from the following sources:

1. a tax of 1·5 per cent *ad valorem*, on all exports of agricultural products;
2. the income derived from sale of produce of the experiment stations, publications and services;
3. subsidies from farming, industry and business;
4. contributions from the provincial governments;
5. interest on capital, legacies and donations;
6. other sources.

In addition, all instruments, machinery, equipment, seeds, chemicals, etc., required by INTA are freed from custom duties and surcharges.

This form of financing the work of a research institution has the following advantages (Herzberg, 1975):

- continuity in the work;
- possibility of relating remuneration of personnel in accordance with professional efficiency and output;
- possibility of adopting less rigid administrative procedures than those of governmental bureaucracy;
- increased motivation of researchers to increase output of crops related to the income of the organisation. (This may, however, have a negative effect on maintaining a balanced research programme.)

The major disadvantage of the system is the fluctuations and unpredictability of income, due to fluctuations in exports and the prices obtained on the world market. For example, during the period 1960–75 income derived from the taxation of export crops accounted for 77 per cent of the approved budget of INTA; this ratio dropped to 51 per cent in 1977 (Trigo *et al.*, 1982).

Extension

Agricultural extension carried out by INTA is concerned not only with agricultural production *per se*, but also includes home economics and work with rural youth. The supervisors of the field extension workers generally have their offices in the experiment stations, thereby facilitating close contact between research and extension.

Farmers participate in the funding of extension activities at the local level. Generally the fixed costs of maintaining the local extension office (mainly salaries and equipment) are borne by INTA, whilst operating costs are shouldered by the villagers' organisation.

Co-operation with the universities

In research
Co-operation with the agronomy and veterinary schools of the universities is encouraged at the national and regional levels.

In post-graduate training
In 1967, following a four-way agreement between the Universidad Nacional de Buenos Aires, the Universidad Nacional de la Plata, IICA and INTA, the Escuela de Graduados en Ciencias Agropecuarios (EPGCA) was established, and continued to function until 1976, when the above-mentioned convention came to an end.

INTA continued to provide courses at the MS level; (a) in animal husbandry at the Estación Experimental Regional de Balcarce, and (b) in plant genetics at the Estación Experimental Regional de Pergamino. These courses have been implemented in association with the National Universities of Mar del Plata and de Rosario, respectively (Trigo *et al.*, 1982).

Ecuador

INTRODUCTION

Organised agricultural research was first initiated in Ecuador in 1942, when the Estación Experimental Agricola de Ecuador was established with the financial and technical aid of the USA. The first 'Centro Experimental' (Pichilingue) was inaugurated a year later in the Litoral, in collaboration with the Development Corporation of Ecuador. During the period 1945–51 seven more experiment stations were established, all of which subsequently ceased to function.

In 1952, the Estación Experimental Pichilingue was transferred to the Servicio Cooperativo Interamericano de Agricultura (SCIA), thereby consolidating agricultural research in Ecuador.

Agricultural production in the 1950s suffered severe reverses which caused grave concerns to the economic and political leaders of the country. The Government therefore decided to establish the National Institute of Agronomic Research (INIAP) which would be responsible for defining the problems of the agricultural sector and to contribute to their solution (INIAP, 1983).

INIAP, as we know it now, was established in 1959, but for budgetary reasons started work two years later on a state-owned farm near Quito, which subsequently became the Estación Experimental 'Santa Catalina'.

In order to enable INIAP to fulfil its obligation to organise a national system for agricultural research and thereby improve the agriculture of Ecuador, the agricultural research activities of various organisations were transferred to INIAP. With the establishment of several new experiment stations in the various agro-ecological zones of the country, INIAP gradually developed a research infrastructure enabling the institute to address itself adequately to the problems of Ecuador's agriculture.

STRUCTURE OF INIAP

Central administration

INIAP is managed by an Administrative Council, of which the Minister of Agriculture (or his deputy) is the Chairman, thereby ensuring effective control of policy of the Institute by the Ministry. The other members of the Council are representatives of the farming sector, the Ministry of Finance, the National Development Council, the National Finance Corporation, the Development Bank, and the Director General of INIAP.

The Administrative Council names the Director General, determines the overall policy of the Institute, approves the budget proposals, follows and monitors its work, approves its regulations and approves the candidates proposed by the Director General for attendance at conferences, seminars, etc., and for post-graduate training.

The General Directorate is the executive and administrative organ of INIAP, with authority over all the dependencies of the Institute. The Directorate generally meets once a month; it is responsible for the planning, organisation, development, direction and administration of all the activities of INIAP. Subject to the approval of the Consejo de Administración, the Dirección General negotiates and implements agreements with national, international, and foreign bodies for co-operation in agricultural research programmes.

The Advisory Committee consists of the Directors at national level, and the Directors of the Experiment Stations. The committee meets generally every three months, or exceptionally at the express request of the Director General, in order to discuss specific technical or administrative objectives.

Research planning

Planning is mainly concerned with the preparation of the research programme in accordance with national and institutional objectives as determined by the National Development Plan: defining alternative policies, establishing priorities, monitoring the implementation of research and evaluating the results.

A Rural Economics Unit supports the technical research carried out by all the research units of INIAP, by analysing the economic implications and cost/benefit ratios of the new technologies proposed by the researchers.

Regionalisation

The research activities of INIAP are carried out in the major agro-ecological regions of the country, at two levels: Regional Experiment Stations and On-Farm research.

Regional Experiment Stations

There are six Regional Experiment Stations, of which four are located in the Litoral, and one each in the Sierra (mountain region) and the Region Oriental (Amazonia).

The results of research carried out at the Regional Stations are not transmitted directly to the farmers, but are first tested on farmers' fields for validation and eventual adaptation.

Experimental Farms

The Experimental Farms have no supporting Service Departments and are therefore dependent on one or more of the Regional Experiment Stations for the disciplinary support for their work.

At the level of the Experimental Farms, the behaviour of new, improved varieties is tested under conditions of soil and climate different to those experienced at the Regional Experiment Stations. Furthermore, the incidence of pests and diseases and their interaction with environmental factors are studied. New cultural practices, proposed by the researchers, are submitted to verification trials.

The land for the trials is made available to the researchers by farmers and ranchers; INIAP supplies seeds, fertilisers and plant protection chemicals and is responsible for the planning, execution and evaluation of the trials. Based on these trials, packages of recommendations are formulated and farmers' field days are organised to familiarise the farmers with the proposed new technologies.

The Experimental Farms are located in the fields of large and medium farms, and research is mainly oriented to the solution of the problems of commercial farming.

Farming systems research

Until 1976, the research work of INIAP was carried out exclusively at the two levels described above. The Institute was becoming increasingly concerned by the finding that its research on basic food crops was having very little impact on the actual productivity of these crops, in particular in the dominant small farm sector.

The situation was most acute in highland maize and wheat production; yields remained low and even tended to decline.

In the search for an appropriate technology for these food producers, INIAP research leaders came to the conclusion that a new research approach had become essential. Therefore, in order to address itself specifically to the problems of the small farmer, a completely new investigational approach was adopted by INIAP in 1977, with the creation of the Programa de Investigación de Producción (PIP), in co-operation with Centre Internacional Mejoramiento de Maize y Trigo (CIMMYT) (Moscardi, 1982).

A special department at the Headquarters of INIAP organises and monitors the experimental units involved in this on-farm research. The Regional Experiment Stations provide technical and administrative support. The research is carried out in close co-operation with extension service personnel, and is conducted according to the principles of Farming Systems Research.

International co-operation

INIAP has established an impressive list of co-operative research projects with various foreign and international institutions, and also receives grants from foreign donors for specific research projects.

Services to farmers

The Regional Experiment Stations provide a number of services to farmers in their respective regions; these include: soil and plant tissue tests, feed analysis, supply of basic and registered seeds to seed farms, and seed-testing. High-grade sires for beef cattle and pigs are made available to farmers for improving their stock.

Resources available for research

Human resources
In the first years of its existence, the major efforts of INIAP were directed to the formation of an adequate cadre of well-trained researchers and technicians (Portilla, 1971). In 1983, the staff of INIAP consisted of 1359 employees, of whom 337 are medium- or high-level researchers, 325 administrative personnel, and 697 field workers, or a ratio of roughly 1:1:2. This staff has made it possible for INIAP to fulfil its obligations in an efficient manner.

Financial resources
Since its inception, government funds made available to INIAP have never been sufficient to enable the Institute to achieve its objectives completely. The Institute has therefore been continuously active in obtaining funds and loans from various national, international and foreign

sources, thereby enabling it to carry out a programme of research out of all proportion to the State funds invested in agricultural research.

Physical resources
Besides the buildings for the Headquarters and Regional offices of the Institute, the major physical assets of INIAP are the large tracts of land made available to the Regional Experiment Stations and Experimental Farms.

Colombia

THE BEGINNINGS

The first agricultural experiment station in Colombia was established in 1879, in the Botanical Garden of Bogotá. The work was mainly concerned with plant introduction, and sporadic crop and livestock experiments. A year later, the Instituto Superior de Agricultura was initiated under Ley 64, which envisaged an institution fulfilling the three functions of research, education and extension. These activities were to be carried out in close co-operation with the farming sector. Though implementation of these concepts was 'sporadic, weak, and frequently changed form and direction', it remained the basic guideline of government policy, until the 1960s (Piñeiro *et al.*, 1982).

From 1914 on, various foreign missions resulted in a number of disparate activities in establishing research centres, experiment stations, educational facilities, etc., in agriculture. The Ministry of Agriculture, and with it, agricultural research, underwent a number of reorganisations.

Ley 132

In 1931, Ley 132 was enacted with the object of 'coordinating the existing disparate attempts at research, experimentation, demonstration, and diffusion' under the direction of the National Government. These efforts were to be part of a Government plan for agricultural development.

Conceptually, Ley 132 was similar to its predecessor Ley 64, enacted half a century earlier. The new law, however, defined a precise organisational framework for implementing the three functions of education, research and extension. It also added another dimension by stressing 'the necessity of these functions being developed within a regional framework, attentive to the needs and problems of the areas under their jurisdiction, but co-ordinated at the national level'.

It was again stressed that the activities of the Agricultural Institutes would be carried out in collaboration with other departments of the Ministry of Agriculture and with the Societies of Farmers and Ranchers. Regarding subsequent developments, Piñeiro *et al.* (1982) write:

> The principles outlined in Ley 132, which envisaged a functional organisation within which a relationship would be established between the State and the farming sectors, on a regional basis, for the development of agricultural research, education and extension, was without doubt, a valuable approach. However, in practice, developments took place in an opposite direction to that envisaged by Ley 132.

Co-ordination between the farming sector and government departments began to disintegrate, giving rise to conflicts and competition. Progressively, several sectorial associations undertook their own research and extension activities, and finally established a number of commodity research institutes (Piñeiro *et al.*, 1982).

The period 1943 to 1963

In the two decades between 1943 and 1963, the Ministry of Agriculture and with it the agricultural research system underwent extensive reorganisations, in 1943, 1950 and 1955.

Throughout this period, Government continued to ensure the participation of all bodies concerned with agricultural development, including the private sector, in defining research policies. During this period, research moved far ahead of agricultural education, and still farther ahead of extension (Streeter, 1972). There was lack of co-operation between research and extension, even though the two functions belonged to the same institutional framework.

THE INSTITUTO COLOMBIANO AGROPECUARIO (ICA)

In order to co-ordinate research, education and extension, a commission, convened by the Rector of the National University, recommended the establishment of an agricultural research and education branch within the Ministry of Agriculture. However, the then President (Illeras), decided that instead of merely co-ordinating the various agencies involved, these should be amalgamated into a single agency which would move out of the Ministry to become a semi-autonomous institute.

In the meanwhile, co-operation with the USA received a new impetus, following the establishment of the Agency for International Development (AID), which aimed at 'helping to develop a more productive and diversified agriculture to supply the essential needs of the nation (AID, 1968). However, AID insisted on a vigorous and effective counterpart. The weak and ineffectual Directorate of Agricultural Research (DIA) of the Ministry of Agriculture was not suitable as a partner for AID.

For these reasons it was decided in 1962 to establish a new institutional structure, with increased autonomy, more flexible administration, and substantial financing: the Instituto Colombiano Agropecuario (ICA).

ICA was established in 1963 to be a 'national institution, under the jurisdiction of the Ministry of Agriculture, with administrative autonomy, and independent patrimony, which would take over the functions of DIA, the training of research personnel in co-operation with the National University, the development of agriculture, and extension' (ICA, 1984).

A Governing Board was appointed, comprising representatives of various ministries, universities, banks and the Director General of ICA. This time, the participation of the farmers' associations was not provided for, the new policy being that all agricultural research should be the exclusive responsibility of the State. By contrast, the role of the universities in determining research policy was strengthened, in line with the concept of integrating agricultural research and education.

Stages of development

There are four clearly defined stages in development of ICA.

1963 to 1967
The major thrust was directed at strengthening agricultural research, with education and extension as auxiliary activities (Ardila *et al.*, 1981).

A considerable effort was invested in widening the agricultural research programme, strengthening the infrastructure of the research centres and experiment stations, and increasing the number of researchers sent abroad for post-graduate training (mainly in the USA). ICA also took over the Extension Service from the Ministry of Agriculture.

In 1967, an agreement was signed between the National University and ICA on setting up the joint *Graduate Study Programme* (PEG), making use of ICA specialists who had been trained abroad. By 1971, PEG had become the major centre for specialised studies, and was providing a steady source of faculty staff.

All these activities in research, education and extension were strongly supported—financially by the Government, financially and technically by various foundations (Trigo *et al.*, 1982).

However, research on the major commodities was being increasingly taken over by the private sector, either independently, or in so-called co-operation with ICA. This tendency was strengthened by the 'cuotas de fomento'—a system of taxation which provided ample funds for research in institutions set up by the private sector.

On the other hand, basic research, which had been largely neglected in the past, was left entirely to ICA, a tendency encouraged by the then incumbent minister of agriculture.

1968 to 1973

In 1968, a general reform (Reforma Instituciónal) of the entire public sector was undertaken. The Ministry of Agriculture was completely restructured; its functions were confined to the direction, planning and evaluation of the activities of the agricultural sectors. Responsibility for the implementation of these policies was transferred from the Ministry to various Institutes, of which ICA was one. To ICA's functions of research, extension and education, were added: responsibility for the promotion of agriculture and animal husbandry; the control and supervision of agricultural inputs and certification of seeds. The rural development activities included: animal and plant health, supervision, technical assistance, and the channeling of credit to rural enterprises.

For the first time, ICA was to undertake special measures specifically oriented to improving subsistence agriculture. A study of production systems in traditional agriculture was to be initiated, as well as studies on mixed cropping.

Concurrently with the above, existing programmes concerned with commercial farming were continued with special emphasis on animal husbandry.

The reorganisation entailed increased budgets and a strengthening of the administrative infrastructure, and special emphasis on planning. Regional decentralisation of ICA's activities was attempted by creating eight 'Gerencias Regionales' for the principal agricultural regions of the country.

The functions, personnel and budgets of a number of institutes that had been previously concerned with single commodities, such as the Institute for Cotton Development, and the Tobacco Institute, or with a single discipline, such as the Zooprophylactic Institute, were transferred to ICA.

In 1970, a change in government occurred, that resulted in a reorientation of policy, whereby major emphasis was shifted from agriculture to the promotion of other sectors of the economy. State interest in ICA declined, and budgets were reduced. The agreements with Foreign Foundations also began to expire. Foreign technical assistance declined rapidly; training of researchers was reduced in scope as the number of stipends were drastically reduced and were not replaced by funds from national sources.

The joint ICA–UN postgraduate programme also encountered great difficulties because of insufficient funds and loss of support from the foundations. This occurred at a time of maximum need for the programme to expand and replace the loss of foreign training opportunities.

In brief, a period in which the scope and complexities of the tasks ICA was charged to carry out increased considerably, coinciding with a drastic reduction in the professional, technical and financial resources available to the Institute.

As difficulties of functioning increased and budgets became smaller, migration of personnel became an increasingly serious problem.

1973 to 1986

In the course of its history ICA, at first a relatively simple and efficient research organisation, with extension and education in limited supporting roles, had changed into a complex institution concerned mainly with rural development, in which research had assumed secondary importance in relation to the other functions undertaken by the Institute.

A breakdown of ICA's total budget for the period 1970–78, shows that ICA had been assigned more and more duties, but had not received a proportionate increase in funds. As a result, research, formerly the most important ICA activity, had been adversely affected (Chaparro *et al.*, 1981).

Not only had the scope of ICA's activities evolved, but the target population towards whom these activities were directed had also expanded: from an almost exclusive preoccupation with the commercial farms and ranches, to a commitment to the numerically very large sector of small, subsistence farms.

These developments have had a number of consequences, described by Trigo *et al.* (1982):

- Government funding, since 1970, has remained fairly constant, notwithstanding the considerable expansion of services ICA was requested to supply.
- As a result, ICA had to make a major effort to increase income from its own sources, such as those derived from the control and testing of agrochemicals. The consequent involvement of scientific personnel in matters unrelated to research increased, leaving only one-third exclusively concerned with research, so that the latter activity became the one which suffered most from these developments.

The complex, diffuse and dispersed organisation of ICA has led to problems associated with administrative gigantism. 'The responsibility for multiple functions of different nature gave rise to conflicts and antagonisms in the implementation of functions; these have become more competitive than complementary' (ICA, 1982).

As the research carried out by ICA became less effective, Government' increasingly encouraged the private sector to engage in agricultural research, thereby reversing the policy which had served as a guiding principle throughout the years since the beginning of agricultural research in the country, namely, maximum concentration of agricultural research in a single national institute.

As the functioning of ICA deteriorated, its image in the eyes of the public was damaged. The formerly vigorous institution had 'changed into a body without impact on agricultural research' (Trigo *et al.*, 1982).

Another indication of the unsatisfactory state of affairs is the numerous meetings, commissions, missions of foreign experts, which follow each other in rapid succession. Their recommendations resulted in a number of reorganisations, which however did not lead to an improvement in the situation.

The Graduate Study Programme (PEG)

The establishment of a joint ICA–national University Graduate Study Programme in 1967 has been briefly mentioned. By 1973, 82 courses were offered, with 59 professors participating. Most of these had joint appointments to ICA and the University. By 1980, more than 650 professionals had received training in the Graduate School, 'an achievement unparalleled in any Latin American country' (Ardila, 1985). With the termination of foreign aid, funding for the programme declined rapidly, and PEG came practically to a standstill by 1985. It is incomprehensible that a programme that is not only in the national interest, but is of advantage to the Faculty of Agriculture, the research institution and the students involved, should have been allowed to lapse. Instead of progressing to a post-graduate school (which should have high priority) the system—which had improved the teaching programme of the faculty through the participation of ICA researchers in professional assignments, ensured a higher quality of undergraduate thesis work under the tuition of research specialists, provided the faculty staff with improved research facilities, and all this at a considerable saving in government expenditure —was inactivated, apparently for reasons of institutional prestige. A new accord is being negotiated between the University and ICA (Arnon, 1986).

Rural development and extension

The extension system has had very little success reaching the small farmers practising subsistence agriculture on minifundias—a sector characterised by high population growth, and limited access to health, educational and financial services.

ICA had therefore found it necessary to devise a new approach to this sector, and created the *Rural Development Service* (Garcés, 1974). This service established the rural development projects with the objective of integrating ICA's efforts to improve the standard of living of the underprivileged rural sector with those of all other agencies engaged in agrarian reform, marketing, credit, infrastructure, education, public health, etc.

RECAPITULATION AND CONCLUSIONS

The functions assigned to ICA, since its inception in 1962, have expanded from the exclusive concern with the three basic functions of research, education and extension, to include responsibility for control, regulatory and fiscal activities as well as rural development. This has resulted in a multiplicity of functions in a single institute, in which research is no longer the priority activity. The diversity and incompatibility of these functions has impaired the efficiency of ICA and has contributed largely to the problems encountered by the organisation (ICA, 1982).

In brief, ICA has been transformed into a miniature Department of Agriculture, without having the political influence of a Ministry. The very situation from which research had attempted to escape, when DIA was replaced by ICA, has been recreated within ICA, with all the problems inherent in such a situation. On one hand, the research activities of ICA were being curtailed by the reduced allocation of resources and by the competition of the private sector, and on the other hand the ability of ICA to conduct a nationwide programme of agricultural research efficiently was constrained by the organisational and administrative framework in which it is confined.

Since 1977, the institutional functioning and structure of ICA in general, and of agronomic research in particular, have been analysed and re-analysed by Colombian and foreign experts. Practically all these deliberations have resulted in the same conclusion: it is necessary to separate the research activities from the functions of regulation, development and control, in which ICA is presently engaged, and leave research as its sole responsibility.

Union of Soviet Socialist Republics

INTRODUCTION

Progress in agriculture in the USSR has lagged far behind achievements in other sectors of the economy. In the same way, the slow progress in agricultural research contrasted with the spectacular advances made in nuclear physics, rocketry, and exploration of space, for example. This in spite of a distinguished Russian tradition in agricultural and related sciences, such as chemistry, zoology, bacteriology, and physiology.

Russian scientists pioneered in soil science. One of the best known Soviet agricultural scientists was N. Vavilov, the eminent plant geneticist, known for his work on the centres of origin of cultivated plants, and his considerable contributions to the stock of germplasm for plant breeding programmes of the major crops, that he personally collected in plant exploration in all parts of the world.

The major reason for the backwardness of agricultural research in the USSR, was the need felt by the political leaders to control scientific development and its application, at least in agriculture, and their inability to distinguish between real science and pseudo-science. This caused a setback of several years in agricultural research, and hence, in progress of Soviet agriculture.

T.D. Lysenko, trained as an agronomist, became the initiator of a movement named agrobiology, better known as Lysenkoism, which arose during a crisis in agricultural production in the USSR. He first proposed to reduce losses of winter wheat by vernalisation, a treatment that had been tried without success in other countries. This, and other proposals, such as the inheritability of acquired characteristics, which claimed to conform with Marxist principles, were endorsed by Stalin, who thought that Lysenko's methods would give more rapid results than the slower capitalistic, and reactionary methods of plant breeding practised in the West, and based on Mendelian principles of heredity.

Symptomatic of the status of eminent scientists at the time, is the treatment meted out to Vavilov. Because he opposed the genetic theories of Lysenko, he was sentenced to death, but was sent instead to a Siberian prison, where he died a few years later. The fact that he was the head of the prestigious Lenin All-Union Academy of Agricultural Sciences for many years, did not protect him against the vindictiveness of a pseudo-scientific charlatan.

The political character of the debate between Lysenko and his opponents became clear at the notorious session of the Lenin Academy of Agricultural Sciences in 1948 (Académie Lénine des Sciences Agricoles de l'URSS, 1949), when Lysenko was able to announce that his agrobiology had the support of the Central Committee of the Communist Party. All geneticists and plant breeders who had opposed Lysenko's spurious theories were obliged to recant 24 hours after having attacked agrobiology, and those who refused to do so were dismissed from their posts and in many cases imprisoned.

Lysenko subsequently became Head of the Lenin Academy of Agricultural Sciences, and Director of the Institute of Genetics. However, agrobiology failed to improve agricultural production, and consequently Lysenko lost his political influence. Subsequently, an investigation disclosed evidence of fraud in the recording of results in an experiment station under his direction. He was dismissed from his posts in 1965.

With the fall of Lysenko and his discredited following, the way was open for renewed progress in agricultural research.

STRUCTURE OF THE AGRICULTURAL RESEARCH SYSTEM
(Casas *et al.*, 1979)

The structure of the agricultural research system in the USSR is extremely complex. It comprises research institutes of the Union Ministry of Agriculture, the Ministries of Agriculture of the Republics, other Ministries at Union and Republic levels that have some involvement with agriculture, the Academy of Sciences and its branches in the Republics and the Universities.

Ministerial research institutes

These institutes are concerned with applied research and are under the jurisdiction of the Lenin All-Union Academy of Agricultural Sciences (the Russian acronym: VASKhNIL).

VASKhNIL was established in 1929, and charged with progressively setting up a network of experiment stations to cover the entire territory of the USSR. Since then, over 700 experiment stations have been established. These comprise 92 Union institutes with 168 experiment stations distributed throughout the USSR and directly managed by VASKhNIL, whose vocation is fundamental research, and 144 institutes directly linked to the Ministries of Agriculture of the 15 Republics, whose research is intended to be of a more applied nature, though this distinction tends to become blurred.

The VASKhNIL institutes are responsible for all areas of agronomic research, excepting technology of plant and animal products, which is the concern of the Ministry of Food Industry and to a lesser degree of the All-Union Academy of Sciences.

Amongst the most important VASKhNIL institutes are: the Vavilov Institute of Leningrad for Food Production, the Moscow Institute for Food Hygiene, the Leningrad Institute for Plant Protection, and the Moscow Institute for Agricultural Economy.

Part of the staff of these institutes is generally distributed among the various branches of the institutes and their experiment stations in the different Republics and some may be stationed at certain production units (sovkhozes, kolkhozes).

Five institutes were established in the 1970s in Novosibirsk, the 'science capital' of Siberia, with the mandate to solve the specific problems of agricultural development of this region.

Ministries other than agriculture

The most important of these for agricultural research are the All-Union and Union Ministries of Food Industry, which carry out most of the research on processing of plant and animal products.

The All-Union Research Institute for Milk Technology, for example, has a staff of about 1200 people, of whom 600 are researchers. Each of the Republics has at least one Food Technology Institute.

Academy of Science of the USSR

The Academy of Science of the USSR is the highest scientific establishment of the Soviet Union. Its declared purpose is to carry out scientific investigations which will further all branches of science. It prepares a programme of research on the major scientific and technological problems according to directives received from the Council of Ministers for Science and Technology (Turkevich, 1960).

This programme, after approval by the Council of Ministers, is carried out in the 367 All-Union institutes of the Academy, and 391 Union Republic Institutes which are the basic research units. The Academy also supervises graduate work at many of its institutes and can award advanced degrees.

The Presidium of the Academy has a special council for the co-ordination of research work of the 13 Union Republic Academies of Science. These Academies elect their own membership and presidium. The President is frequently a member of the Academy of Science.

The Academy also has seven regional branches in the smaller Republics of the Soviet Union, which are mainly concerned with the problems related to the development of the economy and culture of the regions.

Part of the research carried out by the Academy in the fields of biology, technology, economics, etc., is of relevance to agriculture. For example, the All-Union Trimiriazev Institute of Plant Physiology of Moscow, and the All-Union Institute of Genetics and Cytology, as their names indicate, work in fields of potential importance to agriculture.

The universities

By tradition the universities have been mainly concerned with professional education, and to a lesser degree with broad, theoretical research. University-level agricultural education is very developed, comprising over 100 Academies of Agriculture, with a teaching staff of 30 000 and an enrolment of 250 000 students. About 20 000 staff are engaged in research at these Academies, which are under the jurisdiction of the Ministry of Agriculture.

Research by the farming sector

The pioneer in research carried out by the farming sector is the State Poultry Industry, which has established five poultry research centres and five R & D stations; all these work in close collaboration with the official poultry research institutes.

It has been officially proposed to adopt the same production–research system for the vegetable-growing sector.

In order to encourage innovativeness in the State production units, 'Associations for Science and Technology in Agricultural Production' have been established, to promote research and development in the sovkhozes and kolkhozes. Some have been very active; many of the research works published by these Associations are theses for Ingénieur or even DSc undertaken by farm workers, under the auspices of the academic institutions, by correspondence courses. The graduates are frequently offered research positions in the research institutions. The best-known graduate of such a course is Gorbachev.

Research staff

The number of research workers in the USSR increased steeply in the 1970s, at an annual rate of 12 per cent, to reach 80 000 in 1979. At the same time, the academic qualifications of researchers also improved markedly; the number of PhDs increased almost six-fold during the same period.

Recruitment

The Directors of the institutes controlled by VASKhNIL (themselves appointed by the Presidium of the Academy), are free to recruit any personnel they consider essential for the implementation of their research programmes, within the limits of budgetary constraints. Young scientists are appointed on the basis of competitive examinations.

Promotion

The work of individual researchers is evaluated by special committees, every two years for holders of PhD degrees, and every three years for the others.

In addition to scientific achievements, behaviour and ideological commitment are taken into account (Kirillov-Ougrioumov, 1977). Even institute directors are evaluated periodically, by the Presidium of VASKhNIL, on the basis of the scientific achievements of their institutes and the diffusion of the results of their research.

Decisions on the appointment or promotion of researchers are made by the Scientific Council of the Institute. Researchers whose work is not considered to be satisfactory may have their salaries reduced.

Research planning

At Union-Republic level
Overall co-ordination and financing of research activities is the responsibility of the Council for Agriculture and Food of the State Committee for Science and Technology. The Council, comprising representatives of the Union State Ministry of Agriculture and of the research institutes, has a number of sections concerned with specific commodities or disciplines. The principal function of the Council is to establish an overall plan, congruent with the economic objectives of the most recent 5-year Plan; to define the principal research areas and the responsibilities of the different institutes.

Prior to establishing the research programme, the Council evaluates the projects that are being implemented, or have been concluded recently.

The work of the Council is carried out in close co-operation with the All-Union Ministry of Agriculture and the All-Union Academy of Science. The procedures used in the States are similar to those adopted at the All-Union level.

At institutional level
In the case of VASKhNIL, every individual researcher can submit proposals to the Scientific Council of his institute. Each institute transmits its proposals to the appropriate Branch Bureaux of VASKhNIL in Moscow. The proposals are studied, compared and priorities established by each Bureau which transmits them to the Planning and Co-ordination Bureau, which prepares a synthesis programme including approved projects. This is then submitted to the Presidium of VASKhNIL, which finalises its All-Union Programme of Agricultural Research, including

a detailed description of all research projects considered essential, stating their costs, duration and the institutes responsible for implementation. The VASKhNIL programme is then submitted to the Council of Ministers of Agriculture of the USSR for approval.

Implementation of the research programme

After each institute has been informed of the research projects allocated to it, these have priority status and the institutes are under obligation to implement them.

However, each institute also enjoys a certain latitude in the choice of additional projects, in particular contract research for production units. The contracts can be negotiated either directly between the institute and its client, or through the intermediary of the Union-State Ministry of Agriculture. The latter procedure is preferred by the authorities, as it facilitates the avoidance of duplication and the monopolisation of the results of research.

These contracts may account for up to one-third of the cost of personnel salaries of the institute.

Research objectives

Research objectives of all the USSR agricultural research institutions are determined in accordance with the most recent 5-year Plan. This requirement obliges all the research institutions to revise and reorient their research objectives (and sometimes even their structure) in accordance with changes in priorities evidenced in the latest Plan.

In crop production, priority is given to improving the production of cereals, legumes and vegetables. Research is specifically oriented towards achieving higher and more reliable yields, and improved quality. For cereals, priority objectives are improving cold and drought resistance, breeding of triticale and high-protein maize. Other favoured crops are sunflowers, field peas, and soya.

Problems of energy conservation do not figure high in priority, mainly because the use of high-energy inputs per unit area is still very low, in spite of a relatively steep increase in the 1970s; also, the USSR has no shortage of fossil fuel.

Considerable research efforts are invested in other areas of importance such as biological control of plant pests and biological fixation of nitrogen by legumes and cereals.

Most of the Soviet research objectives are similar to research problems studied in the developed market economies of the West.

Publications

The Institute of Agricultural Information, Economic and Technological Studies of the Union Ministry of Agriculture, with a staff of 500 people, handles the flood of scientific and technological publications, both Russian and foreign. The Institute publishes bibliographic information in the form of indices, yearbooks, and translations.

The Institute is supported in its work by other institutions with a more general vocation, such as the Union Institute for Scientific and Technological Information, the Scientific and Technological Library and the National Centre for Translation of Scientific and Technological Publications.

Vietnam

Agriculture is the most important economic sector of Vietnam. Seventy per cent of the population derive their livelihood from agriculture. The predominant crop is rice. The productivity of the agricultural system is extremely low and unstable.

HISTORICAL BACKGROUND

During the French colonial period, an Institute of Agronomic Research was founded in 1920 in Saigon, with some laboratories in Hanoi. Research work was carried out on soils, diseases of plantation crops, and botany. Some research on rice was done at the Rice Office, in Saigon.

After the 1945 revolution, the agricultural research system was modified seven times, between 1952 and 1975.

STRUCTURE OF THE NARS (Tuan, 1985)

The Ministry of Agriculture is responsible for the agricultural research institutes. The Department of Agricultural Science and Technology of the Ministry has established a *Scientific Council* which is the co-ordinating body, and formulates five-year research and development programmes. Since 1980, National Commodity Research and Development Programmes have been established for rice, secondary food crops, pulses, pigs, cattle, poultry, etc. Each Programme is headed by a Board of Scientists. Transfer of research results to farmers is allowed only after approval by the Scientific Council.

Research institutes

The first agricultural research institute, for crops, was established in 1953, and later became the *National Institute of Agricultural Sciences* (INSA). The major objective of INSA was to undertake long-term, strategic research to support and improve agricultural production (Hoan, 1987).

In 1967, several research departments of INSA were separated from the main institute in order to undertake expanded programmes of applied research in various commodities and disciplines, leaving INSA to work on more long-term research. Ten new institutes were formed (Hoan, 1987):

- The *Food Crop Research Institute* for research on important food crops, including rice, sweet potatoes, potatoes, pulses, and vegetables.
- The *Industrial Crops Research Institute* for crops like tea, citrus and lychee.
- The *Animal Husbandry Research Institute* is responsible for research on genetics and cross-breeding, husbandry of pigs, buffaloes, cattle, horses, poultry, and other livestock. The Institute maintains seven research stations in different regions.
- The *Veterinary Research Institute* works on the control of animal diseases, and production of vaccines.
- The *Soil and Fertiliser Institute* conducts research on soil management and optimum fertiliser utilisation.
- The *Plant Protection Institute* studies pests, diseases of plants, and their control.
- The *Agricultural Economics Research Institute* studies production economics, methods for improved farm management.
- The *Institute of Agricultural Machinery and Mechanisation*, is re-

sponsible for research on the manufacture, operation and maintenance of agricultural machinery.

- The *Agricultural Planning Institute* studies soil resources, the regionalisation of crop production, optimal resource utilisation.
- The *Institute of Agricultural Construction and Design* carries out research on capital construction and agricultural design.

Two additional institutes were established after the reunification of Vietnam: the *Agricultural Technology Institute* and the *Rice Research Institute*.

Three Regional Research Centres and four Commodity Research Centres (cotton, maize, pulses, sericulture) have also been established.

Funding

Besides routine funding of research by Government, the raising of research funds for specific projects through contracts between the research institutes and co-operative farms is encouraged. It is considered that this system provides an incentive to develop better and more appropriate technologies.

Staff

The technical staff engaged in agricultural research numbers about 2500 of whom 6 per cent are seniors, 66 per cent are juniors and 28 per cent are technicians. It is recognised that the proportions of senior researchers and of technicians in relation to junior researchers are too low.

The National Institute for Agricultural Sciences provides postgraduate training for a limited number of students. In-service training for junior researchers is provided at the Institutes, as well as in-service training for farm labourers, to enable their promotion to technicians.

Priority areas for research

Priority areas for research identified by missions visiting the country and by local staff, are: management of acid sulphate soils, water management, rainfed farming systems, animal disease control.

Research planning

The following procedure has been established for planning and approving the research programme (Hoan, 1987):

1. Each research institute, together with a national Programme director, formulates the research programme of the institute for the coming year, in accordance with the goals and objectives of the current Five-year Plan.
2. The proposed programme is submitted to the Department of Agricultural Science and Technology (DAST) of the Ministry of Agriculture (MOA). DAST makes an initial scrutiny of the proposal and holds discussions with the relevant institutes for any necessary clarification.
3. DAST integrates the research proposals into an overall agricultural research plan, with priorities based on the urgency of the problems and the economic potential of the projects.
4. The research plan prepared by DAST is vetted by the Council of Agricultural Science and Technology; with the concurrence of the Minister of Agriculture, it is then sent to the State Committee for Science and Technology.
5. The State Committee for Science and Technology examines the plan in consultation with the State Committee for Planning, the Ministry of Finance, and the MOA.
6. The State Committee for Science and Technology, on behalf of MOA, announces the approved research plans, including programmes and topics at national and ministry levels, together with the resources allocated for their implementation.

Programme implementation (Hoan, 1987):

On the basis of the approved plan, MOA allocates specific responsibilities for the implementation of the various components of the plan, to the institutes, centres and agricultural units. These in turn, assign specific projects to the appropriate research departments. These are required to submit reports every three, six and twelve months.

Agricultural research is mainly carried out through national commodity programmes, thereby drawing research workers into integrated teams and forming links among units. Each programme has a Board of Directors responsible for formulating and monitoring the programme. The

Board of a National Programme is appointed by the State Committee for Science and Technology, and its chairman by the Council of Ministers, indicating the importance attached by the political leaders to this form of organising research. The chairman of the board is responsible for the management of the programme.

Monitoring and evaluation of research programmes (Hoan, 1987)

At the *State level*, inter-ministerial programmes are evaluated by the State Committee for Science and Technology. Following the comments of the Committee, MOA takes the required action.

At the *Ministerial level*, DAST carries out periodic evaluations following which projects may be terminated, modified or intensified. Following these recommendations, the Minister gives directives to the institutes concerned for implementing the required changes.

At the *Institutional level*, each Institute has its Council for Scientific and Technological Research, which supervises the research work, evaluates progress, examines reports, and submits its observations through DAST, to the National Council for Science and Technology.

Shortcomings

Hoan (1987) comments: 'Research planning has not involved the participation of the majority of research users or the collaboration of research institutes and non-agricultural institutions. Hence, many targets have not been achieved, and some topics have overlapped, causing a waste of investment and resources. No standard criteria for more accurate evaluation of programmes have as yet been developed'.

Non-government research

An interesting feature of agricultural research in Vietnam is the number of companies and enterprises which have established their own research centres, combining production, research and technology transfer. Hoan (1987) mentions the following examples: the Coffee Research Centre of the Union of Coffee Enterprises, the Cotton Research Centre of the Cotton Company, the Vegetable Research Centre of the Vegetable Company, the Chicken Research Centre of the Union of Poultry Enterprises, etc.

People's Republic of China

THE DEVELOPMENT OF THE AGRICULTURAL RESEARCH SYSTEM

The development of the agricultural research system and of the transfer of technology in China have occurred in the context of a unique political system (Sprague, 1975).

In the short period since the establishment of the Peoples' Republic of China, the leadership's policies towards agronomic development has followed a zig-zag course, characterised by repeated major reversals of direction. The manner in which research in general, and agricultural research in particular, was to be structured was a political issue, and therefore underwent the same veerings in course dictated by the prevailing policies.

The early days of the Republic

Prior to the Revolution, the elite of Chinese scientists had been trained abroad, mainly in the USA, Japan and Europe. They valued academic training, foreign contacts, and independent basic research. Subsequently, in the 1950s, most of the scientists were trained in the USSR. Both groups found it difficult to accept Mao's concept that the primary function of science was in the service of production.

China, however, had no choice at the time but to rely on these foreign-trained scientists, working within the Institutes of the *Chinese Academy of Sciences*. The Academy maintained its position of chief policy maker and implementing agency for research until the late 1950s.

The *'Great Leap Forward'* (1958–59) was the first reaction against the neglect of agriculture and called for the simultaneous development of agriculture and industry—'Walking on two legs'.

The new leadership created the State Scientific and Technological Commission, through which primary control over research was gradually wrested from the Academy of Sciences.

During the 1960s, the authorities diverted many of the scientific activities from the academically-oriented institutes to the rural areas.

The *Cultural Revolution* (1966–68) continued and amplified the general decentralisation policies of the Great Leap Forward movement. From 1966 to 1976 the system went through what the Chinese call 'struggle, criticism, transformation'. The rural bias was strengthened, and a drastic reorganisation of agricultural research was implemented. Many of the research institutes, in particular those related to specific crops or specific regions, were removed from the aegis of the Academy of Agricultural Sciences and placed under the supervision of the provinces, in order to make them more responsive to local needs. Extension stations and personnel were transferred to the communes. Agricultural college curricula and student selection procedures were transformed. Publication of scientific journals was suspended, and little research was done during this period. Agro-technical experiment stations were established at county, commune, brigade, and team level. The new system was called 'open door research' and designed to assure free exchange between farm and laboratory (Stavis, 1978).

Period of restoration and consolidation, 1976–80

In the early 1970s, the basic responsibility for science and technology started reverting to qualified scientists and to the academies. Many of the research institutions were reopened, and by 1976, 70 institutes were again functioning under the Academy of Science; by 1979 the number had grown to 90, and continued to increase subsequently (Orleans, 1979).

A similar trend was evident in relation to the Academy of Agricultural Sciences, renamed Academy of Agricultural and Forestry Sciences.

Whilst Chinese scientists had learned to accept the ultimate authority of the Party in establishing the general orientation of research, they started to react strongly when politicians attempted to dominate science. This opposition appears to have been recognised by the Party as justified, as evidenced in a speech by Teng Hsiao-Ping to the National Science Conference in which he stated: 'The Party leadership must ensure the correct political orientation, they must guarantee supporting services and supplies, they should get acquainted with the work of the scientists, but the scientific leadership of the research institutes should have a free hand in the work of science and technology' (Orleans, 1979).

HUMAN RESOURCES

To develop a cadre of research workers amenable to the orientation by the Party, important changes were made in agricultural education. Agricultural colleges were moved from the cities to the rural areas and the curriculum made shorter and more oriented to practical matters. Formal post-graduate training became virtually non-existent.

There was a determination to prevent the emergence of an elitist class of scientists. All agricultural scientific societies were disbanded, but all research workers were organised into an Association of Agriculture (Sprague, 1975).

College education seems to have been oriented towards the very practical. To make up for possible deficiencies, senior scientists at some of the institutes took the initiative to organise a kind of 'apprentice training programme', something similar to post-graduate training (Sprague, 1975).

College-trained personnel were reinforced in their work in the communes to which they had been assigned by experienced farmers, as well as urban high school graduates who were required to live in the countryside since the Cultural Revolution.

In many cases, not more than one-third of the scientific staff of an institution were working at headquarters at any given time, the rest working in communes, or travelling to study conditions in different communes.

The scientists are required to recognise innovations that have been made by farmers, organise their testing and eventual adaptation so that large-scale adoption can be promoted (Stavis, 1978).

Incentives

An attempt was made to minimise basic research, and political–moral incentives were stressed. Small study groups, organised by the Communist Party, engaged in criticism and self-criticism. Scientists were encouraged not to be concerned with promoting their personal interests, but to work on problems of national economic importance.

Status

Before and during the Cultural Revolution, scientists were suspect, and were accused, indiscriminately, 'of being elitist, living in ivory towers, seeking personal fame and gain, on persisting with theoretical research unrelated to production needs' (Orleans, 1979). Re-education was required for all of them.

AFTER THE DEATH OF MAO TSE-TUNG

Since the death of Mao Tse-tung, the emphasis is on rapid economic development. Science and technology have been designated 'leading factors' in the four 'modernisations' of agriculture, industry, national defence, and science and technology. Agricultural research therefore belongs to two of the leading factors.

As a result, scientists are no longer under suspicion of being politically unreliable, but have even been elevated to an elite position within society (Orleans, 1979).

PRESENT STRUCTURE (Zhou, 1987)

The Government of China is now paying much attention to agricultural research and has undertaken the organisation of an integrated national research system. Science and Technology Commissions have been set up at state, province, prefecture and county government levels, and are responsible for overall co-ordination and management of all aspects of science and technology. The departments responsible for agriculture at each of these four government levels have management mechanisms for the implementation of the guide-lines provided by the Commissions.

There are four subsystems for agricultural research, each with specific responsibilities:

(1) *Chinese Academy of Science* is affiliated with the State Council. Its mandate is basic research, part of which relates to agriculture. It has 122 research institutes distributed throughout China.

(2) *Ministerial level:* The *Chinese Academy of Agricultural Science* (CAAS) is under the Ministry of Agriculture, Animal Husbandry and Fisheries (MAAF). Its mandate is to solve key technological and scientific problems in crop production and animal husbandry of significant economic importance. It concentrates on applied research, mainly in crop breeding and cultivation, plant genetic resources, soils and fertilisers, plant protection, agroclimatology, and veterinary science.

CAAS has a network of 35 national research institutes; each institute has one or two main research functions. CAAS does most of the research on key projects with national priority. It also co-ordinates the provincial programmes. Research programmes have to be approved by MAAF.

The *Chinese Academy of Forestry* (CAF) is under the Ministry of Forestry. CAF is a national centre for applied forestry research. It has 11 research institutes and pilot farms, and is managed in the same way as CAAS.

Other research institutions at ministerial level include the *Chinese Academy of Agricultural Engineering Research and Planning* (under the MAAF) and the *Chinese Academy of Water Conservation*, under the Ministry of Water Resources and Electric Power.

(3) *Provincial and prefectural levels*: each province or autonomous region has its own Academy of Agricultural Science, of Forestry, etc. They are controlled by the respective Departments of Agriculture or Forestry, etc. Research is focused on solving provincial production problems, but they also participate in national and regional problems.

Prefectural research institutions, under the prefectural departments, are smaller than the provincial institutions, and their role is more circumscribed. There are about 1400 research units in and above prefectural level, in the two sub-systems of MAAF and Ministry of Forests.

(4) *University-affiliated institutions* specialise in basic theory and applied research in accordance with educational requirements. China has 59 agricultural universities and colleges.

Funding

Funds for CAAS and CAF are allocated, after approval by the planning and financial sections in the Ministry, from the national budget of the Treasury.

Funds for university research flow from the State Education Commission through the State financial budget.

An additional source of funding is contract research for the production sectors. Institution-initiated research can be carried out in a large range of areas in which the researchers are interested and competent, provided the funds are found by the research institutes themselves, from foundations, loans, or foreign assistance.

Research planning (Zhou, 1987)

Research planning is exemplified by the procedure followed by a typical

Provincial Research Institute:

- the Provincial Department of Agriculture produces guidelines for an annual programme and assigns specific projects;
- several months before the new financial year, the research institute makes a socio-economic survey of the provincial development requirements;
- projects are selected which conform with the provincial development plan, the development of the institute and its resources;
- a research proposal is prepared, indicating research objectives, progress schedule, budget estimate, etc.
- an internal academic committee evaluates and approves the proposal;
- the programme is submitted to the Provincial Department of Agriculture, which decides which projects will be included in the provincial research programme. Projects of wider socio-economic significance are sent for consideration to a higher level.
- the approved programme is then allocated funds through the provincial department of the Treasury.

Major deficiencies (Zhou, 1987)

The decentralisation of research according to administrative divisions, and not to agro-ecological regions is a source of overlapping and duplication. There are no formal relationships with the producers. Each institution reports its activities to the next higher level in the hierarchy, resulting in over-control by administrative authority.

Programmes do not generally reflect accurately the needs of agricultural development, and the funds allocated for research are inadequate.

These deficiencies have been recognised by Government, and steps are contemplated to correct them.

International co-operation

China's NARS has established relationships with over 10 International Research Centres, and is co-operating with the agricultural research systems of several countries.

Chapter 2

National Agricultural Research Systems: An Overview

INTRODUCTION

In the preceding chapter we represented case histories illustrating the development of the national agricultural research systems (NARSs) in a number of selected countries.

These countries were not chosen at random, though availability of reliable documentation and the personal acquaintance of the author with many of the situations described, were factors in deciding which countries to include. An effort was made to choose case histories which would reflect most, if not all, the situations encountered in the field in developed countries and developing countries; free marketing systems and centrally planned systems; large and small countries in each of the former systems; countries that had, under similar eco-agricultural, political and social systems, adopted entirely different solutions to research organisation, etc.

In the developing countries, are represented NARSs from Asia, Africa, and Latin America, with contrasting colonial pasts. In the centrally planned economies, we have been limited in the number and choice of case histories, because of the paucity of information available.

Though the topics analysed in this overview are not based exclusively on the case histories presented in the previous chapter, the latter do illustrate the different problems and topics that will be presented and discussed in this book.

This overview raises many questions, to which we still do not have all the answers.

191

NATIONAL AGRICULTURAL RESEARCH SYSTEMS IN THE DEVELOPED MARKETING SYSTEMS

The changing circumstances of agriculture

From the end of World War II and until the early 1970s, the developed countries with free marketing systems enjoyed a period of strong economic growth. However, the after-effects of the energy crisis in 1973, resulted in lower economic growth from then till the 1980s. Since the mid-1970s, agriculture has experienced the following major changes (Stevenson, 1984; Le Gouis, 1986; Santucci, 1986).

Self-sufficiency and overproduction
Not only has self-sufficiency in the major commodities been achieved, but overproduction has resulted in surplus commodities with limited markets because of the lack of purchasing power of many of the potential importers.

Lower income from agriculture
In a buyer's market, farmers have become weak in bargaining power for prices for their products, especially as two-thirds of their produce goes to food processors. The low prices obtained for agricultural products contrast with the steep increase in the cost of energy-rich inputs essential for achieving high yields. The combination of low prices and high costs of production have led to heavy expenses, debts and, frequently, near-bankruptcy.

Despite income support measures implemented by the European Community, the per capita income of farmers has remained virtually unchanged from 1974 to 1984. In the USA, government farm support expenditures have increased from $4 billion in 1981, to almost $32 billion in 1987. Almost one-third of the cash income of the American farmers is derived from government subsidies (Myers, 1987).

Part-time farming
Many small farmers have attempted to keep their farms by supplementing their income by off-farm work. In 1975, only 37 per cent of the farms in the EEC were run on a full-time basis. In Italy, the proportion of full-time farms is only 16 per cent.

Smaller numbers of farms, and fewer workers in agriculture
Overall, in many developed countries, small farms are increasingly going out of business. In the USA, some 30 million people migrated from the countryside to the urban areas between 1940 and 1973, and 2000 farms went out of business every week (Wade, 1973e). The trend continues, and between 1981 and 1987 the number of farms decreased by 14 per cent (from 2·43 million to 2·1 million) (Myers, 1987).

Even in countries like Holland and Denmark, traditionally supportive of family farms, from which most of the agricultural production is derived, small farms are disappearing at an increasing rate.

In the European Community as a whole, there was a decline of six million in the number of workers in agriculture during the period 1974–84. At the same time the average age of the farmers increased, with only 9 per cent of them under 34 years of age.

Decline in political power of the farming community
As the number of farmers and farm workers becomes smaller, there have followed changes in the power structure of the Farmers' Associations, and their political power has declined significantly. There is also less public sympathy for farming and rural problems.

Deterioration of natural resources
The conservation of natural resources has not been as successful as their exploitation. Though the USA, for example, has been a world leader in research on soil erosion and in the development of methods for its prevention, half the original top-soil has been lost, and this loss continues on about one-third of the farmlands. Ground-water supplies have diminished and the quality of the water impaired (USDA, 1983).

Agricultural policies
The targets of the agricultural policies of the EEC countries (Skovgaard, 1984) are fairly typical of those of other developed countries. These are: (a) to increase agricultural productivity by rationalising production methods, without further increasing yields; (b) to assure the farmers of a reasonable standard of living; (c) to ensure supplies and to stabilise markets by avoiding overproduction; (d) to ensure reasonable prices for agricultural commodities to the consumers.

Unlimited guarantees for unlimited quantities of agricultural produce are no longer to be given.

Implications for agriculture research

Funding
Many NARSs in the developed countries expanded rapidly and considerably after World War II. Towards the 1970s, the increase in funds devoted to agricultural research began drawing to a close.

Governments were having increasing doubts about the benefits that their countries were deriving from agricultural research; the public was becoming increasingly concerned with pollution resulting from intensive use of agrochemicals and the whole trend of technological societies (Ulbricht, 1977).

As a result of the foregoing, funding for agricultural research is now generally decreasing in real terms; the demands for evidence of efficiency of the research service is increasing as well as demands for privatisation (Stevenson, 1984).

Interest groups
New interest groups have emerged, which have markedly changed demands made upon agricultural research (Busch & Lacy, 1983). For example:

• the consumer movement is concerned with proper diet and food safety;
• the environmental movement is pressing for more research on farm run-off, pollution from animal wastes, and soil erosion;
• farm labour organisations show an increasing awareness of the role of agricultural research in changing the labour needs of farming.

These and other groups have potentially conflicting objectives, with which many researchers have difficulty in identifying. However, it will be essential for agricultural research to widen its base in order to maintain wide public support.

Research objectives
Crop yields and animal productivity appear to have reached a plateau. The percentage of annual increases in agricultural production and technology, in the USA, for example, has declined in recent years (USDA, 1980).

The major challenge facing agricultural research in the developed countries is to provide answers to farmers who desire an adequate return for their labours by developing technologies that increase efficiency and productivity, through decreasing costs of production, rather than by increasing yields (Skovgaard, 1984).

Environmental groups are increasing their interest in the acceptability of the production methods used, and are demanding intervention against some of them through legal prohibitions and regulations. Since the 1970s, *protection research* has been playing an increasingly prominent role. The objective is to ensure that production methods are environmentally acceptable in the long run (Skovgaard, 1984).

To offset increased production, over half a million hectares will have to be taken out of production every year. For the EEC countries, it is estimated that up to 30–40 per cent of the land will no longer be cultivated by the year 2000 (Röling, 1986). Research will have to concern itself with the use of these, mostly marginal, lands for the maintenance of landscapes, protection of ecosystems, recreation, etc.

Societal

Hightower (1978) states that agricultural research has been, and still is, 'committed to the technological and managerial needs of the largest-scale producers and of agricultural business corporations and . . . to omit those most in need of research assistance'.

Whether one agrees or not with Hightower's generalisation, there is no doubt that 'socio-economic research will have to apply itself to devise ways for limiting the social costs of the mass loss of rural livelihoods' (Röling, 1986).

Increasingly sophisticated research

Recent breakthroughs in the basic sciences, such as have led to genetic engineering, have opened the way for increasingly sophisticated research. Examples of the kind of research already on the programmes of agricultural research institutions are the most recent objectives of the British Agricultural Research Council (AFRC, 1987):

. . . a co-ordinated programme on cell signalling and recognition—a topic of crucial importance to the understanding and exploitation of many biological processes in animals and plants, including fertilisation, cell division, growth and development, host–pathogen interactions, symbiosis and stress response; the uptake and utilisation of mineral nutrients in plants; robotics in agriculture; novel vaccines and vaccination; transgenic animals; manipulation of growth regulation; the control of the functional properties of proteins and carbohydrates in food, etc.

Busch and Lacy (1983) point out that this type of research may increase the fragmentation of research along disciplinary lines, as the researchers typically receive all, or most, of their training within the same discipline to which they are going to devote themselves. 'Each discipline develops its own jargon, making cross-disciplinary communication difficult.'

Institutional
The overriding impression from the case histories presented in the previous chapter, is that almost all the developed countries had established over the years complex national agricultural research systems, characterised by the fragmentation of research among a number of bodies, with much duplication of effort, and lack of success in achieving co-ordination between them.

The undeniable success of these NARSs in promoting agricultural development, the impressive increase in yields and the parallel decrease in human effort required, have ensured very high returns on investments in agricultural research, but have obscured for many years the fact that most NARS had been highly inefficient in achieving their objectives.

Lack of national planning and co-ordination of the complex NARSs has led to high costs, which did not cause any special difficulties during the period of economic prosperity following World War II. However, the slowdown in economic growth, beginning in the 1970s, the agricultural overproduction and environmental stresses, have caused a gradual decline in public support for agricultural research.

All these factors have led to devastating criticisms of agricultural research, more stridently in some countries than in others.

The 1971 Royal Commission in the UK, headed by Lord Rothschild (Rothschild and Dainton, 1971), was the first in Europe to stress the need for a contractor–client relationship in agricultural research, whereby research services, specifically created by government as an instrument of its policy for agricultural development, should conduct only that kind of research that leads to the solution of high-priority problems of the farming sector; a concept that is still not willingly accepted by many agricultural scientists.

Funding for agricultural research is on the decline in most developed countries, and this trend appears to be accelerating. Demands for rationalising NARSs are becoming increasingly harsh. The UK is one of the first countries in Europe to have started closing down redundant research institutes, amalgamating others, and selling still others to private interest (privatisation).

A similar process of rationalisation of the research system appears to be inevitable, sooner or later, in most developed countries.

Another weakness of the NARSs of the developed countries, pointed out by Ruttan (1985) are the rudimentary linkages amongst them. He writes: 'For example, there has not yet emerged any institutional capacity to rationalise or co-ordinate agricultural research among EEC member countries. There is a modest programme of information exchange among OECD countries, but its activities appear to be more ceremonial than substantive.'

Private sector research

A major change in the NARSs of the developed countries is the ever increasing involvement of the private sector in agricultural research and diffusion of technology.

Certain advances made in basic research have made it possible for private enterprises 'to appropriate the economic benefits generated by new technology, and the trend in market economies is for technology to lose its nature as a public good and become a merchandise' (Piñeiro, 1986).

In the USA, already, the private sector invests more in agricultural research than the USDA and the state experiment stations together.

THE DEVELOPING MARKET ECONOMIES

The World Bank (1983) defines 94 nations as developing (or less developed) countries, of which 35 are considered low income countries (average per capita GNP below $400) and 59 middle income countries. They include all of Latin America, all of Asia except Japan, and all of Africa and the Middle East except for the high-income oil exporters. Three-fourths of the world's people live in these countries.

Typically, the poorest countries are the most dependent on agriculture. In 1981, the low income countries employed 73 per cent of their labour force in farming, compared with 44 per cent in middle income countries and 6 per cent in industrialised countries (2 per cent in the USA). In 1983, low income countries obtained 37 per cent of their GDP from agriculture, compared with 15 per cent in middle income countries and 3 per cent in industrialised countries (2 per cent in the USA).

Food production in the LDCs has expanded 38 per cent in the decade 1974 to 1984, but rapid population growth has offset much of the improvement, so that the per capita gain, starting from a very low base, was only

8 to 9 per cent. This improvement was very uneven, and in the case of Africa (excluding the Republic of South Africa) there was even a decline of 6 per cent for this period.

It is a tragic paradox, that the countries in which 3 per cent of the active work force are engaged in agriculture, produce food in excess of their requirements, whilst it is in nations in which over 70 per cent of the labour force are engaged in agriculture, that most of the world's hungry or near-hungry people are to be found.

Population growth and urbanisation are causing a considerable increase in food requirements. Many countries, and in particular those in the low income category, have since 1970 been meeting their food deficits by increasing imports of cereals and other foodstuffs, at an unsustainable rate, so that there is no practical alternative to increasing domestic food production (ISNAR, 1984c).

Neglect of agricultural research

In most LDCs, until the 1970s, politicians did not recognise the importance of agricultural research for development, and consequently, investment in research was not given high priority. Links between whatever research was being done and the majority of the farming sector were practically non-existent (Rivaldo, 1987).

These attitudes could find some justification as long as the growth that had taken place in the agricultural sector had arisen from further increasing the area under cultivation, without changing the low-technology methods in use. However, possibilities of further increasing the cultivated areas are becoming increasingly limited, and, generally, there is no longer an alternative to increasing productivity of the land already being farmed.

Neglect of natural resources

However important the objective of increasing yields, the LDCs cannot continue to neglect, as in the past, the impact of agricultural practices on the environment and on the loss of natural resources.

> Unsuitable agronomic practices are leading to massive soil erosion; in the drier areas, overgrazing is leading to a progressive degradation of the grasslands and finally to desertification; where irrigation is practised, the addition of large amounts of salt and waterlogging is taking huge areas of the most productive soils in the world out of cultivation;

conversely, overexploitation of ground-water resources with their consequent depletion is also causing salt-water intrusion in the aquifers and land subsistence, resulting in irreversible damage to the aquifers. The rate of destruction of forests, has been increasing at an alarming rate, causing irreversible processes in the denuded soil (laterisation), that make the land unfit for production and causing erosion, siltation of reservoirs, devastating floods, reduced underground water resources, loss of wildlife, and more.

For the LDCs, the breakdown of ecosystems and the resulting loss of land constitute a grave threat to their very survival. Hence the importance of intensive research programmes to devise appropriate measures for preventing in the first place, and remedying where necessary, the negative impact that has already taken place. However, the main constraints are not only technological, but also social and political; there is generally a lack of political awareness of the social, economic and environmental consequences of the abuse of the natural resources, which are not always immediately apparent and remedial measures are therefore not given high priority in the past by most politicians (Arnon, 1987).

The problems of increasing food supply, and of preventing the loss or degradation of natural resources, cannot be solved without an adequate national agricultural research capacity, able to address itself to problems of high-priority. Many of the LDCs have reacted to the situation by investing a major effort in the reorganisation of their national agricultural research systems, as shown in many of the case histories of selected countries presented in Chapter 1.

Historical development of the NARSs in the LDCs

In most developing countries, organisational structures for agricultural research were established by the colonial powers. The earliest measure taken to improve agriculture in the colonies was the establishment of botanical gardens during the 18th and 19th centuries. Through these gardens, crops and crop varieties were exchanged between continents and countries, and when successful, became major components of agricultural production in their new environments.

Towards the end of the 19th and beginning of the 20th centuries, agricultural experiment stations were established in an increasing number of these countries. Research was mainly devoted to export crops of economic interest to the colonial powers and significant results were achieved. Research on food crops was generally neglected, and in many of the countries this one-sided orientation of research continued unchanged after independence. In the early 1970s, several LDCs were still spending more than half their annual research funds on export crops (FAO, 1984).

The result in most countries has been a dualistic structure of the agricultural sector, consisting of a relatively progressive sub-sector of mainly large production units for export, and a stagnant sub-sector of small, subsistence farmers using traditional methods of production.

One common feature of the research systems of all the colonial powers, was that the researchers were mostly recruited in the respective home countries, and little or no effort was made to develop a local research capacity. The result was that after independence, the research system either continued to be manned by expatriates (former French and Dutch colonies), or broke down completely when the foreign researchers left *en masse* (former English and Belgian colonies). India is one of the few exceptions.

In the Latin American countries, which had gained their independence a century earlier than those in Africa and Asia, it was the pervasive influence of the aid provided by the USA, and the commercial interests of the large American companies that determined, to a large extent, the form and content of agricultural research, until the beginning of the 1960s, when the NARSs of most of these countries started to undergo extensive reorganisations.

Funding of agricultural research

In all the Third World countries, expenditure on agricultural research in the 1970s, accounted for US$ 240 million, constituting only 15 per cent of the total world outlay for agricultural research; of this sum, slightly more than one-third was contributed by foreign aid (Casas, 1977).

These limited funds were not even devoted to the problems of major importance to the economies of the countries concerned. Historically, the financial interests which invested heavily in plantation crops, also financed and organised research on these crops, usually by setting up specialised institutes for this purpose. This trend continued after independence because of the shortage of foreign exchange faced by the countries. Food production, the major consideration for the vast majority of farmers, was neglected, and research on food crops, still more so.

The World Bank (1981) describing the situation of agricultural research in the LDCs writes:

A pervasive problem is the inadequate appreciation by governments of the role that effective research plays in agricultural growth and development. This attitude has several consequences. Low priority, in terms of budgetary and manpower support, is given agricultural research by governments. Deficient organisational structures for research and extension result in fragmentation of effort, weak linkages between research and extension, and the generation of research results that have little practical value to farmers. Training programmes for scientists and technicians are generally deficient.

The situation described by the World Bank explains the generally poor performance of many national research services. These have been plagued by 'problems such as low salaries, poor promotion prospects, vulnerable research budgets, inadequate provisions for the maintenance of facilities, weak management leading to the inadequate targeting of research on key issues, low overall productivity and high staff turnover' (Farrington & Howell, 1987).

Factors inducing change since the 1970s

The pressing need to increase food production as a result of the steep increase in population, precipitated a number of developments which fundamentally changed the situation in the LDCs and, subsequently, caused a belated impact on the NARSs of those countries.

Among these factors, Von der Osten (1982) mentions:

- a concerted, world-wide effort in support of agricultural development in Third World countries;
- the adoption of national policies (in some) LDCs in support of agricultural development;
- the gradual build-up of a global international research system.

Together, these efforts brought about the 'Green Revolution', based in the development and diffusion of high-yielding varieties of wheat and rice, which has resulted in an impressive increase in production and a reduction of the costs of these two major cereals.

Yet this impact has been highly uneven, and has largely by-passed the countries in Africa, and the less favoured agro-ecological zones, and the most underprivileged sectors of the rural population in all Third World countries.

The situation in the 1980s

Many developing countries have started to reorganise and strengthen their NARS, and to reorient their research programmes with more emphasis on food crops. Of the developments since 1965 into the 1980s, Jain (1987) writes: 'more investments have been made in agricultural research in many developing countries during this period than in the entire history of scientific support for their agriculture. The fundamental changes have been in respect to the administrative framework, governance mechanisms, linkages and relationships with policy-making bodies and development departments.'

The number of agricultural researchers in the LDCs of Asia, Africa and Latin America has risen from about 14 000 in 1969, to 63 000 in 1980. Even so, all the LDCs together had only slightly more agricultural researchers than the USA federal and state systems, and fewer than Japan. Also, most of this growth is restricted to a few countries, mainly India, China, Brazil and the Philippines (Ruttan, 1985).

The overall growth of agricultural research capacity has been very uneven. The World Bank (1981) divides the LDCs into four categories in relation to their research assets and needs:

(1) Countries with generally adequate levels of manpower available for both the management and conduct of research and a fairly well developed research infrastructure. Less than 10 per cent of the LDCs belong to this category.

(2) Countries that generally have well-trained staff, but their research activities are frequently fragmented, unco-ordinated and isolated from the development process. These countries frequently require assistance in organising an appropriate administrative structure for research, in managing the national research effort. The countries in this category account for about 10 per cent of the total.

(3) Countries lacking an essential research infrastructure. The needs of these countries are twofold: to develop an effective organisation and management system for research, and strengthen their scientific personnel base to conduct research. These countries account for almost half of the total.

(4) Countries with very limited research resources, and no single crop

of sufficient importance to warrant a comprehensive research system. The major need of these countries is to develop a limited research capability to carry out mainly adaptive research, for a small number of their most important commodities. About 30 per cent of the LDCs belong into this category.

Organisational structures

The period of development of the NARSs has also been a period of constant changes in the apex authorities for public-funded research systems. The overall lack of structural stability is exemplified by the fact, that in a review of 12 LDCs by FAO (1984), only half had their present system of research organisation in place for more than four years. This flux does not always indicate a search for greater efficiency, but is frequently due to changes in government, resulting in a reallocation of agricultural research units to different ministries, or its fragmentation between a number of different units within a ministry. These institutional changes are frequently accompanied by the replacement of top research management personnel.

Despite the need for efficiency and economy, there is a tendency to establish additional institutions, or small stations, for each research topic or commodity. This proliferation of research units leads to replication of administrative services and physical facilities, to shortage of funds required to provide the critical mass of staff required to maintain viable research programmes (FAO, 1984).

In practical terms, FAO (1984) recommends that expenditure on agricultural research in the LDCs should reach 10–20 per cent of the total funds committed to agricultural development, and at least 1 per cent of agricultural GDP.

FAO (1984) summarises the major requirements of the LDCs in relation to agricultural research: (a) an increased share of funds for research on problems of rural areas, including socio-economics; (b) better adaptation of research activities to rural needs; (c) intensification of research on the special problems of less favourable production conditions, such as rainfed agriculture in the drier areas and shifting cultivation; on the prevention or reduction of post-harvest losses; on energy use in agriculture; and on a balanced approach to the exploitation of natural resources.

Ruttan (1985) expresses his concern about 'the cycles of development and erosion of capacity that have characterized a number of national agricultural research systems', in particular those that have received external assistance for long periods. When the external assistance comes to an end, research capacity is eroded or collapses, when funding from national sources is not made available to continue the former initiatives.

Strengthening the NARSs of the developing countries

In describing the work of ISNAR to strengthen the NARSs of the developing countries (the mandate for which ISNAR was established), Von der Osten (1982) writes: 'We have found that, in order to be productive, research systems need adequate capacity in the areas of research policy, research organisation, and research management.'

Research policy
The strength of NARSs is largely determined by their ability to mobilise political support and funding. This requires: (a) that the basic orientation of the research programme is in accordance with national goals; (b) that the choice of priorities is well-based; and (c) that there is potential for impact on agriculture.

Research organisation
Most NARSs of the LDCs require strengthening if they are to achieve their objectives in meeting present and future challenges. Though the resources allocated to the NARSs, from national and foreign sources have increased markedly since the 1970s, they are still far from achieving the same levels of investment that have enabled the developed countries to achieve the high productivity of their agriculture. 'Even to approach parity with the agricultural research situation in the scientifically advanced countries would require increasing overall 1980 expenditure by two and one-half, and the number of scientists by more than three.' For many poorer countries these multiples would have to be greater (Madamba & Swaminatham, 1983).

Strengthening the NARSs is, however, far from being exclusively a matter of providing greater financial resources. The NARSs of many LDCs are characterised by an even greater degree of fragmentation, duplication and lack of co-ordination than the developed countries on which they are modelled, though they are far less able to afford these deficiencies.

In many cases, they have conserved intact the systems inherited from the former colonial powers, and to these have added layers of additional structures and models, frequently on advice from foreign donors or experts.

It is the Latin American countries which have made the greatest advances in reorganising their NARSs on a rational basis, and it is the African countries which have made the least progress in this field. The need for the LDCs to foster the involvement of the faculty staff in the national research programme cannot be overstressed.

Research management
Even among the better endowed NARSs, deficiencies in research management greatly reduce the effectiveness of agricultural research. This is not only due to the heavy hand of bureaucracy, but also to the lack of well-trained managers. Scientists who are charged with managerial responsibilities are rarely professionally or psychologically prepared to undertake a type of work for which they have not been adequately prepared. The perception that such preparation is important is also very rare.

Regional workshops on research management, where those responsible for research management can meet and discuss their common problems could play a very useful role.

International co-operation
The developing countries cannot, nor do they have reason to, rely exclusively on their own research efforts. Much of the information they need on agricultural technology can come from the International Research Centres, from scientifically advanced countries, and even from neighbouring countries facing similar problems. This information can then be adapted by their own research system to their specific conditions. That this need is being realised by many countries is evidenced by the growing interest shown by developing countries and the international community in horizontal co-operation between countries and the increasing attention shown to regional approaches in the form of networks (ISNAR, 1987).

Information exchange has also improved substantially among NARSs, International Research Centres and International Associations (Nestel, 1983).

AFRICA

Socio-economic background

About 70 per cent of Africa's population still live in the rural areas. Population growth is proceeding at a faster rate than agricultural production, resulting in an increase in malnutrition, underemployment and poverty, desertification and soil degradation.

In most areas in which shifting cultivation is practised, the critical population density is being exceeded, resulting in a shortening of the fallow period; regrowth of an adequate vegetative cover becomes impossible, soil fertility is not restored and finally, the soil itself is lost by accelerated erosion.

Until the mid-1980s, most African States were still assigning low priorities to agricultural development, even though agriculture is the dominant sector of their economies, not more than 5 to 15 per cent of public funds being assigned to the development of this sector. At the 1985 OAU meeting of African Heads of State, it was agreed that public expenditure for agriculture would have to account for 20 to 25 per cent (ISNAR/SPAAR, 1987).

Eicher (1970) states that three generally accepted assumptions regarding agricultural development in Africa are myths: (a) that farmers are lazy and not responsive to economic development; (b) that certain crops are by definition plantation crops, implying economies of scale which make it extremely difficult for small farmers to compete with large, commercial units; and (c) that communal land tenure, land fractionisation and general land tenure problems are significant obstacles to progress.

Though these comments were made in relation to West Africa, they apply in large measure to East Africa, too.

Regarding the first myth, that farmers are lazy and not responsive to economic development: Undernutrition, endemic diseases and difficult climate are the reasons for the apparent laziness. An avoidance of effort is the body's natural defence mechanism under these conditions. The traditional African diet may be more or less adequate for tribal life requiring little sustained effort, but is insufficient for a worker from whom a regular, sometimes considerable, output of energy is required.

As to the lack of response to economic development, the resistance to change of the African farmer should not be exaggerated. Given the proper incentives and supporting services, and using appropriate extension methods, these farmers are more ready to adopt improved practices than is generally assumed (Arnon, 1987).

Regarding the second myth, experimental evidence shows that small farmers can produce practically all crops as efficiently as commercial plantations or State farms, provided they have equal access to new technology, credit, extension assistance and central processing that is critical for such crops (Eicher, 1970).

It is doubtful whether land problems as a constraint to development is really a 'myth' in Africa. It is true that land tenure patterns in Africa can be considered, in a world perspective, as egalitarian. Under traditional tenure, most households enjoy secure usufruct rights to communally owned land, and the herds of all households usually have unrestricted access to the grazing land (Eicher, 1970). However, increased population densities have resulted in deforestation, land degradation and desertification on increasingly expanding areas.

Major problems facing agricultural development

From 1974 to 1984, food production in Africa grew at half the rate of population increase, which averaged 3.2 per cent per year, whilst food demand was increasing by 3–5 per cent. Only very few countries in the world (none in Africa) have been able to achieve such an increase in production in recent years (World Bank, 1983).

No less critical than food supply, are the problems of providing productive employment to the rapidly increasing work force. In Kenya, for example, it is estimated that between 1980 and 2000 the labour force will double. The industrial and service sectors, at the expected rate of development, will not be able to provide more than 15 per cent of the jobs needed, so that 85 per cent of the increase will have to be provided by a labour-intensive agriculture for a long time into the future (World Bank, 1983).

Therefore, a major objective of agricultural development, and hence of research, is to devise labour-intensive, productive farming systems. Unfortunately, at the time the problems described above were becoming acute, development strategies were mainly concerned with urban and industrial development; 'political support for agricultural research was declining precisely when there was a rising need to address the more complex research problems of improving food productivity and incomes in diverse farming systems' (World Bank, 1983).

The benefits of technological progress have, so far, bypassed Africa. Von der Osten (1982) lists some of the major constraints in the continent to agricultural modernisation:

- *Agroecological:* the limited irrigation potential mitigates against the widespread adoption of high-yielding varieties.
- *Technological:* the knowledge base of African dryland farming is limited.

- *Socio-economic:* the complexity of farming systems and the great diversity of crops complicates the transfer of technology.
- *Institutional:* most national research systems are in the early stages of their development, and staffed with largely inexperienced research workers.
- *Policy-derived:* weak national support of agricultural research.

The present situation calls 'for a concerted effort by national, regional, and international research systems to arrest the trend . . . proper planning and development of agricultural research systems at the national levels, co-operating with regional and international research centres' (ISNAR/SPAAR, 1987).

Special programme for African agricultural research (SPAAR)

It is in order to address the problems outlined above, that a group of fifteen international donors have set up SPAAR with the mandate of strengthening the research capacity of African agricultural research systems; to increase the effectiveness of funds contributed by donors, and to provide for a gradual increase in external assistance to African countries.

The organisation of agricultural research

Before independence
Anglophone Africa. In the British colonies, commodity experiment stations were set up to study the requirements of a few major crops that had achieved considerable economic significance, mainly for the Colonial powers themselves. These stations were usually established and maintained by the commercial interests concerned and were each devoted to research on a single crop. The Empire Cotton Growing Association established a large cotton research station at Namulonge in Uganda, financed mainly by the cotton spinning interests in Lancashire. An experiment station for sugar cane was set up in Java, a Rubber Research Institute in Malaya, a Tea Research Institute in Ceylon, Coffee and Sisal Experiment Stations in East Africa, an Oil Palm Experiment Station in Nigeria, and a Cocoa Research Station in Ghana.

Only a single research centre of importance for the study of certain basic aspects of agriculture was set up: The East African Agriculture and Forestry Research Station at Muguga, near Nairobi.

After independence
In 1961, the author assisted as an observer at a conference on agricultural

research held in Muguga, Kenya. The only African delegate present was from Ethiopia. The British participants presented the long-term research programmes of their respective colonies—Kenya, Tanganyika, Rhodesia, Ghana, etc.—on the assumption that the current situation would continue indefinitely; the French were more realistic, presenting their contingency plans for the time when their colonies would inevitably become independent; describing how they intended 'entering through the window after having left through the door'.

The present structure of agricultural research in many African nations still strongly reflects the divergent attitudes of the British and French 'delegates' at the pre-independence conference (CCTA-CSA, 1961).

The anglophone countries. Under British colonial rule, the general responsibility for research was vested in the Department of Agriculture of each colony. However, British Planters' Associations and various commercial interests established a number of autonomous experiment stations for research on export crops. A few 'regional' (i.e. serving several countries within an eco-agricultural zone) research institutes had also been established: for cocoa in Ghana, for oil-palm in Nigeria, for rice in Sierra Leone, etc.

After independence, the British left without having trained native-born staff to take over the research infrastructure. Their former role in agricultural research was partially taken over by expatriates from various countries, and a start was made to train national research workers.

Francophone Africa. In the French colonies, private interests were less closely associated with agricultural research, which was carried out mainly under the direction of scientific institutions located in France itself, in particular the Institut National de l'Agronomie Coloniale.

Important centres for research were established on the Ivory Coast (Bringersville) and in Senegal (Bambey). High-level agricultural research was carried out in Algeria (Maison Carrée), Morocco (Rabat) and Tunisia.

In summary, the main characteristics of agricultural research in the colonial dependencies were that:

(a) they were concerned almost exclusively with cash crops (mainly plantation crops),† and made very little attempt to conduct research directly aimed at improving native farming *per se*;

†An apparent exception: agricultural research in the North African countries was on a par with that of metropolitan France, and was concerned with all aspects of farming. This research, however, was mainly aimed at solving the problems of the 'French colon' rather than of native farming.

(b) it was conducted almost entirely by European scientists, whilst only technicians were natives of the countries concerned; and

(c) the areas covered by many of the research institutes had no relevance to political frontiers, but were interregional in nature.

Over time, the technologies developed by research in the colonies gave rise to the highly profitable production of a range of export crops: coffee, cocoa, tea, cotton, sisal, groundnuts, oil-palm, bananas, etc. Advisory, input supply and marketing and transport services were developed to support the production of these crops, and agricultural research was oriented almost entirely to the same purpose. Communication on these crops among researchers was practically world-wide.

The research stations remained largely unchanged, under the Ministries of Agriculture, with the major effort invested in export crops. It is only in recent years that the emphasis is being shifted to food crops.

The francophone countries. After independence, responsibility for agricultural research in several francophone countries was vested in a Ministry of Science. However, the former institutes continued their independent existence, reporting directly to their parent institutions in France, and only through them to the Ministry of Science. The institutes continued to be funded by France (some co-funded by the host country), and to be staffed by expatriate French workers of whom over 600 remained on permanent assignment in the francophone countries south of the Sahara (Casas, 1977).

A major effort was undertaken, and is still underway, in the former French colonies to replace expatriates by nationals, a process that may take many years, because of financial restrictions. To achieve independence in agricultural research involves forgoing the massive financial and technical help provided by France to her former colonies. The influence of the continued French presence on the national research systems, is demonstrated by the case history of the Ivory Coast (cf. p. 100–7).

In the former *Belgian dependencies* (Belgian Congo, Rwanda, Burundi) research had been vested in the Institut National pour l'Etude Agronomique du Congo Belge (INEAC) which had established an enormous research complex in Yangambe—probably the biggest in the World—with experiment stations in all parts of the vast country.

After independence, the whole research complex was divided between Zaire, Rwanda and Burundi. Zaire in particular was unable to make use of the vast research infrastructure left by the Belgians.

As to the expatriate staff, the Belgians acted more like the British than the French, withdrawing their researchers, without having prepared local cadres to take over—one characteristic common to all the colonial powers.

Trigo (1986) summarises the present situation of the structure of agricultural research in the African countries as follows:

The post-colonial structure of agricultural research in Africa appears to be characterized by the existence of a vast array of organizations, which mostly correspond to what was in place at the time of independence. The 'nationalization' of those research structures has undoubtedly been the main task of the last 20–25 years. This process has taken place against the background of different colonial heritages, which has affected the types of institution established in the newly independent countries and the decolonization strategies, which influenced the nature and pace of the nationalization.

Funding of agricultural research

The substantial differences in GDP per capita among tropical African nations, ranging from US$ 100 to 1000 (World Bank, 1984), is reflected in the allocation of resources to agricultural research by the different countries (Torres, 1987). Allocations range from 0·1 per cent of the agricultural GDP to over 2 per cent (SPAAR, 1986). On the basis of certain criteria (budgetary allocations, number of research workers, number of operational research stations, working relationships with international research systems, etc.), USAID (1983) identified eight countries as capable of generating new technologies, 13 as having the minimum resources required for adaptive research, and 10 as not having reached even this stage.

After independence
The situation in agricultural research. The organisation of agricultural research in Africa is generally still in the transition period between the situation as it existed before independence and the establishment of a new structure designed to answer the needs of an independent, developing economy.

In most cases, agricultural research is poorly organised and poorly managed, the organisation is in continual flux, the research activities are hampered by bureaucratic procedures, low financial resources, inadequate facilities, poor research-extension linkages, and above all, much of the research has little relevance to the problems of the small farmers. There is a lack of programming and decisions on priorities, little or no evaluation of the work being done, little effort at diffusion of the results. In the francophone countries, much of the research is planned and implemented by expatriates.

Casas (1977) points out that in recent years, a large number of countries have reacted to this situation 'by applying two types of measures, often in combination: stricter control of the research activities and goals being pursued on their territory . . . and the rapid mobilisation of a national research potential which will gradually replace foreign technical assistance'.

There have been significant increases in agricultural research personnel, much support has been provided by the international research centres, and considerable investments have been made in extension. Notwithstanding the efforts made since the 1960s, 'the output from national research and its impact on agricultural productivity in sub-Saharan Africa have been disappointing (ISNAR/SPAAR, 1987).

The World Bank (1984) reports: 'agricultural research is failing to provide adequate support to producers of both food and export crops. Major advances like those that revolutionised wheat and rice cultivation in Asia have not been made since the 1960s, when new maize hybrids were adopted in southern and eastern Africa. No major breakthroughs have been achieved in genetic improvement of rainfed millet or sorghum, which account for 80 per cent of the cultivated land in the Sahel and other areas of low rainfall. Nor can rapid progress be expected.'

Additional weaknesses mentioned by ISNAR/SPAAR, 1987 are:

- A shortage of experienced programme leaders, and a high proportion (up to 50 per cent and more) of research personnel with less than three years' experience.
- Dispersion of research efforts and frequent changes in research personnel, leading to slow completion of research tasks.

It is important to stress that all these shortcomings are not shared by all African countries; some have been successful in generating and adapting food-increasing technologies.

What is important is that from the deficiencies and from the successes it is possible to derive some important lessons. In particular, successes have arisen where there has been a concentration of human and other resources on priority problems; teamwork, continuity of research leadership, and stability of funding. Where research results are efficiently transferred to farmers; where the preconditions essential to enable the farmer to adopt new practices are provided, and above all, where policy leaders are committed to promoting agricultural progress.

ASIA

According to the Asian Development Bank (1977), the Asian countries can be divided into three broad categories in terms of their performance over the decade 1965 to 1975: (a) those that have been able to maintain a fairly rapid growth of 2 per cent or more, by combining expansion in irrigation with modern inputs (e.g. India, Indonesia, Malaysia, Pakistan, the Philippines, Sri Lanka); (b) those whose growth rates have been slowed down, in part by the inability to boost yields through the extensive adoption of the modern seed-fertiliser technology and in part due to unfavourable policies (e.g. Bangladesh, Burma, Thailand); and (c) those whose growth rates have slowed down because they appear to be approaching a ceiling in terms of the yield potential of current technology (e.g. China).

Historical development of agricultural research

The first step made to improve agriculture in Asia, was the establishment of botanical gardens in the former colonies: in the West Indies, Mauritius, India, Jamaica, Trinidad, British Guiana, Ceylon, Singapore, Java. The main contribution of these gardens was the introduction of new crops and the replacement of disease-stricken varieties by disease-resistant varieties, thereby saving important branches of agriculture from extinction in the countries involved.

The first research institutes were devoted to important export crops, beginning with sugarcane in Indonesia. This was followed by eight commodity research institutes in India, seven in the Philippines, six in Indonesia, and one each in Malaysia and Thailand.

This tendency of research institutes to work on a single crop has continued to the present time, and has been an important factor impeding the co-ordination of agricultural research into an integrated national research system (Mosher, 1971).

Subsequently, agricultural colleges and universities have been established in increasing numbers in all the countries of the continent, and have made important contributions to agricultural research.

After independence

A common feature of most, if not all NARSs in Asian countries, is that they have generally maintained the structures inherited from the colonial period; reorganisation has therefore generally consisted in adding new structures, frequently inspired by the various donor agencies, resulting in multiple institutions with overlapping functions and requiring special bodies for co-ordination between the disparate elements of the system.

This situation often resulted in 'wasteful duplication and overlapping of research efforts, as well as the wastage of scarce resources in terms of manpower, costly scientific equipment, and financial allocation' (RAPA/ FAO, 1986).

Even a country like Pakistan, which started with a relatively clean slate, having inherited a single agricultural research institute from its colonial past, has, in the course of four decades, established a complex system of agricultural research, with a large number of institutions working under different federal and provincial agencies, with little co-ordination between them.

All efforts of various national and international agencies to promote a unified national agricultural research system, have been frustrated for years by provincial politics, because the provincial governments wanted to maintain their exclusive rights to undertake agricultural research.

Another country that started with a clean slate, the Philippines, was able to build in 20 years a research 'system' in which responsibility and resource allocation were fragmented, an excessive number of research stations and facilities had been established, planning of research was inadequate and ineffective. Though large sums had been allocated annually, research had not made a substantial impact on agricultural production. It required international intervention and support, and over a decade of efforts by a co-ordinating body, to bring order into this chaos.

Present structure of the NARSs

The NARSs in Asian countries, most of which were initiated in the 1960s and 1970s, are at present in different stages of development. Some have remained weak, others have been more successful (Nestel, 1983).

A review of the case histories of the Asian countries highlights certain common constraints to the development of an effective and efficient NARS.

Lack of overall co-ordination of rural development efforts. In Asia, there are generally many agencies involved at the national, district/provincial, township and village levels for promoting development and assisting the rural population. The unco-ordinated and often competitive functioning of these agencies has resulted in the country's not achieving the goals it has set itself.

Fragmentation of the research effort. The unco-ordinated rural development effort is reflected in the fragmented NARSs. Traditionally, the NARSs of the Asian countries have been more fragmented than in other regions. The usual pattern has been for research to be undertaken by several agencies within each ministry or department. Research was conducted in ten different bureaux in Ceylon (Sri Lanka), by 15 separate institutes managed by five different directorates within the Ministry of Agriculture in Indonesia, and by nine different agencies within the Department of Agriculture and Natural Resources in the Philippines (Mosher, 1971).

Venezian (1981), describing the situation ten years later, writes: 'In general, national research systems are not cohesively structured. Linkages and communications among the institutions are weak, and thus national programmes may duplicate efforts, leave significant gaps, and become routine and repetitive or slow to respond to changing situations. This again lowers the systems' overall productivity.'

Lack of adequate programme planning. It is only in a very few countries that the planning of the research programme has moved beyond the setting of the individual institutes, to encompass a national perspective (Nestel, 1983).

A general insufficiency of manpower. Physical infrastructure has frequently outstripped available qualified manpower. Many NARSs in the region do not yet provide an attractive career structure for well-qualified research personnel. Rarely are there incentives to encourage staff to reside at, or near, research stations, rather than at central headquarters (Nestel, 1983).

Inadequacy of financial resources. Expansion of the NARSs has, to a large extent, been funded by external agencies, without there being enough financial support from national sources. This overdependence on foreign support results in up-and-down cycles of development and erosion in the national research institutions.

Imbalance in research orientation. Most countries in the region have long continued to support the pre-independence preponderance of research on export crops and concurrent neglect of food crops. This trend was only reversed following the success of the new high-yielding varieties of rice and wheat. However, a change is again evident in the countries that have achieved self-sufficiency in food grains and started experiencing acute foreign exchange problems in the late 1970s, when there was some re-emphasis on research on industrial crops which were being increasingly processed locally (FAO, 1984).

Integrated national commodity programmes
The major and most effective instrument for co-ordinating the disparate elements of the NARS, ministerial and independent institutions, research council institutions, faculties of agriculture and agricultural universities, have been the National Commodity Programmes, initiated in India, and subsequently adopted in most countries of the region.

LATIN AMERICA

The evolution of the agricultural research systems

The agricultural research systems have generally passed through the same three phases:

(a) An initial period of unstructured agricultural research, frequently initiated by private bodies.

(b) From the 1940s onwards, many Latin American countries made efforts to develop a structure for agricultural research and extension within their ministries of agriculture.

(c) From the late 1950s, the general tendency has been to transfer responsibility for all government-funded research from the Ministry of Agriculture to a National Agricultural Research Institute.

The beginnings
Agricultural research was initiated in a few Latin American countries by the beginning of the 20th century. Early research efforts were undertaken by individuals from Europe and the USA, and were naturally influenced by the concepts current in their countries of origin. Typically, they tended to approach agricultural problems through scientific disciplines, in particular descriptive botany or zoology, and in isolation from other disciplines. The research was considered as a goal in itself, and not as a means to solve problems of food production (Marcano, 1982).

Research in the ministries of agriculture
As the need for agricultural development became more and more acute,

and the importance of research for furthering this process became increasingly evident, the major responsibility for planning, financing and implementing agricultural research was taken over by government, in all the countries of the continent.

The first official experiment stations were established in the 1930s and 1940s. By the 1960s, agricultural research had expanded considerably, mainly as a result of significant support received from foreign agencies; these in turn encouraged greater participation by the national budgets. One of the main beneficiaries of these trends were the training programmes for research personnel, including in some countries the development of a national capacity for post-graduate training (Trigo & Piñeiro, 1982).

The concentration of agricultural research activities within the ministries of agriculture generally gave disappointing results. The major potential advantages of having responsibility for research and extension vested in the same body responsible for agricultural development did not have the expected results. It did not produce research programmes congruent with the priority needs of agricultural development. Lack of understanding of the potential contribution of agricultural research to development by the policy-makers resulted in inadequate and unstable funding. Rigid bureaucratic procedures prevented the efficient implementation of research programmes.

The national agricultural research systems were not cohesively structured; links and communication between units were weak or non-existent, resulting in much duplication whilst important problems remained untreated. Co-operation between research and extension, though operating within a common framework, was unsatisfactory. Research was also slow to respond to changing situations and needs (Marcano, 1982). As a result, agricultural research was often not well aligned with national priorities.

Besides a chronic shortage of capital and operating funds, research was also plagued by low salaries and migration of workers.

The national institutes of agriculture
Until the end of the 1950s, the leaders of the predominantly agrarian countries of Latin America had been obsessed with the need to concentrate almost all available resources on industrial development. Only after it was realised that the modernisation of agriculture was an essential prelude to industrial development (Arnon, 1987), both as a source of financing investments and the creation of a significant local market for industrial products, was more understanding evidenced by the policy-makers for

the need for advanced agricultural technology which only an effective research organisation can provide.

Much thought was therefore given to a restructuring of the existing research system; this thinking was evidently influenced by the experience of the developed countries and the research structures they had developed.

The solution adopted finally by most Latin American countries was to concentrate all government-funded research in an autonomous or semi-autonomous public institution, free of the bureaucratic constraints experienced within the Ministry, but firmly linked to the Ministry of Agriculture in matters relating to research policy.

All the existing organs carrying out agricultural research operating in a government framework and their assets were transferred to the new institutes, which became the major bodies for agricultural research in their respective countries. This process was considerably strengthened by the support given by foreign sponsors, who provided the significant financial means and technical aid required to implement the structural transformation.

The first country to establish a national agricultural research institute in Latin America was Argentina in 1957: the National Institute of Agricultural Technology (INTA), followed by the National Institute of Agricultural Research (INIAP) of Ecuador in 1959; the National Foundation for Agricultural Research (FONOIAP) of Venezuela (1959–61); the National Institute of Agricultural Research (INIA) of Mexico in 1960; the Columbian Agricultural Institute (ICA) in 1963; and the Agricultural Research Institute (INIA) in Chile in 1964.

Relative latecomers were: the Bolivian Institute of Agricultural Technology (IBTA), the Institute of Science and Technology (ICTA) of Guatemala, and the Agricultural Research and Development Institute (IDIAP) of Panama, in 1975; the National Institute of Agricultural Technology (INTA) of Nicaragua. One of the latest countries to adopt the new system was Costa Rica in 1987.

Common characteristics of the national institutes
In all the new institutes, policy decisions and overall administration are centrally directed by a national headquarters, and implementation is decentralised and carried out by a hierarchy of regional research centres, experiment stations and experiment fields, established in accordance with the agro-ecological regions of the country.

The National Institutes were vested with responsibility for research on all commodities produced in the country, supported by all the necessary discipline research. Much thought was given to the adoption of a suitable method of planning a research programme, to be based on priorities in accordance with the National Development Plan.

As a suitable operational framework for commodity-oriented research, most of the Institutes adopted National Programme Networks, which encompassed all the country-wide research on: (a) a commodity, (b) a group of related commodities, or (c) subjects that related to agricultural production as a whole, such as farming systems, problem soils, water management, etc.

It will be noted that this model is extremely similar to the model structure for national agricultural research proposed by the FAO Panel for Agricultural Research in 1965 (cf. p. 223).

Differences between the national institutes
Within the overall framework described in the previous section, significant differences occurred in the final form and content of some of the National Institutes.

Governance structure. While all the institutes have their own legal status, under the aegis of the Ministry of Agriculture or its equivalent, they differ in whether they have a board of governors or trustees responsible for policy guidance and management control (INTA of Argentina, ICA of Columbia, ICTA of Guatemala and INIA of Chile belong to this category) or that they do not have such a body and the Director General of the Institute reports directly to the Minister of Agriculture.

Functional differences. It is mainly in the functions vested in the Institutes that these differ from each other. These range from exclusive concern with research (INIAP in Ecuador, IDIAP in Panama, and others); research and extension (INTA of Argentina, FONAIAP of Venezuela and others); research, extension and education, in Argentina, Colombia, Peru, Uruguay, Mexico and Brazil (the efforts at education have generally been short-lived, excepting in Mexico and Brazil) (Trigo, 1986).

ICA of Colombia and EMBRAPA of Brazil are special cases. In ICA, research, extension and education were combined with responsibility for agricultural development. EMBRAPA 'is a multi-organisational model, involving separate levels of administration in the public sector (federal and state) and in the private sector' (Trigo, 1986).

Funding. INTA of Argentina has a special funding arrangement, whereby most of its resources derive from a special 2 per cent tax on agricultural exports.

Human resources
The chief limitation faced by the young national research institutions has generally been the shortage of trained research personnel. Therefore, a major effort has been made by all these institutes to provide adequate graduate and post-graduate training for their research personnel, an effort which received considerable support from international sources and universities.

During the 1960s, when the reorganisation of agricultural research began to accelerate, post-graduate training was given high priority. There were no post-graduate programmes at national level and all the training programmes were based on foreign universities.

In time, several countries developed their own national training programmes (at Master's level). The most important: the ICA Graduate Programme in Colombia, the Graduate School of Agricultural Sciences in Argentina, and the Graduate Programme of La Molina, in Peru. In Peru, the training was provided by the University itself in Argentina and Colombia, the training programmes were located within the research institutes.

In the early 1970s, most of the foreign support for these programmes had been terminated; concurrently, the national budgets for the institutions concerned were cut. As a result, all three programmes were affected: in Colombia and Argentina they were practically terminated, leaving these two countries without effective alternatives for graduate training, and in Peru, it was severely curtailed (Ardila *et al*, 1981).

Policy issues
The major change in policy in recent years is the emphasis on the need to divert more attention to the problems of small-scale farmers. One country, Guatemala, has gone to the extreme of establishing a national research institute exclusively devoted to the interests of the small farmer.

The importance of farmers' participation in the formulation of research policy is now becoming increasingly recognised, and farmers' representatives are now generally included in the decision-making bodies of the NARS.

Shortcomings and problems
Some of the problems that plagued the research institutions when they were still part of the Ministry of Agriculture have not completely disappeared with the restructuring of agricultural research.

Public recognition. Publicly, the importance of agricultural research is still weak. This is the consequence of the low priority often accorded to scientific activities in general, and, until the 1960s, to agricultural development in particular. Agricultural research, until recently, has also been of little benefit to the campesinos (small farmers) who form the majority of agricultural producers in the Latin American countries.

Venezian (1981) writes: 'agricultural research is not sufficiently understood by its direct beneficiaries, the farmers, and much less by the general public. Policy makers and governments reflect this undervaluation by low expenditure for national agricultural research.'

As a result of the lack of public recognition 'many of the research systems were plagued by cyclical sequences of development and erosion of capacity as budgetary priorities responded to changes in political regimes' (Ruttan, 1982).

Other constraints. Those common to many NARSs were: overconcentration of the research efforts on commercial farming; poor programme planning without concern for priorities; neglect in transferring the results of research to farmers (see below); scarcity of skilled research management professionals; difficulties in training and keeping research and technical staff; and absence of a congenial research climate (Venezian, 1981).

Transfer of technology. Research results frequently gather dust in the files of the research institutes, or take very long to reach the farmers. One of the major reasons for this situation is deficient links between research and extension. Marcano (1982) points out that 'interaction is assumed between the research and dissemination organisations, but often there is no such interaction; in some cases there is actual antagonism'. This is true, even when both services form part of the same framework, whether a ministry or a research institute.

These are problems that are not confined to Latin American countries, and will be treated in detail in Chapter 3.

Chapter 3

Components of the National Agricultural Research Systems

In all the case histories presented in Chapter 1, various institutions or organisations are involved in agricultural research, each contributing in a greater or lesser measure to the solution of the problems of agriculture of the country. There is probably no country in the world in which a single institution undertakes all national agricultural research.

The great variety of combinations of different research institutions which together constitute the national agricultural research system (NARS), the role each of these components in the system of different countries, and the relative importance of their contributions, makes attempts to propose a typology of NARSs difficult and inaccurate. It is therefore preferable to consider each of the components of a NARS individually, analyse its advantages and weaknesses, examine how it interacts with the other components of the system, and what is its potential to contribute to the objectives of the national development plans.

INSTITUTIONS INVOLVED IN AGRICULTURAL RESEARCH

The following institutions are generally involved in the NARSs:

Ministries: The research is conducted as part of the line activities of (a) a single ministry—usually of agriculture (or rural development), less frequently a ministry of scientific research or of education. Within the ministry, there may be a single department for research, or the research may be divided among different commodity departments, or (b) responsibility for agricultural research may be allocated to several ministries.

Institutes can be: *Semi-autonomous* —linked to a ministry in the determination of policy, but independent in the execution of the policy; or *autonomous*—single commodity-oriented or single disciplineoriented or polyvalent.
Research councils may have purely advisory and co-ordination functions, or may have their own research institutions.
Statutory or commodity boards responsible for a single commodity, or for the development of a specific region.
Universities with one or more faculties carrying out agricultural research according to the choice and interests of the faculty staff; or agricultural universities (modelled more or less on the Land Grant College model).
Private sector, in which the most common are farmers' associations and industrial or commercial corporations.

In most countries, one of the components of the NARS has the major responsibility for nationally funded agricultural research, whilst the others make their respective contributions either in co-ordination with the central institution or in isolation therefrom.

Agricultural research carried out by one or more ministries of state

National agricultural research services can be the responsibility of a single ministry: agriculture, rural development, science, education, or fragmented among various of these ministries, or even within a single ministry, when each commodity department (livestock, plantation crops, field crops, etc.) has its own research unit.

Each of these options has its specific characteristics, which we will examine briefly. However, they all have one major disadvantage in common. All ministries are structured for planning, regulatory and control activities. Research cannot function effectively within a framework which is not adjusted to deal with its specific requirements and needs. Operational and implementation problems due to the general inflexibility of ministerial budgetary, administrative and personnel procedures frequently frustrate the research process. In particular, procedures for the appointment and professional advancement of research personnel are completely inappropriate.

The problem is further compounded when the research activities are fragmented among a number of commodity departments, or worse still, among several ministries. In addition to the organisational and administrative disadvantages inherent in a research unit which forms part of a ministry, the division of authority due to the fragmentation of research among different departments makes co-ordination of the research effort impracticable and the creation of a research climate impossible.

Another drawback of a ministerial framework for research is its lack of stability. Changes in government usually involve major changes in the administration: ministries have to be added, or consolidated, in response to political requirements; the new ministers frequently reallocate their respective 'fiefs'. Research units are separated or joined together in accordance with these arrangements. This lack of continuity is certainly not conducive to a good research environment or to the morale of the research staff.

The result is 'a mediocre government research establishment, with the best workers migrating to the universities or the private sector' (Webster, 1970).

Ministry of agriculture (or of rural development)
The major advantage of this framework for agricultural research is the close linkage between the research and the people responsible for planning and implementing the national rural development plan, working within the same ministry.

On the other hand, the ministry will mainly demand and foster 'brushfire' research, and accord very low priority to long-term and basic research.

Another advantage claimed for having an agricultural research department within a Ministry of Agriculture, is that it then operates within the same framework as the Extension Service. Unfortunately, experience has shown that a common organisational framework is no guarantee for satisfactory collaboration between the two services. In view of the importance of this subject, it will be discussed in detail in Chapter 22.

Ministry of science (or of scientific research)
This structure has the advantage of providing a more congenial environment for agricultural research, and a better understanding for long-term and basic research. However, it does suffer from all the bureaucratic constraints of the Ministry of Agriculture model. In addition, when agricultural research is the responsibility of a ministry that is not directly concerned with agricultural development, the almost inevitable consequence is a lack of congruence between the research programme and the priority problems of the agricultural development plan. Consequently, there is a tendency for the Ministry of Agriculture, and for parastatals to set up

their own programmes of research in accordance with their special needs (ISNAR/SPAAR, 1987).

Research carried out simultaneously by a ministerial department and a number of autonomous specialised institutes

This solution has usually evolved as the result of various sectorial or political interests having established, in the course of time, commodity or discipline-oriented institutes. The major formal justification for this approach is the possibility to avoid 'monopoly' in agricultural research. It is legitimate to question the advisability of concentrating all planned agricultural research within a single institution. Lack of competition between institutions, due to monopoly in research, may result in organisational rigidity, in complacency, lack of initiative and absence of a feeling of urgency which could be extremely detrimental to research output. The monopoly is, however, more apparent than real. There are usually a number of academic institutions, such as faculties of agriculture, technological colleges, etc., in which individual scientists, though enjoying academic freedom, are concerned with various aspects of agricultural research to which they may make important and substantial contributions, as well as competing with the governmental institution for research funds.

Parastatals or statutory boards

These are vested with responsibility for the development of a number of key crops, or less frequently, for the development of a specific region. These autonomous bodies are governed mainly by representatives of private sector producers whose interests, by definition, are not necessarily coincident with national interests. These boards, though heavily dependent on government subsidies, carry out their own planning, set their own salary scales, negotiate crop-specific international loans, handle their own local and overseas marketing, do their own or contract farming, offer special extension and credit services to farmers, and establish their own research facilities.

Many of these parastatals or commodity boards carry out their activities with dedication and efficiency and achieve impressive results.

Inevitably, there is severe competition with the government research service for staff, facilities and funds. The boards are concerned with the major commodities; therefore the government service is relegated to the financially less important crops, or must engage in duplicating the work on commodities already being investigated with far greater resources by the boards.

Duplication of effort is wasteful, and few countries can afford this luxury, nor can they afford a weak and ineffective national agricultural research service.

A multiplicity of autonomous organisations creates vested interests that are not responsive to objective requirements. Co-ordination is essential in order to ensure that the solution of problems is undertaken according to priorities, with a minimum of overlapping and a maximum of teamwork. Ensuring co-ordination between a large number of autonomous bodies is time-consuming, effort wasting, and rarely effective. The difficulty is usually compounded because the research institutions are controlled by different authorities and have different types of status (academic, governmental, private, etc.). Really effective co-ordination can be achieved only by a combination of centralised decision-making and executive authority (Wansinck, 1964).

Worst of all, even after Government has taken a decision to restructure its agricultural research system, so as to include all government-funded research, the boards are generally sufficiently powerful to stall all efforts to implement the reorganisation, even when the latter is supported by World Bank or other international aid that is made contingent on the integration of several components of the national research effort.

A national agricultural research institute (or service)

In this model, responsibility for all government-funded agricultural research is vested in a single, semi-autonomous organisation, whose policy is dictated by the Minister of Agriculture, either directly or through a board of governors, appointed by him, in which the ministry's interests are predominant.

The institute is, however, autonomous in the implementation of the research plan, as approved by the Minister or the board. It administers its budget independently of the public system.

The National Research Institute is far less influenced by political changes than the former models. Typically, it has several sources of funding: the national budget, the Research Council, international and bilateral aid, and from the 'clients' themselves.

A central national organisation for the implementation of all government-funded research has the advantage that it can best respond to the policies determined by national interests, co-ordinate countrywide programmes effectively, ensure teamwork between disciplines and commodity-oriented research, and integrate the research capabilities of the academic institutions in a national research effort.

In view of its importance, this model will be discussed in detail.

Agricultural research councils

Agricultural Research Councils (ARCs) have been established for several reasons:

- because of the view that all scientific effort, including agricultural research, should be insulated against political interference;
- in response to a chaotic situation, where many individual institutions engage in agricultural research with resultant duplication, overlapping, waste of resources, and ineffective coverage of the priority problems of the country;
- a critical food shortage induces Government to make a major effort in strengthening the national agricultural research system, and adopts the Research Council as the most effective means to achieve this aim.

The last two factors, single or in combination, have been the main reasons for the establishment of ARCs in many Asian countries (Trigo, 1986).

The ARCs have the power to discuss and implement measures for rationalisation and co-ordination of research in all fields of agriculture. These powers are exercised by vesting the Council with the right to scrutinise departmental estimates for research and to advise the Government on areas which could be usefully rationalised. The Council also dispenses a research budget of its own, to be used at its discretion for strengthening special areas, which deserve support either for answering urgent needs or for correcting inbalance.

In certain countries, the national research council functions in an advisory capacity only, and is responsible for co-ordinating research carried out by various agencies; in others, the Councils also have executive functions and establish and administer national research institutes and laboratories.

One of the first duties of a scientific council is to determine whether all fields of scientific effort are adequately covered, and to provide the means for eliminating any existing gaps. This requires an inventory of the country's scientific resources, in particular of its scientific manpower, and a study of the distribution of the research effort in relation to the needs of the country's economy.

Other functions of the ARCs are:

- The co-ordination of interdisciplinary research efforts;
- the establishment of major research facilities;
- the organisation of scientific information centres;
- the review of scientific performance in different sectors of research;
- stimulating the development of research findings into technical innovations;
- improving methods of research organisation and administration;
- conducting surveys of available scientific manpower and estimating future needs.

Characteristics of the ARCs

A feature common to ARCs in most countries is that the Council is an autonomous body, with a board of directors, a majority of whose members are scientists, appointed as individuals, and not representatives of institutions. The board is assisted by an executive office, with a permanent staff. In order to be effective in setting policy, ready access to political leaders, and even their participation in decision-making, is essential. It is therefore necessary to include in the board representatives of the Ministers of Agriculture, of Financing, and of the Prime Minister's office. The Council then provides a framework within which top scientific and political leaders work together on a continuous basis.

There are four different concepts regarding the functions of the ARCs:

(a) *The ARC has no executive functions:* It is a central body with an advisory role in national agricultural research policy. This type of ARC has generally been established in response to the proliferation of individual research institutes that define their own policies and pursue their own interests.

Its basic functions are planning and co-ordination. These consist in: developing a long-term research plan; co-ordinating the activities of all government-funded research institutions; advising on the allocation of funds among the various research institutions; co-ordinating foreign financial and technical assistance for agricultural research; monitoring and evaluating existing programmes.

Though this type of ARC has no executive functions, it can influence research policy by allocating funds according to priorities it has established, by having representatives on the boards of the research institutions, and by the recommendations it makes to the policy-making bodies at the national level.

(b) *An ARC with executive functions:* Experience has shown that ARCs with only planning and co-ordinating functions are rarely effective, and the general tendency is to enable them to assume at least partial control over the implementation of government-funded research. This can take one of two forms: (1) responsibility for oriented basic research is vested in the ARC, whilst the National Agricultural Research Institute remains responsible for applied research; (2) the ARC becomes the body responsible for the entire national research system.

In both cases, concern with the technical and administrative problems involved in administering a number of specialised research institutes, diverts the attention of the ARC from the basic and long-range problems which should be its principal concern. Also, by becoming directly involved in the implementation of research, it forfeits its status as an independent, objective body, which is essential if it is to fulfil its basic roles of determining policy and of ensuring co-ordination.

(c) *An ARC which is responsible for the national agricultural research but is independent of the Ministry of Agriculture:* This option will undoubtedly favour basic and long-range research. However, the very detachment of the Council from the Ministry of Agriculture and from the immediate problems of the farming community will frequently bring in its wake the neglect of their legitimate needs.

A Ministry of Agriculture, faced with the responsibility for implementing the national agricultural development plan, will, sooner or later, have to establish a research establishment of its own, thereby recreating the dichotomy that the ARC was supposed to overcome. An example of this process is what occurred in Great Britain, a country which was one of the first to entrust its national agricultural research to an independent ARC (cf. p.3).

(d) *An ARC which is responsible for the national agricultural research, and is part of, or responsible to, the Ministry of Agriculture:* The main argument advanced in favour of this model is the same mentioned in relation to the semi-autonomous research institute, namely, that the usual ministry department cannot provide a desirable research climate, has little understanding of the problems of research, is not prepared to invest in exploratory or long-term research and is affected by political change.

This type of ARC has been adopted by several Asian countries. Actually, the justification for this model, and its objectives and functions, are practically identical in all but name with those described for the semi-autonomous National Institute of Agricultural Research, and all that has been stated, or will be stated further on, in relation to the one, applies equally to the other.

National academies of science

In most centrally planned economies, it is the National Academy of Sciences that is recognised as the supreme scientific body and is responsible for co-ordinating the research activities of universities, academies, technological and scientific institutes, in addition to its own research activities (Turkevich, 1960).

According to Starnovsky (1962), the following are the three main fields of activity of the Academy as a co-ordinating body:

(a) determination of research tasks of crucial importance, including supervision; of the necessary provision of personnel and funds, and control of the rate of fulfilment of these tasks;

(b) organisation of a network of scientific and research centres and planning of their development;

(c) planning of the training of scientific workers and evaluation of their scientific proficiency.

Research by the extension services

Research carried out by the extension services can take several forms:

(a) the extension service sets up its own research unit, operating independently of the ARC;

(b) extension workers are allowed to allocate a certain proportion of their time for on-farm experimentation, carried out at their own initiative;

(c) extension workers participate in research teams organised by the national research institute, in Farming Systems Research, verification trials, adaptive research etc., on the farmers' fields.

The drawbacks of the first approach are costly duplication, and, in addition, all the disadvantages of a research unit operating within a government department (cf. p.223). Vesting responsibility for research in a service structured for completely different activities—extension work— can be counterproductive for both research and extension.

The implications of involving extension workers in agricultural research under options (b) and (c) will be discussed in Chapter 22.

AGRICULTURAL RESEARCH IN THE UNIVERSITIES

The way and the degree in which Universities engage in agricultural research differ between and within countries. Fundamentally, a Faculty of Agriculture can choose one of three different approaches:

(a) In tradition-oriented universities, academic freedom is a prerogative of faculty that is jealously guarded. The faculty generally choose research projects according to their personal interest and/ or the educational value of the project. Occasionally, contract research is undertaken. Research planning is practically non-existent.

(b) The Land-Grant College approach, whereby the Colleges (actually Faculties of Agriculture) assume, statewise, full responsibility for agricultural education, research and extension. This system, developed in the USA, has been a great success there, but has generally failed in most other countries which have attempted to adopt this system. The reasons for the success in the USA and lack of success elsewhere will be discussed below.

(c) A third approach is for the university to establish a special mechanism to channel the research efforts of the faculty into research of national significance, in co-ordination with the national agency responsible for agricultural research.

The tradition-oriented university

Universities in many developing countries have taken as their model the old-established European-type universities and have adopted their traditions.

Darling (1970) describes very aptly the traditional 'ivory tower' approach of the average English university (not very different to that of most European universities). 'There was an insistence that Universities were part of the universal world circuit, that their responsibilities were far above the needs of the communities which supported them, and that in the interests of academic freedom a university should not allow itself to become involved in any way in local affairs.'

While agricultural faculties never subscribed wholeheartedly to the ivory tower approach, they were affected by it; in the years following World War II, particularly, the natural science departments in Faculties of Agriculture threw themselves into the pursuit of science for its own sake with considerable abandon. Darling writes 'that the effects of these attitudes have been unfortunate in creating a sense of isolation from the industries which the Faculties serve', but adds: 'I am happy to state that in recent years this trend has been reversed.'

The university schools of agriculture, in common with all institutes of higher learning, have two basic functions—teaching and research —both relevant to their contribution to agricultural research.

In teaching, their aim is to prepare agronomists for the functions of planning, production, technology, research and extension in the field of agriculture. Research at institutes of higher learning has two objectives: to extend the field of human knowledge and to contribute to the training of their students. Traditionally, research objectives are followed without consideration of practical implications, or restrictions on the liberty of the researcher in the choice of his problems. It is called academic, pure, fundamental, or basic research—these terms are considered interchangeable when applied to university research. Whilst it is true that much of this research was carried out in the past on a shoestring, it was entirely free of outside pressure.

During World War II, many faculty staff members in different countries felt the obligation to respond to the call of their respective governments and to devote their research activities to the national war effort. This research received massive governmental financial support to which faculty members became accustomed. They also became partly accustomed to the concepts of teamwork, directed research, the need for administrative controls, etc. Research itself became increasingly sophisticated and expensive. Faculty members became aware that 'freedom of research' was usually associated with lack of funds for research whilst there was no difficulty for a good researcher to obtain ample financial support for projects that could be 'sold' because of their economic potential. There is every evidence that research would become even more complicated, require more team effort, and involve even more expensive equipment in the future than in the past.

After the war years, universities continued to be involved in 'contract' research financed by outside sources, usually with well-defined objectives.

Universities are traditionally concerned with the source of financial support they receive, with safeguarding the freedom of the individual to select the research areas in which he is active, and are allergic to the controls invoked by outside sources.

Many universities became alarmed at their increasing involvement with 'applied' research and thought that this interfered with their basic education functions and their traditions of free research. Many university scientists were also alarmed by the tendency of the university authorities no longer to feel under an obligation to provide research funds to those members of the staff who were unwilling or unable to commit themselves to contract research, whilst those who were willing to forgo their freedom of action had no shortage of funds for research. This situation 'reduces the scholarly atmosphere of the university, and makes more difficult the creativity of the scholars' (Simons, 1960).

For all these reasons, many universities decided against further involvement and adopted a policy of refusing further financial support for research with economic objectives.

In many universities, research does not generally relate to national problems, even in countries in which universities constitute practically the sole centres of agricultural research activities (Nestel & Franklin, 1981). Even a very high percentage of post-graduate theses bears no relationship to the realities of agriculture in their respective regions. This 'ivory tower' attitude also affected the first generation of universities developed after World War II in anglophone countries in the former colonies, which refused to become involved in 'local affairs', and this was a hindrance to their integration in the national research effort.

Universities which foster mission-oriented research

By contrast to the attitudes adopted by the universities which fear that mission-oriented research will adversely affect the scientific level of their research and also restrict academic freedom, other academic institutions attempted to solve the problem by setting up, within or in proximity to the university campus, affiliated institutions or 'research companies' to handle contract research, with close ties to the university. These affiliated institutes formed a framework within which those members of the university who desired to contribute to applied research, could do so without violating the basic concepts of the university. In addition to enabling the universities to dissociate themselves from direct connection with contract research, the affiliated research institutes helped solve the problem of research direction. Traditionally the university organisation 'is not a very satisfactory model of how to organise and manage research' (Simons, 1960). The desire for individual freedom is not usually compatible with research co-ordination and direction. With contract research, the need to stay within well-defined objectives, and to keep to a schedule, requires a more formal organisation than is possible in the university.

However, in certain specialised fields, of which medicine and agriculture are typical, the 'conflict' described above between pure and applied research has never been relevant. In these fields, applied research is closely related to the educational objectives; research cannot be divorced from the subject taught, and hence medical or agricultural research in the respective faculties cannot be 'pure'.

In practice, less basic research is done in Faculties of Agriculture than is generally assumed. Darling (1970) mentions three reasons for this: (a) basic research is expensive, and faculties of agriculture are generally short of money; (b) faculty staff do not wish to get too far from problems relevant to agriculture in their country; and (c) science faculties are better equipped to do basic research and have better chances of competing for grants to finance this type of work.

The establishment of separate research institutes for applied research, recently adopted by the universities to solve the conflict in which they found themselves involved, is not a new approach so far as medicine or agriculture is concerned; these have traditionally carried out their research in establishments separate from the faculties themselves: hospitals in the case of medicine, experiment stations in the case of agriculture. Whether these were established by the universities, by government or other agencies is irrelevant; they have always served the faculties as research establishments. The agricultural institutes of research have a separate administration, their own staff—usually mainly, if not exclusively, concerned with research. They also own special facilities, including experimental farms, in various parts of the country, and they have their separate sources of funds. If they so wish, faculty members can undertake research in the institutes, using their facilities and supported by their funds.

Agricultural research in faculties of agriculture—academic freedom

The nature of agricultural research carried out in faculties of agriculture is not necessarily different from that in government agricultural research institutes. There is, however, one characteristic that distinguishes the two: the freedom of choice of research subjects. The faculty member has the privilege of absolute discretion in the choice of the problem on which he and his students will be engaged, whilst the government researcher has the moral obligation to undertake research on subjects according to priorities that are binding on him. These are usually determined by public or governmental bodies, albeit with his participation, but in which his own desires are not decisive and may be overridden.

It is only natural that these decision-making bodies tend to give priority to problems of immediate significance to the agricultural economy, and are not always able to recognise the potential contribution of basic research to the solution of practical problems. Their greatest weakness lies in their inability to assess the potential value of entirely new approaches, which are the source of major breakthroughs in agriculture. Even scientists would have been hard put to foresee the connection between the discovery and isolation of plant hormones and wide variety of practical applications that have resulted from this discovery and have revolutionised modern agriculture.

Few people will question the need for a planned, and hence directed, programme of agricultural research to which the research institute is dedicated; hence the greater is the need for faculty research which preserves the 'freedom of research' of agricultural faculty members. This provides a haven in which researchers with original ideas, when forthcoming, are not limited by formal or other restraints from devoting their time and whatever facilities they have available, to any research they feel is of potential interest, whether short- or long-term, and without regard to whether it is possible to define *a priori* the area of possible application or not.

This does not imply that faculty members should retire to 'ivory towers'. Their thinking, as unconventional as it may be, should still be related to the realities of the agricultural economy and to the solution of its problems.

Choice of research topics—influence on students
Faculty members should at the same time, be just as free to become involved, if they so wish, in the 'planned' programme of research, either by undertaking individual research projects or by fully committing themselves to a field of research in the agricultural research institute.

In common with all university research, agricultural research in a School of Agriculture is not only an end in itself but is an important part of the educational system. As such, the main objective of research is to train the student and to teach him a systematic approach to the solution of problems. Problems should be chosen 'that encourage students to think imaginatively, reason scientifically and gain new understanding of principles—convert a worker into a thinker' (Bailar, 1965). These objectives can be attained equally well if the subjects chosen for the research have worthwhile practical implications, as when they are esoteric they have no relation to agricultural practice. The choice of research subjects has, however, a considerable effect on the shaping of the future agronomist's predilections and motivations. If the research subjects usually chosen contribute little or nothing to worthwhile knowledge or practical objectives and their only 'practical' application is to provide material for a 'paper', this will strengthen the latent careerism of the student. Research subjects that have no other objective but to satisfy scientific curiosity will tend to form a graduate divorced from agricultural reality. However, it is possible and desirable to choose subjects that provide scientific training and, at the same time, attempt to solve problems of significance to agriculture—thereby showing the young graduate or post-graduate student that basic and applied research are interwoven and that scientific curiosity can serve agriculture. This will encourage devotion to, and involvement with agriculture.

Darling (1970) wrote: 'Fortunately a strong wind of change is blowing and the modern Faculty (in the LDCs) is usually only too keen to play its full share in the development of national research in agricultural science.'

There are three forms of pressure to encourage Faculties of Agriculture to contribute to national interests (Darling, 1970): (a) The need to win grants from outside the university; (b) the need to attract students who wish to work on subjects that are relevant to the existing situation; and (c) involvement of heads of departments as members of national councils, committees and commodity panels, which influences their attitudes.

The land grant college model

Of all the structural types of research organisations described above for the different countries, the one that at first view appears the most rational and convincing is the land grant college model, in which responsibility for research, education and extension is vested in the colleges of agriculture, which, in turn, form part of the state universities. This approach, adopted by the individual states of the USA, has ensured close links between education, research and extension and has been exceedingly successful in transforming American agriculture. And yet the many attempts to transfer this model to other countries have not been successful; and in those countries in which the establishment of land grant type colleges is in the early stages of development, serious doubts may be entertained as to their suitability.

Reasons for success of the land grant model in the USA
The success of the land grant colleges in the USA is due to a number of factors specific to that country, and which are generally lacking elsewhere:

1. The land grant colleges were extremely dependent on the goodwill and support of the farming communities in which they were established. This made them sensitive to the needs of agriculture and responsive to the demands from farmers to solve their problems.
2. The teaching and research staff were only a single generation removed from the pioneering farmers of the frontier days; they themselves were usually of farming stock, eager to dedicate themselves to the solution of farmers' problems, in which they saw a worthwhile and interesting challenge. The majority were close to the land and to the farmers and had considerable practical knowledge and experience in farming. The manual labour involved in field experimentation was not considered demeaning for a scientist; on the contrary, familiarity with farm operations and ability to operate farm equipment and use tools was a source of pride.
3. The large number of land grant colleges ensured complete coverage of practically all the problems encountered even when no careful programming or planning of research was carried out. Research results from a single experiment station were mostly applicable to a number of neighbouring states, thereby making an 'overlapping' of research programmes possible.
4. Wherever shortcomings in the research programme remained, these were usually covered by the additional research structure of the federal government.
5. This system entailed a considerable amount of duplication and repetition, but the country was sufficiently rich to afford this and sufficiently aware of the importance of research for the development of agriculture to agree to the necessary financial appropriations.

Inapplicability of the land grant model in the LDCs .
The number of countries in which a similar set of circumstances exists is

few indeed! Universities are generally concerned with maintaining their academic freedom, and hence are not usually amenable to an oriented research programme which is a *sine qua non* of agricultural planning and development in most countries. Their interests naturally lie in the more basic aspects of research, to the detriment of the applied research essential in agriculture.

Similar views have been expressed by Ruttan (1968):

> The United States is characterized by a highly developed institutional infrastructure linking the university to other private and public institutions involved in technical, social and economic change. In societies where such an infrastructure has developed, research and education within the framework of the traditional academic disciplines and professions have represented an effective link in a larger system devoted to the production, application and dissemination of new knowledge.
>
> The same pattern of academic and professional organization, when transplanted into societies where the institutional infrastructure which it presumes does not exist, rarely performs as an effective instrument of technical, social or cultural change. In my judgment this is one of the major factors responsible for the substantial frustration involved in attempting to utilize the 'university contract model' as an instrument to induce technical, social or cultural change in developing economies. The institution building approach to the replication of either the 'land grant' or the 'classical' university in developing countries has rarely been productive in terms of either technical or cultural impact. The frequent result is to burden the developing country with an over-extended academic bureaucracy which is unable to make effective use of the limited professional capacity available to it.
>
> If developing countries are to overcome the technical and institutional limitations that separate the performance of the world's low and high income economies they must make efficient use of the professional competence which represents their single most limiting resource. This implies a pragmatic search for patterns of institutional organisation which permits a nation to have access to the professional competence available to it and to focus this competence directly on the critical barriers to technical, social, and cultural change.

Notwithstanding the inappropriateness of the land grant college model for most developing countries, United States technical assistance personnel has been extremely active in getting the model adopted in the countries in which they are active.

The report of the Committee on Institutional Co-operation points out that, 'Insistence on a US organizational form . . . has impaired many technical assistance institution building activities' and suggests that the land grant model can be made more relevant if US technical assistance personnel can become 'less doctrinaire in assisting host nationals to find an organizational structure for teaching, research and extension that is politically feasible and operationally efficient'.

The agricultural universities in India

There is no gainsaying the attractiveness, at first sight, of the land grant college model, and the efforts made to propagate the idea in developing countries; it is therefore worthwhile examining in some detail, what may occur when this system is applied indiscriminately and under inappropriate conditions.

It has already been mentioned that, on the basis of the recommendation of a Joint Indian–American committee, the Government of India decided to establish autonomous 'agricultural universities' in each of the States, and to transfer to these all the existing experiment stations as well as responsibility for extension (cf. p.777).

As a concrete example, let us consider the developments that have occurred in the State of Tamil Nadu, in which an agricultural university (TNAU) was established. Agricultural research and extension carried out in the past in Tamil Nadu have to their credit some outstanding achievements.

The Ministry of Agriculture had established 30 agricultural experiment stations in all parts of the State. The central experiment station at Coimbatore had established an international reputation, mainly for its breeding work, and the extension service was the first to apply, on a large scale, in the Tangore District, the principle of a 'Package Programme', with considerable success.

With the establishment of TNAU, it was decided to transfer responsibility for research and extension from the State Department of Agriculture to the University, in accordance with the policy of the Government of India, as expressed in the Tamil Nadu Agricultural University Act.

It was soon realised that the University was not in a position to effectively assume responsibility for extension, and it was therefore decided at an early stage that the State Department would continue to discharge this function, leaving the University with the responsibility for teaching extension methods, and training extension workers.

However, the State Department did not accept the decision to transfer agricultural research to the University, and fought a rearguard action to maintain its activities in this field. By 1974, only a few of the experiment stations under its control had been transferred to the University, causing a considerable amount of overlapping and duplication in the research programme.

As a similar situation had arisen in a number of other States (High-Level Committee, 1973); the National Commission on Agriculture studied the situation and found that with the varied patterns of research administration and execution prevalent in different States, there was a growing tendency for recrimination between University and State departments. Such a lack of co-operation and clear definition of responsibilities and duties was leading to a situation which could be potentially very harmful to agricultural research, education and extension. The National Demonstration Programme of the Indian Council of Agricultural Research with which the Farmers' Training Programme had been linked, was not achieving the anticipated results owing to lack of adequate co-operation between the research staff under the control of the University and the extension staff under the control of the State Departments of Agriculture and Animal Husbandry.

Ironically, the co-ordination between research, education, extension and development was far better before the establishment of the agricultural universities. Because of the lack of understanding, the two organisations instead of being complementary and supplementary to each other have unfortunately involved themselves into unhealthy rivalries. This fissiparous tendency is detrimental to the scientific development of agriculture and must be nipped in the bud. There is enough scope for both the organisations to purposefully serve agriculture (High-Level Committee, 1973).

The Commission therefore concluded that: 'a clear-cut definition of the respective responsibilities of Government Ministries and Departments on one hand, of the Agricultural Universities on the other, is an essential prerequisite—not only for the avoidance of overlapping and wasteful duplication, but also to facilitate harmonious co-operation between the two bodies'.

The National Commission therefore recommended that 'while the Agricultural Universities would be responsible for basic, applied and operational research, the State Department of Agriculture should assume responsibility for adaptive research'.

The Commission apparently overlooked its own statement, expressed elsewhere in the report, that 'Taking an overall view of agricultural research, one must recognise that it cannot be compartmentalised artificially into one kind of research or the other', a statement with which the author fully agrees.

Taking the long view, there is no doubt that the TNAU, notwithstanding the problems aroused by its establishment, will make important contributions to agricultural progress in the State of Tamil Nadu, but that in its final form, any resemblance with a land grant college will probably be purely coincidental.

PRIVATE SECTOR RESEARCH

Private sector research generally falls into two categories: research carried out by agro-industrial enterprises, and that undertaken by commodity-oriented organisations.

Agro-industrial enterprises

Research on manufactured inputs used by farmers and in which the investors can capture all or most of the benefits of research: such as development of hybrid varieties (sale of hybrid seed), improved agricultural machinery (patentable innovations), animal breeding (mating fees), development of new chemicals for plant protection, weed control, animal health, plant hormones, etc. (patentable). This type of private sector research is perfectly legitimate and deserves encouragement. The firms that undertake this research wish to make profits and their research is aimed mainly at the requirements of the developed countries in which most of them are located and in which they have their major markets. The products produced by the agro-industrial enterprises, in as far as they are exported to developing countries, mainly serve the commercial farmers. Little emphasis is therefore given to the needs of the small farmers by private sector research (Yudelman *et al.*, 1971).

Many of the new technologies, based on genetic manipulation of plants, bioconversion, the production of metabolites for producing antibodies to detect virus infections, and the extraction of useful compounds, such as drugs, pigments and insecticides give patentable results (Piñeiro, 1983).

Many of the technological innovations that are being developed by the private sector are dependent on basic research. This dependence may lead to a division of labour with the Universities and the National Research Institutes. If these institutions undertake the basic research required by the private sector, the former are justified in charging the cost of the research to the firms that request specific projects (Piñeiro, 1986).

As agriculture becomes increasingly modernised in the developing countries, the role of the agro-industrial and related enterprises in research becomes more important, gradually taking over some activities formerly undertaken by the national research organisation.

Farmers' associations

Commodity-oriented farmers' organisations frequently undertake their own research on the grounds that public-funded research is too understaffed or has inadequate infrastructure to undertake effectively the research programme they consider essential. These associations frequently finance the research by a tax on the product for which they are responsible. The result is a multiplicity of separate specialised research institutes in many developing countries, usually commodity-oriented.

The research work in the private sector institutes is often of excellent quality: it concentrates far greater efforts on a single commodity than the public institution can afford; funding is generally adequate, salaries and working conditions of the scientific staff are far better than in the civil service. Some of these private research institutes, such as the sugar and the pineapple research institutes in Hawaii, set up by their respective industries, have achieved a well-deserved international reputation. The Escuela Agricola Panamericana, founded and supported in part by the United Fruit Company, has been concerned with training in addition to research. 'The role played by its graduates throughout Latin America in agricultural research, education and policy has been remarkable' (Venezian, 1981).

The maintenance of specialised private research institutes may be justified as long as the specific sectoral needs of the farmers' associations are not being adequately treated by the official research institutions, and as long as no serious effort is made to develop a national agricultural institute capable of responding to the needs of all the sectors involved. However, the multiple single-commodity agricultural institutes of the private sector, as is the case with the parastatals in many LDCs (cf. p.225–6) leads to a fragmentation of the research effort, especially in the disciplines that cut across branches of production (crop protection, irrigation techniques, soil fertility, etc.). On the other hand, important areas of research are neglected, in particular those of concern to small farmers (food crops, farming systems, mixed cropping, small livestock, etc.).

When Government resolves to establish or strengthen a national research organisation, and to undertake a comprehensive and balanced research programme, it finds itself hampered by vested interests. As long as the parastatals or the associations refuse to integrate their institutions, including infrastructure, manpower and financial resources, into a common effort, the Government simply cannot implement its good intentions. The situation of the national research organisation remains unsatisfactory, and public support deteriorates, because the national research institute is no longer considered the only possible source of new technology.

In some cases, the private sector (agro-business and farmers' associations alike) provides funds for research it is interested in to the public research institution, avoiding thereby the possible lack of confidence in research results produced by an interested party.

Conversely, as long as the national research institution remains incapable of providing the required services, the private organisations have a good excuse to refuse to integrate their research into the national system, and a vicious circle is the result.

Efforts to overcome the duplication, the unbalanced research programme, and the differences in conditions of service between the private sector research and the national system, by appointing co-ordinating committees, have generally been ineffective and have almost always resulted in no more than window-dressing. Time-consuming meetings are held, but the effects are generally negligible.

A possible solution under these circumstances consists of: (a) a government decision to unify all agricultural research into a single national system; and (b) ensure that the bodies which maintained the specialised institutes would have a full voice in the decision-making process of the national institute and would therefore be sure of receiving the research support they required and deserved.

Country experiences

The contrasting approach of three Latin American countries to sectoral research is of interest:

Colombia
Developments in Colombia provide an excellent example of how narrow

sectoral interests can transcend and even subvert national policy and objectives (Piñeiro *et al.*, 1982).

In 1931, a law was enacted, whereby all national agricultural development activities, including research, extension and education, would be planned and implemented in co-ordination and co-operation with the Sociedades de Agricultores y Ganaderos, whose representatives were appointed to the Consejo Nacional de Agricultura. Despite this law, the Federación Nacional de Cafeteros established in 1940 its own experiment station at Chinchina, later to become the Centro Nacional de Investigaciónes de Café (CENICAFE), in order to be completely independent of government institutions, a position that was jealously maintained and protected. This was the first step in invalidating the concept of national authority in matters of agricultural policy, in co-ordination with the farming sector—to be replaced by competition and conflicts between the private sector and government departments for control of research and transfer of technology. The cafeteros were followed by the textile industry which founded the Instituto Nacional de Algodon. Subsequently, an accord was signed between Government and the association of cotton producers, whereby the Instituto de Fomento Algonodero obtained the status of a semi-official body. Though enjoying strong governmental support, the cotton sector jealously retained control of the functioning of the institute, through its majority representation in the Executive Council. Funding for research was obtained through a tax on cottonseed and cotton fibre, in addition to the government subvention.

In the same period, the rice industry established the Federación de Arroceros, which signed in 1950 an agreement with the Government similar to that obtained by the IFA. The rice growers were followed by the tobacco growers, who in 1954 established the Instituto Nacional de Fomento Tabacero.

Though Decreto 3092 of 1950 explicitly re-stated that 'the functions of experimentation remain exclusively in the hand of the Ministry of Agriculture in all topics of national interest', research in various commodities remained firmly in the hands of the specialised institutions established by the private sector, thereby achieving what Trigo *et al*, (1982) called 'the dismemberment by commodities of the research system'; the Ministry of Agriculture had in effect renounced the prerogatives accorded to it by law.

With the establishment of the Instituto Colombiano Agropecuario (ICA) in 1963, the concept that the state would be exclusively responsible for all agricultural research in the country was not only restated, but also institutionalised. Representation in the Executive Council of ICA by the private sector was cancelled. And yet, IFA, INTABACO and CENICAFE continued to maintain research institutes for cotton, tobacco and coffee, respectively. In the 1950s, the 'cuotas de fomento' were initiated. These were taxes derived from the producers of the major commodities for the promotion of their respective products.

In practice, control of the funds enabled the private federation to exercise complete control over experimentation in their respective commodities, while leaving basic research and the problems of the subsistence farmers as the exclusive domain of ICA.

The number of private-sector and public-independent institutes for research and extension further increased in the 1960s: the Instituto Zooprofiláctico established so-called diagnostic centres with research programmes, and provided extension under the aegis of the Banco Ganadero and the Federación de Ganaderos; the Instituto Colombiano de Reforma Agraria (INCORA) carried out socio-economic research and provided extension services; regional organisations such as the Corporaciones de los Valles del Cauca and del Magdalene y la CAR provided extension and education services in their respective areas. A research institute for sugar cane was established in 1977—CENICANA.

All this led to a considerable fragmentation of agricultural research and extension, and a consequent weakening of ICA, leading to the 'Reforma Institucional de 1968', which again reaffirmed the exclusive jurisdiction of ICA over all agricultural research and extension activities. A number of the autonomous institutes, with all their assets—financial, material and human—were then transferred to ICA.

This trend was again reversed following a change in government in 1970, which led to a gradual transfer of most agricultural research of concern to the commercial farming sector back to the various institutes, each concerned with one of the major commodities.

As a result of this zig-zagging policy, and lack of resolve in implementing government decision, the state lost its ability to direct the technical development process in its entirety, as part of a global strategy for the development of the agricultural sector (Piñeiro *et al.*, 1982).

Argentina also has very strong farmers' organisations, and yet, in contrast with Colombia, these do not maintain their own experiment stations, but influence the orientation and programming of research of INTA, through their strong representation in the Executive Council of the Institute. Trigo *et al.* (1982) ascribe this difference in attitudes to the different nature of the 'gremios agropecuarios' (farmers' associations) in the two countries. The gremios in Colombia are commodity-oriented, their members form a homogeneous group with common interests, and they are concentrated in circumscribed areas of the country. They also have a long tradition of providing services to their members and have considerable political strength.

Argentina

By contrast, the gremios in Argentina are each concerned with a variety of commodities, and have a very diffuse and heterogeneous membership. Because of the diversity of interests of their members, the Argentina gremios cannot take the same activist stand as can their Colombian counterparts, when the latter are defending the narrow sectoral interests of their members.

Peru

In Peru a different approach has been adopted: Several farmers' associations have established small experiment stations to study the specific problems of their respective valleys. The work at these stations is fully coordinated with INIA, especially within the framework of the national commodity programmes, many of which are of considerable importance for the valleys.

In this way mutual benefit is achieved: the private experiment stations enjoy the support of the national research institute which even contributes funds; the associations obtain valuable information that it would have been difficult to generate by their own means, and INIA can expand its regional work by using the facilities of the 'private' experiment stations.

The Sociedades Agricolas de Interés Social (SAIS) are autonomous bodies that have developed agricultural activities. They encourage researchers from the national research organisation to carry out experiments on the SAIS estates, and these also serve as demonstration fields.

FUTURE TRENDS

According to Trigo *et al.* (1982), the trend in most Latin American countries is towards a complex system comprising public, semi-public and private institutions which participate in specific aspects of the process of generation and transfer of agricultural technology, creating together a multi-organisational system, replacing the state monopoly in agricultural research and extension.

This raises two questions: (a) is the trend towards a fragmentation of research inevitable, so that the nature and objectives of the national research institution have to be adapted to the changed situation; (b) is this trend desirable, or should every effort be made to maintain the integrity of the national research institute and its overall responsibility for all aspects of agricultural research?

Trigo *et al.* (1982) consider the trend towards multi-institutional agricultural research systems as inevitable; this trend being an integral part of the market economies which characterise the majority of Latin American societies. These authors also view these trends as favourable developments, because of the increased involvement of wider economic and social sectors in the technological process and their greater contribution to the resources available to research.

The involvement of the private and semi-public sectors, such as producers of agricultural inputs, farmers' associations, commodity boards, etc., in the agricultural research process is not a future development or even a new one. Already in the past, fragmentation of agricultural research among many bodies was a regular feature in many countries.

Nor is it exclusive to developing countries. In the developed market economies, private sector research has already established itself as a major contributor of new technologies, and its role is expanding. In the USA, for example, the private sector already contributes more than half the funds invested in agricultural research (Fig. 7).

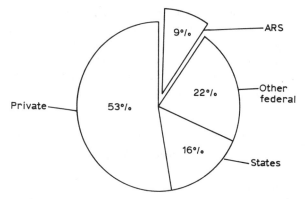

Fig. 7. Public and private funds for US food and agricultural research. (From USDA (1983), reproduced by permission of USDA.)

In drawing conclusions on the desirability of the process, one must distinguish between research initiated by agro-business, and the proliferation of sectorial and semi-public research institutes.

Research by agro-business interests is perfectly legitimate, and can make valuable contributions to agricultural progress. An accommodation between private and public research can, and should, be promoted.

The situation is different in regards to sectorial and semi-public institutions.

Experience in many of these countries has even shown that research carried out by sectorial interests is frequently more effective in responding to the needs of these sectors than the research carried out by the national research institution, especially when the latter is starved for human and financial resources.

The advantages that accrue to certain narrow sectorial interests when agricultural research is fragmented among a number of semi-public and sectorial bodies are offset by the damage that inevitably is caused to the overall national research effort, due to reduced public and political support, competition for human and financial resources, and the inability of the authorities to define and implement a research policy that is consistent with overall national interests. It is therefore in the national interest to maintain the integrity of the national research system. This objective should be possible to achieve and maintain, if all legitimate sectorial interests are taken into account by ensuring adequate representation of all sectors concerned, in the process of defining research policy and priorities.

Whether the trend towards the fragmentation of research is inevitable, as postulated by Trigo & Piñeiro (1982), is more difficult to predict. It will depend largely on whether government policy is aimed at strengthening public research, and its ability to implement its decisions. It will also depend on the measure of trust that the private sector feels it can place in the desire and ability of the national research institution to provide answers to the technological problems with which they are faced.

EVOLUTION OF NATIONAL AGRICULTURAL RESEARCH
SYSTEMS (NARSs)

Although countries differ significantly in available resources, cultural traditions and political objectives, the processes that have led to the establishment of the NARSs and their subsequent evolution often show remarkably similar trends, and generally pass through a number of stages (Fig. 8).

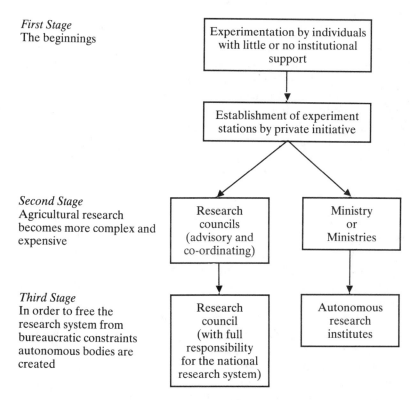

Fig. 8. General evolution pattern of national agricultural research system.

First stage

Research—as distinct from trials and perceptions—is undertaken by qualified and talented individuals, such as Van Theunen and Liebig in Germany, Johnstone in Scotland, Young in England, etc., with little or no institutional support. This was followed by the establishment of experiment stations or agricultural laboratories, by the private initiative of individuals or farmers' associations: Boussaingault in Alsace, the Highland

Agricultural Society in Scotland, a landowners' association in Saxony, Lawes and Gilbert in England, etc.

Second stage

An awareness of the need for agricultural research as a pre-requisite for increasing agricultural productivity, leads to the establishment and funding by Government of agricultural experiment stations, first in a few pioneering countries, and then worldwide. These stations were generally single-commodity or single-discipline units, with little or no linkages between them. In an effort to achieve some semblance of co-ordination and orientation of the research effort, responsibility for agricultural research is vested in one or more ministries, or alternatively in a research council.

Third stage

In the long run, the 'Ministry model' proves to be inadequate as a framework for agricultural research, because of the inflexibility of ministerial budgetary, administrative and personnel procedures, and other reasons (cf. p.223). The solution adopted to remedy this situation is the establishment of a semi-autonomous institute (or service) with strong linkages to the Ministry responsible for agricultural development: the Ministry reserves for itself the prerogative of ensuring that the research work carried out by the Institute is congruent with the national interests and requirements; the Institute having complete independence in the implementation of a mutually agreed research programme.

In the case of the 'ARC model', most ARCs have evolved, or are evolving, along similar lines. First established as advisory, planning and co-ordinating bodies, they gradually assume greater directional and executive control over funding and the executive control over the actual implementation of research. Trigo (1986) writes: 'The force behind this trend appears to be the increasing conviction that without at least partial control over funding and the capacity to actually implement certain strategic components of the research programme, the co-ordination function cannot be properly performed.'

Subsequently, the ARS is formally charged with the responsibility for the organisation and implementation of the national agricultural research programme.

Because of the basic concept 'that the ARC should be shielded from political interference', the Ministry of Agriculture has only minimal influence on the research programme. This leads easily to a situation in which the research work becomes increasingly isolated from the pressing problems of the Ministry and of the farming community. This in turn leads to a curtailment of the independence of the ARC in its research programming, through such devices as channeling the funds for research through the Ministry, appointing the Minister or his representative as chairman of the Board of Governors of the ARS, or making the ARS directly responsible to the Minister of Agriculture. Following these developments, the ARS model becomes practically identical to the semi-autonomous National Institute of Agriculture, in all but name.

It should be stressed, that in almost all countries, universities and private sector interests have made significant contributions to agriculture research, and are therefore an integral part of the national agricultural research system.

Dual systems

In countries such as England, in which the responsibility for co-ordinating the work of the autonomous institutes of agricultural research was assigned to a research council, and the latter either did not try, or did not succeed in overcoming the resistance of the many research institutes to their integration into a mission-oriented research system. Consequently, a parallel research system was established by the Ministry of Agriculture that was responsive to its requirements.

Responsibility for the national research vested in a faculty or agricultural university

In the few countries in which faculties of agriculture or agricultural universities are responsible for the national research effort, a second system has developed, superimposed on the former: a federal system in the USA, a NRC in India.

Countries consisting of a federation of states

In all such countries, there is a federal system that either co-ordinates and complements the state systems, or establishes a research system superimposed on that of the individual states.

Arrested evolution

In many of the *developed countries*, the national research system did not progress from the second stage to the 'climax'; the main reason being that the systems in place had proved to be very effective, and had made considerable contributions to agricultural development in their respective countries. That they were very costly because of the considerable amount of duplication, was not a sufficient incentive for the governments concerned to attempt to overcome the resistance to change of vested interests. This situation may change, in view of the retrenchment that has become necessary in many of these countries.

At the end of the colonial period, most of the *developing countries* inherited a typical second-stage research system, which for many years remained unchanged. However, under pressure of the severe constraints they face in the human, financial and institutional resources available to them, as well as the need to modernise their agriculture as rapidly and efficiently as possible, they had to improve the ability of their NARS to make a significant contribution. In the 1950s there were only very few countries that had what we have described as the climax model. In the 1980s, most countries in Latin America had already adopted the National Agricultural Research Institute model or were in the process of doing so; in Asia the tendency has been to adopt the National Agricultural Research Council model; in Africa, where the influence of the ex-colonial powers is still very strong, change has been slower in coming. However, many countries are now contemplating organisational change, as evidenced by the numerous missions that have been invited in recent years to advise on reforming their respective NARSs.

Recapitulation

The national agricultural research effort is generally fragmented among many bodies: Ministries, research councils, research institutes, universities, extension services, farmers' associations, parastatal agencies, and other components of the private sector.

The original rationale for this multiplicity of research agencies was the assumption that there would be a clear division of work between them: ministries would concern themselves with short-term, pressing problems; universities would devote themselves to academic research without thought of practical application; research councils would engage in basic, mission-oriented research, free from political interference; research institutes would undertake applied research in accordance with national priorities; extension services and farmers' associations would carry out verification and adaptive trials, and the private sector would engage in research the benefits of which could be 'captured' by those undertaking the research. In the case of Federal–State systems (USA, Australia, Brazil) the national research effort was to be directed to the problems of nationwide significance and that of the individual States, to their respective specific problems.

Actually, all these differences have largely disappeared, if ever they existed, and the research carried out by most of the agencies engaged in public-funded research activities is very much the same.

Paradoxically, it is the developed countries which have generally shown the least inclination for radical organisational change, because the resultant high cost of the system is generally masked by the high returns of investment in agricultural research. There are already many indications that economic constraints are making close scrutiny of the present system essential in many industrialised countries.

In the developing countries, there is little evidence that the fragmentation of agricultural research between ministries, parastatal bodies, and universities is resulting in healthy competition leading to higher research productivity. An FAO survey has shown that this situation 'had induced inefficiency in the use of scarce manpower, physical facilities and equipment. The dispersal of responsibility has led to obstacles in adopting an intersectoral approach and the neglect of research on farming systems. It also made it difficult to curtail the duplication of effort' (FAO, 1984).

However, most of the Third World countries have shown an awareness of the need to rationalise their respective NARSs, and many have already undertaken important restructuring of their research organisation.

A NORMATIVE MODEL FOR THE RATIONAL STRUCTURE OF A NARS

'No single model can be recommended as the most effective approach to the organisation of a national research effort, as situations in individual countries are unique' (World Bank, 1982).

Similar statements, often in identical terms, have been made by various authors. Trigo (1986), for example, states that: 'in organisational format matters, there is no way of organising agricultural research systems, and not all formats are equally effective.' The different models of research organisation originated as coherent responses to the conditions that existed at the time of their inception.' '"Optimality" results from political and technical fit within a given environment. An optimal format is one that gets the job done.'

The fact that it is possible to associate success stories with each of the four main types of research systems prevalent in the world is seen by Trigo (1986) 'as proof, albeit inadequate, that there is no one best way to organise'.

A final quote, from Ruttan (1982): 'the organisation of a productive national agricultural research institute is not independent of the broader historical forces that have fashioned a nation's political, cultural and economic system'.

No one will dispute the fact that the organisation of agricultural research in each country has been fashioned by its political, cultural and economic system. However, this does not necessarily imply that the resultant system is a coherent response to the conditions prevailing in the country.

Even a superficial examination of the case histories presented in Chapter 1 will show that very rarely, if ever, has the national agricultural research organisation developed according to a planned blueprint, that takes into account the political, cultural and economic factors prevailing in the country. In most cases, clashes of interests among ministries, departments, institutions, and personalities engaged in fief building, or group and sectorial pressures, have had more influence in shaping the organisation than has objective planning in accordance with the perceived needs of the country. How can one explain otherwise the fact that countries in geographical proximity to each other, with similar political, social and economic systems, have developed entirely different patterns of organisation that cannot be related to differences in their individual requirements? How can one explain the fragmentation, duplication and overlapping of research apparent in many of the case histories? How can one explain the neglect of areas of research of major political, cultural and economic importance as evidenced in the neglect of the small farmers' sector and research on food crops in many developing countries?

Factors that have shaped NARSs

The *political factors* that have influenced the type of structure adopted in a country, have often been:

    (a) changes in government involving the need to increase or, on the contrary, to consolidate ministries; clashes of interests between personalities engaged in empire or fief expansion. All this results in a volatile situation conducive to lack of continuity in research programmes and low morale of the researchers;

(b) influence of pressure groups—'pork-barrel politics which paralyse research planning' (Wade, 1973e);

(c) sectoral interests which result in a proliferation of single commodity institutes that weaken the national research institute;

(d) regional interests that lead to the establishment of many unnecessary regional research centres, or locate them where they are not needed;

(e) lack of stability, as reorganisation follows reorganisation with each change in the political constellation.

Occasionally, political factors may have beneficial effects on the research system, as when there is a perceived need to improve agricultural productivity and hence to strengthen the research structure, or to promote research programmes aimed at improving the situation of disadvantaged sectors of the rural population.

Cultural factors such as the *laissez-faire* approach typical for England and the Netherlands, has resulted in a complex, fragmented research structure, that has produced excellent work with a considerable impact on the national and world agriculture, but at a high cost, which few countries can afford.

Need for a rational and stable structure

A national programme of agricultural research that covers both urgent, day-to-day problems, as well as long-term topics; that is congruent with the needs of the National Rural Development Plan, and that ensures the most efficient use of scarce resources, is of major importance for all countries. And yet, it is practically impossible to achieve the rational planning and implementation required for such a programme, within many of the existing structures, even if they result 'from political fit within a given environment' and even if 'they get the job done'.

One must therefore conclude, that a rational and stable structure is an essential framework for effective planning of research. The question naturally follows whether it is possible to design a prototype of research organisation, that can, with minor modifications, serve as a normative model, applicable to most countries.

The optimal format of a NARS

First, it is necessary to clarify what should be understood by an optimal format of the research structure. 'One that gets the job done' as proposed by Trigo (1986) appears to be irrelevant, as it does not consider the cost involved in getting the job done. A more relevant definition, which will serve as a basis for further discussion is that 'an optimal format is a structure that achieves its objectives as economically, efficiently, and equitably as possible'.

This again brings us to the question of whether the type of organisation that has evolved historically in each country is really the one that is best adapted to the needs of the country (as frequently stated) and the one that gives optimal results.

Effective versus efficient

The political, cultural and economic factors in the USA have resulted in a dual agricultural research structure—federal and state—which has produced some of the most dramatic breakthroughs in agricultural science, and has been a major factor in the development of the most productive agriculture in the world.

The special characteristics of the Dutch people have resulted in a NARS, the complexity of which is in inverse proportion to the size of the country. This system has also contributed to the development of a highly modern, intensive, diversified and prosperous agriculture.

The *laissez-faire* approach of the British has enabled the development of a dual-system of agricultural research, ARC and Ministry systems respectively, one arm of which (the ARC) has been severely criticised by the agricultural sector as being alienated from the realities and needs of agriculture. And yet, this system has produced high-quality research second to none in the world, and a number of breakthroughs in basic and applied research of international significance.

Indisputably, these NARSs 'have got the job done'—but 'optimally'? Notwithstanding their diversity, the three NARSs we have brought as examples, have one characteristic in common: they are extremely expensive. Duplication and *laissez-faire* can be extremely effective, by promoting competition, attacking problems from different angles, encouraging personal initiative—but at a cost most countries cannot afford.

Even in the countries mentioned, it is becoming more and more difficult to justify the high cost of these systems. Decreased funding for research, reduced political support, competition from private sector research, are obliging most developed countries to take a hard look at their NARSs, and to draw the necessary conclusions.

Obviously, these systems cannot, and should not, be transferred to other countries (as demonstrated by the ill-fated attempts to apply the Land Grant College model, historically the most effective of all, in many developing countries). But can one conclude from the foregoing, that 'no single organisational model can be recommended as the most effective way to organise a national research effort' and that 'a model will need to be developed or modified to meet the special needs of each country' (Elz, 1984)?

The following section will address itself to this question.

A MODEL STRUCTURE FOR A RATIONAL AGRICULTURAL RESEARCH ORGANISATION

We will now consider whether it is possible to design a prototype of organisation within which most, if not all, national agricultural research structures can be accommodated.

The main requirements for such a model structure would be:

(a) Commitment to solving the problems of the agricultural community.

(b) Ability to carry out a balanced programme of research, covering both the urgent and day-to-day problems of agriculture as well as the long-term problems. This involves:

- close contact with the ministry responsible for agricultural policy and development;
- readiness to carry out both applied and basic research according to the type of problem to be handled;
- sufficient detachment from immediate pressure to be able to devote the necessary efforts to long-term and exploratory research.

(c) Ability to make the most efficient use of research personnel, equipment and funds. This involves:

- a central organisation that can effectively co-ordinate its work with a minimum of outside interference, and the avoidance of needless duplication of effort;

- autonomy in the implementation of its research programme in order to ensure flexibility of administration with a minimum of outside bureaucratic interference and red tape;
- a system of remuneration, grading and promotion appropriate for researchers;
- ability to respond to the requirements of the different ecological regions of the country.

As these prerequisites for an efficient national agricultural research organisation are the same for all countries, it is possible to conclude that any model that can provide all these requirements, is by definition a model applicable in all countries.

Proposed model

At a meeting of an international panel of experts on the organisation and administration of agricultural research, called at the initiative of FAO in 1965 (FAO, 1965), the unanimous conclusion was reached that really effective co-ordination of the national research effort could only be achieved by establishing a national agricultural research system (NARS) in which overall responsibility for all government-funded research was vested. Such an organisation would be in a position to integrate the research work of the faculties of agriculture and other bodies, into the national research programme, by means of joint planning, research contracts for specified projects, and by co-opting faculty staff into interdisciplinary research projects.

Consequently the type of research organisation that could most effectively serve as a general adoptable model is a central national research institute (or service, or research council with the same characteristics), closely linked to the Ministry of Agriculture in matters of policy, research planning and transfer of technology, but independent in the implementation of the approved programme. This organisation would have the following characteristics (Fig. 9):

- responsibility for implementation in the whole field of governmental agricultural research;
- semi-autonomous status within the framework of, or linkage with, the Ministry of Agriculture;
- funds and research directives derived from a plurality of sources:
  1. the Ministry of Agriculture for general financing and for programmes adjusted to development plans of the Ministry and the most pressing needs of the farming community;

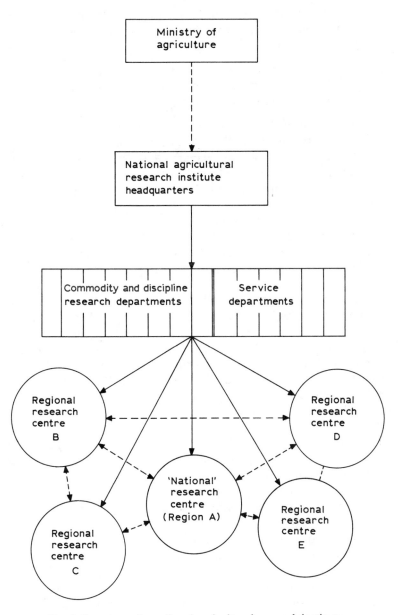

Fig. 9. Structure of a national agricultural research institute.

2. the national research council, for long-term and basic research;
3. uncommitted institutional funds for exploratory research, doctorate theses, etc.; and

- centralised organisation and administration of research, and partly decentralised implementation of research based on a network of regional polyvalent experiment stations.

The maintenance of close links with the Ministry of Agriculture is essential: (a) in order to ensure that the research programme is oriented towards the solution of problems related to the implementation of the development plans for which the Ministry is responsible, and (b) to ensure effective co-operation in the transfer of research findings to farmers, through the Extension Service.

A semi-autonomous status, as a public body, is essential, because a ministry or a government department is usually structured for regulatory and control activities, whilst research cannot function effectively within a rigid framework which cannot adjust to the specific requirements of research.

The research organisation needs freedom from political and outside administrative control in its day-to-day operations, whilst the ministry justifiably requires that control is maintained over the policy and general research programme of the research institute and its dependencies. This control can be achieved by vesting full powers of orientation, review and evaluation in a high-level committee, chaired by the Minister of Agriculture, in which the research organisation, the farmers' interests, the universities and other interested bodies are effectively represented (cf. p.389).

The research organisation should have control over its budget, subject to certain administrative procedures. It should be exempted from the general career-service rules that govern the recruitment and promotion of civil service personnel, and be allowed to develop its own system, with promotion based on research achievements, and a salary scale preferably modelled on that of faculty staff.

Centralised organisation and administration, combined with the decentralised implementation of research, are essential because of the special nature of agricultural research. The centralisation is essential in order to ensure that the entire organisation works in accordance with a research programme oriented towards the needs of the country as defined in the national development plans, and to provide the supporting services required for an effective research effort.

The actual research in the field must be decentralised, for several reasons. The obvious reason for establishing a network of regional experiment stations or centres, is that a substantial part of agricultural research is highly location-specific: the results and subsequent recommendations depending to a considerable degree on environmental conditions. In addition to ecological considerations there are a number of other reasons justifying the need for the regionalisation of research:

* It is important, for political and psychological reasons that research should have a grassroot orientation.
* A co-ordinated programme of research carried out in a number of regional stations is effective in shortening the time required before conclusions can be drawn and practical recommendations formulated.
* A pattern of regional stations can dovetail with the regional organisation of extension, thereby facilitating the transfer of research findings and shortening the time-lag in their adoption.

An appropriate system for policy guidance, planning the research programme, and defining priorities based on the development needs of the country, is another essential feature of the model. Further requirements are: a research structure consisting of a combination of research units based on commodities (crops or animals) and on disciplines (agrometeorology, plant physiology, genetics, soil science, sociology, etc.) respectively and research projects based on interdisciplinary teamwork.

The size of the organisation, the number of regional centres or experiment stations, as well as details of the infrastructure, can be adjusted to the specific needs of each country (cf. Chapter 12).

Adoption of the proposed model
In the 1960s, when the FAO panel proposed the National Agricultural Research Institute (or Service) as a model that can be generally adopted, there were only a few countries in which such a model was operating. In Europe, France was probably the only country that had adopted such a system; however, it was not transferred to any of the French colonies.

Since then, many countries that have felt the need to reorganise their NARS have adopted this model. In Latin America, the changeover from the Ministry of Agriculture model to the National Institute model has taken place in country after country, and the few laggards are following suit. In Africa, it is mostly the francophone countries that are changing over, or already have national institutes of research, but these are generally linked to Ministries of Scientific Research, instead of the respective Ministries of Agriculture, a serious drawback (cf. p.224). In several Asian countries, the responsibility for agricultural research has been vested in ARCs; these, are however, practically identical in their functions, linkages and organisation to the national institute model, and the major difference is in name only.

Conditions for effective functioning of the proposed model
Even an ideal organisational structure does not ensure that it will be effective and efficient. A number of conditions have to be met to make this possible:

- Government commitment to agricultural development in general, and to support of agricultural research in particular.
- A clear statement of research goals, and appropriate procedures for defining priorities with participation of political leaders, planners, researchers, extensionists and the clients.
- A clear mandate vesting responsibility for all public-funded research, and a legal status ensuring autonomy in administrative matters, and in the implementation of the research programme.
- Sufficient and stable financial support, consistent with the tasks required to implement the approved research programme.
- Procedures that ensure orderly administration but reflect the needs of a research organisation: minimum of bureaucracy and paper work.
- Authority based on scientific leadership and not on formal administrative hierarchies.
- An adequate number of qualified research workers, and supporting personnel; adequate on-the-job training programmes.
- Adequate supporting services.
- Effective linkages with politicians, the general public, other research institutions, national and international; the extension service, and the users of technology. An effective flow of information to farmers.
- An adequate network of interlocking regional experiment stations.

All these requirements will be addressed in detail in the chapters that follow.

Modifications of the basic model

The main modifications of the basic model that have been attempted have consisted in assigning one or more functions, besides research, to the responsibilities of the institute: extension, education and development.

Extension. The rationale for the research and extension functions in the same institution is discussed in detail in Chapter 22. The advantages of such a solution are more apparent than real; the major disadvantages of this solution result from including in the same organisational unit two services that require entirely different organisational and administrative frameworks for them to function effectively. Their respective professionals have different aptitudes, motivations and goals; different criteria have to be applied for the professional advancement of individuals belonging to the two groups.

The inclusion of two services with such different characteristics within a single institution will more probably be a source of friction and alienation than of understanding and co-operation.

Education. The proper venue for the formation of professionals who require an academic degree is the university, which has the recognition and the organisation for this purpose. This is its *raison d'être*, and an agricultural research institute should not compete with the university in this area. However, the research institute can make an important contribution to agricultural education, which benefits all the parties involved, by providing the institutional framework in which graduate and postgraduate students can carry out their research on subjects relevant to the objectives of the institute. This training should be under the aegis and responsibility of a university, and supervised by senior research personnel accredited by the university for this purpose (see Chapter 9). This training is therefore an adjunct to the research and does not require any special organisational modifications of the model proposed.

Development. Only in one country in Latin America has a research institute evolved into a complex institution concerned not only with research, education and extension but also with agricultural development (ICA in Colombia). The difficulties resulting from this type of organisation have been described in detail in Chapter 1. It is sufficient to state here that after 20 years' experience with this model, it has been decided to reassign these functions to separate frameworks.

This brief analysis of the different modifications leaves us with the basic model described above as the preferred alternative.

A RESEARCH INSTITUTION ORIENTED EXCLUSIVELY TOWARDS THE SMALL FARMER

Guatemala has established the Institute of Agricultural Sciences and Technology (ICTA) with small farmers as its exclusive clients (McDermott & Bathrick, 1981), and is directing all its efforts toward generating technology relevant to small farm systems.

The major activity is on-farm research; with minimal experiment station research. ICTA has no central research station. Its regional stations are neither large nor elaborately equipped, and are mainly concerned with maintenance of genetic purity and with varietal crossings. Almost everything else is done on farms, so that 75 per cent of the research is on-farm, and only 25 per cent is carried out at the regional centres (called 'production centres'), and laboratories with meagre facilities. All innovations are tested by farmers in their own system before being recommended for use on small farms.

Concentrating on the small farmers, in order to compensate for years of neglect, is a very laudable objective; it is, however, doubtful whether absolute neglect of commercial farmers' problems is feasible, or economically justified in the long run.

Nor is it probable, that ICTA, as it is presently conceived, can be a really effective research organisation. Farming systems research depends on a continuous flow of new technology from the research centres, to be tested, adapted, and integrated into modified farming systems. Once the available backlog of technologies is exhausted, it is improbable that the limited research capacity of the few production centres and laboratories with 'meagre equipment' will be capable of producing an adequate flow of innovations. In short, the ICTA model may be justified in the short term, but does not appear to be a viable long-term proposition.

DUAL NATIONAL AND STATE RESEARCH SYSTEMS

Beyond a certain size, distances within a country may become a physical factor that affects the ability of the organisation to carry out its responsibilities. In addition, in very large countries, such as the USA, Australia, USSR and India, a political factor intervenes: in addition to physical dimensions, a national government is superimposed on a number of states or provinces.

These countries have national interests that transcend the interests of the individual states or provinces and are therefore a legitimate concern at national level. The national interests may lie in promoting research aimed at expanding the production of export crops or commodities required for processing by industries in other states than those producing the raw material; conversely, there may be a conflict between national and statal interests. For example, a state may wish to promote labour-saving technologies which cause an influx of unwanted migrants to the cities of other states.

It is therefore generally accepted that there is a legitimate role for agricultural research at national level, even when each of the component states has an adequate agricultural research infrastructure. What is less clear is the nature of this role. Should the national level confine itself to planning, promoting and co-ordinating the agricultural research nation-wide, or should it also have a role in implementation?

Most of the large countries in the world, faced with this problem, have opted for the second solution, and have a national infrastructure for agricultural research superimposed on the state research structures. Before considering whether this approach is desirable for developing countries, it may be useful to examine the situation in the country which has the longest experience with a dual national–state agricultural research system—the USA.

The formula that was adopted by consensus: the Agricultural Research Service (of the Department of Agriculture) was 'to be primarily responsible for research on problems of national and regional concern to agriculture,' whilst the State Agricultural Experiment Stations would be primarily responsible for research on problems within the borders of their respective states. The following statement 'that basic research is a responsibility both of the ARS and of the State Experiment Stations and should be advanced in each institution and in each area so far as feasible' made any distinction between the work of the federal and state organisations practically unenforceable. To agree that both were to be engaged in basic research, namely: 'discovering, verifying and establishing the practical bearings of scientific facts and principles of agriculture', and at the same time attempt to distinguish between subjects of national interest and those of state interest only, was completely illusory. Work on streptomycin at the New Jersey experiment station, on the development of hybrid corn at the Connecticut Station, on male sterility of sorghum at the Texas station, on rust research at the Minnesota Station, on controlled atmosphere for fruit storage at Cornell, to mention only a few examples, was of an importance which not only transcended the borders of the states concerned, but was also of immeasurable benefit to all nations in the world.

In a recent assessment of the US Food and Agricultural research system, it is stated that with its present structure 'there is some question as to whether the USDA has a national research program or merely a series of local and regional activities. Consequently, the USDA and the state agricultural research experiment stations appear to be working on seemingly indistinguishable problems' (Konkle, 1982).

It is indisputable that the dual system for agricultural research in the USA has been extremely effective and has played a major role in developing one of the most productive and sophisticated agricultures in the world. However, effective is not necessarily efficient, and there is an enormous amount of duplication between national and state research activities and even between states within a common agro-ecological region.

The assessment mentioned above comes to the conclusion that much of the nation's research efforts 'are dissipated in operational activities at the expense of policy-level activities'. The results of this dual activity have been: (1) to reduce the time and attention that can be given to determining policy, planning and co-ordination; (2) to reduce the authority of the administrators of the national research organisation; (3) to reduce the operational efficiency of the latter; and (4) to increase bureaucratic delays in decision-making (Konkle, 1982). To these deficiencies the author would add: to reduce the human and financial resources available to the state research systems.

The basic conclusion from the foregoing is that the most efficient approach for large nations comprising several states or provinces, is that the national agricultural research service should concentrate on policy and co-ordination: planning, promoting, co-ordinating and funding agricultural projects of national significance and strengthening the research structure of the states. A major objective at national level would be the promotion of research on regional problems, e.g. problems that extend over a major portion of two or more states.

This solution provides government with a single national agency to hold responsible and accountable for the use of all national funds allocated to agricultural research, and for the co-ordination of national and regional research programmes.

On the other hand, unnecessary expense and duplication are avoided by entrusting the implementation of the regional and national co-operative programmes to the state (provincial) research institutes, whether working individually or on joint programmes sponsored, co-ordinated and funded at national level.

CO-ORDINATED NATIONAL RESEARCH PROJECTS

The major and most effective linkage between the national (federal) and state (provincial) research systems is provided by the Co-ordinated National Research Projects (CNRPs).

This system has been adopted in both developed and developing countries, usually with considerable success, on condition that the necessary prerequisites are forthcoming.

Advantages

Moseman (1977) assigns the following advantages to the CNRPs:

- they make the most effective and efficient use of research resources;
- they facilitate prompt and continuous interchange of new knowledge and materials among research workers of the Central Government, the States (Provinces, Prefectures), and the private sector;
- they ensure effective programmes of broad national or regional concern;
- they facilitate the introduction and testing of knowledge and materials introduced from abroad.

To these advantages, the author would add that they provide a common framework for national and state research, with a common purpose —avoiding competition and duplication.

Conditions for success

Experience has shown that the CNRPs cannot be successful unless the following conditions are met:

- national government leaders understand fully the objectives of the programmes and provide the full support needed;
- the project is established within the institutional framework of the national research institute; linkage with the ministry responsible for agricultural development is adequate;

- there are adequate organisational and institutional facilities in the participant states;
- all complementary resources in the colleges of agriculture and private sector are mobilised and integrated into the project;
- the project is multidisciplinary, with participation of agronomists, plant breeders and scientists from allied disciplines concerned with disease and insect control, plant nutrition, water requirements, weed control, etc.;
- identification of the nature and relative importance of the constraints faced in the production of the commodity with which the project is concerned;
- the selection of capable leadership.

Duration

Experience has shown that a 10-year period is the minimum required for implementing a successful CNRP.

VERY SMALL COUNTRIES

The size of a country to be served by a NARS is a factor that cannot be ignored. Though even very small countries can have a great variety of ecological zones, this is not a hindrance to the functioning of a national research institute, provided it relies, for the execution of adaptive research, on a suitable network of regional experiment stations. However, the size of a country, the importance of agriculture in its national economy, and the availability of resources, determine the size of the investment possible in the establishment and maintenance of the NARS.

Gamble & Trigo (1984) point out that there is no standard definition of what can be considered a small country. Countries may be large in some aspects (population, area) and small in others (income, growth rate). The authors define 'smallness' in terms of their agricultural resources.

Taking the bottom half of a ranking based on this variable for different regions, the countries in Central America and the Caribbean are considered small in the Latin American context; and countries such as Benin, Guinea-Bissau, Liberia, Sierra Leone, Togo, Somalia, Burundi, Gabon, Congo, Rwanda, and Swaziland would be small in the African context.

A criteria proposed by Trigo & Piñeiro (1984) for estimating the resources which need to be invested in agricultural research, is the number of important commodities produced. They estimate that the minimum research module for one commodity would require a team consisting of four researchers, complemented by eight specialists, plus a complement of support personnel. The cost for a single commodity would amount to about US$ 500 000 (1984 figures). For a small country with 6–10 major commodities and several agro-ecological zones, this could imply a research budget of 5–8 million US dollars. When this effort is complemented by cross-commodity research, such as water requirements, plant protection, farming systems, and socio-economic research, the authors estimate a budget in the 12–15 million dollar range.

Ruttan (1985) formulates three generalisations:

- small research systems have higher research investments per hectare than large ones to achieve an equal level of effectiveness;
- small countries, with great agroclimatic variations, will face higher costs to develop productive farming systems than more homogeneous countries;
- small countries cannot avoid being dependent on others for much of their agricultural technology.

The international research centres are especially important to the small countries, as a source of technology and direct support of their NARS, as well as the training provided by the centres.

The World Bank (1983) recommends a gradual development of the NARS in a small country. A first step would be the development of one or two experiment stations in the most important agro-ecological regions of the country, with a small, multidisciplinary team of researchers, together with adequately equipped support staff, working on the major problems facing the farmers of the region(s). If agricultural experiment stations already exist, these should be upgraded and possibly augmented by another one or two, if justified by the potential importance of additional regions.

Many of these countries will depend on external help in establishing their NARS, and for at least a few years of implementation.

Regional collaboration in research between neighbouring small countries, or groups of small countries, could help them to overcome, at least partially, problems of scale. An example of such an approach has been proposed for the Caribbean region, with its many small countries, by Wilson (1984), of the University of the West Indies. It is based on the development of a NARS in each country in the region, and a strategy of collaboration in agricultural research, training, and development between the states and their institutions, which were non-existent at that time (1984). By aggregating their efforts in these functions, it would become possible to achieve a level of research and education capacity which each country separately is not capable of achieving.

Chapter 4

International Agricultural Research Systems and Co-operation

THE INTERNATIONAL RESEARCH CENTRES

The first two international centres: the International Rice Research Institute (IRRI) in Los Baños (Philippines) and the International Maize and Wheat Improvement Centre (CIMMYT) in El Batan (Mexico), were established by the Rockefeller Foundation and the Ford Foundation in 1959 and 1966, respectively.

The considerable achievements of these two centres which culminated in the 'green revolution', contrasted so strongly with the relative lack of impact of the national research establishments, that a change of policy of the international organisation towards the latter occurred.

It was felt that the autonomous international centres, sheltered from the vagaries of national politics, able to attract highly competent professionals from all over the world, and enjoying ample financial backing, would be far more effective in contributing to a solution of the world food problem than the ineffectual national systems, plagued by limited human and financial resources and constantly changing political climates.

For these reasons, in the mid-1960s, international efforts were directed away from improving the research capacities of the LDCs themselves, and concentrated on establishing and reinforcing a network of international agricultural research institutes (Ruttan, 1982).

The Consultative Group on International Agricultural Research (CGIAR)

In 1971, the Consultative Group on International Agricultural Research was formed, an association of mainly donor agencies. The 40-strong

271

membership consists of 12 'donor' countries, UN agencies, a number of countries representing the developing countries, the Regional Development Banks, the Ford, Kellog and Rockefeller Foundations, and various International Development Agencies. The CGIAR is advised by a Consultative Technical Committee (TAC), comprising 13 scientists (Fig. 10).

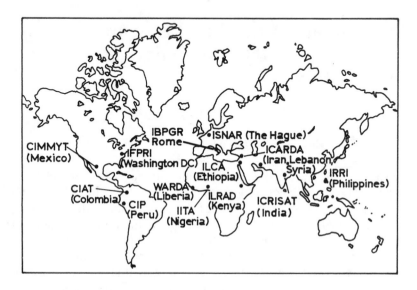

Fig. 10. The geographical location of the CGIAR-supported institutes.

The major objective of CGIAR is to use the funds provided by the members to ensure a co-ordinated programme of research by the international centres, and to support research in the LDCs in accordance with internationally agreed priorities.

In addition to the Centro Internacional Mejoramiento de Maize y Trigo (CIMMYT) in Mexico, and the International Rice Research Institute (IRRI) at Los Baños in the Philippines, other international centres have been established, whose activities will be described below.

The following services are normally provided by these international centres to the national agricultural programmes in the regions for which the respective centres are relevant (Hardin, 1979).

(1) Developing of and training in modern, problem-solving research procedures;

(2) generation of complete commodity 'production packages' with suggested procedures for adaptive testing and modification to fit local conditions;
(3) provision of genetic materials;
(4) international testing of materials and practices with associated data retrieval and analysis;
(5) consulting services;
(6) seminars and workshops;
(7) publications; and
(8) direct technical assistance through contracts for special out-reach projects.

The CGIAR system of research centres is conducting research on most of the important foods of the LDCs. All of the major cereals, which as a group provide 60 per cent of the energy and 55 per cent of the protein consumed by people in the LDCs, are included. Four centres work on pulses, and two on livestock (Sawyer, 1987).

In the majority of the centres the research is carried out by multidisciplinary teams. These teams are frequently organised around a single commodity or commodity system. The primary purpose of research on a single commodity is to develop a package of technology which should contribute to increased yields, reduced production risks, or improved quality. The research is mission-oriented, and in many cases plant breeding plays a major role (Pinstrup-Andersen, 1982).

Research programmes of the international centres

Planning a national agricultural research programme which will make the most effective use of the technology developed at the international centres requires an awareness by the national planners of the activities carried out by each of the centres.

International Rice Research Institute (IRRI)
Rice is the central concern of IRRI. Interdisciplinary teams of agronomists, physiologists, plant pathologists, entomologists and geneticists work together to produce high-yielding rice varieties and study the cultural methods required to achieve their potential. Top priority is given to plant protection in the search for genetic resistance to a wide range of diseases and insect pests. The institute maintains a genetic bank with over 30 000 accessions; in a typical year 10 000 seed samples are sent out in

response to requests from national programmes in different countries (CGIAR, 1978).

The institute has also developed extension and demonstration techniques and pioneered the concept of a 'package of practices'. The socio-economic and technological barriers preventing the small farmer from adopting the new technology are studied at 14 locations in six countries. Farming systems for a variety of environments with special emphasis on multiple cropping and focused on the small farmer, are being developed. Investigations on a pilot basis are carried out for increasing the efficiency of water use on small farms. Small-scale machinery is being designed with simple replacement parts that can be manufactured by local enterprises.

International Maize and Wheat Improvement Centre (CIMMYT)
Crop improvement of maize and wheat is the primary objective, complemented by the refinement of management technique. CIMMYT maintains a systematic international programme of 1429 trial nurseries in which 91 countries co-operate (CGIAR, 1978).

The maize programme has concentrated on developing plant populations that can either be used directly in co-operating countries (in which case they are released by the national programmes after screening and given local names) or incorporated into local synthetic varieties. Early work was aimed at developing distinct maize types for narrowly delimited growing environments: varieties for high and low altitudes, for early, intermediate and late maturity, resistance to disease and pests prevalent in given areas, the incorporation of a high-quality protein trait into tropical varieties. More recently work has concentrated on breeding dwarf varities *with wide adaptation* and a broad genetic background, incorporating multiple disease and pest resistance and high protein quality (high-lysine).

These dwarf varieties are relatively insensitive to day-length; they are characterised by low ear placement and upright leaves making possible a doubling of the usual population density.

The wheat programme produced the first high-yielding dwarf wheats in the early 1960s. Since then, experimental programmes have been intensified, aiming for greater stability of yield, broader disease and insect resistance, and adaptation to cold, drought and other stress conditions. A programme for improving durum wheats is under way. Barley and a triticale cross between wheat and rye are being improved along the same lines as wheat.

The International Institute for Tropical Agriculture (IITA)
This centre, established in Ibadan, Nigeria, concentrates on developing improved farming systems as alternatives to shifting cultivation. The work is mainly concerned with the major food crops of the humid tropics; cowpeas, yams and sweet potatoes, but also includes maize, pigeon peas, soybeans, cassava, rice and forage crops. The structure, chemistry and physics of tropical soils are being thoroughly studied. Soil management practices, including tillage and mulching techniques, have been designed for topsoil retention, moisture conservation and improved soil temperature regimes.

Weed control research is carried out to identify the most effective herbicides for root and tuber crops, rice and cowpeas.

IITA is also co-operating with IRRI and CIMMYT in developing rice and maize varieties more specifically adapted to ecological zones in Africa.

IITA has major world responsibility for cowpea improvement, the most widely grown legume in the humid tropics, and on a regional basis for several other food legumes that could be fitted into farming systems of the humid tropics: limabeans, African yam beans, pigeon peas and soybeans. A systematic collective effort aimed at assembling all available germplasm of indigenous legume species in western and central Africa is being undertaken.

Work is being done on the improvement of productivity, adaptation and food quality of the three major energy foods of the humid tropics: cassava, sweet potato and yams. In cassava, the main objectives are to achieve resistance to mosaic and bacterial blight; and to produce varieties with a low level of cyanide. In sweet potatoes, major objectives are resistance to sweet potato weevil, and a lowering of sugar content.

In both cassava and sweet potatoes, disease-free vegetative material for distribution to national programmes is being produced in a tissue-culture programme.

Centro Internacional de Agricultura Tropical (CIAT)
This centre was established in Colombia, with the objective of increasing agricultural productivity in the lowland tropics of Latin America. Research is being conducted on cassava, forage crops, beef cattle and pigs. CIAT also has a small rice research programme.

The establishment and organisation of germplasm banks for cassava, and subsequent screening and multiplication, provide a wide range of genetic material to plant breeders throughout the tropics.

A Small Farm Systems Programme has been established which is attempting to integrate biological, engineering and social science expertise to study the complexity of small farmers' problems and offer relevant alternatives within the current context on the farm which will best help the farmer increase production (Francis, 1979). The work is being carried out in stages as follows:

(1) An analysis of present traditional systems;
(2) the synthesis of prototypical systems;
(3) the design of improved technology;
(4) the validation of this new technology on the farm;
(5) implementation of this process in specific zones; and
(6) evaluation of the impact of the intervention on farm income and human nutrition.

International Crops Research Institute for the Semi-Arid Tropics (ICRISAT)

This centre was established in 1972, with headquarters at Hyderabad, India. In the initial stages, ICRISAT's research programme has three main objectives:

(1) To serve as a world centre to improve the genetic potential for grain yield and the nutritional quality of sorghum, pearl millet, pigeon peas, chick peas, and groundnuts.
(2) To develop farming systems which will help to increase and stabilise agricultural production through better use of natural and human resources in the seasonally dry, semi-arid tropics.
(3) To assist national and regional research programmes through co-operation and support and contributing further by sponsoring conferences, operating training programmes on an international basis, and assisting extension activities.

It is envisaged to test and evaluate the results of the research conducted at Hyderabad in the entire semi-arid tropical regions of the world, by an extension programme of co-operation with national research institutes. In the plant breeding programmes, for example, promising material is made available to the relevant national research institutions at early generation stages, so that local plant-breeders can make the evaluation and selection on the basis of their local experience and judgement. Therefore, the benefits of the crop improvement programme that can be obtained by different countries will depend on the competence of the national research centres (Cummings & Kanwar, 1974).

In Latin America, parts of Brazil and Argentina, and a substantial area in Mexico and Central America fall in the semi-arid tropical region to be served by ICRISAT, and it is envisaged that appropriate co-operation links with the national research institutes of these areas will be established.

Centro Internacional de la Papa (CIP)

The International Centre for Potato Research was established in Peru with the mandate of improving potatoes and potato growing in LDCs in general, and to expand the area of adaptation of this crop, including the lowland tropics.

Only a relatively small proportion of the resources available to the institute is invested in developing its headquarters facilities, and most funds and efforts are devoted to research carried out in a network of co-operating national and regional centres. Key problems investigated are late blight, resistance to bacterial wilt, to potato cyst nematode, to cold; and to improvement of potato quality.

A major effort is being made in the collection, classification, maintenance and distribution of all tuber-bearing *Solanum* species.

CIP has used contract linkages with institutions in developed countries for work on fundamental problems related to the CIP projects more extensively than any of the other centres (Ruttan, 1982).

International Livestock Centre for Africa (ILCA)

ILCA was established in 1974 in Ethiopia. The declared purpose of ILCA is 'to assist national efforts which aim to effect a change in the production and marketing systems in Tropical Africa so as to increase the total yield and marketing systems in Tropical Africa so as to increase the total yield and output of livestock products and improve the quality of life in organisational constraints to improved livestock production in the tropical countries south of the Sahara. It is planned to co-ordinate the work of animal scientists, ecologists, sociologists, economists and others in the investigation of the man–animal–environment complex. Associated projects will be carried out by co-operating national research centres in various African countries.

Animal research will include genetics and physiology, nutrition, reproduction, adaptation to climate and epidemiology. Environmental studies will involve climate, soils, ecology of range lands and pasture improvement. Socio-economic research includes the geography of livestock raising; the social structures, cultural attitudes and value systems of pastoral societies, economic analyses of livestock operations, marketing systems, and the relationships between producers and consumers. Data from these studies are used in the development of models for different types of range management.

International Board of Plant Genetic Resources (IBPGR)
In 1974, the International Board of Plant Genetic Resources was established under CGIAR auspices, 'to encourage and assist in the collection, preservation and exchange of plant genetic materials' (Ruttan, 1985).

The National Seed Laboratory at Fort Collins and the N.I.Vavilov Institute in the USSR have large collections which conserve a considerable proportion of the genetic variety of many crops.

IITA in Nigeria, and IBRI in the Philippines have large collections of genetic resources of the crops included in their breeding programmes.

Most of these collections consist exclusively of seeds; no solution has yet been found for long-term storing of genetic material of crops propagated vegetatively, such as yams and cassava.

International Food Policy Research Institute (IFPRI)
IFPRI was established in 1975 in Washington, DC and formally incorporated into the International Agricultural Research System. This institution is devoted to research on agricultural production, price and trade policy issues.

Impact on national research

The steep increase in funds made available to the International centres—from US$ 1·1 million in 1965, to US$ 120 million in 1980—is an indication of the high regard with which these institutions were considered by their sponsors.

The international centres have given an entirely new dimension to the national research programmes of the LDCs, which have to contend with limited financial resources and a lack of trained personnel, and cannot possibly, on their own, afford the comprehensive research required to solve all the problems faced by their agriculture.

By contrast, the considerable resources, excellent equipment and above all the interdisciplinary teams of high-level scientists attracted to these centres make possible work on a scale and level that cannot possibly be matched on a national scale, except by the richest countries.

By providing consulting services to governments of LDCs and assigning staff members as residential specialists, the international centres help to plan, organise and implement local programmes. By taking advantage of these services, the national research systems of the LDCs can obtain valuable back-stopping for their own programmes. They get access, rapidly and without undue expense, to much of the basic information that they lack, to technology that is relevant to their needs, to seeds of improved varieties for screening at their regional stations and to genetic material for their breeding work. By orienting the national programmes to adapting the findings from the work of the international centres to their own conditions, so that they can be applied at farm level, rapid progress can be made on many problems with relatively modest resources and a modest research system. In the plant-breeding programmes, for example, promising material from the international centres is made available to the relevant national research institutions at early generation stages, so that local plant-breeders can make the evaluation and selection on the basis of their local experience and judgement. Therefore, the benefits of the crop-improvement programme that can be obtained by different countries will depend on the competence of the national research centres (Cummings and Kanwar, 1974). 'The more successful are the International Centres, the more productive is the national research work' (Kislev, 1977).

During the 1970s more than 130 developing countries joined in international networks for testing experimental crop varieties. By 1979, the wheat network involved 115 collaborating countries, the maize network 84, the rice network over 60, and other groups of countries were testing varieties of sorghum, millet, potato, cassava and pulses. Each network created links among the participating developing countries and with the international centres (Hanson, 1979).

However, the ability to evaluate and adapt the technology provided by the international centres depends on the existence of a well-planned, effective national research system in each country, staffed by trained and committed researchers.

Consequently, a shift in the emphasis of the international organisations has again taken place in the late 1970s. It is now generally accepted that 'continued effectiveness of the international centres requires both suitable mechanisms to transfer the newly created technology to the LDCs *and* strong national research systems to absorb and complement the centres' work' (Venezian, 1981).

This renewed concern with strengthening national agricultural research systems has led to the establishment of specialised institutions such as the International Service for National Agricultural Research (ISNAR) (cf. p. 289) and the International Agricultural Development Service (IADS) (cf. p. 295).

The activities of the international agricultural research institutes are expanding; they are becoming primary centres for genetic resources conservation in their respective fields; they are expanding their research into cropping and farming systems; developing research programmes in pre- and post-harvest technology and are placing greater emphasis on small-farm technology.

In recent years the tendency is for the international institutes to complete the evolution from single-commodity improvement centres to broad-based rural development centres (Ruttan, 1985).

Considerable attention is being given to the problems of the farmer with scarce resources, whose interests had been largely neglected in the past by the NARS (Pinstrup-Andersen, 1982).

The international centres have established 'outreach' programmes with the intention of reinforcing their co-operation with the NARSs. Scientists are assigned to different countries, with the following objectives (Ampuero, 1981).

- to obtain genetic material and collect information on national and regional problems that can serve to orient the centres;
- to carry out certain kinds of research (e.g. farming systems research) which require specific environmental conditions;
- to evaluate genetic material in areas with a high incidence of plant disease;
- to validate the transfer technologies developed at the centres;
- to promote regional co-operation.

Deficiencies in the relationships between the IARCs and NARSs

Farrington & Howell (1987) underline some of the major deficiencies in the relations between the IARCs and the NARSs. These relations 'bear similarities between those of the NARS and the farmers: whilst nominally client-focused, in practice, the IARCs' research agenda is inherently biased by top-down factors such as the availability of donor funding for research into specific areas, and the spheres of interest of IARC scientists themselves'. These authors also state that there is a bias in favour of the larger LDCs, whilst it is precisely the smaller countries which are at least able to finance the 'critical minimum mass of effort necessary for any research other than the most directly adaptive'.

There is also the real danger that government leaders will see the IARCs as satisfactory alternatives to the NARS, and thereby weaken their support for their own national research effort.

It must be stressed that the efforts of the IARCs will have only marginal effects on the NARSs, unless the authorities in the LDCs themselves make an all-out effort to this end, by insulating their research systems from the vagaries of political change, providing stable financial support, and promoting the professional development of their own scientific and technical staff.

Future developments

Ruttan (1985) points out that when the system of IARCs was established in the 1960s, it was generally assumed that after several decades, the management and support of the centres would be transferred to the host countries. Ruttan considers this assumption to be unrealistic, and that the system should be viewed as a permanent component of the global agricultural support system, though this does not preclude the phasing out of certain programmes and the initiation of new ones.

The emergence of new natural resource and basic science research centres would require an appropriate governance system, and Ruttan (1985) suggests the establishment of new supervising bodies, such as a Consultative Group for Natural Resources Research and a Consultative Group of Biological Sciences for Tropical Agriculture.

There will be an increasing demand that the international centres concern themselves more with the difficult areas and undertake breeding work on hitherto neglected crops.

Greater efforts are expected in co-ordinating the work of the centres with the national research programmes. 'Such a co-ordination constitutes the best possible solution to the real needs of these countries and the most effective way to overcome existing constraints' (Trigo *et al.*, 1982).

OTHER INTERNATIONAL CENTRES

In almost all the CGIAR-supported centres, research is limited to food commodities. A number of international research institutes have been established, outside the CGIAR system, in areas not covered by the latter.

These include (Ruttan, 1985): an International Fertilizer Development Centre (Alabama, USA) for research on resource development; an International Irrigation Management Organisation (IIMI) in Sri Lanka; the United Nations Industrial Development Organisation (UNIDO) which has sponsored exploratory studies leading to the establishment of an International Centre for Genetic Engineering and Bio-Technology (ICGER); ICIPE (Nairobi) for research on insect physiology and ecology; AVRDC (Taiwan) for research on tropical vegetables; ICLARM (Philippines) for research on living aquatic resources; INTSOY (Illinois, USA) for soybean research; ICRAF (Kenya) for agroforestry systems.

EXTERNAL ASSISTANCE FOR NATIONAL RESEARCH

Aid required by developing countries

External assistance to strengthen the NARSs of the developing countries must take into account the size of a country's agricultural sector and the current state of its research system.

The World Bank (1981) estimates that about 10 per cent of the LDCs already have adequate research skills, good national programmes, and effective links with the IARCs. These countries may even be able to provide assistance to other LDCs.

Another 10 per cent of the LDCs have adequate research expertise, but the system is poorly organised and managed. For this group of countries, the main need is for assistance in improving research organisation and management.

Nearly 50 per cent of the LDCs are large enough to support a NARS, but lack the essential research infrastructure and trained manpower. The needs of these countries are to develop an effective NARS, to acquire adequate research facilities and to train a sufficient number of researchers and technicians.

Forms of aid provided to developing countries

The aid being provided by various countries to the developing countries in the field of agricultural research usually follows the same general lines: provision of staff, equipment and financial aid to research centres in the developing countries; research carried out in specialised institutes of the donor countries, aimed at solving problems of significance to the recipient countries; provision of fellowships for special courses and organising seminars for the training of research workers from the developing countries at all levels.

In many of the countries involved in giving aid to developing countries, special units or bodies are responsible for this activity: In the United Kingdom there is a special Ministry for Overseas Development; in France, responsibility is vested in the State Secretariat for Foreign Affairs, with a special secretariat for co-operation with African countries south of the Sahara; in Belgium, Netherlands, Denmark and Israel, special units function within the Ministries of Foreign Affairs (in Israel they work in close co-operation with the Ministry of Agriculture); in the USA responsibility is vested in the Agency for International Development (AID); in Canada, the International Development Research Council, etc.

The provision of aid to *independent* countries for the improvement of their agricultural research is a relatively new concept. It is therefore not surprising that no standard methods for providing this aid in the most efficient manner possible, without impinging on the sovereignity of the recipient countries, has yet been devised. France and the UK, the two countries whose former colonial empires have largely achieved independence, have naturally attempted to find formulas acceptable to the newly independent countries for transferring the research efforts previously carried out by them in their former colonies. The solutions adopted by each of these donor countries are entirely different from one another.

Much of the assistance provided by bilateral donors supports research programmes in certain commodities or in a particular discipline. About one-third of the support provided by major bilateral donors goes to strengthen NARs or IARCs, while two-thirds are dedicated to research in selected disciplines. France is an exception; much of its external support for research, particularly to francophone Africa, goes to basic research and commodity-oriented research. An important part of this support takes the form of paying the cost of French researchers who work in the experiment stations in the former colonies (World Bank, 1981).

Organisation and implementation of aid to LDCs

The needs of the developing countries are so great, and the means needed to provide assistance so considerable, that much thought must be given to the planning, co-ordination and execution of this aid.

At a meeting of an *ad hoc* group on measures for close co-operation in the provision of aid in agricultural research to African countries, organised by FAO in 1966, the following recommendations were made (FAO, 1966).

Planning
In an effort to determine priorities which would make it possible to channel aid into the areas of greatest need, a general survey of the agricultural research requirements of the developing countries should be carried out by an international organisation such as FAO.

International agencies and donor countries should give assistance in agricultural development plans and the preparation of agricultural research programmes adjusted to the development plans. Objectives and programmes should be established on a sufficiently long-term basis to ensure sufficient continuity and stability of the research work.

Organisation
Co-ordination between donor countries. The ideal would be for an international organisation such as FAO to take the initiative in co-ordinating aid given by the developed countries. There are, however, considerable political difficulties which have to be overcome in order to achieve this aim. FAO could, at least, serve as a clearing house for supplying information on the work already done in the field, to countries contemplating an aid programme.

There are cases of co-operation between donor countries, in specific research programmes, such as between Belgium and the Netherlands in Burundi, but they are rare. Co-ordination between donor countries is particularly pertinent when the aid given can serve a whole region and not only a single country.

Implementation
Training of research workers. First priority should be given to establishing or strengthening the teaching and research requirements of agricultural education at all levels.

Provision of fellowships for post-graduate training appears to be adequate as far as numbers are concerned. However, there is a need to ensure that the training is related to the requirements of the developing countries. It is desirable to give priority to candidates who have had a few years' experience in research in their own country, and who undertake to return home and work in research for at least 3–5 years after the completion of their fellowship.

Provision of experienced research workers from developed countries. Recruiting of staff for research work in the developing countries is becoming increasingly difficult. In addition to the general shortage of trained research personnel, it is difficult to find the people with the necessary dedication who are prepared to work under conditions which are infinitely more difficult than those at home, on assignments which by their very nature are short-term and do not provide for a life-time career.

Though these general recommendations are still as valid as when they were formulated, they have mostly been either ignored, or only partly implemented by the donor countries.

Several governments have however recognised the need to make arrangements specially adjusted to this situation and designed to overcome the inherent difficulties. Probably the most practical and promising of the approaches is to create 'pools' of research workers in different fields, by expanding the establishment of appropriate organisations in the donor country, so that a number of workers are always available for work in developing countries.

There is an increasing tendency to provide research teams to research institutions in developing countries, rather than individual research workers. This approach helps solve many of the social and technical problems that burden the lone expatriate scientist.

Another approach is the creation of a corps of specialists in the direct employ of the ministry concerned with international co-operation.

A very promising initiative has been taken in the establishment of partnership arrangements between individual research institutions in developed countries and those in the developing countries. Research workers from the sponsoring country can carry out research assignments for shorter or longer periods, initiate research projects, and train replacements from native staff to carry on after their departure, without being in danger of losing their positions or status in their home institution. The direct contacts between the foreign specialists and their colleagues at home, and after their departure, between the specialists and their native replacements, adds a personal dimension that enhances the level of the research and makes it far more effective.

CIP has taken the initiative of relocating a number of headquarter scientists to the regions where a problem can be addressed best in association with national scientists. Other headquarter scientists have taken sabbaticals within developing country programmes (Sawyer, 1987).

Provision of research facilities in the donor countries

Frequently, the most practical *short-term* solution for certain problems, especially of a fundamental nature, which require highly specialised staff or very expensive equipment, is to carry out the necessary investigation in the donor country itself. Examples of this approach have been given above. Even countries that have no direct experience in tropical research can provide valuable aid by making available laboratory, computer and library services.

Financing research in the recipient countries '

Financing fundamental and applied research in the developing countries on subjects of mutual interest to donor and recipient countries, following the example of the research sponsored by Public Law 480 of the USA, can strengthen the research infrastructure of these countries, provide opportunity and challenge to senior research workers, training for young research workers, and at the same time provide solutions to problems of scientific and practical importance.

Toulmin (1966) writes that 'farming out research problems, to be worked on in developing countries, will be one of the cheapest, most effective and mutually beneficial ways in which the advanced nations can contribute to the development of poorer countries'.

Inventory of research results already available

In many of the developing countries there is a wealth of data resulting from research work carried out over the years by expatriate staff. During the interregnum that occurred after the departure of the expatriate staff, and before it is found possible to continue the work by trained nationals, much of the information in the files and station records may be lost. Even for published data, a review of that which is pertinent, its classification, indexing and, in particular, interpretation, is essential if valuable information is not to be lost and is to remain available to future research workers. This work could well be carried out by qualified workers from developed countries prepared to accept a one- to two-year assignment for this purpose.

Drawbacks of external assistance

External assistance 'provides a relatively easy alternative to the development of internal political support for the national research system, causing cycles of development and erosion inherent in the traditional external aid approach' (Ruttan, 1985). Most existing aid systems 'therefore have built-in incentives for NARS' leadership to direct entrepreneurial efforts towards the donor community rather than toward the domestic political system (Ruttan, 1985).

This results in the 'cycles of development and erosion', corresponding to the funding of projects initiated by foreign aid, and the termination of the foreign assistance, when the national government is unwilling or unable to provide continuing support.

Gamble (1984) adds another major drawback. He writes:

Each aid agency (multilateral, regional, bilateral, or private) that provides such assistance has its own priorities, capabilities, and modes of operation. These may, or may not, mesh well with the needs, priorities, procedures, and absorptive capacities of recipient countries. Donor countries may use foreign aid as an instrument of national foreign policy and diplomacy, an area in which goals are not necessarily compatible with those of development assistance itself.

Gamble (1984) concludes that foreign assistance has been subject to the ups and downs of political fortune and to economic cycles in donor countries. In the LDCs, priorities have often been determined by the availability of foreign aid rather than by the real needs of the country.

SOURCES FOR AID TO AGRICULTURAL RESEARCH IN DEVELOPING COUNTRIES

The major sources for aid to agricultural research in the developing countries are international organisations, national foreign aid agencies, private foundations and institutions.

International organisations

Many international organisations provide support to the agricultural research in the developing countries, either within an overall framework of encouraging development or exclusively concerned with strengthening research. These include:

- World-wide organisations, such as the specialised institutions of the UN: FAO, UNDP, ISNAR;
- multi-government organisations, such as the World Bank, the Interamerican Development Bank, the Instituto Interamericano de Cooperación en la Agricultura;
- Regional Associations, such as CATIE;
- The International Research Centres.

Closely tied with one or more of these organisations are various international programmes and campaigns, such as the World Food Program, Freedom from Hunger Campaign and the Interamerican Committee for Agricultural Development. Sometimes, large-scale trials and surveys are carried out in the framework of these trials, such as the fertiliser trials on major food crops initiated by FAO.

International organisations have made considerable contributions to the building and strengthening of national research institutions, and their potential is far from exhausted.

This support has been in the form of expert advice on the organisation of agricultural research systems, provision of long-term visiting scientists, construction and equipment, libraries, fellowships, investment and operating funds.

The Food and Agriculture Organisation of the United Nations
The first world body to deal with the general problems of agriculture was the International Institute of Agriculture, established in Rome in 1905. It was to serve as a clearinghouse of world agricultural information, by assuming responsibility for the collection, analysis, publication and distribution of information on the economic and technical aspects of crop and livestock production in the world. In 1945, after World War II, it was replaced by the Food and Agriculture Organisation of the United Nations (FAO), which took over its functions and assets. FAO is concerned with a broad spectrum of agricultural problems: land and water use, plant production, animal husbandry, fisheries, forestry, rural institutions and services, the use of atomic energy in agriculture, agricultural economics and statistics, human nutrition.

Whilst as a major international agricultural organisation, FAO is mainly concerned with the application of improved agricultural practices, particularly in the developing countries, it is also concerned with assisting member countries in the organisation and development of agricultural research.

FAO takes the initiative for organising international and regional forums for the discussion of scientific, technical and economic problems. An example is the Panel of Experts on organisation and administration of agricultural development, research and extension, organised in 1965, to review policies and working methods, and to discuss appropriate systems of research, extension and development organisation, with particular reference to developing countries (FAO, 1965).

The FAO has also been active in initiating working parties on various subjects and promoting international co-operation in these fields. Examples are the working groups of hybrid-maize breeders in Europe, the co-operative work on the improvement of grasslands in the Mediterranean region and on wheat breeding in the Near East, etc.

In 1977, FAO provided support to 58 projects designed to establish or strengthen national research institutions. Many of these projects were funded by the United Nations Development Programme (UNDP), with the FAO providing the technical personnel (World Bank, 1981).

International Service for National Agricultural Research (ISNAR)
ISNAR was established by the Consultative Group on International Agricultural Research (CGIAR), for the purpose of assisting governments of developing countries to strengthen their agricultural research. ISNAR began operating at its headquarters in the Hague, Netherlands, in 1980. It is a non-profit, autonomous agency, international in character, and non-political in management, staffing, and operations. It is the only international service which focuses primarily on national agricultural research issues in the developing countries, the ultimate goal being 'to enable these countries to plan, organise, manage and execute research more effectively from their own human, natural and financial resources' (ISNAR, 1982*b*).

ISNAR works in close co-operation with all international organisations, complementing but not competing with the work of other programmes and sources of technical assistance related to agricultural research.

ISNAR gives emphasis to the generation, introduction and use of adapted technology suitable to resource-poor farmers and local farming conditions.

Up to the present, ISNAR has been mainly active in African and Asian countries. It may be assumed that Latin American countries have not yet shown sufficient awareness of the aid potential of this new agency. Whilst 'ISNAR's services are available to any developing country, ISNAR will provide assistance to a country only at the country's request' (ISNAR, 1982*b*).

If such a request is forthcoming, ISNAR is mandated to provide or arrange for assistance to developing countries in:

- identifying needs for planning and carrying out agricultural research;
- determining research priorities;
- formulating overall research policies and strategies;
- elaborating programmes of action including specific projects for external financing;
- designing necessary institutional and organisational arrangements for carrying out research programmes and projects;
- identifying the necessary resources for the execution of such activities;
- contacting potential external sources for financing research programmes and projects;
- promoting effective links between research organisations, extension services and the farming community;
- determining the basic facilities required to conduct research (laboratories, equipment, experiment stations, adequate staffing, finance, etc);
- establishing and strengthening links to existing information systems in order to speed up exchange of information on research results, ongoing research, and training opportunities at international and national institutions;
- organising appropriate flows of information within a geographic region so that interested countries may arrange to co-operate on specific research efforts;
- arranging training programmes for research and research support staff;
- organising and supporting symposia and seminars for the interchange and dissemination of ideas and information useful in the development and operation of national research systems;
- evaluating the effectiveness and suitability of various forms of research organisation and activity.

After ISNAR accepts a request from a government for its aid, the approach and procedures to be adopted are determined and mutually agreed upon. Normally, the first stage is a review of the existing system by a team consisting of ISNAR staff and consultants.

Recommendations are then formulated 'on issues such as structuring, programming, financing, staffing and linking of the research system to policy-makers, extension services, producers and other groups' (ISNAR, 1982*b*).

Subsequently, ISNAR provides the links between the national system investigated and the international agricultural research centres.

The World Bank

In the early years after its inception, the World Bank concentrated on infrastructure projects—transportation, ports, power, etc. Recently, however, there has been a growing awareness of the need to increase the productivity of agriculture in the developing countries as an essential prelude to their overall development. As a result of this change in policy, the World Bank has become the largest single multilateral financing agency supporting NARSs. The World Bank's cumulative lending for research and extension, as of June 1982, amounted to US$ 1126·3 million.

World Bank support for agricultural research is provided in three ways: for research components within agricultural development projects; for projects solely in support of NARSs and extension; and for research activities at agricultural universities, through educational projects (CGIAR, 1978).

For this purpose, the Bank examines the adequacy of the national research effort of a country requesting aid, in terms of the size, scope, research philosophy, and priority accorded by the government concerned.

The financial resources of the Bank are then combined with the expertise of other bodies with whom the Bank works closely: ISNAR, FAO, UNDP, and other multilateral and bilateral donors, as well as the international agricultural research centres.

When a developing country is prepared to follow mutually acceptable research strategies and related national policies, the Bank is prepared to consider providing continued technical and financial support for a programme of 10–15 years' duration.

Such a programme is envisaged as comprising several phases. The first phase 'would be largely a technical assistance project, involving assistance for national research planning, preparation, staff recruiting and training, and the maintenance of current work. This phase would prepare plans for future research and extension activities, construction designs, and feasibility studies that would be required to implement a second phase project. A third-phase project would follow if it was needed' (World Bank, 1981).

The World Bank has recognised that special consideration needs to be given to the arrangements by which research projects are financed in the low-income countries. For this purpose, the International Development Association (IDA) was established in 1960, as an integral part of the World Bank, to serve as a channel for assistance to countries, which could not afford to accept loans from the World Bank under the same conditions as the better-off developing countries (World Bank, 1982).

The various regional banks, such as the Interamerican Bank of Development, have the same functions and act according to the same principles as the World Bank, for their respective areas of influence.

International associations (IAs)

A significant development in world agriculture since the 1970s is the establishment of international associations (IAs), whose objectives involve agricultural research (Ruttan, 1985).

There are three groups of IAs: those that perform agricultural research themselves, such as the Centro Agronómico Tropical de Investigación y Enseñanza (CATIE); those concerned with promotion and co-ordination of agricultural research, such as the International Federation of Agricultural Research Systems for Development (IFARD); and others providing support for NARSs, such as the Inter-American Agricultural Information Service (AGRINTER).

There are some 50 international associations operating in different regions. Some are world-wide in scope, some are limited to specific regions.

Examples of IAs with global operations are IFARD, already mentioned, and the International Association of Agricultural Economists (IAAE).

The regional associations encompass a variety of international bodies, that originate in the countries themselves, and are generally located in one of the participating countries.

The greatest concentration of these associations is found in Latin America. The reasons attributed by Venezian (1981) for this concentration are: 'the urgency of common problems, country size, common language and geographic proximity'.

Venezian (1981) gives a detailed list of regional associations in Latin America, Africa, the Middle East, Asia and the Pacific, to which the reader interested in this particular subject is referred.

Since IAs have been first established, they have, to varying degrees, contributed to the development of agriculture of the countries within their sphere of operations, in providing training, carrying out research, technical assistance, diffusion of information, co-ordination and planning (Ruttan, 1985).

Foundations as a factor in international aid programmes

Private foundations, such as the Rockefeller Foundation and the Ford Foundation, have made striking contributions to the agricultural development of certain countries, by supporting agricultural research. They have the advantage of considerable flexibility in action as compared to official sources of aid.

An outstanding example is the technical assistance provided by the Rockefeller Foundation to Mexico, at the request of that Government (Richardson, 1964). Initiated in 1943, the co-operative programme was originally aimed at improving the quality and quantity of the basic foods of Mexico. Initially, maize and wheat breeding projects were undertaken, in view of the considerable chronic shortage of these basic foods. Substantial increases in maize production were achieved, which enabled the diversion of a large proportion of this crop for feed grain. A feed manufacturing industry was established, and thriving poultry and pig production made possible. Research on sorghum grain made possible a more than four-fold increase in yields in five years. Striking successes were also achieved in breeding rust- and lodging-resistant semi-dwarf varieties of wheat, which were grown in 75 per cent of the cultivated wheat area. National average yields were increased from 7½ bushels per acre in 1945 to 32 bushels per acre in 1963.

As the breeding programme developed, projects were added on soil fertility, bean improvement, plant pathology, entomology, potatoes, vegetables, sorghum, forage legumes and grasses, soybeans and agricultural economics.

The agricultural research programme made it possible to achieve diversification of agricultural production and to reduce the conomic dependence on export crops, with its inherent dangers. A high degree of self-sufficiency in foodstuffs was also achieved.

Mexico is an outstanding example of success in improving agricultural productivity. During the decade 1954–64, national and agricultural output rose about 7 per cent a year as compared with a population growth of about 3 per cent. The labour force in agriculture has continued to increase but has declined in relation to the total labour force. Net migration out of agriculture has averaged about 300 000 annually in the 1950–60 period (Christanssen & Yee, 1964).

All these achievements were accomplished mainly through research resulting in the development of improved crop varieties, the adoption of improved methods of cultivation, pest and disease control, the large-scale use of fertilisers, and increased areas under irrigation.

However, the greatest and the most lasting benefit from the foreign assistance programme was the development of *national competence* to continue and expand agricultural research. From the beginning, the programme was based on co-operation between foundation field staff and Mexican agronomists. The maximum number of foundation staff never exceeded 18, whilst more than 100 Mexican scientists were directly associated in programme research projects, out of a total of over 700 who received in-service training in the course of 20 years. As local competence and leadership developed, the number of foundation personnel was reduced, whilst Mexican scientists assumed complete responsibility for most of the programme. Many of the young scientists trained by the programme assumed major responsibilities in agricultural extension and in seed production enterprises. More than 125 became staff members of Mexico's National Institute of Agricultural Research and of Livestock Research. Others have assumed positions in teaching, research and extension in other institutions throughout the country.

Concurrent with the operating of research and scholarship programmes, a series of grants assisted over the years in developing institutions and enabling outstanding scientists to study abroad.

With the reduction of the foundation staff in Mexico, a gradual shift in emphasis of the foundation's effort was made towards the international application of research results from the Mexican programme.

Satellite programmes for research on maize were carried out in Thailand and in Nigeria, with extensions into Senegal, Liberia, and the United Arab Republic. These and other programmes, though they were administered separately, they remained conceptually part of the centre in Mexico, which evolved in form and function so as to cope with the new responsibilities (Rockefeller Foundation, 1965).

Finally, it became one of the first International Agricultural Research Centres under the aegis of CGIAR.

The International Rice Research Institute is another example of a research centre established by private foundations. In view of the fact that half of the world population depends on rice as a staple food, the Rockefeller and Ford Foundations decided jointly to establish a rice research and training centre in southeastern Asia. A site was chosen in the Philippines, whose government co-operated by supplying land for laboratories, housing and experimental fields in proximity to the College of Agriculture of the University of the Philippines (Moseman & Hill, 1964). The institute was established as a registered, non-profit Philippine corporation with an international board of trustees.

Some 400 persons, more than 90 per cent of whom are Filipinos, worked at the institute. The professional staff included scientists of seven nationalities. The work was concerned with basic research on the rice plant, and all phases of rice production, management, distribution and utilisation. It included research in plant breeding, genetics, chemistry, agronomy, soil chemistry, soil microbiology, agricultural economics, agricultural engineering, communication and statistics. The institute promoted and co-ordinated a growing number of co-operative research projects in other countries (Rockefeller Foundation, 1965).

The institute also gave advanced training to young researchers, mainly from south and southeastern Asia, through a resident training programme. It operated a library which served as a world centre of information on rice.

The International Rice Research Institute was a partner of CIMMYT in initiating the 'green revolution', and became an important member of the CGIAR group of International Research Centres.

The International Agricultural Development Service (IADS) was established in 1975, with funding provided by the Rockefeller Foundation 'to provide contract research management and development services to NARSs'.

IADS offers assistance in agricultural planning, strengthening the NARS, training personnel, and implementing production programmes. Through meetings and publications, IADS also contributes to an understanding of development programmes (IADS, 1982).

Certain universities in developed countries have established a special relationship with counterparts in the LDCs. For Latin American and Asian countries, it is the USA universities that have been the most active, and have often attempted to 'sell' the land grant college model (cf. p. 236–9).

NATIONAL FOREIGN AID AGENCIES

Bilateral assistance has been implemented in a number of ways: Great Britain, France, and the Netherlands have continued to maintain the research institutes they established in their respective countries during the colonial period, which now address themselves to the problems of the countries in the sub-tropics and tropics. These activities have led to a dearth of international support for developing the basic research in the LDCs themselves.

Ruttan (1985) argues correctly that 'intellectual commitment to the solution of scientific problems is enhanced when scientists working on these problems are located in the environment in which the problems exist. Basic research capacity in the tropics will also facilitate more effective dialogue with the research community of the developed countries.'

Although most LDCs will continue to rely for some years on developed countries 'for research leading to a better understanding of basic processes this does not imply a continuing scientific and technical neocolonialism' (FAO, 1972). The needs of the LDCs for basic research must be identified, and resource allocations will have to be made, sooner or later, to meet these needs (FAO, 1972), so that in the course of time, the LDCs should be fully capable of undertaking the basic research they require.

Many other countries, in addition to those mentioned above, have been active in providing aid to the agriculture of the LDCs in general, and to the strengthening of the NARSs in particular. The USA, Sweden, West Germany, Canada, Australia, Israel and others have participated in bilateral agreements.

The largest single source of bilateral aid was from the USA, which contributed about 40 per cent of the total bilateral aid to the NARSs of the LDCs, followed by France (19 per cent), Canada (12 per cent), and West Germany (9 per cent) (Oram, 1982).

The major support of bilateral agencies has been in promoting collaboration between the NARS of the donor and of the recipient countries.

CO-OPERATION BETWEEN DEVELOPING COUNTRIES

Research efficiency could be enhanced by research linkages among countries situated in the same agro-ecological zone.

Co-operation between NARSs in the LDCs is still very limited. Even between individual researchers, communications are generally very poor; 'a researcher is more likely to know more about research in a European rather than in a neighbouring country' (Qasem, 1985).

Many attempts have been made to ensure inter-state co-operation in jointly maintaining commodity research institutes of a regional nature, established by colonial powers, such as the Cocoa Institute in Ghana, the Oil-Palm Institute in Nigeria, and the Rice Institute in Sierra Leone, and which, after independence, were to serve the needs of their respective countries. These attempts at international co-operation in agricultural research have generally disintegrated as a result of political forces. There was also the perception that the host country benefited out of proportion to its financial inputs in a common enterprise. All the regional research institutes, mentioned above, whose research cut across national boundaries, have by now been nationalised.

In view of the lack of success of eco-zone institutes, the concept of *regional programmes* has been proposed by Devred (1966). This concept is based on the fact that rarely do political boundaries coincide with broad ecological zones. Countries that are situated within the same ecological zone could, by pooling their efforts in solving common problems, make more efficient use of their limited resources and personnel. The steps involved, proposed by Devred, were:

(a) defining the ecological zones of the continents in which developing countries are situated. This has already been done by Devred for Africa;

(b) preparing an inventory of research results already available for each ecological zone;

(c) each group of countries within an ecological zone would define the principal problems of the region and decide on priorities.

A co-operative research programme would then be established, its execution based on the existing national research institutions, each of which would make its contribution in carrying out the programme, according to a mutually agreed upon plan. Each country would, of course, be free to carry out its own national research programme, according to its own needs and priorities, in addition to its contribution to the regional programme.

AGRICULTURAL RESEARCH NETWORKS

An approach that is well suited to problems of common interest to a large number of countries (including LDCs, developed countries and international centres), which may not require a large central staff or much sophisticated equipment, is the *research network*.

Such a network requires a central board of governance and a small core of staff to ensure continuity, to plan and to co-ordinate activities. Some research might be centralised, such as on methodology, information and training activities, and special equipment such as a computer (FAO, 1972).

Research networks can be used to promote regional, agro-ecological, or commodity-oriented research of an applied nature, preferably with links to an international centre working in the same field.

Networks are also particularly suited to supporting activities covering several countries. These could include surveys, information collection, storage and retrieval systems and the collection, conservation, storage and exchange of genetic materials.

During the 1970s more than 130 developing countries joined in international networks for testing experimental crop varieties. By 1979, the wheat network involved 115 collaborating countries, the maize network 84, the rice network over 60, and other groups of countries were testing varieties of sorghum, millet, potato, cassava, and pulses.

Each network created links among the participating developing countries and with the international centres (Hanson, 1979).

Network goals vary considerably, from the distribution and testing of germplasm, to a pooling of research resources among nearby countries with similar problems. They all have certain characteristics in common (Nogueira, 1987): they arise from a mutual agreement among participants; there is collaboration and complementation among participants; a central core co-ordinates contributions and manages overall interactions; resources are assigned specifically to finance activities. These mechanisms transcend their member organisations, and they differ from unscheduled interactions between centres and institutes, however frequent and productive these may be.

History (Plucknett & Smith, 1984)

Whilst there has been considerable activity in networks in recent years the concept is far from new. During the colonial period, the European powers each established a number of research institutions, mainly for investigating export crops, which shared the information obtained, but only among compatriots. In the 1920s, the USDA established a regional maize network in the Mid-West in which the participants exchanged ideas and information on maize hybrids, so as to avoid duplications; similar networks were formed among winter wheat breeders in the northern Great Plains and in the Northern Central Region.

In the 1950s, an international network was established following the outbreak of stem rust strain 15B epidemic in the US; as the problem might have hemispheric consequences, Argentina, Canada, Chile, Colombia, Ecuador, Mexico and the USA collaborated in setting up nurseries to identify sources of resistance to race 15B. The success of this common effort prompted about 100 countries to join the wheat network.

At about the same time, various kinds of networks were established in India. These became the All-India Co-ordinated Crop Improvement Programmes. These networks were for maize, millet, sorghum, rice, wheat and other crops.

By the 1980s, networks of different kinds had become commonplace. FAO, for example, reports that there are more than 100 international agricultural research networks (Winkelmann, 1987).

Advantages

The major advantage of the networks is the efficiency achieved in the use of resources, by avoiding duplication, attainment of critical mass, and exchange of ideas. For donors, networks can be an efficient framework for supplementing national support for research (Winkelmann, 1987).

Networks are also suitable vehicles for collective interaction of NARSs with the international centres. As the networks become stronger, they may assume responsibility for certain regional programmes currently funded or managed by the international centres (Piñeiro, 1987).

Conditions for success

Winkelmann (1987) mentions the following prerequisites for success: a clearly defined common goal or problem; strong self-interest of the participants in an important problem; strong and effective leadership; resource commitments by the participants; access to outside funding and an effective advisory group.

To these conditions for success, Bonilla & Cubillos (1987) add: 'Homogeneity of the scientific and technological competence of participant programmes'. They mention that owing to the large number of participants in some FAO networks, 'a common language to speak about the same problem is missing'.

Disadvantages

In a large measure, the disadvantages experienced by the networks are the result of the complexity of managing them. With many participants of uneven experience and with different levels of commitment, co-ordination and direction become difficult. Also, it is costly to bring participants together (Winkelmann, 1987).

Kinds of networks

SPAAR (1986) identifies three major categories of networks. The first aims at facilitating information exchange; it requires little integration, and consists mainly in a sharing of ideas, methods, results, and germplasm. The second adds to the information exchange by the organisation of meetings at which participants exchange ideas directly. The third adds to the two former activities, the joint setting of priorities, planning, implementing, and monitoring of defined undertakings, probably with some assignment of important tasks.

Each step involves a greater commitment of national resources, a greater reliance on the efforts of the network, greater complexity of co-ordination and higher costs (Winkelmann, 1987).

Network Activities (Nogueira, 1987)

Research inputs and outputs
Exchange of information relating to programmes and projects; transfer of methodologies, materials (such as germplasm), and resources (availability of infrastructure for research); training, through exchange and visits, courses, and post-graduate; technical assistance in design and execution of projects.

Research work
Co-ordinated research, involving identification of common problems, independent project execution, and co-ordination at different stages of the project; *research collaboration* consisting in the distribution of responsibilities within a common programme with shared objectives, joint review and adjustment in the course of implementation; *joint research* shared management of project implementation.

Conferences

A special type of conference has been initiated by the International Agricultural Research Centres, with the specific purpose of co-ordinating a research network for a specific commodity or discipline of a commodity. At these conferences, information is exchanged among members on recent advances in research strategies and actions to be followed in the next agreed period are planned. IRRI holds such a conference each year on rice research for a world-wide membership of co-operating researchers from twenty countries. CIAT holds one conference of this type every two years for each of its commodity programs (beans, cassava, tropical pastures and rice).

International network of plant genetic resources

An example of an important network which links institutes in the developed and developing countries and the international centres involved in plant breeding work is the International Network of Plant Genetic Resources. The conservation of primitive types of crop plants and related wild species is an expensive task, that requires exploration of these resources, their conservation, their evaluation, and their utilisation in crop improvement programmes. This world-wide objective requires the collaboration of every country, but can only be successful if there is an international co-ordinating centre, with a central fund for exploration, conservation, training and information (FAO, 1972).

References to Part One

Académie Lénine des Sciences Agricoles de l'URSS (1949). *La Situation dans la Science Biologique*. Editions en Langues Etrangères, Moscow.

AFRC (1987). *Report of the Agricultural and Food Research Council for the Year 1986–87*. AFRC, London.

AID (1968). *Building Institutions to Serve Agriculture: A Summary Report of the Committee on Institutional Co-operation*. Agency for International Development, Purdue University.

Alves, E.R. de A. (1983). Desafíos de la investigación agricola en el Brasil. In *Seminario Internacional Sobre Generación de Información y Cambio Tecnologico en la Agricultura*. IICA, Montevideo, pp.15–37.

Alves, E.R. de A. (1984a). Management and development of an agricultural research institution. In *Brazilian Agriculture and Agricultural Research*, ed. L. Yeganiantz, EMBRAPA, Brasília. DF, pp.129–49.

Alves, E.R. de A. (1984b) Notes on dissemination of new technology. In *Brazilian Agriculture and Agricultural Research*, ed. L. Yeganiantz, EMBRAPA, Brasília. DF, pp.221–46.

Ampuero, E. (1981). *Organización de la Investigación Agrícola para Beneficiar a los Pequeños Agricultores en America Latina*. Cornell International Agriculture, Mimeo 85, Ithaca, NY.

ARC (1963). *The Agricultural Research Service*, London.

Ardila, J. (1985). *Marco Doctrinario y Politicas de la Subgerencia de Investigacion y Transferencia de Tecnología*. ICA, Bogotâ.

Ardila, J., Trigo, E. & Piñeiro, M. (1981). Human resources in agricultural research: three cases in Latin America. *Resource Allocation to Research*, ed. D. Daniels & B. Nestel. Int. Development Centre, Ottawa, p.151.

Arnon, I. (1976). Rapport d'une *Mission Consultative au Senegal*. ISRA, Dakar.

Arnon, I. (1986). *Conceptualización sobre el Modelo Institucional del ICA para la Investigación Agropecuaria*. ICA, Bogotà.

Arnon, I. (1987). *Modernisation of Agriculture in Developing Countries: Resources, Potentials and Problems*. Second Edition. John Wiley & Sons, Chichester.

302

Asian Development Bank (1977). *Asian Agricultural Survey 1976: Rural Asia: Challenge and Opportunity*, Manila.

Baharsjah, S. (1985). Improving research-extension linkages: the Indonesian experience. In *Research–Extension–Farmer*, ed. M.M. Cernea, J.K. Coulter & F.A. Russell. A World Bank and UNDP Symposium, Washington, DC, pp.28–33.

Bailar, J.C. (1965). The evaluation of research from the viewpoint of the university professor. *Research Management*, **8**, 133–39.

Beringer, H. (1988). *Personal communication*. Landwirtschaftliche Forschungsanstalt Büntehof, Hannover.

Bonilla, S.E. & Cubillos, A.G. (1987). Chile's experience in agricultural networks. In *International Workshop on Agricultural Research Management*. ISNAR The Hague, pp.131.

Brown, J.R. (1965). *Increasing World Food Output*. USDA Foreign Agricultural Economic Rep. No.25.

Busch, L. & Lacy, W.B. (1983). *Science, Agriculture, and the Politics of Research*, Westview Press, Boulder, Colorado.

Carter, Mary, E. (1987). R & D management of a mission-oriented government agency. US–Malaysian Workshop on *Management of Research & Development*, Kuala Lampur.

Casas, J. (1977). *Reflexions sur la Recherche Agronomique dans les Pays Sous Développés et Moyennement Développés*. Série Notes et Documents. Ecole Nationale Supérieure Agronomique, Montpellier.

Casas, J, Carrière, P. & Lacombe, P. (1979). *La Recherche Agronomique et la Diffusion du Progrès Technique en Union Soviétique*. Ecole Nationale Supérieure Agronomique, Montpellier.

CCTA–CSA (1961). Colloquium on the *Organisation of Agricultural Research*, Muguga, Kenya.

CGIAR (1978). *Report of the Task Force on International Assistance for Strengthening Agricultural Research*. CGIAR Secretariat, Washington, DC.

Chang, C.W. (1964). *Present Status of Agricultural Research Development in Asia and the Far East*. FAO Regional Office for Asia and Far East, Bangkok.

Chaparro, F., Montes, G., Torres, R., Balcázar, A. & Jaramillo, M. (1981). Research priorities and resource allocation in agriculture: the case of Colombia. In *Resource Allocation to Agricultural Research*, ed D. Daniels & B. Nestel Proc. Workshop held in Singapore. Intern. Dev. Research Centre, Ottawa, pp.68–96.

Christanssen, R.P. & Yee, H.T. (1964). The role of agricultural productivity in economic development. *J. Farm. Econ.*, **46**, 1051–61.

Chung, Y.B. & Dong, Y.M. (1984). *A Study of Agricultural Extension in the Republic of Korea*. FAO Regional Office for Asia and Far East, Bangkok.

COA (1987). *Agricultural Science and Technology Development in Taiwan*. Council of Agriculture, Executive Yan, Taiwan.

Conseil Fédéral Suisse (1975). *Ordonnance sur la Recherche Agronomique*, Berne.

Cummings, R.W. & Kanwar, J.S. (1974). Transfer of technology outreach of ICRISAT. In *International Workshop on Farming Systems*. ICRISAT, Hyderabad, pp.487–97.

Darling, M.S. (1970). The place and function of the University Faculty in the national research programme. *The Organisation and Methods of Agricultural Research*, ed. J.B. Robinson, Ministry of Overseas Development, London, pp.22–28.

De Zeeuw, D. (1985). Research policy linkages: a case in developed countries. In *Agricultural Research Policy and Organisation in Small Countries*. Report of a Workshop. Agric. Univers. Wageningen, pp.53–57.

Devred, R. (1966). *Organisation of Regional Agricultural Research Programmes for Agricultural Development within the Framework of Ecological Zones*. FAO Meeting on Savana Development, Khartoum. FAO, Rome.

Drilon, J.D.L. & Librero, Aida R. (1981). Defining research priorities for agriculture and natural resources in the Philippines. In *Resource Allocation to Agricultural Research*, ed. D. Daniels & B. Nestel, Proceedings of Workshop held in Singapore. Int. Research Centre, Ottawa, pp.97–103.

Eddy, E.D. Jr (1957). *Colleges for Our Land and Time*. Harper, New York.

Eicher, C.K. (1970). *Research in Agricultural Development in Five English-speaking Countries in West Africa*. Agricultural Development Council, New York.

Elz, D. (1984). (ed.). *The Planning and Management of Agricultural Research*. The World Bank, Washington, DC.

Evenson, R.E. (1984). Observations on Brazilian agricultural research and productivity. In *Brazilian Agriculture and Agricultural Research*, ed. L. Yeganiantz, EMBRAPA, Brasilia, DF, pp.247–76.

FAO (1965). Report on First Session of the Sub-Panel of Experts on the *Organisation and Administration of Agricultural Research*, Rome.

FAO (1966). Ad hoc Working Group for *Closer Co-operation in the Provision of Aid to African Countries in Agricultural Research*, Rome.

FAO (1972). Accelerating agricultural research in the developing countries. In *The State of Food and Agriculture 1972*, Rome, pp.141–54.

FAO (1984). *National Agricultural Research*. Report of an Evaluation Study in Selected Countries, Rome.

Faraj, N. (1986). Planning and programming in Morocco. In *Improving Agricultural Research Organisation and Management: Implications for the Future*. ISNAR. The Hague, pp.61–62.

Farrington, J. & Howell, J. (1987). *The Organisation and Management of Agricultural Research: Current Research Issues*. ODI, Disc. Paper 20, London.

FONAIAP (1981). *Plan Indicativo para la Investigación Agropecuaria del FONAIAP* durante el Periodo 1981–85, Caracas.

FONAIAP (1983). *Memoria 1981*, Maracay, Venezuela.

Francis, C.A. (1979). Small farm cropping systems in the tropics. In *Soil, Water and Crop Production*, ed. D.W. Thomas & M.D. Thomas, Avi Publishing Co. Inc, Westport, Conn., pp.318–48.

Gamble, W.K. (1984). *Improving the Global System of Support for National Agricultural Research in Developing Countries*. ISNAR, The Hague.

Gamble, W.K. & Trigo, E.J. (1984). Establishing agricultural research policy: problems and alternatives for small countries. In *Agricultural Research Policy and Organisation in Small Countries*. Report of Workshop. Agricultural University, Wageningen, pp.36–47.

Garcés, C. (1974). The experience of the ICA in agricultural development through education and research. In *Strategies for Agricultural Education in Developing Countries*. The Rockefeller Foundation, New York, pp.141–70.

Gardner, R. (1970). The organisation, administration and function of the national advisory service experimental centres. In *The Organisation and Methods of Agricultural Research*, ed. J.A.D. Robinson, Ministry of Overseas Development, London.

Gerente General (1978). *Informe para Junta Administradora del FONAIAP*. Caracas.

Gerente General (1980). *FONAIAP, Informe de la Gestion Realizada Durante el Periodo Junio de 1979–Junio de 1980*, Caracas.

Gunawardena, S.D.I.E. (1986). Organisation and structure in Sri Lanka. In *Improving Agricultural Research Organisation and Management: Implications for the Future*. ISNAR, The Hague, pp.46–52.

Halstead, R.J. (1988). *Binational Agricultural Research and Development Fund (BARD)*. USDA, Washington.

Hanson, H. (1979). *Biological Resources*. Report prepared for the Conference 'AgriculturalᵢProduction: Research and Development Strategies for the 1980s'. German Foundation for International Development, Bonn.

Hardin, L.S. (1979). *Emerging Roles of Agricultural Economists Working in International Research Institutions such as IRRI and CIMMYT*. Paper presented at the 17th International Conference of Agricultural Economics, Banff, Canada.

Hayami, Y. & Akino, M. (1977). Organisation and productivity of agricultural systems in Japan. In *Resource Allocation and Productivity in National and International Agricultural Research*, ed. T.M. Arndt, D.G. Dalrymple and V.W. Ruttan. University of Minnesota Press, Minneapolis, pp.29–59.

Hayes, T. (1987). AFRC man savages R & D policies. *Farmers' Weekly*, 13 November, England.

Herzberg, J. (1975). *Staatliche Landliche Beratungsdienste als Instrument der Entwicklungsförderung in Südamerika*. D.L.G. Verlag, Frankfurt (Main).

High-Level Committee (1973). *Reorganisation of Agricultural Education and Research in Uttar Pradesh*. University of Agriculture and Technology, Pattnagan.

Hightower, J. (1978). *Hard Tomatoes, Hard Times*. Shenkman, Cambridge.

Hoan, N. (1987). Agricultural research planning, monitoring and evaluation in Vietnam. In International Workshop on *Agricultural Research Management*, ISNAR, The Hague, pp.59–62.

Hunter, G. (1978). Report on administration and institutions. In *Rural Asia, Challenge and Opportunity*, Vol. IV, Asian Development Bank, pp.1–26.

IADS (1982). *Report 1981*. International Agricultural Development Service, New York.

ICA (1982). *Comision sobre Politicas y Prioridades para la Investigación Agropecuaria y sobre la Reestructuración del ICA*, Bogotá.

ICA (1984). *El ICA: su Organización, Realizaciones y Recursos*. Informe presentado para la Comision Octava de la Camara de Represantes, Bogotá.

IESA (1980). *FONAIAP—Estudio de Organización*, Instituto de Estudios Superiores de Administración, Caracas.

INIAP (1983). *Como Nacio INIAP*. Dep. de Communicación del INIAP, Quito.

Isarangkura, R. (1981). Inventory of agricultural research expenditure and manpower in Thailand. *Resource Allocation to Agricultural Research*, ed. D. Daniels and B. Nestel, Proc. Workshop held in Singapore. Intern. Development Research Centre, Ottawa, pp. 32–41.

ISNAR (1981). *Kenya's National Agricultural Research System*. The Hague.

ISNAR (1982a). *A Review of the Agricultural Research System of Malawi*. The Hague.

ISNAR (1982b). *The Role of International Associations in Strengthening National Agricultural Research*. The Hague.

ISNAR (1984a). *L'Institut National de La Recherche Agronomique du Maroc: Bilan et Perspectives*. The Hague.

ISNAR (1984b). *Agricultural Research in Ivory Coast: Presentation, Evaluation, Proposals for Improvement*. The Hague.

ISNAR (1984c). *Considerations for the Development of National Agricultural Research Capacities in Support of Agricultural Development*. The Hague.

ISNAR (1985). *Kenya Agricultural Research Strategy and Plan*. The Hague.

ISNAR (1986). *Improving Agricultural Research Organisation and Management: Implications for the Future*. The Hague.

ISNAR (1987). *International Workshop on Agricultural Research Management*. The Hague.

ISNAR/SPAAR (1987). *Guidelines for Strengthening Agricultural Research Systems*. World Bank, Washington, DC.

Jain, T.C. (1985). Constraints on research–extension linkages in India. In *Research–Extension–Farmer. A Two-Way Continuum for Agricultural Development*. World Bank and UNDP, Washington, DC.

Jain, H.K. (1985). *India's Co-ordinated Crop Improvement Projects: Organisation and Impact*. ISNAR, The Hague.

Jain, H.K. (1987). Structure and organisation in national agricultural research systems. In *International Workshop on Agricultural Research Management*. ISNAR, The Hague.

Jogaratnam, T. (1971). Agricultural research organisation and operations in Ceylon. In *National Agricultural Research Systems in Asia*, ed. A.H. Moseman, Agricultural Development Council, New York, pp.20–22.

Kanwar, J.S. (1971). Soil research organisation in India. In *National Agricultural Research Systems in Asia*, ed. A.H. Moseman, Agricultural Development Council, New York, pp.207–21.

Karama, A.S. (1986). Planning and Programming in Indonesia. In *Improving Agricultural Research Organisation and Management: Implications for the Future*. ISNAR, The Hague, pp.72–5.

Kim, Y.S. (1987). The progress and contribution of research projects for agricultural development in Korea. In *The Impact of Research on National Agricultural Development*. ISNAR, The Hague, pp.211–24.

Kirillov-Ougrioumov, V. (1977). Le système d'évolution des cadres scientifiques en URSS. *Le Progrès Scientifique*. **191**,71–83.

Kislev, J. (1977). A model of agricultural research. In *Resource Allocation and Productivity in National and International Agricultural Research*, ed. T.M. Arndt, D.G. Dalrymple & V.W. Ruttan. University of Minnesota Press, Minneapolis.

Knoblauch, H.C., Law, E.M. & Meyer, W.P. (1962). *State Agricultural Experiment Stations: A History of Research Policy and Procedure*. USDA Misc. Publ. 904.

Konkle, W.W. (1982). *An Assessment of the United States Food and Agricultural System*. Office of Technology Assessment, Washington, DC.

Le Gouis, M. (1986). Role of Advisory Development Services in the Adjustment of Agricultural Production to the Present and Potential Market Requirements. Paper presented at the Seminar on the Future of the Advisory Services, Louvain.

Lopes, J.R.B. (1985). Agricultural research in Brazil. In *Strengthening National Agricultural Research*, ed. B. Bengtsson & G. Tedla, SAREC, Stockholm.

McDermott, J.K. & Bathrick, D. (1981). *Guatemala, Development of the Institute of Agricultural Science and Technology (ICTA) and its Impact on Agricultural Research and Farm Productivity*. US Agency for International Development. Washington, DC.

Madamba, J.C. & Swaminathan, M.S. (1983). Co-operation between national research systems and the international research support community. In *Agricultural Research for Development: Potentials and Challenges for Asia*, ed. B. Nestel. ISNAR, The Hague, pp.4–10.

Mahapatra, I.C. (1971). The All-India Co-ordinated Agronomic Experiments Scheme. In *National Agricultural Research Systems in Asia*, ed. A.H. Moseman, Agricultural Development Council, New York, pp.225–29.

Maida, J.M.A. (1982). *Development and Administration of Agricultural Research and its Contribution to Agricultural Development in Malawi* (mimeo).

Malawi Government (1982). *Mid-year Economic Review 1981–82*. Government Printer, Zomba, Malawi.

Mangundojo, S. (1971). In *National Agricultural Research Systems in Asia*. ed. A.H. Moseman, Agricultural Development Council, New York, pp.39–49.

Marcano, J. (1982). Latin America and Caribbean. In *The Role of International Associations in Strengthening National Agricultural Research*. ISNAR, The Hague, pp.58–60.

Menon, K.P.A. (1971). Building agricultural research organisations—the Indian experience. In *National Agricultural Research Systems in Asia*, ed. A.H. Moseman, Agricultural Development Council, New York, pp.23–38.

Morillo, A.F. (1971). Organización de la investigación agricola en Venezuela. In *Seminario sobre la Administración de Instituciónes de Investigación Agricola*, IICA, Zona Andina, Quito, pp.60–80.

Moscardi, E. (1982). *The Establishment of National On-Farm Research Entity in Ecuador*. CIMMYT. Mexico City.

Moseman, A.H. (1977). Co-ordinated national research projects for improving food crop production. *Resource Allocation and Productivity in National and International Agricultural Research*, ed. T.M. Ardnt, D.G. Dalrymple and V.W. Ruttan. University of Minnesota Press, Minneapolis, pp.367–80.

Moseman, A.H. & Hill, F.F. (1964). Private foundations and organisations. In *Yearbook USDA: 'Farmers World'*. Washington, DC.

Moseman, A.H., Robins, J.H. & Wilcke, H. (1981). *The Role of the Federal Government, State Agricultural Experiment Stations, and the Private Sector in Research*. Office of Technology Assessment, Washington, DC.

Mosher, A.T. (1971). *To Create a Modern Agriculture.* Agricultural Development Council, New York.

Muturi, S.N. (1981). The system of resource allocation to agricultural research in Kenya. *Resource Allocation to Agricultural Research,* ed. D. Daniels & B. Nestel, Proceedings of Workshop held in Singapore, International Research Centre, Ottawa, pp.123–28.

Myers, M. (1987). Challenges of the US fertilizer industry. *Fertilizer International* No.249.

NAS Committee (1972). *Report of the Committee Advisory to the USDA.* National Technical Information Service, US Department of Commerce, Springfield, Va.

Nestel, E. (ed.) (1983). *Agricultural Research for Development: Potentials and Challenges for Asia.* ISNAR, The Hague.

Nestel, P.L. & Franklin, D.L. (1981). Summary of workshop decisions. In *Methods for Allocating Resources in Applied Agricultural Research in Latin America,* ed. P. Pinstrup Andersen and F.C. Byrnes, CIAT, Cali, Colombia, pp.13–21.

Nogueira, R.M. (1984). Comments by participants. In *The Planning and Management of Agricultural Research,* ed. D. Elz. The World Bank, Washington, DC, pp.72–83.

Nogueira, R.M. (1987). Agricultural research networks: an analytical framework. In *International Workshop on Agricultural Research Management.* ISNAR, The Hague, pp.119–30.

OECD (1965). *Survey of the Organisation of Agricultural Research in OECD Member Countries.* Paris.

OECD (1981). *Agricultural Advisory Services in OECD Member Countries.* Paris.

Oram, P. (1982). *Strengthening Agricultural Research in The Developing Countries: Progress and Problems in the 1970s.* World Bank, Washington, DC.

Orleans, L.A. (1979). Science in China. In *China's Road to Development,* ed. N. Maxwell. Pergamon Press, Oxford, pp.219–29.

OTA (1981). *An Assessment of the US Food and Agricultural Research System.* Office of Technology Assessment, Washington, DC.

Panabokka, C.R. (1985). Agricultural research policy and organisation in small countries: scientific linkages in a small country. In *Agricultural Research Policy and Organisation in Small Countries.* Agricultural University, The Hague, pp.63–64.

Pastore, J. & Alves, E.R. de A. (1984). Reforming the Brazilian agricultural research system. In *Brazilian Agriculture and Agricultural Research,* ed. L. Yeganiantz, EMBRAPA, Brasilia, DF, pp.117–28.

Phillips, R.W., Moskowitz, I. & Lininger, F.F. (1953). *The Organisation of Agricultural Research in Europe.* Dev. Paper No. 29. FAO, Rome.

Piñeiro, M. (1983). *El Sector Privado en la Investigación Agropecuaria: Reflexiones para su Analisis.* PROAGRO, Documento 1.ISNAR, The Hague.

Piñeiro, M. (1986). *The Development of the Private Sector in Agricultural Research: Implications for Public Research Institutions.* PROAGRO Paper 10, ISNAR, The Hague.

Piñeiro, M. (1987). Agricultural production in Latin America and the Caribbean: international organisations and regional programmes. In *The Impact of Research on National Agricultural Development*, ISNAR, The Hague, pp.89–95.

Piñeiro, M., Fiorentina, R., Trigo, E., Balcazar, A. & Martinez, Astrid (1982). *Articulación Social y Cambio Tecnico en la Producción de Azucar en Colombia.* IICA, San José.

Pinstrup-Andersen, P. (1982). *Agricultural Research and Technology in Economic Development.* Longman, London.

Plucknett, D.L. & Smith, N.J.H. (1984). Networking in international agricultural research. *Science,* **25**,989–93.

Portilla, F. (1971). Organización de la investigación agricola en el Ecuador. In *Seminario Sobre la Administración de Instituciónes de Investigación Agricola,* IICA, Zona Andina, Quito, pp.43–59.

PSAC (1962). *Agricultural Panel Report on Science and Agriculture.* Washington, DC.

Qasem, S. (1985). Research policy linkages. In *Agricultural Research Policy and Organisation in Small Countries.* Rep. of a Workshop Agricultural University, Wageningen, pp.58–63.

Ranhawa, M.S. (1958). *Agricultural Research in India.* Indian Council for Agricultural Research, New Delhi.

RAPA/FAO (1986). *Report of the Expert Consultation on Linkages of Agricultural Extension with Research and Agricultural Education.* FAO Regional Office for Asia and the Pacific, (RAPA) Bangkok.

Rebishung, J. (1964). *Relation entre la Recherche et la Vulgarisation.* Information presented to meeting on the Organisation of Agricultural Research, European Commission of Agriculture, Berne.

Review Committee on Agricultural Universities (1978). *Report on Agricultural Universities,* New Delhi.

Richardson, R.W. Jr (1964). The Rockefeller Foundation's Mexican Agricultural Programme. *Agric. Sci. Rev.,* **2**(1), 12–20.

Rivaldo, O.F. (1987). Strategies for strengthening the Brazilian agricultural research system. In *The Impact of Research on National Agricultural Development.* ISNAR, The Hague, pp.161–81.

Rockefeller Foundation (1965). *Programme in the Agricultural Sciences.* Annual Report, 1964–65.

Röling, N. (1986). *The Structure of Advisory Services in the Context of the Future Evolution of Agriculture: Conceptual Aspects.* Paper presented at the Seminar on the Future of the Advisory Services, Louvain.

Rothschild, L. & Dainton, E. (1971). *A Framework for Government Research and Development.* Her Majesty's Stationery Service, London.

Russell, Sir E.J. (1966). *A History of Agricultural Science in Great Britain, 1620–1954.* Allen & Unwin, London.

Ruttan, V.W. (1968). *Organising Research Institutions to Induce Change, The Irrevelance of the Land Grant Experience for Developing Economies.* Minn. Agric. Expt. Sta. Misc. J. Paper Series 1313.

Ruttan, V.W. (1982). *Agricultural Research Policy.* University of Minnesota Press, Minneapolis.

Ruttan, V.W. (1985). Toward a global agricultural research system. In *Agricultural Research Policy and Organisation in Small Countries*. Report of a Workshop. Agricultural University, Wageningen, pp.19–35.

Santucci, F.M. (1986). *Prospects for Advisory Services in the Context of the Future Development of Agriculture*. Paper presented at the Seminar on the Future of the Advisory Services, Louvain.

Sawyer, R.L. (1987). The CGIAR and the national agricultural research programmes: a maturing partnership. In *The Impact of Research on National Agricultural Development*. ISNAR, The Hague, pp.47–52.

Sibale, P.K. (1986). *The Reorganisation of the Department of Agricultural Research in Malawi*. Paper presented at International Workshop on the Role of Applied Research in Agricultural Development, Ruppin Inst. of Agriculture, Natania, Israel.

Simons, H.A. (1960). *The New Science of Management Decision*. Harper & Row, New York.

Skovgaard, I. (1984). *Research in Support of Agricultural Policies in the Nordic Countries*. Working Document for the 14th FAO Regional Conference. Rome.

Skovgaard, I. (1987). Agricultural research and advisory services. In *Agriculture in Denmark*. The Agricultural Council of Denmark, Copenhagen, pp.2–13.

SPAAR (1986). *Report of the Technical Group on Networking*. Special Programme for African Agricultural Research, Brussels.

Sprague, G.F. (1975). Agriculture in China. *Science*, **188**, 549–55.

Starnovsky, B. (1962). *Planning of Science and the Resources of Research*. Paper prepared for the UN Conference on the Application of Science and Technology for the Benefit of the Less Developed Areas. Geneva.

Stavis, B. (1978). Agricultural research and extension services in China. *World Development*, **6**, 631–45.

Stevenson, R. (1984). Agricultural extension at a crossroads? In *Proceedings 6th European Seminar on Extension Education*, ed. R. Volpi and F.M. Santucci. Centro Studi Agricoli, Borgo a Mazanno, pp.36–39.

Streeter, C.P. (1972). *Colombia. Cambios en la Agricultural, el Hombre y los Métodos*. The Rockefeller Foundation, New York.

Torres, F. (1987). Agroforestry research networks in tropical Africa: an ecozone approach. In *The Impact of Research on National Agricultural Development*, ISNAR, The Hague, pp.105–23.

Toulmin, S. (1966). Is there a limit to scientific growth? *Sci.J.* **2**(8),80–85.

Trigo, E.J. (1986). *Agricultural Research in the Developing World: Diversity and Evolution*. Working Paper No. 6. ISNAR, The Hague.

Trigo, E.J. & Piñeiro, M.E. (1982). Institutional aspects of agricultural research organisation in Latin America. In *Proceedings Caribbean Workshop on the Organisation and Administration of Agricultural Research*, ed. W.M. Forsythe, A.M. Pinchinet and L. McLaren. IICA, San José, pp.47–78.

Trigo, E.J. & Piñeiro, M.E. (1984). Funding agricultural research. In *Selected Issues in Agricultural Research in Latin America*. ISNAR, The Hague, pp.76–98.

Trigo, E.J., Piñeiro, M. & Ardila, L. (1982). *Organización de la Investigación Agropecuaria en America Latina*, IICA, San José.

Tuan, Dao. T. (1985). *Agricultural Research System in Vietnam.* A Consultancy Report. FAO/RAPA, Bangkok.

Turkevich, J.O. (1960). Organisation of science in the Soviet Union. In *Proc. Conf. Acad. Ind. Basic Research.* National Science Foundation Princeton University, New Jersey, pp.56–65.

Ulbricht, T.L.V. (1977). Contract agricultural research and its effect on management. In *Resource Allocation and Productivity in National and International Agricultural Research,* ed. T.M. Arndt, D.G. Dalrymple and V.W. Ruttan. University of Minnesota, Minneapolis, pp.381–93.

USAID (1983). *Strengthening the Agricultural Research Capacity of the Less Developed Countries: Lessons from AID Experience.* AID Programme Evaluation Report No.10. Washington, DC.

USDA (1980). *The Mission of Science and Education. Agricultural Research.* Washington, DC.

USDA (1983). *Agricultural Research Service Programme Plan.* Misc.Publ.1429. Washington, DC.

Van de Zaag, D.E. (1985). Organisation of agricultural research in the Netherlands with special reference to potato research and farmers' participation. In *Agricultural Research Policy and Organisation in Small Countries.* Report of Workshop, Agricultural University, Wageningen, pp.76–82.

Venezian, E. (1981). International associations and national agricultural research. In *The Role of International Associations in Strengthening National Agricultural Research.* ISNAR/IADS, The Hague, pp.10–57.

Venkatesan, V. (1985). Policy and institutional issues in improving research-extension linkages in India. In *Research–Extension–Farmer. A Two-Way Continuum for Agricultural Development.* World Bank/UNDP Washington, DC, pp.13–27.

Vez, A. (1987). *Agriculture et Environnement.* Paper presented at a meeting of SAGUF, Changins, Switzerland.

Von der Osten, A. (1982). The impact of research on development—needs and potentials. In *The Impact of Research on National Agricultural Development.* ISNAR, The Hague, pp.41–46.

Wade, N. (1973*a*). Poor research, *Science,* **179**,45–57.

Wade, N. (1973*b*). Agriculture: critics find basic research stunted and wilting, *Science,* **180**, 390–93.

Wade, N. (1973*c*). Agriculture: signs of dead wood in forestry and environment research, *Science,* **180**, 474–77.

Wade, N. (1973*d*). Agriculture: social sciences oppressed and poverty stricken, *Science,* **180**, 719–22.

Wade, N. (1973*e*). Agriculture: research planning paralysed by pork-barrel politics, *Science,* **180**, 932–37.

Wansinck, G. (1964). *The Relationship between Technical Agricultural Research and the Economic Aspects of Farming.* Paper presented at Meeting on the Relationships between Agricultural Research and Agricultural Practice, Berne.

Wansinck, G. & Ulbricht, T.L.V. (1972). *Mechanisms for Adapting Agricultural Research Programmes to New Goals.* Paper presented at 2nd Working Conference of Directors of Agricultural Research. OECD, Paris.

Wapakala, W.W. (1986). Improving agricultural research organisation and management in Kenya. In *Improving Agricultural Research Organisation and Management: Implications for the Future*. ISNAR, The Hague, pp.37–41.

Watts, L.H. (1984). The organisational setting for agricultural extension. In *Agricultural Extension: A Reference Manual*, ed. B.E. Swanson, FAO, Rome, pp.20–30.

Weber, A. (1979). *Some Preliminary Notes to the Early History of the German Agricultural Research System*. Christian-Albrechts University, Kiel.

Weber, A. (1986). *Auftrag, Aufstieg, and Ausbreitung der Agrarwissenschaften*. Paper presented at the Welternährungstag. Christian-Albrechts Univ., Kiel.

Webster, C.C. (1970). The organisation, place and function of the agricultural research council in the national research programme. In *The Organisation and Methods of Agricultural Research*, ed. J.B.D. Robinson, Ministry of Overseas Development, London, pp.53–60.

Webster, B.N. (1970). Restraints in building national research capabilities. In *National Agricultural Research Systems in Asia*, ed. A.H. Moseman. Agricultural Development Council, New York, pp.106–17.

Wilkinson, B. (1987). Letter to the Editor, *Farmers' Weekly*, 20 November.

Wilson, L.A. (1984). *Toward the Future: An Alternative Framework for Agricultural Research, Training and Development in the Caribbean*. The University of the West Indies, St. Augustine.

Winkelmann, D.L. (1987). Networking: some impressions from CIMMYT. In *The Impact of Research on National Agricultural Development*. ISNAR, The Hague, pp.125–34.

World Bank (1981). *Agricultural Research*. Sector Policy Paper. Washington, DC.

World Bank (1982). *World Development Report 1982*. Washington, DC.

World Bank (1983). *World Development Report 1983*. Washington, DC.

World Bank (1984). *Toward Sustained Development: A Joint Programme of Action for Sub-Saharan Africa*. Washington, DC.

Yudelman, M., Butler, G. & Banerji, R. (1971). *Technological Change in Agriculture and Employment in Developing Countries*. OECD, Paris.

Zhou, F. (1987). Organisation and structure of the national agricultural research system in China. In *International Workshop on Agricultural Research Management*, ISNAR, The Hague, pp.219–23.

Zuurbier, P.J.P. (1983). *De Relatie tussen het Landbouwkundig Onderzoek, de Landbouwvoorlichting en de Boer in Nederland*. Minis.van Landbouw en Visserÿ, The Hague.

Zuurbier, P.J.P.(1984). *De Besturing en Organisatie van de Landbouwvoorlichtingsdienst*. Agricultural University, Wageningen.

Part Two

AGRICULTURAL RESEARCH—FORM AND CONTENT

Chapter 5

The Conceptual Framework

THE CHARACTERISTICS OF AGRICULTURAL RESEARCH

Agricultural research involves, by its very nature, the application of the principles of basic sciences to the solution of problems of immediate or prospective usefulness to agriculture. It draws heavily on various scientific disciplines, mainly, but not only, on the biological sciences: plant and animal physiology, genetics, microbiology, entomology, serology. As many of the practical agricultural problems depend on the interrelationships between plants (or animals) and soils and climates, chemistry and biochemistry, physics, geology and meteorology are also involved. Research findings in any of these sciences may find entirely unforeseen applications in agriculture, hence, the work of the agricultural research worker provides ample scope for the application of the scientific method in a wide spectrum of disciplines.

Agricultural research as a public responsibility

The role of agriculture is as a basic industry that not only supplies food and fibres to the population but is also the essential base that can make possible an industrial 'take-off'. In view of its wide implications of national significance and its service to the community as a whole, it is fully justified that agricultural research should be the concern of a government, and be sustained by national funds and not be a burden to be borne by the farming community alone. It would also be unrealistic to expect that an industry such as agriculture, however large and important its role in the national economy, fragmented as it is into numerous small production

315

units, even if they are organised into co-operatives, would be able to maintain by itself a viable research organisation. Therefore, in contrast to industry, government financing generally constitutes the major source of support for agricultural research.

However, as the economy develops, the industries which supply agriculture with a variety of inputs, such as pesticides, fertilisers, agricultural machinery and equipment, etc., usually make important contributions to agricultural research. In the United States, investment in agricultural research by industry had in the 1960s already exceeded that by government, both federal and state (Brown, 1965).

Whatever the source of the research funds, results (in as far as they do not relate to the industrial process proper) are made *freely available* to all farmers; a considerable effort is even expended on attempting to make the results *acceptable* by the farmers. This, of course, precludes any possibility of secrecy; the free publication of research results makes them available on an international basis, so that the benefits derived from successful research, financed and developed by one country, are generally freely accessible to other countries. In this respect, agricultural research resembles academic research more than it does industrial research.

Farmers' influence on research policy

Another characteristic of agricultural research is the considerable influence that the sector of the electorate most directly concerned, e.g. the farmers, has on research policy. Farmers, especially in the developed countries, are often able to evaluate the significance and importance of the problems with which they are faced, and as they are usually well organised, are able to exert pressure on the government agencies responsible for research. The nature of agricultural research in a country will greatly depend on whether farmers insist on research of immediate economic benefit, or realise the importance of long-term research on problems of basic importance.

KINDS OF RESEARCH

Definitions

The word 'research' has been defined in many ways. Klopsteg (1945) gives the following comprehensive definition:

Research is original and creative intellectual activity, carried out in the laboratory, the library or the field, which endeavours to discover new facts and to appraise and interpret them properly in the light of previous knowledge. With constantly increasing understanding, it revises previously accepted conclusions, theories and laws, and makes new applications of its findings. Whether it seeks to extend knowledge for its own sake or to achieve results with specific economic or social value, its *raison d'être* is its contribution to human welfare.

Hertz (1957) is more succinct in defining research as 'the application of human intelligence in a systematic manner to a problem whose solution is not immediately available'.

When attempts are made to distinguish between various *kinds* of research (Fig. 11), ambiguity and contradictions take over. These are due mainly to confounding the nature of the research with the motivations of the research worker. The National Science Foundation of the USA (NSF, 1959) defines 'basic research' as that which is 'motivated primarily or exclusively by intellectual curiosity and an interest in the study of the laws

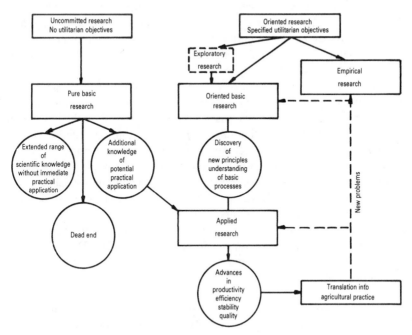

Fig. 11. Kinds of research and research objectives.

of nature for their own sake without regard for the immediate applicability of any finds he may reach'. Whilst 'applied research' is 'research projects which represent investigations directed to discovery of new scientific knowledge and which have specific commercial objectives with respect to either products or processes'; 'technical activities concerned with non-routine problems encountered in translating research findings or other general scientific knowledge into products or processes' is called development.

Many versions and variants (sometimes conflicting) of these definitions are to be found in the literature. In some of these, the idea that basic research is uncontaminated by any thought of practical application or material gain is the recurring theme: basic research is 'that carried on only to extend the range of scientific knowledge' or 'free enquiry into nature' (Kornhauser, 1963). Basic research is 'uncommitted research, prompted by disinterested curiosity, and aimed primarily at the extension of the boundaries of knowledge. Applied research, by way of contrast, is usually committed to the search for a solution to a specific problem' (NSF, 1959).

Schmookler (1964) goes even further and makes the definition of research contingent on the organisation which sponsors the research. 'Viral or nuclear research is applied research if the sponsors pursue the research with potential economic or military uses uppermost in their minds', whilst 'a census of butterflies in a region' is pure research if carried out by a university entomologist!

Some authors are more specific, and distinguish between two kinds of basic or fundamental research, according to motivation: Zuckerman *et al*. (1961) call *pure basic research* 'research that is selected by the individual to satisfy his own tastes and intellectual curiosity'; and *objective basic research* 'research in fields of recognised potential technical importance'. In this case the distinction is not between 'basic' and 'applied', but between 'pure' and 'applied'. At a conference on academic and industrial basic research, sponsored by the National Science Foundation, the term 'basic basic research' was used by participants in the discussion! (Schmookler, 1964).

Within the broad definition of fundamental research, further distinctions have been made. In a report prepared for the OECD on 'Government and allocation of resources to science' (Brooks *et al*., 1961), distinctions are made between:

> *Academic research*, that is 'free-ranging and essentially individualistic in approach, requiring relatively inexpensive equipment, so that cost per investigator is not high.

Programmatic fundamental research, which is 'organised around highly expensive equipment and for a critical mass of skilled manpower which are often multi-disciplinary'; and

Establishment fundamental research, 'undertaken in a departmental research establishment whose programme is aimed primarily at a practical objective'.

It does seem surprising that scientists should define their basic activity —research—not objectively, according to its inherent characteristics, but according to extraneous, subjective factors. That the same research can be differently defined according to the motivations of the researcher, the wish of the sponsor or the nature of the organisation in which it is carried out, in itself underlines the inherent weakness of this approach.

Judging the value of research

Whilst there is ambiguity in *defining* research, scientists are much clearer in their criteria when *judging* research. When judging the research work of their peers, the criteria used are: its explanatory value—generality and span of relevance; the clarifying power; and the degree of originality of the research. Scientists may also be impressed by the elegance of the solution, but the purity or 'usefulness' of the research does not enter into the judgement at all.

Barnard (1957) writes that whether a given research project is classified as 'basic' or 'applied' inevitably reflects the bias of the person who does the classifying. As an illustration, he cites a report in which university administrators estimated that they were awarded 885 million dollars for basic research by the Federal Government of the USA, whilst the public officials who disbursed these funds report that they provided about half the amount to the universities for the same purpose! The reason for the discrepancy lies in the subjective approach used for defining the research: basic research enjoying high status in the university, whilst the federal agencies, which have to justify the allocations in practical terms, played down the basic aspect! The confusion and ambiguity are further increased by using interchangeably the terms 'pure', 'fundamental', 'basic', 'theoretical' and 'academic'.

Mission-oriented research

In 1967, a working party was appointed by the British Agricultural Research Council to advise the Council on agricultural research policy for the next ten years. In a statement of general principles (ARC, 1967) the distinction is made between 'mission-oriented' research and 'speculative' research. The former includes both 'applied' research and basic research that aims at contributing to the solution of practical problems, and the term 'mission-oriented' research will be used in this book as defined by the British ARC.

Speculative research includes 'truly speculative or intuitive investigations, and also objective research which is of no foreseeable practical importance, although designed to increase understanding of scientific phenomena'. The justification for speculative research is that it is 'the only source of new ideas on which the practical advances of the future can be based'.

In this context, speculative research is simply another term for pure basic research. Speculative research carried out by agricultural institutions is not, and *cannot be* the 'only source of new ideas on which the practical advances (of agriculture) in the future can be based' (ARC, 1967). These advances, in the future as in the past, are dependent not only on agronomic research of all kinds, but also on research findings in practically all fields of scientific endeavour, in particular in the biological sciences. Speculative research as defined by the working group is characterised by the very fact that in the planning stage, no foreseeable implications for agricultural progress can be envisaged. It is therefore difficult to visualise how such research projects, whose implications for agriculture cannot be guessed at, will be evaluated in a mission-oriented agricultural research institution when determining research priorities.

Exploratory research, on the other hand, has a completely different meaning: it relates to original ideas or approaches which, *a priori*, appear to have agricultural implications, but which require a period of exploration before it is possible to decide whether the idea has merit or not. This kind of original thinking is not only legitimate in an agricultural research organisation, it should be actively encouraged.

A missed opportunity is generally a greater loss to society than the risk involved in making a relatively small investment in the exploration of new ideas. Exploratory research is needed not only to solve existing problems of agriculture, but also to make possible future breakthroughs and progress. It can be of either the basic or the applied kind.

The classic example of exploratory basic research is the work of F.W. Went (1957) on isolating a growth factor in oats. The identification of indole-acetic acid and its subsequent synthesis led hundreds of other workers to do empirical research on the effect of applying growth factors, by different methods and at different stages of development, to a large variety of plants. It was found that certain concentrations promoted root production and affected fruit set and fruit drop, whereas at higher concentrations the dormancy of bulbs and tubers was affected; at still higher concentrations the life processes of the plant were affected adversely — the different plant types showing differential responses in this respect.

The empirical research that followed Dr. Went's exploratory work has resulted in truly revolutionary developments of incalculable value in agriculture; rooting of cuttings, control of flowering and fruit drop, defoliation, and selective and general weed control, to mention but a few of the applications on a wide variety of crops.

Agricultural research is, by its very definition, research whose objective it is to apply a wide variety of scientific disciplines to the development of new approaches to agricultural production, and to the solution of the problems besetting the farmer, and is, therefore, essentially applied research in the wider sense. Logically, there can be no such thing as *agricultural* research that is *not* oriented into research activities that may find useful application in agriculture. In this context, however, there is no contradiction whatsoever between what is commonly called applied research and basic research. If we define research by its characteristics, and not by extraneous factors, we can call *basic research* investigations that extend scientific knowledge by the search for new principles and the understanding of underlying processes, whilst *applied research* is the use of these principles and techniques to the solution of specific problems. It is clear that both basic and applied research are complementary and essential stages in planned or oriented agricultural research.† It is the nature of the problem, and not the motivations of the researcher, that determines whether basic research or adaptive research is necessary or preferable for the solution of a specified problem.

Norman Borlaug did not break new paths in fundamental science or in the basic theory of plant breeding. He applied well-known techniques of plant breeding, along with a few innovations in testing, to create a line of improved wheat varieties that were used to increase food production dramatically in many countries, including LDCs. Also, he did not publish

---

†'Objective' basic research is mainly distinguished from 'pure' basic in that, being stimulated primarily by technological needs, it calls for a planned approach. Other terms that are frequently used for basic research that seeks new knowledge required for practical purposes are oriented, or mission-oriented basic research (Zuckerman *et al.*, 1961).

very much before receiving his Nobel Prize, and those papers he did write were not classified as basic research (Konkle, 1982).

Kind of research in relation to the problem to be investigated

The kind of research to be adopted in relation to a specific problem should be dictated by what will 'achieve the best results, both in quality and quantity, in relation to the available time and the effort and money applied to it' (Hertz, 1957). If the basic information needed is already available—whether through local research or research carried out elsewhere—then an adaptation of this available knowledge to the specific conditions pertaining is indicated. Much effort is devoted to basic research of sorts, with minor and unimportant variations, on subjects that have already been investigated, an activity that has been irreverently called: 'transferring bones from one graveyard to another'.

Nor can it be denied that many discoveries in agriculture have been made with little or no basic research. A few of these are listed by Salmon and Hanson (1964): the value of lucerne and other legumes in improving soil fertility was known to farmers in ancient Greece; a method of protecting sheep from blowflies was developed in Australia by a shepherd, after scientists had failed to solve the problem; the relation between barberries and stem rust was known long before it was understood; zinc deficiency as the cause of little-leaf disease of peaches was discovered by accident when a galvanised iron bucket was used as a container for spray material used on a few trees, instead of the customary wooden bucket. Sometimes, basic research has quite unintended results of considerable practical importance. The classic example is the work of Went on isolating a growth factor in oats mentioned above.

In many cases, obtaining 'an understanding of the underlying processes' is the shortest and most efficient way of solving the problem, and in that case basic research is clearly and unequivocally indicated for practical purposes. Frequently, several parallel approaches are justified. Who would venture to state, *a priori*, that in research concerned with acquired resistance of insects to insecticides, a study of the mechanism whereby the insect achieves its acquired resistance is less justified, from the *practical* point of view of solving the problem, than a programme of empirical research, based on screening countless new formulations of insecticides? Certain problems require a combined approach, solving certain aspects of the problem by *basic research* to be followed by the application of these findings. In this approach, we are following in the footsteps of Pasteur, who said: 'There are not (pure) sciences and applied sciences, but simply science and the applications of science.' It is time that one should accept the notion that the *practical importance* of a problem has no bearing on the scientific character of the research carried out to solve it.

Research and academic snobbishness

Unfortunately, there has developed in the academic world a snobbishness that gives a higher rating to 'pure' research, that is 'prompted only by curiosity and that has no practical motivation or immediate application' over research that is planned with the declared aim of solving problems of practical or economic significance.

Simons (1960) is of the opinion that the word 'research' has been 'debased' and 'has lost its stature, and we in the Universities would do well to replace it'. The term he proposes is 'creative scholarship'.

Carew (1964) lists the following consequences to agricultural research resulting from the lower status accorded to planned research which is directed towards practical results:

(a) 'Basic' projects take precedence over 'applied' ones — preference being given to a study of single cell infection mechanisms over 'practical' disease control work; fertiliser studies are abandoned in favour of the study of ion exchange phenomena, etc.

(b) 'Basic' plants are favoured over economic crops: instead of using a sugar beet for studies on sugar synthesis by the plant, some exotic weed is preferred for this purpose.

(c) The use of cell tissues is favoured over that of whole plants for the study of nutrition and other problems.

(d) And last, but not least, papers in what Carew calls 'prestige' professional journals, are favoured over bulletins or articles in popular magazines.

Carew does not question the need for basic research, but underlines the need for 'a good, imaginative, original research that will provide the knowledge to improve food and fibre production abilities'. This includes both basic and applied research.

The snobbish attitude to planned research is not only unfortunate, but also unjustified. There is little doubt that planned research provides a more serious challenge to the research worker than 'pure' research. In pure research, every contribution to knowledge is an achievement, while in planned research appropriate solutions to definite problems must be found, and found within a reasonable time limit. There are no alibis for failure in this respect. 'Objective basic research can be as exciting, rewarding and intellectually demanding as pure basic research. It requires for its success the highest intellectual qualities: imagination coupled with doubt, flexibility with persistence and precision with daring' (Zuckerman *et al.*, 1961). 'The opportunities for creative intellectual work are probably greater today in applied than in pure science' (Dancy, 1965).

It must be stressed that there are no actual differences in principle between the two types of research. Pure research and planned research require the same scientific methods and the same objective and critical approach. The only difference is in the motivation: 'Looking at an experiment, you cannot tell whether it is pure or applied research' (Hertz, 1957). In developing countries, however, considerable relevance must be attached to the relative importance of basic research and applied research in agriculture in the overall planned programme. The former is, by its very nature, universal in its application, while the latter has mainly regional significance.

Research in developing countries

There is a great amount of basic agricultural research going on in the world. In a developing country, where one starts almost from scratch, it would appear both wasteful and illogical not to draw first of all on the enormous fund of basic knowledge available, and to concentrate the main efforts on determining the proper application of this knowledge to local conditions. This is what should be done in the early stages, as has already been discussed above. However, in the long run, this has certain drawbacks which may not be apparent at first sight. In agricultural research, too, the law of diminishing returns is operative. At first, relatively simple and empirical experimentation may give spectacular results; however, a stage is soon reached where further progress is dependent on a greater research effort. Certain regional problems may be so specific in nature that no appropriate basic research has been carried out elsewhere.

Many problems of significance to the tropics, in which most of the developing countries are situated, have not yet been studied sufficiently. Basic studies of soils and their fertility, the ecology of major pests and diseases, animal physiology, and soil–water–plant relationships under tropical conditions, are only a few examples of the scientific problems whose solution is essential to the sustained progress of agriculture in tropical countries.

Institutional separation of applied and basic research

Basic research may be needed not only to solve existing problems of agriculture but also to provide for future breakthroughs and progress, in particular by exploring original and unconventional approaches and ideas.

Certain countries have attempted to solve these problems by concentrating basic research in a central institution and carrying out applied research in district experiment stations according to ecological zones. While this may be the most 'practical' approach, it is far from being an ideal solution. It is important to avoid segregating research workers into two castes. This can be achieved by finding a suitable balance between basic and applied research—not between institutions or departments, but within institutions and departments; not between research workers, but between the tasks allotted or undertaken by each individual. This is not always possible, even though it should be the declared policy and objective, because the characteristics and abilities of the individual are also a factor to be taken into account.

In agricultural research there is, therefore, need and justification for basic research as well as applied research. However, the primary consideration of the organisation is to ensure that the work carried out under its auspices should, at least potentially, be of benefit to the farming community in particular, and to society in general. The means available for research therefore have to be channelled into the appropriate direction and control exercised over their proper use. This signifies that in agricultural research, the effort is directed towards mission-oriented research, whether basic or applied.

Adaptive research is carried out in order to make research recommendations more appropriate for specific conditions, or to intended users; this kind of research is being increasingly emphasised, as research and extension face the challenge of being relevant for farmer sectors with very small margins of potential and high vulnerability to risk (Röling, 1984). It is the major kind of research carried out in the early stages of agricultural modernisation. It does not usually require interdisciplinary teams or sophisticated equipment (Piñeiro, 1986).

Insurance research is aimed at *preventing* a given situation from occurring. Such research, like insurance, is undertaken in the hope that it will never need to be implemented. Consider, for example, a country in which citrus is an important industry. It is known that in certain other countries the crop as a commercial undertaking has been completely wiped out because a particular virus disease, tristeza, assumed epidemic proportions. There is no certainty that this will occur in the country under consideration; it may even by highly improbable. But, still, should by remote chance an epidemic occur, could the research organisation possibly justify not having undertaken, in time, research for the prevention or control of this eventuality? In this particular case, efforts were made to develop a rapid method for identifying propagation material carrying the disease. The economic justification for this type of research is the same as that for any type of insurance policy.

Education or 'learning' research

The development of cadres of qualified researchers is an essential prerequisite for effectively carrying out a national research programme. Much of the publicly supported research is organised around graduate education programmes so that the investment in research simultaneously generates new knowledge and trains additional research personnel (Tichenor & Ruttan, 1970). A very large proportion of the basic research carried out in any special field throughout the world may be relevant for a worker in applied research. However, he will generally be unable to realise the implications of this reservoir of information and its potential value to his work unless he, too, engages in basic research. To be really productive, he will have to piece together information and ideas from several sources, frequently from separate fields of research, and apply them to the solution of a specifically local problem.

Nor should the human factor be overlooked. Sole concern with the routine solution of 'practical' problems may cause the research worker to lose interest and enthusiasm and to become stale. This is a calamity to be avoided. Therefore, it is quite legitimate to select certain projects, not for their merits in solving problems of practical importance, but because they provide research workers with valuable experience, thereby developing useful capabilities and increasing the professional competence of promising young researchers. A pay-off for this kind of research occurs when problems arise in the area in which competence has been gained.

Sociological and economical research

Anderson (1972) has correctly stressed the need for an investment of effort in social science research to parallel investment in agricultural production and biological research. The translation of research findings into agricultural practice, particularly in developing countries, is highly dependent on social attitudes and motivations, as changes in the organisation of farming are generally necessary for the successful adoption of new practices and fields of production, and changes in productivity may cause increases or decreases in labour requirements with far-reaching social consequences.

The value of agricultural research findings is in many cases greatly enhanced if accompanied by economic research. Agricultural research workers should plan their experiments so that the results are amenable to economic analysis (Dillon, 1966), and research teams working on certain types of problems should preferably include an economist if meaningful results that can serve as practical guidelines to extension workers and farmers are to be obtained.

For all these reasons, a balanced programme of agricultural research should include sociological and economic research, with a proper balance between short- and long-range planning.

Summary

Authors are increasingly adding new names and definitions for different kinds of research. To avoid confusion, the following are the most generally used definitions; they are also the ones adopted in this work.

Basic (fundamental, theoretical, pure, academic) *research* studies the laws of nature, without regard for the immediate applicability of the findings.

Mission-oriented basic research is directed to the discovery of new scientific knowledge of potential economic importance and/or needed for the solution of a specific problem.

Applied research consists in the application of existing knowledge to the solution of practical problems.

Adaptative research is designed to adapt technology to the specific needs of a particular set of conditions.

Developmental research is the use of scientific knowledge to produce new or substantially improved materials, devices, products, programmes, processes, systems or services.

Insurance research is aimed at preventing an undesirable situation from occurring, or preparing a remedy for such a situation in time.

Exploratory research is based on original ideas which might have practical implications, but requires preliminary work in order to ascertain if the idea has merit.

Strategic research is designed to generate new knowledge and new methodologies needed for the solution of specific research problems (ISNAR/SPAAR, 1987).

The *basic characteristics of agricultural research* can be summarised as follows:

1. Its dependence on a wide variety of scientific disciplines;
2. its regional character: Not only does the application of basic research findings require investigation under a wide variety of ecological conditions, but basic research, too, is often required for problems of regional significance;
3. its international character, resulting from the exchange of ideas, information, expertise, genetic material, etc.;
4. the interdependence of the various fields of agricultural research, which is a source of organisational difficulties that can only partially be overcome by teamwork.

ROLE OF AGRICULTURAL RESEARCH

The mission of agricultural research has been defined by Aldrich (1966) as:

to apply all possible sources of scientific discovery to the solution of the technical and practical problems of agriculture;

to engage in basic research where the lack of fundamental knowledge may impede progress; and

to solve the specific problems with which agriculture is faced.

The primary objectives of agricultural research are:

to increase productivity by increasing production per unit of area (or animal), or in irrigation agriculture per unit of water, if water is the limiting factor;

to increase efficiency by reducing the input of labour in relation to production or by making the work less onerous;

to increase the stability of production: by breeding varieties of crops and breeds of animals that are more disease-resistant or more immune to unfavourable environmental conditions, by improving methods of crop protection against diseases, pests and weeds;

to improve quality by breeding varieties with inherently higher nutritive values, improved flavour or eye-appeal; improving production techniques that affect quality, improving post-harvest techniques;

to produce the type of products required for consumption, industry and export. This frequently involves the introduction of new crops or methods of production, with the attendant required research, increased control of environmental factors, investigating new uses for established crops, etc; and

to avoid environmental pollution and prevent soil erosion.

The relative emphasis placed on each of these objectives depends mainly on the stage of development of agriculture in each country and its economic requirements.

Increased productivity is usually the first demand to be made on agricultural research, during a period when agricultural methods are still primitive, the population is increasing rapidly, and demand elasticities are still at levels which justify and allow increased agricultural output (Christanssen & Yee, 1964).

For example, in Israel, immediately after independence, the first objective of agricultural policy was to make the country self-supporting in food production. The research programme was therefore geared mainly to the problem of increasing yields of the principal food crops, and productivity of animal husbandry in the supply of milk and dairy products, meat and eggs.

As the result of an intensive research programme, yields were increased three- to ten-fold and more: yields of grain under dryland conditions rose from 600 to 5000 kg and more per ha; under irrigation, from 3000 to over 10 000 kg. The previous primitive breeds of cattle and sheep were replaced by locally developed or selected breeds; milk production increased from 800–1500 kg to 8000 kg per cow per year, placing Israel in the first ranks of dairy producers in the world. An essential element in the transformation of Israeli agriculture, was the gradual change from dryland to irrigated agriculture. When it became evident that the amounts of water that could be made available to agriculture would not suffice, even for the limited areas of land available, emphasis on maximum production per unit area was transferred to the search for ways and means to achieve maximum productivity per unit of water.

In western Europe, in the decade following World War II, increased production was imperative: grain yields were increased 35–40 per cent, milk yield per cow 12 per cent. Total output per unit of input increased by 2 per cent per year in the 1950s, through the use of fertilisers, improvements in plants and livestock, control of pests and diseases, better animal management practices, etc. (ARC, 1967).

Different countries may change the emphasis on the various objectives of agricultural research at different stages of their development. In the United States, with its vast reserves of land, emphasis passed, at a very early stage, from maximum productivity per unit area to increased output per person in agriculture. During the period from 1900 to 1960, maize output per man-hour increased 13-fold (Brown, 1965).

In Japan, with its fixed-land economy, and a farm population that has not yet declined markedly, efforts have continued to be concentrated on expanding output per unit area. Japan's rice yield, which averaged 3064 kg per ha in 1900, reached 5740 kg per ha in 1960 (Barnard, 1957).

Role of agricultural research in developing countries

A formidable task faces the developing nations. The economic and social problems that have to be overcome have been described by Myrdal (1965). The most important social and economic change that has taken place in the developing countries since they achieved independence is a population explosion that makes it much more difficult to solve their economic problems. The flow of financial resources from developed countries is not increasing, but has levelled off; there is also a reduction in private long-term investment, due to a slowing down of development caused by internal political instability and to boundary conflicts.

This situation has been aggravated further because the demand of the developed countries for traditional export commodities of the developing countries has fallen off whilst outlets of new exports have not developed sufficiently. The considerable income gap between developed and developing countries is widening, and food production per head is declining; two-thirds of the people in developing countries suffer from malnutrition or undernutrition; disease and illiteracy, in addition to certain social customs, generally handicap any programme aimed at increasing productivity in agriculture. There are many opinions as to the 'preconditions' which are essential if a 'take-off' into 'self-sustaining growth' is to be achieved by developing countries, but two essential facts are generally acknowledged:

(a) Agriculture has a crucial role to play, and 'take-off' in agriculture is the first essential step. In addition to improving the existing nutritional levels both in quantity and quality, and providing for a rapidly increasing population, agriculture must make substantial contributions which will enable national economic growth on a wide front (Hertz, 1957), including:

- provide raw materials for industrial processing;
- produce export crops as a source of foreign currency;
- increase efficiency, so as to free labour for the expansion of industrial and other economic activities.

(b) These requirements cannot be achieved by reliance on traditional agriculture. In a detailed analysis of the factors involved in achieving a transformation of traditional agriculture, Schultz (1964) reaches the conclusion that traditional agriculture, as such, is in a state of equilibrium, in which the traditional factors of production available are used as efficiently as possible under the circumstances.

Traditional agriculture is characteristically poor 'because the factors on which the economy is dependent are not capable of producing more under the circumstances'. Hence, real economic growth cannot be achieved by simply increasing the input of traditional factors of production. What is required are 'improvements in the quality of the inputs', namely, new agricultural inputs with a relatively high pay-off.

Virtually all these 'new' inputs of potential promise must come from *outside* traditional agriculture—improved varieties, fertilisers, equipment, pesticides, etc., and their success is dependent on their being used efficiently.

Christanssen and Yee (1964) estimated the minimum required increase of agricultural productivity that must be achieved by developing countries to enter the take-off stage of economic development at 4–5 per cent per year. How difficult this is to achieve is illustrated by the fact that in Japan, with its spectacular success in increasing output per unit area, the long-term annual rate of yield increase usually ranged between 1·0 and 1·5 per cent per year; that it *can* be achieved is shown by the fact that more recently agricultural productivity in Japan increased at the rate of 4·6 per cent per year (Myrdal, 1965).

Certain techniques, when applied in combination, can give rapid and spectacular results in a very short time. A combination of improved variety, appropriate fertilisation, adjusted plant population, efficient weed control and plant protection can give increases in yield that range from fifty to several hundred per cent. The expenditure required from the farmer for inputs is low in relation to the additional yield produced, provided the prices charged for fertilisers and pesticides are not inflated as a result of deliberate policy, dependence on unscrupulous middle-men, unrealistic distribution costs, or other man-made factors that disrupt the cost ratio of crop and input factors.

Is a national agricultural research effort justified?

A tremendous amount of agricultural research is being carried out in all parts of the world, and there is already an enormous fund of knowledge available on how to increase the productivity and efficiency of agricultural production. This knowledge is freely available to all and at practically no cost. The argument that most of this research has been carried out in developed countries, i.e. in physical, economic and social environments totally inappropriate to the developing country, has lost much of its plausibility following the establishment of the International Agricultural Research Centres in the tropical and sub-tropical regions, whose research is carried out under environmental conditions that are largely representative of those of the developing countries.

The need for 'own' research in developing countries themselves can therefore be legitimately questioned. In countries which are struggling to establish a sound economy, which lack trained personnel, and in which most of the farmers are illiterate, research may appear to be a luxury which poor countries can ill-afford. It may well be asked whether elementary logic does not compel a developing country to concentrate on disseminating and applying knowledge already available in other countries, or in the international institutes. In other words, should the available limited resources and trained personnel be devoted to extension instead of research.

That this has been the policy followed in many developing countries is indicated by the finding that whilst the highly developed regions invest considerably more intensively in research than in extension, the developing countries in Africa and Asia are several times more 'extension-intensive' than the developed Western countries. A negative correlation was actually found between the level of development and the propensity to invest in extension. The lower the level of per capita income, the higher the proportion of agricultural product spent on extension (Boyce & Everson, 1975).

Even the World Bank gave low priority to agricultural research until the mid-1960s, on the assumption that the already available technology was adequate to achieve production objectives in the developing countries. Leaders in developing countries still feel that research is too costly and appropriate only for wealthy countries (IADS, 1982).

The apparent logic of down-playing the importance of agricultural research in favour of other activities is a dangerous fallacy—and as a policy it is self-defeating.

Basic principles can be established anywhere in the world, but their application to a specific environment requires a local research effort. Not only each country, but several regions within a country, have unique combinations of soils, climate, social, economic, and other conditions which are not duplicated elsewhere.

There is no gainsaying that the cost and duration of research can generally be significantly reduced by transfer of technology from one country to another, or from an international research system to a national research organisation. However, in all cases, adaptive research is required in the actual region in which the innovation is to be introduced before large-scale adoption can be considered. Without adaptive research, costly mistakes and disappointments cannot be avoided.

Quite apart from the need for adaptive research, it has been shown conclusively that the extent to which a country can benefit from the research findings of other countries (or those of the international institutes) depends on its own investments in research (Evenson, 1973).

In an intensive study on the subject of technology transfer from developed to developing countries, Kislev and Evenson (1973) specified an international model in which productivity (in wheat and maize production) was related not only to the research programme of the country in question, but the research programme in other countries located in similar geo-climatic zones. The idea was to determine how much of the research discoveries of other countries could be borrowed by, or transferred to the country in question. The basic relationship found is shown in Fig. 12. This figure shows that a country does 'borrow' or benefit from the research findings of other countries, but that the extent to which it does depends on its own indigenous research capability. This makes policy and planning of research in LDCs even more important.

The international agricultural research institutes can make major contributions to strengthening national research systems and improving their efficacy, but they cannot replace them. The adoption of the high-yielding varieties of wheat was relatively rapid in countries that had their own research capacity, and had only a short-term impact in countries that lacked the research capability to fit the high-yield potential of the new high-yielding varieties to local growing conditions (IADS, 1982).

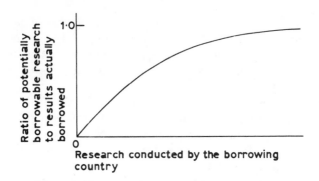

Fig. 12. Basic relationship between 'own' research and research adopted from other countries. (From Kislev & Everson (1973), reproduced by permission of the authors.)

Hence, an essential element in the ability to transform traditional agriculture is an appropriate agricultural programme and the necessary organisation to carry it out. Research on production factors that may have a rapid and considerable impact on yields, does not require overly expensive or sophisticated research in the early stages. It is well within the capabilities of developing countries to establish the *minimum research infrastructure* that is essential, including a network of regional experiment stations with testing facilities and competent staff. Actually, they cannot afford not to do so.

Let us consider plant breeding as an example. Thousands of new, improved varieties of the principal crops are developed and released yearly in the world. It is a common fallacy to believe that such varieties can be successful only in climates identical, or at least very similar to those in which they have been developed. Pragmatic experience in many countries indicates that it would be an unnecessary limitation to confine oneself to crop introductions from the relatively narrow limits of homoclimes.

Some of the most conspicuous successes of plant introduction in Israel have been with introductions from very dissimilar climates, and some of the most marked failures from homoclimes with a typical Mediterranean climate. Excellent results were obtained with potato varieties from Ireland, peas and tomatoes from the UK, sugar beets from Holland, grasses from tropical Africa. The same holds for animal breeds: Friesian milk cattle from Holland, merino sheep from Germany, horses from

Scotland, poultry from New Hampshire have produced outstanding results. An outstanding example in beef cattle is the Hereford breed, which, when transferred from the humid climate and lush pastures of England, was highly successful in the arid, hot plains of Texas, with their sparse and parched vegetation.

One cannot, however, simply go shopping for improved crop varieties — an Introduction Service must be established and the promising introductions tested for adaptability, disease resistance and technological suitability, according to the most scientific procedures. Even such a simple 'application' of scientific achievements developed elsewhere is therefore far from simple. In addition, introduction of new varieties of crops, if carried out without the necessary scientific safeguards, may cause untold damage by introducing pests or diseases previously nonexistent in the country.

Even the best bred varieties are practically useless unless placed under appropriate ecological conditions: nutrient and water requirements, day length, crop sequence, weed, pest and disease control, and many other factors need elucidation. These cannot be determined by reading textbooks published in foreign countries but require appropriate research under local conditions.

However good imported varieties may be, they will soon show one or other deficiency which seriously limits their usefulness. These can be overcome by an appropriate breeding programme carried out by local scientists working under local conditions.

Sooner or later, emergency situations will develop; an unexpected invasion of insects, an unexplained epidemic of disease, or one of the other numerous emergency situations which normally appear in progressive agriculture. These cannot be tackled simply by consulting textbooks or journals, but require teams of research workers who are experienced in their respective professions and fully acquainted with local conditions.

It would be a mistake to conclude from the foregoing that the duties of research workers in a developing economy are limited to 'troubleshooting'. The research worker must pioneer new developments, must think ahead of farmers and planners, and spearhead agricultural progress.

Not only is there a need for the LDCs to adapt research results developed elsewhere, but there are specific problems that can only be investigated in the tropical and subtropical regions in which most of the LDCs are located. Ruttan (1985) stresses the 'very strong need for capacity to conduct research about problems of tropical soils in many parts of Africa, lack of knowledge about soil fertility represents a serious constraint on the ability to design viable short rotation systems, to replace the more extensive slash-and-burn or other long rotation systems now in use'.

Some problems are not region-specific but are even almost entirely country-specific. For example, teff is a major cereal in Ethiopia, but is hardly grown elsewhere; currants are of great importance in Greece; llamas and alpacas are important in the Andes; jute is crucial to the economy of Bangladesh (FAO, 1972).

RESEARCH POLICY

Agricultural research as a client–contractor relationship

Agricultural research workers the world over will readily agree that the basic function of agricultural research is to solve the problems of the farming community; many act, however, as if they feel they are the best, if not the only, arbiters of what requires to be done.

This attitude has been self-defeating, because it has generally alienated the research workers from the farming community and in consequence has resulted in lack of support from the latter for agricultural research. This lack of support inevitably leads to reduced funds for agricultural research, constraints on the ability of the research organisation to achieve the required objectives, and further alienation from the public sectors the research is supposed to serve.

Wherever an effort has been made to break this vicious circle, and to ensure that research should make its proper contribution to agricultural development, it has been found that the only way to achieve this purpose is to establish a formal client–contractor relationship between the research organisation and its clientele, whereby the clients indicate what they require, and the contractor undertakes to implement a research programme in accordance with these requirements.

This interaction between clients and contractor should result in (a) a formulation of research policy, (b) a budgetary framework for implementing this policy, and (c) a research programme based on priorities. These topics will be dealt with in detail in the appropriate chapters.

Relationships of research to users

The researcher–user partnership is an expression of the principle that publicly financed research workers should be accountable for what they do and that they have a responsibility to meet the needs of their country.

This principle has gained wide acceptance as a basis for mission-oriented research.

It is useful to remind planners, administrators, or researchers involved that the objective of agricultural research is to serve national interests —including the interests of producers and consumers of agricultural products.

However, certain differences between this and the ordinary client–contractor relationship need to be considered:

- A research system is not simply a contractor; it is actively involved, on an equal footing, in defining the objectives and requests of the 'clients'. In this sense, it is more that of a partnership than a client–contractor relationship.
- Researchers must frequently engage in 'exploratory' investigations before it is possible to assert what lines of inquiry may be of future use to the clients.
- A research organisation has the responsibility to engage in long-term research for which clients have little sense of need.

The research clientele

The major 'clients' of agricultural research are: the *ministries* concerned with agricultural planning, development, and implementation; the *producers* of all sectors; the *consumers*, including businesses in export or manufacturing based on agricultural raw materials; other *researchers*; *teachers*; and *extension personnel*. Agricultural research policy must take into consideration the varied needs of the several 'clients', which may differ considerably.

Research planners may be subject to pressures arising from conflicts of interests between clients: certain short-term aims of farmers' groups may be opposed to the requirements of the national economy (for instance, government policy may be intended to promote the expansion of crops for export, while certain farmers' organisations may be more interested in growing subsidised food crops). Conflicts of interest may also arise between the national objectives and the wishes of the research workers themselves, who may give preference to subjects of general scientific interest.

Accommodating conflicting demands of clients

An effective research programme attempts to accommodate the different, and possibly conflicting, demands of its clients within the constraints imposed by national goals and the socio-economic and ecological environments. It thus becomes essential to establish a framework within which the 'clients' interact with each other.

A major goal: improving the productivity of the small farmers

A major change has taken place in recent years in agricultural development policy in many developing countries, and even in some developed countries. Formerly, the dominant policy was to favour the agricultural sector that could respond most efficiently to economic inputs (economic efficiency), and there has been little responsiveness to sectors of agriculture with poorly organised-small farmers, part-time farmers, minority groups. However, since the 1970s, the realisation has grown in many LDC's that a tax-supported research system has the obligation, both for reasons of equity and economic efficiency, to respond to the needs of the deprived rural sectors in planning the research programme, and not only to the clients who are most affluent and best organised (Konkle, 1982). At present, many development plans state explicitly that the top national priority is to improve the productivity of the small farmer (social efficiency). This emphasis reflects an awareness that 'without rapid progress in smallholder agriculture throughout the developing world, there is little hope either of achieving long-term stable economic growth or of significantly reducing the levels of absolute poverty' (McNamara, 1972).

A policy discriminating against the backward sector, though it may possibly be based on short-term economic considerations, leads, in the long run, to a widening of the economic and technological gap between the two sectors.

Overall national development requires a more equitable distribution of income which makes possible a wider and more effective market. In a predominantly agrarian economy, it is likely that the overall level of saving and investment will be higher if the productivity of land and labour is increased by the widespread adoption of new technology on a large number of family farms, than by concentrating resources on a relatively restricted scale in an advanced sub-sector of farming. It is also probable that the latter will, in any case, adopt new practices even when most government efforts and incentives are concentrated on the more backward sectors of the rural economy. There does not *have* to be a conflict between equity and productivity.

Equitable income distribution due to increased productivity of small farms also favours the growth of small scale agro-industries in the rural regions, which can further improve income distribution (Schutyer & Coward, 1971).

Elz (1984) argues that the emphasis on technology for the small farmer may have certain short-term disadvantages under certain conditions: 'these farmers produce little, if any, surplus for the urban sector. If the larger part of the total population, including most of the low-income groups, is located in urban areas, the supply of technology to the low-income farmers may further reduce the food supply and increase the poverty in the cities.' However, the lack of a surplus for marketing is precisely the direct result of the subsistence-type farming practised by the low-income farmers. Improving the technology of this sector aims at creating a surplus for marketing, and thereby augmenting food supply to the urban sector.

A further argument proposed by Elz, is that: 'there may be so much migration from rural to urban areas that by the time technology for the small farmers has been created, they may already have migrated to the city, and there may be no more small farmers in a particular area to use the new technology developed especially for them. Or these small farmers may have formed a co-operative to use modern large-scale technology and some may have grown into larger farmers and may no longer be interested in small-scale technology. In this case, the resources allocated to create technology for the small farmer may not be of any use and may be considered to have been wasted.

It is the poverty of the small farmers that drives them to migrate to the cities; only the improvement of their farming methods can counteract the tendency to migrate to the cities, where they are neither needed or wanted. There is not the slightest prospect of small, subsistence farmers growing into larger farmers, or even forming co-operatives using large-scale technology, unless their productivity is first increased by the adoption of improved technology.

Implications for agricultural research

Discrimination against large and medium farmers?
Does a policy oriented towards small farmers necessarily result in discrimination against the more favoured agricultural sectors? Such a discrimination could prove to be costly in terms of economic development, in view of the important share in agricultural production of the advanced sectors.

Schultz (1977) writes: 'Important as it is that economic policies do not bypass and not discriminate small farmers, agricultural scientists who are endeavouring to develop more "efficient" plants (and animals too) in terms of their genetic capacities, and chemists who are engaged in developing cheaper and better chemicals, should *not* be placed under the constraint that the fruits of their research be applicable only to small farms.'

There is no reason to assume that most of the results of biological research cannot be adapted and adopted by large farmers. It is only a few special objectives, such as breeding and management of small stock (sheep, goats, rabbits, etc.), special equipment for small farms, storage methods, mixed cropping etc., that may be of significance to small farmers only.

Scope and nature of agricultural research
Though concern for the small farmer does not involve major changes in the traditional topics of biological research, it does entail a revision of basic concepts of the responsibilities of the research organisation, and on how and where research should be implemented.

Role of agricultural research reappraised
The formerly widely accepted premise that the role of the agricultural research worker is concluded when a solution to a given problem is achieved in the laboratory and/or the fields of the experiment stations, and that 'adaptation' of these results to actual farming conditions is the responsibility of the extension service, is no longer tenable.

The net result of such a policy was the formulation of technological proposals that frequently could not be adopted by the majority of the farmers, and consequently caused an alienation of the research worker from the real needs and circumstances of the majority of the farming community.

The realisation that the responsibility of the research worker is involved right up to the stage at which a proposal can be implemented by the farmer, is gradually gaining ground.

Another major implication of the new approach relates to the way research is implemented. Results of the research carried out at the experiment station have to be tested under the conditions faced by the farmer, and the recommendations subsequently have to be adapted to these conditions. This has led to a far greater dependence on on-farm research (cf.

(cf. p. 345), adaptive research, verification trials, farming systems research.

Farmer involvement. Testing technologies developed at the experiment station by on-farm trials is not a new approach. What is new, is that in the past these experiments were carried out in the farmer's fields, with very little involvement by the farmer. The new approach stresses participation of the farmer in diagnosis of the problem, choice of the variants to be included in the trial, in the actual field work, and in the evaluating of the results.

Farming systems research. Frequently, on-farm trials reveal that the adoption of a new technology is not possible without major dislocations of the existing farming systems. In these cases, new farming systems have to be evolved and tested. Furthermore, certain goals cannot be achieved by research on individual commodities, but require a 'systems' approach. For example, a stable farming rotation to replace shifting cultivation in the humid tropics, a more intensive system of rain-fed cropping by integrating arable cropping and animal husbandry in the Near East.

These requirements have led to the now widely accepted concept of Farming Systems Research (FSR).

INTEGRATIVE RESEARCH

Most research, in developed and developing countries alike, has been mainly concerned with the investigation of individual components of production-improved varieties, fertilisation, soil management, irrigation, control of diseases, pests, weeds, etc.

Even factorial designs can only cope with a very limited number of interacting factors. Also, experiment station research is not usually conducted under the prevailing farming conditions, especially if the farming is largely traditional.

Whilst farmers in developed countries can easily overcome the problems involved in the introduction of a new crop or the adoption of a new technique into their farming systems, this is not the case with the subsistence farmer, who faces numerous economic, social and institutional constraints impeding change. Also, a dramatic improvement in the productivity of subsistence farming cannot be achieved by the piecemeal introduction of single practices (cf. p. 776).

For all these reasons, an intermediate testing stage is essential that can bridge the gap between the experiment station and the farm. This role has been allocated to farming system research. This approach was first applied in Senegal (Unités Expérimentales) where it was found to be very effective, and was called pre-extension research.

An example of the need for such an intermediary testing stage is the attempt to apply station experiment results to farm practice in the case of an experiment on the effect of depth of ploughing on the yields of groundnuts. This experiment was carried out at Bambey, the Central Research Station of Senegal. It was found that by increasing the depth of ploughing the seedbed by a few centimetres, it was possible to achieve a dramatic increase in yield of groundnuts, a major food and export crop in Senegal and other African countries.

What can be more simple than increasing the depth of ploughing by a few centimetres, especially if this makes possible a significant increase in yields? However, when the extension agents instructed the farmers to set their oxen-drawn ploughs, mounted on tool-carriers, so as to achieve the required depth of ploughing, the animals were unable to budge the plough. What had been easy work for the improved breed of well-fed oxen in the research centre, proved to be an impossible task for the small, emaciated oxen of the farmer, poorly fed during the months of the dry season on sparse, overgrazed range, and required to make a major effort at the beginning of the rainy season.

In order to incorporate an apparently simple change in the existing technology of a traditional crop, the following steps were found to be essential:

- breeding a larger and stronger ox, entailing a cross-breeding programme;
- providing forage reserves to supplement range during the dry season, so as to have the animals fit for heavy work at the beginning of the rainy season. This in turn entails:

- research on appropriate forage crops;
- changes in the cropping system, to include one or more forage crops;
- research on appropriate forage conservation methods;
- research on feeding rations;
- affecting a complete change in the mentality of the farmer, for whom it has been axiomatic that his oxen work for him, and not he for the oxen, as the growing and feeding forage implied.

This example shows how even a simple change in a traditional operation, can involve far-reaching changes in the technology, crops, farming system, and attitudes of the farmer.

Safrenko (1984) mentions some of the reasons why the transfer of research results to the majority of the farmers has been hampered:

- the research did not take into account the complete farming system;
- it was difficult to identify the social and institutional constraints faced by the farmers;
- the capabilities of the small farmers were not taken into account;
- the research was often carried out with equipment and labour not available to the farmers.

The need to overcome these and other constraints to the efficient transfer of research results has been increasingly accepted worldwide since the 1970s, and has resulted in various methods aimed at testing research proposals under conditions relevant to actual farming conditions.

This has involved the use of testing and evaluating systems, from simple on-farm trials, verification trials on representative sites, to the more comprehensive pre-extension research and farming system research.

Farming Systems Research (FSR)

Farming Systems Research in its modern form was pioneered by R. Bradley, a soil scientist as well as a practical farmer, who on retirement from Cornell University, moved to the International Rice Research Institute (IRRI) in the Philippines. He investigated agronomic practices to increase the total annual production of rice, using carefully planned farming systems in which rice was intertilled with upland crops or followed quickly by dry season crops.

He was able to increase the traditional yields from rice land from around eight tonnes per hectare to more than 22 tonnes equivalent grain for the full growing period, whilst maintaining soil structure unimpaired.

After Bradfield's pioneering work, interest in what is now called FSR intensified in the decade that followed. Many research teams in different countries have addressed themselves to the study of traditional farming systems and to the design of new farming systems adapted to the various combinations of environment, production factors and technological levels under which various categories of farmers operate. Various formal procedures have been elaborated and there is now an extremely rich literature on the subject.

There is therefore no need to describe the FSR in detail here, and we will confine ourselves to a brief description and discussion of certain salient points of the system.

Basic factors in FSR
The basic factors of FSR have been identified by Hildebrand & Waugh (1983) as being:

- concern oriented to the disadvantaged rural sectors;
- recognition that thorough understanding of the farmers' situation, gained first-hand, is critical to enable an increase in their productivity and form a basis for improving their welfare;
- the need for a multidisciplinary approach in order to understand the farm as an entire system rather than the isolation of components within the system.

FSR 'requires researchers to examine more closely the components of a farming system, and their interactions; to identify constraints to, and opportunities for, increased productivity; and to design, test, and evaluate technical innovations under actual farming conditions. The farmer, the farm, and the surrounding physical, biological, and socio-economic environments become the foci of research efforts'. An overview of such an integrated approach is shown in Fig. 13.

FSR techniques
Various techniques for FSR have been developed and described in the literature. They all have certain basic approaches in common.

The research team. The farming system studies are carried out by teams comprising several of the following disciplines: agronomy, animal husbandry, agricultural engineering, extension, economics, sociology, anthropology.

The core of such a team generally consists of a team leader, an agronomist, an economist and an extension specialist.

Steps in FSR. FSR involves the following stages (Kellogg, 1977):

1. diagnosis of farmers' circumstances and actions in a defined 'target area';
2. planning and design of technological adaptation;
3. on-farm testing and verification;
4. multi-locational field trials (Fig. 14).

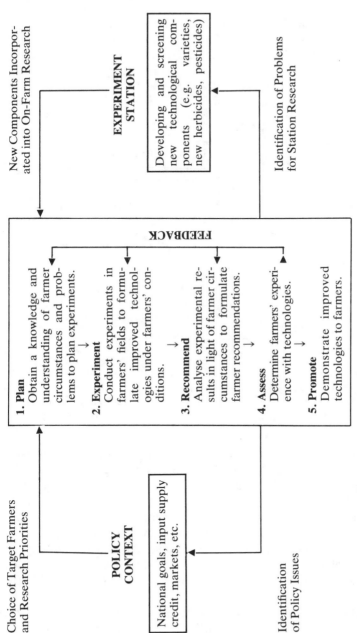

Fig. 13. Overview of an integrated research programme. (From Byerlee *et al.* (1984), reproduced by permission of CIMMYT, Mexico.)

ON-FARM RESEARCH

FEEDBACK

1. Plan
Obtain a knowledge and understanding of farmer circumstances and problems to plan experiments.
→

2. Experiment
Conduct experiments in farmers' fields to formulate improved technologies under farmers' conditions.
→

3. Recommend
Analyse experimental results in light of farmer circumstances to formulate farmer recommendations.
→

4. Assess
Determine farmers' experience with technologies.

5. Promote
Demonstrate improved technologies to farmers.

New Components Incorporated into On-Farm Research

EXPERIMENT STATION

Developing and screening new technological components (e.g. varieties, new herbicides, pesticides)

Identification of Problems for Station Research

Choice of Target Farmers and Research Priorities

POLICY CONTEXT

National goals, input supply credit, markets, etc.

Identification of Policy Issues

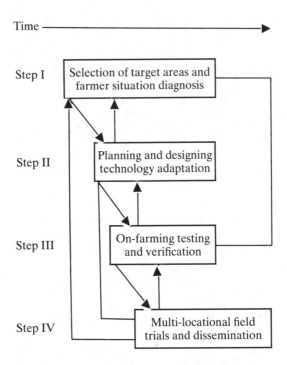

Fig. 14. Steps in a general approach to adapting and extending new agricultural technology. The arrows indicate that steps overlap in time and results in later steps may require going back to previous steps for further analysis. (From Johnson & Kellogg (1984), reproduced by permission of FAO, Rome.)

Some authors (e.g. Johnson & Kellogg, 1984) assign the responsibility for stages 1 and 2 to the research organisation, and stages 3 and 4 to the extension service. This division of responsibility is unacceptable if one accepts the premise that the function of agricultural research does not end until its recommendations have been tested and adapted under the conditions in which the farmer, for whom the recommendations are intended, operates. Under this conception, on-farm testing and multi-locational field trials are an integral part of the research process, in which extension workers participate and can make valuable contributions to a joint effort.

It is at the diffusion stage at which the extension service should take over responsibility. In view of the importance of strong linkages between research and extension, the subject will be treated in detail in Chapter 22.

Drawbacks in FSR

An overly-detailed collection of socio-economic and environmental data may be academically excellent, but it is also time-consuming and costly. There is the danger that FSR may become an objective unto itself, and no longer fulfil its role as an essential link in the transfer process from the experiment station to the farmer.

There is an enormous amount of information that can be collected in a 'target area' and the essential information may become 'lost in a mass of environmental, social, economic, institutional, technical and individual detail'. Hence the importance of Safrenko's (1984) insistence that 'a choice must be made as to what is relevant and important'. One must focus on some of the more obvious major constraints farmers perceive and experience and the environmental factors affecting agricultural decision-making.

It is in the light of these considerations that simplified, but satisfactory, methods have been developed for FSR, which are based on two principles (Chambers, 1980):

Optimal ignorance: attempts to avoid 'overkill' of information and to be satisfied with sufficient information to start development planning, whilst remaining conscious of the need to gradually build up more information through a process of doing by 'learning'.

Proportionate accuracy: attempts to obtain a degree of accuracy of the data collected which is sufficient to orientate the research programme.

Chapter 6

Allocation of Resources for Agricultural Research

SOURCES OF RESEARCH FUNDS

The major sources of funding for agricultural research are:

- government allocations from the national budget;
- grants, loans and technical assistance from foreign sources;
- taxes on agricultural commodities; and
- private sector contributions.

The source of the funds available for research evidently has considerable bearing on budgeting policy. Usually, only part of the funds needed for agricultural research are obtained from government sources on a more or less permanent basis. The budget allocated by the government is usually adequate mainly for maintaining the organisational framework of research: permanent staff, laboratories and field stations. The actual operation of this framework is largely dependent on outside allocations, derived from farmers' organisations, foundations, foreign interests, etc., which by their very nature are fluctuating, impermanent and usually earmarked for specific purposes. This results in the highly unsatisfactory situation of planning by the purse, instead of the other way around; objective planning under these circumstances is extremely difficult. Probably the least scientific aspect of agricultural research is the actual planning of the research programme, based on the budget made available.

348

GOVERNMENT ALLOCATIONS FOR AGRICULTURAL RESEARCH

Why should funding for agricultural research be a public responsibility? Since the 1970s, the question of who benefits from agricultural research has been increasingly debated, and it is far from clear who benefits most: the farmer, agribusiness, or the consumer.

The distribution of economic benefits from publicly funded research among different sectors of society will depend to a large degree on the major goals and strategies adopted (Pinstrup-Andersen, 1982). Interest groups from the producer, consumer and marketing sectors will try actively to influence agricultural research programmes. Considerable conflict of interest may exist between these sectors, and even within a sector (e.g. commercial farmers versus subsistence farmers).

Whoever benefits most does not affect the justification for governmental funding of research, because: (a) the benefits that accrue from the research are generally shared by the majority of the population, and (b) most results of agricultural research cannot be appropriated by those who finance the research.

The competition for national funds for research

Research resource allocation from the national budget involves several categories of decisions, made at different levels:

(a) What proportion of the national budget should be devoted to research in general and to agricultural research in particular;

(b) how the total allocation to agricultural research should be apportioned between problem areas in which research is required to achieve the stated goals;

(c) how to divide research resources between the different ecological regions of the country;

(d) what proportion of the research budget should be reserved for research that is not directly linked to the short-term goals of the development plans—such as long-term research, basic research, exploratory research, etc., essential for ensuring a balanced programme;

(e) how to determine priorities between research projects within each area of research.

This grouping of decisions is necessary if one is to adhere to the principle that the decisions have to be made between *commensurable alternatives* (Toulmin, 1966) e.g. the respective importance of different commodities, between ecological regions, between long-term and short-term research, etc. Choices that are based on the national interest, as perceived by the decision makers, are *political decisions*, whilst choices that are concerned with implementing the political decisions are *professional* or *administrative* decisions.

Decision-making should be confined to making choices between political alternatives or between professional alternatives. When attempts are made to decide between an alternative that is an end in itself (a political choice) and an alternative that is a means to an end (a professional choice) confusion is bound to arise. Professional considerations may of course influence political decisions and vice-versa.

Allocation of funds between different fields of science

Allocating funds earmarked for research among different fields of science is essentially a political decision. One of two paths is generally followed:

- An overall sum is allocated to a special Ministry of Research, which will have to make decisions on the relative shares of: academic and applied research and subsequently, within the latter between industrial, medical and agricultural research.

  A Ministry of Research will be hard put to find a rational basis for making decisions on how to divide funds between academic research and mission-oriented research; or between mission-oriented research aimed at improving health of the population or that required for increasing agricultural production.

- A different and less frequent path used, mainly in some developed countries, is that of deciding on the amount to be allocated to uncommitted research at the cabinet level and leaving to an appropriate scientific body the decisions on its allocation to different fields of academic science. Funds earmarked for technological and economic-oriented research in different fields are included in the budgets of the ministries concerned. These then decide on the relative amounts to be allocated to development, regulatory, extension and research activities. In this case there is a clear distinction between political decisions and administrative decisions.

Whatever pathway is followed, in all countries in which agricultural research is a public sector activity, and is funded on the basis of an annual allocation from the national budget, it has to compete for funds with other public sector activities.

In view of the chronic financial difficulties faced by most LDC's, this competition is generally detrimental to agricultural research, and support for it has fallen considerably in many counties in the 1980s (Trigo & Piñeiro, 1984).

Allocation of resources to agricultural research in the aggregate

The determination of total national expenditure for agricultural research presents many problems.

Ideally, the national budget for agricultural research would be calculated after the research programme has been established. It would consist of the cumulative amounts necessary to undertake all the research projects that have been found, after careful evaluation, to have sufficient merit for inclusion in the national programme. This approach would be satisfactory if the proposed budget were adequate for funding all *essential* research. This situation hardly pertains to agricultural research.

In general, the total budget allocated to research is determined pragmatically, as a result of negotiation and bargaining, under the influence of the following factors:

(a) the size of the budget allocated in the previous year;
(b) changes in the cost-of-living index;
(c) government policies in relation to retrenchment or expansion;
(d) changing trends in agricultural production;
(e) the influence of pressure groups;
(f) the size of allocations from outside sources.

Seiler (1965) writes that even in industrial research, in which a number of highly refined formulas and models have been developed for determining optimum research and development budgets, such as those based on 'return-on-investment', the determination of the actual budget is also in most cases a mixture of intuition, judgment and experience.

However, as the World Bank (1981) points out 'it is ironic that despite the promise of high returns, investment in agricultural research continues to be accorded low priority in many developing countries'.

In a study by Boyce and Evenson (1975), the authors come to the conclusion that developing countries were allocating insufficient resources to agricultural research, notwithstanding 'the extraordinarily high rates of return that have been measured in virtually all of the studies of agricultural research productivity'. The authors indicate that there are two major factors limiting the responsiveness of policy-makers to allocating funds to agricultural research: (a) the limited availability of qualified researchers; (b) the inability of the investing country to appropriate for itself all the gains from research, as these may flow to other countries. The first constraint mentioned by Boyce and Evenson is primarily a consequence of limited allocation of funds to research and not the basic cause; as to the second constraint, most of the adaptive research carried out in the developing countries is extremely location-specific and the results would hardly be 'captured' by other countries. Far more convincing is the assumption that the allocation of funds to agricultural research is limited because of the scepticism of policy-makers regarding its potential contribution to development. This view is largely due to the lack of impact of past research in their respective countries on the agriculture practised by the majority of the farmers. The policy-makers do not generally recognize that the major reason for this situation is the *inability* of the farmers to adopt research recommendations, *mainly* because of constraints due to economic factors, inadequate infrastructure and general institutional neglect of the small farmer. It must also be admitted that the research itself was generally costly, because of its fragmentation among many authorities, the dispersal of its meagre resources among a large number of institutions, the lack of planning and defining of priorities, and other shortcomings.

In brief, there can be no convincing justification for an increase in allocations to agricultural research unless measures are simultaneously taken: (a) to organise the national research effort so as to make it productive; (b) to ensure that the research findings are transmitted to the majority of farmers; and (c) making it possible for farmers to adopt new technology by appropriate economic and institutional measures.

CURRENT INVESTMENT IN AGRICULTURAL RESEARCH

The developed countries, accounting for less than one-third of the world's population, spend three-quarters of the global expenditure on agricultural research (which approached $5000 million in 1980 (World Bank, 1981)). As a result, it is the developed nations that reap most of the direct economic, political and social benefits from agricultural research.

Though the expenditure on agricultural research in Third World countries has risen moderately in the 1970s, it still does not amount to more than about 25 per cent of the world total, and the sums spent are markedly below those justified by economic and social requirements. Pinstrup-Andersen (1982) writes: 'that it is not surprising that poor countries emphasise investment with expected short-term pay-off, something that generally cannot be expected from agricultural research. Politicians faced with a situation of extreme poverty, starvation and related existing or prospective social and political unrest, may have little choice but to attempt short-term solutions, even with certainty of high costs in the long term'.

ISNAR data show that about 65 per cent of NARSs are facing a decline in real terms, in their per-scientist financial support, leading to low productivity, lack of staff motivation, and high rates of attrition as staff leave for more attractive opportunities elsewhere.

A World Bank study (1981) indicates that in order to provide an adequate amount of food to their rapidly increasing populations, the yield of land that is already being farmed in the LDCs must increase by about 2 per cent yearly.

In a review by the International Food Policy Research Institute of the resources allocated to agricultural research in 65 developing countries, it was found that there are serious problems of underfinancing and concomitant understaffing of agricultural research in low-income countries, which are incidentally the countries in which agriculture is the dominant economic sector. The proportion of the agricultural GDP spent on agricultural research was 0·26 per cent in the low-income countries and 0·33 per cent in the middle-income and high-income countries. This compares with 1–2 per cent range in the developed countries (World Bank, 1981 (Fig. 15).

Determining the financing of research on the basis of a given percentage of the agricultural GDP can be a highly misleading criterion (cf. p. 354). It may, however, have some use for pointing out to policy-makers that a country is not allocating sufficient funds to agricultural research. An allocation of less than 0·5 per cent is certainly too low, and a target figure of at least 1·0 per cent of agricultural GDP is desirable (ISNAR/SPAAR, 1987).

A more logical approach is to consider research expenditures in relation to *per capita* expenditure of those directly dependent on agriculture. The spread between developed and developing countries is very wide. In a comparison of five developed countries (USA, UK, Germany,

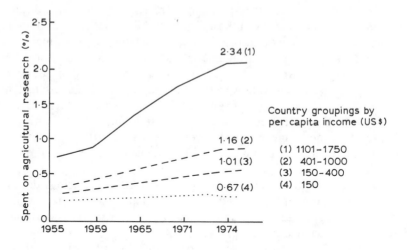

Fig. 15. Expenditures on agricultural research as a percentage of the value of agricultural product (1951–74). (Based on data from the Asian Development Bank, 1977.)

France and Holland) and five developing countries (India, Malaysia, Sri Lanka, Chile and Tanzania), the expenditure was in the mid-sixties between $20 to $50 per farm family per year in the developed countries and between 0.5 *cents* and $2 in the developing countries (Yudelman *et al.*, 1971). Matters have not changed appreciably in recent years.

The major drawback in associating investment level in research with the value of the agricultural product is that according to this criterion the more advanced the agricultural productivity of a country (and hence the larger the value of the agricultural product) the higher the investment in research. It is, however, those countries in which the majority of the farmers still practise subsistence agriculture that have the lowest value of agricultural production and the greatest need for technological change. It is therefore the countries in which the need is greatest that suffer from a chronic shortage of funds to establish and operate effective national research programmes and whose investment is lowest, both in absolute figures and in relation to the agricultural GNP. It is these countries which are in the greatest need for external support, both for the establishment of an efficient, if small, research infrastructure and for operating expenditure. If such aid is not forthcoming, the vicious circle of low-productivity

– inadequate investment in research – stagnation in agricultural productivity – will be perpetuated. The international research centres should also give preferential attention to these countries, and serve them by providing improved practices, materials and training.

In many countries the national system is so weak as to impede the utilisation of the new technologies being developed by the international agricultural research centres. The present levels of investment in agricultural research are so low that the prospects for rapid technological advances and for achieving the target proposed by the World Bank are seriously jeopardized.

In the mid-1960s, the average annual expenditure for agricultural research of 41 developing countries (23 in Africa, 9 in Asia and 9 in Latin America) was about $2 million which is less than the estimated annual expenditure in one experimental station in California (FAO, 1970).

The World Bank considers a four-fold increase in expenditure on agricultural research as a desirable long-term target in countries where agriculture is the key economic sector, but is aware that such an increase is not always immediately feasible. Four aspects must be taken into consideration in evaluating an appropriate investment target for a country:

- the size and scope of its existing agricultural research programme;
- the present share of the agricultural GDP spent on research;
- the availability of qualified research workers and support staff;
- the capability of the country to support financially a larger research effort.

ISNAR (1987) points out that there are often inconsistencies between the objectives fixed for agricultural research, the policies designed to realise these objectives, and the level of resources allocated to research. The rates of growth in national support for agricultural research in real terms are declining; at the same time governments are placing increased emphasis on agricultural development.

The World Bank (1981) points out that financial aid to the developing countries for an expanded research effort, that promises attractive economic and social benefits will be of limited usefulness 'unless borrowing countries accord high priority to national agricultural research by providing adequate financial support and appropriate economic and other policies that encourage the adoption of improved technologies and that, above all, establish a hospitable work environment for national research scientists and administrators'.

Allocation of resources to different areas of agriculture research

Every country must define the research areas that are indispensable for the achievement of the goals defined by the competent bodies. These research areas can be broken down into four categories (Ruttan, 1981):

(a) *activities*: which specify the kind of activity or its purpose; e.g., improving biological efficiency of plants and animals, protection of plants and animals from losses or damage due to insects, diseases, weeds, etc.; efficient marketing. Socio-economic research, such as identification of constraints to adoption of new technology, and the impact of adoption on different sectors is becoming increasingly recognised as an essential complement to biological research;

(b) *commodities*: on which research is required and their relative importance;

(c) *resources*: indicating the resources—soil, water, forest, range, wildlife, etc.—that require research. The establishment and maintenance of an inventory of the country's resources is an essential element in developing an adequate research programme. Developing countries are becoming increasingly aware of the need for research aimed at conserving the productivity of their natural resources and, in particular, the ecological consequences of sustained high levels of application of chemical inputs.

(d) *field of science*: which designates the scientific disciplines to be employed in solving the problem: entomology, genetics, nematology, biochemistry, etc.

To these the present author would add:

Regions: which indicates how the total research efforts should be apportioned to the different regions of the country, with special emphasis or farming-system research in relatively resource-poor regions.

This grouping of decisions is necessary if one is to distinguish between political and professional decisions. For example, the decision to orient research primarily to small farmers, to food crops or to a marginal region is a *political* decision which may legitimately override economic considerations; priorities between projects within a given field should definitely be a *professional* decision; others, such as apportioning funds between short-term and long-term research, may involve both political and professional considerations.

Basically, the political decisions define the broad social goals and priorities, and research administrators and scientists allocate the research resources according to professional criteria so as best to meet these goals and priorities.

How decisions on priorities within areas of research are made is treated in detail in Chapter 7.

CRITERIA PRESENTLY USED

All research efforts have to operate within budgetary constraints, and hence the great importance of using rational methods for deciding on the relative shares to be allocated to different areas of research. In practice, this is rarely done, and the following criteria, based on 'rule of thumb', are generally used.

Precedent

The previous year's allocation for each field is used as a base, to which a fixed percentage is added or subtracted. 'The very nature of the research process, its continuity and the fact that to a large extent research feeds on itself, means that the funds allocated at any one time will depend on the amount that has been spent in the past' (Horowitz, 1960). This past expenditure represents not only projects that must be continued until completion, but also knowledge that has been gained and gives rise to future research. The extent to which the people responsible for policy-making in a ministry of agriculture are research-oriented and appreciate the value of research will have influenced past expenditure. The present budgets will, in turn, be influenced by these expenditures. However, it is not logical to assume that the starting level was rational or that there is justification for perpetuating it. Despite this serious drawback, allocation of research funds to research institutions, departments, and other bodies tends to be governed by precedent. Once a research unit is allocated a certain amount of funds, it becomes difficult to change this allocation, other than to increase it.

The main obstacles to change and adjustment of allocations to research areas are the following:

- The historical pattern in a given situation gives certain research areas special emphasis. Thus training and experience, resulting in special competence, have been greatest in those areas. This tends to perpetuate past emphasis, even after conditions have changed.

- Pressure groups may occasionally help bring about a change in emphasis, but they most usually play a conservative role.
- The personal biases of administrators, research workers, and other interested persons are also influential.
- There is a tendency to favour short-term problems, as well as 'safe' problems, and to avoid untried areas.
- Tenure and flexibility of specialists are other contributing factors.

Increases or decreases in allocations are generally shared more or less equally by all departments, thus accounting for the relative stability of shares. This suggests that even in a country which is establishing a modern dynamic agriculture, tradition and rule of thumb predominate, while the process of problem-oriented decision-making is still only marginally applied. Generally, a change in the relative shares of different research areas is dependent on some significant crisis, such as the retirement of a key figure or the dying out of a branch of production. As a result, new research fields can usually be developed only when additional funds, specially earmarked for this purpose, are made available. By contrast, in crops in which much research has been invested for a number of years, the potential for further progress may be largely exhausted.

The rigidity in allocation patterns and the perpetuation of traditional research activities, described above, raise doubts as to whether the research organisation is sufficiently sensitive to the problems of society, particularly of a rural society, and whether it adequately reflects the changes that occur in a dynamic architecture.

Percentage of gross value

Frequently a certain percentage, usually between 0·5 and 1·5 per cent of the total gross value of the annual production of a commodity, is devoted to research. This solution favours the branches of production already well established, whose research needs are not necessarily greater than those of the less developed branches. In particular, new branches of production—the potentialities of which cannot even be guessed at without preliminary research—are the most discriminated against in this approach.

Another bias, due to this approach, is in favour of export and cash crops, discriminating against the traditional food crops. Other fields of research that tend to be neglected, due to this approach, are livestock, tropical pastures and agro-forestry.

The method also has serious methodological limitations. Production values for food crops, which are largely consumed on the farm, are generally grossly underestimated in national accounts, resulting in a downplaying of their importance when establishing the criteria for priorities. The prices of the commodities may be artificially distorted for reasons of policy, further discriminating against them.

Matching funds

Research funds allocated by the industry itself are in certain cases matched in a fixed ratio by government funds. The industry funds may derive either from voluntary contributions made by the farmers' organisation concerned or from a special tax imposed on the product itself— either by the government, by the farmers' organisation or by marketing or production boards. This approach has the same weakness as the previous one, in as much as it also favours strong fields of production over weak ones, and certainly does not generate new possibilities.

Ad-hoc basis

Research funds may be allocated on an *ad-hoc* basis, in which case an overall *policy* favouring certain sectors of production or research areas at the expense of others will be the basic guideline. Useful guidelines followed for determining policy in relation to the different sectors are (a) growth potential, (b) improving income distribution, (c) potential contribution to improving the trade balance (increasing exports, reducing imports), (d) influence on other fields of production (e.g. producing raw materials for industry), and (e) efficient use of input (land, water, labour, etc.). Rule of thumb is still in evidence when decisions are based on expectations of future developments; yet, such decisions are probably more rational than those based on the other yardsticks mentioned above.

GRANTS, LOANS AND TECHNICAL ASSISTANCE

Research institutes in Third World countries augment the funds available from government with grants, loans, and technical assistance from multilateral credit organisations and bilateral aid programmes of developed countries.

An idea of the magnitude of the sums involved is provided by the following figures: from 1971 to 1980, the Interamerican Development Bank (IDB) granted 13 loans to eight Latin American countries for a total value of about US$138 million. Non-reimbursable grants for technical co-operation totalling US$25 million were provided by the World Bank for 20 agricultural research projects in 13 countries, and two research loans totalling US$96 million. In 1980, the USA had 25 projects under way in Latin American countries and nearly US$70 million committed until 1985 (Trigo & Piñeiro, 1984).

In most cases, financial assistance to LDCs is made contingent to the concurrent allocation of national resources: monetary, technical, human, etc. As a result, national resources are allocated to projects whose priorities are largely dictated by outside interests. Furthermore, these commitments reduce flexibility in national research planning (Elz, 1984).

Pinstrup-Andersen & Byrnes (1975) point out correctly that 'this situation places great responsibility on the external agencies to ensure that their priorities do in fact correspond to national needs'.

NON-GOVERNMENT LOCAL SOURCES OF FUNDING

Bredahl *et al.*, (1980) argue that researchers are not likely to search for financial support from interest groups if their needs for research funds can be met adequately by the research institution. However, with increasingly reduced budgets, annual appropriations are channelled more and more into salaries and overhead expenses, leaving little for implementing specific research projects. 'The interest group that provides the marginal addition of spendable cash can influence the entire research activity including substantial magnitudes of permanent state and federal funds' (McCalla, 1977).

Special granting agencies

In many developed countries, research workers can go 'shopping' for research funds made available by different public, semi-public and private agencies, such as research councils, foundations, etc. In countries where competitive *research grant systems* have been adopted, these are often used to redress the bias against research in the basic natural sciences, the social sciences and the humanities and to provide incentive for many creative scientists in the universities (ISNAR, 1987b).

Special taxes (cesses)

A different form of support is the imposition of special taxes on specific commodities, mainly export crops. In some rare cases, these taxes may be a major source of funding for the NARS. For example, the Argentine NARS (INTA) derives most of its funding from a tax on agricultural exports. The major drawback of this system is that the value of the exports may fluctuate considerably from year to year, being influenced by factors over which the NARS has no control, such as volume of production or market prices.

The private sector

In many countries, farmers' associations have provided financial support to the NARS. This support is generally directed to specific commodities, generally produced by a small, but advanced sector.

The private sector is also becoming increasingly involved directly in agricultural research, especially in the developed countries, in those areas in which it can 'capture the benefits of research'. In the USA, over half the agricultural research is funded by the private sector (cf. p. 247).

Shortcomings of dependence on outside sources of funding

Since the 1970s, traditional sources of support for agricultural research have begun to decline in many countries, developing and developed. Broad farm groups have become increasingly overshadowed by many powerful sectoral interest groups. This has led to a greater fragmentation in the overall support for research. At the same time, national funds for research, measured in constant dollars, have stagnated in many cases. Administrators, however reluctantly, have felt compelled to encourage researchers to seek external funds (Busch & Lacy, 1983).

We have already pointed out how this can influence the entire process of allocating resources according to national priorities.

Bush & Lacy (1983) further point out that 'time spent on searching for grants is time that might otherwise be used for research'. As funds become scarce, the probability of the individual researcher obtaining a grant is reduced and the costs of grant-seeking increase.

Over the years, research administrators have courted various groups to ensure continued funding of research. In so doing, 'perhaps they have become the servants of power' (NAS Committee (1972)). Scientists are generally unhappy with this role.

Researchers who seek to free themselves from the constraints of this system, have sought support from various granting agencies. Many have, however, found that such support entailed severing one's ties with the problems of agriculture and rural life. This has tended to fragment the agricultural research system (Busch & Lacy, 1983). If this is true for the rich USA, how much greater can the power of interested groups to orient the research effort to their particular benefit be in less rich countries?

The potential conflicts arising from pressure by various interest groups, and in particular from funding by commercial organisations, were recognised in the USA already in the 1920s. A special committee on 'Organization and Policy for the SAEs' recommended the following guidelines in determining when grants could be accepted from outside sources (Barre *et al.*, 1928):

(a) if the research to be funded was of general public importance;
(b) the funds were to be made available through the research institution and not directly to the researchers;
(c) all patent rights were to be reserved to the research institution;
(d) results were to be published through the regular channels, whether or not they were favourable to the organisation that provided the funds.

These guidelines are just as valid today, as when they were formulated more than half a century ago.

In addition to the conditions outlined above, an excellent way to avoid undue influence on the research programme is to submit research projects funded by interest groups to the same procedure of evaluation and priorisation as all other research projects, irrespective of source of funding (see Chapter 7).

The research organisation's own resources

A major source of income for most NARSs has been the sale of produce from experimental station farms. In recent years, following the stagnation in government funds for research, several countries have started charging their 'clients' for services for which a charge can be collected (e.g. England, The Netherlands, etc.).

The national research organisation of Brazil, EMBRAPA, generates more than 40 per cent of its operating costs by selling various services, including computer services, seeds and patents; it also buys and sells farms and cattle (Alves, 1984).

BUDGETING FOR RESEARCH

Budgeting can be defined as 'the formulation of plans for a given period in the future in specific numerical terms (Koontz & O'Donnel, 1955).

Objectives and usefulness of budgeting

The budgeting process fulfils several roles (Koontz & O'Donnel, 1955):

(a) it is an instrument of planning;
(b) it makes it possible to delegate authority broadly without loss of control;
(c) it is an important instrument of control of research activities; and
(d) it obliges the project leader to plan his work carefully, to give realistic estimates of costs and to become 'cost conscious'.

As an instrument of planning

The budget is the main instrument with which the research effort can be planned and directed into those channels which accord best with the overall research policy; it enables research management to strike a balance between the various areas of research, as well as between short-term and long-term research, basic and applied research within each area. It also makes it possible to plan the effective use of the scientific and technical manpower, and the research facilities in laboratory and field, available to the research organisation.

A number of factors, however, limit the possibilities of using the budget as an instrument of planning; a large proportion of the budget has to be allocated to fixed costs (such as salaries, maintenance and other services) and to maintaining a minimum of activity in each of the research areas of the organisation.

As an instrument for broad delegation of authority

The budget, by approving in advance expenditure for specific activities, makes it possible to delegate authority to the project leader to dispose of the funds earmarked for his research without need for further reference to his superiors, provided this is done in accordance with the approved budget. Only deviations from the allocations made for individual items will require consultation and approval.

As an instrument of control
By requiring automatic reviews at fixed intervals, the budget makes it possible to compare the actual performance of the research project with the anticipated expenditure, and therefore provides a yardstick of the accuracy of the planning; at the same time, within certain limits, it keeps expenditures in line with allocations as they were approved originally.

In research budgeting, there is no need to provide extremely detailed and itemised information on expenditure. The information provided should be in sufficient detail to show whether the expenditures are according to the approved plan and do not exceed the limits of the budget. Meticulous details will only obscure the picture by providing a mass of data that serves no practical purpose.

As an instrument for providing factual information on the costs of research
When evaluating the results of research, an important aspect is information on what the research has cost. The careful planning involved in preparing the budget, the need to be as economical as possible in order not to jeopardise the allocation of funds to the research project, the need to 'stretch' funds and facilities as far as possible, the continuous negotiations and bargaining that accompany the 'battle of the budget', 'cause awareness of the resource scarcities that confront the research organisation as a whole and cause the researcher to take a deeper interest in efficiency' (Gross, 1964).

Weakness of budgeting as a tool for planning and control

Bureaucratic procedures
Ruttan (1982) writes of his concern about the excessive administrative burden that stifles both routine investigations and research entrepreneurship. 'It appears that a concern for fiscal responsibility has often been carried to the point where it becomes an excessive burden on research productivity.'

Inflexible financial regulations, excessive paperwork by slow moving bureaucrats, delayed approval for essential items or operations may cause disruptions in the operation of costly research projects.

Budget control, applied dogmatically by administrative personnel with overly perfectionist tendencies, can also be a cause of tension and conflicts, besides causing difficulties in the implementation of projects.

In many LDCs professional support for financial management is inadequate. Operational responsibility for the day-to-day disbursement of funds and for the maintenance of records is passed to clerks who are in no position to introduce flexibility in government rules.

Release of funds
The way money is made available for research operations is very important. Release of funds in monthly or quarterly instalments is a bureaucratic device that may play havoc with field experiments. Agricultural operations are not amenable to an equal distribution of funding over the months of the year.

Foreign exchange
Procedures for the allocation of foreign exchange for essential research requisites are frequently cumbersome and often result in depriving research stations of basic supplies (FAO, 1984).

Inefficiency and waste
Many countries require funds to be used within a specific time frame. This can cause special problems. Researchers are thereby motivated to adjust their records of expenditure in accordance with the provisions of the budget, thereby destroying its usefulness as a source of information for the cost of estimates of future projects. Budgets may also be a source of inefficiency and waste, when allocations have been made for items of expenditure which subsequently prove less expensive or even unnecessary, and the project leader does not want to 'lose' the money allocated. Budgets depend largely on precedent, and economising one year may easily result in the allocation being reduced the following year, hence the tendency to spend the full amount that has been allocated, whether it is needed or not. Attention should therefore be given to allowing carry-over of funds to the next year.

Lack of balance between fixed recurrent costs and operational budget
A large proportion of the budget (the 'core' budget) has to be reserved for fixed costs, such as salaries, overhead and other services. A great matter for concern is the high proportion of salaries to operating costs. There are agricultural research organisations in which up to 90 per cent of the budgetary resources are allocated for salaries (World Bank, 1984). Any

reductions in the budget affect, first and foremost, the operational funds for research. This leads to a situation whereby a reduction of only 15 per cent in the annual budget can immobilise half or more of the research programme.

A number of factors contribute to this situation: the reluctance of donor countries to provide funds for operational expenditures, inflation, inadequate planning, etc. The shortage of operational funds causes 'both a high degree of inefficiency in the use of the scarce resource of trained scientists and a high level of frustration among those scientists keen to do a good job' (Oram & Bindlish, 1981). The situation tends to become worse as a result of grade inflation due to automatic promotion, which further increases the imbalance between salaries and operating funds.

In brief, inadequacy of operating funds in relation to salaries, makes it impossible to carry out a research programme that is commensurate with the resources invested in research personnel.

Lack of balance between capital investment and funds for maintenance
Considerable investment has been made in many countries in infrastructure and equipment, not least by donors, without the concomitant provision for maintenance (ISNAR/SPAAR, 1987). Costs of maintaining the physical plant and equipment further exacerbate the problem. Donors are generally hesitant to finance maintenance expenditures and some impose an absolute sanction on such support (Oram & Bindlish, 1981). Recipient countries are loth to forego any possible source of outside funding and the result is the establishment of an infrastructure without the means for its efficient use. Funds for maintenance are usually the first to suffer in budget cuts, as can be observed from the large number of non-functioning facilities at experiment stations across Africa (ISNAR/SPAAR, 1987).

As a matter of course, capital investment should be accompanied by an increase in the operational budget sufficient for maintenance of the new facilities.

Lack of stability in funding
The budget cannot serve as an effective tool for planning if the yearly allocations fluctuate widely. Lack of stability in funding is particularly detrimental to research programmes given the long-term nature of the research process. 'There is no real substitute for government commitment to a stable research budget. Only this can ensure that funds are available for both staff salaries and operational expenditures at a

sufficient level to safeguard the high-priority programmes' (ISNAR/ SPAAR, 1987).

Trigo & Piñeiro (1984) have analysed the evolution budgetary resources for agricultural research, from 1960 to 1980, in Latin American countries for which information was available. From 1960 to 1970, budgets rose steadily, but during the 1970s, as a result of persistent economic crises, they became extremely irregular, with year-to-year variations often surpassing 50 per cent.

Among the countries most affected by budgetary instability were some of those with the longest established research organisations: Argentina, Colombia, Ecuador, Peru and Uruguay.

Such instability, exemplified in Fig. 16, creates special problems for agricultural research organisations, because of the long-term nature of research, and the impossibility of making frequent and sudden adjustments in research programmes.

Summary
The challenge faced by the direction of the research organisation is how to obtain increased funding for the research programme and ensure its stability.

Continuity and stability of the programme, a proper balance between short-term and long-term research, between applied and basic research, is only possible when funding comes from a plurality of sources; institutional funding from government for maintaining core personnel and maintenance costs; project funding by the private sector for applied and adaptive research on a contractor–client basis; funding from a research council or national agricultural research fund for basic and long-term research; and a contingency fund at the disposal of the director of research to meet unexpected situations and to encourage exploratory research.

ALLOCATION PROCEDURES

In many countries, the processes whereby resources are allocated to agricultural research are generally ill-defined and depend more on historical, political or personal influence than on any other criteria (Daniels & Nestel, 1981).

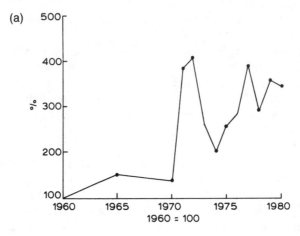

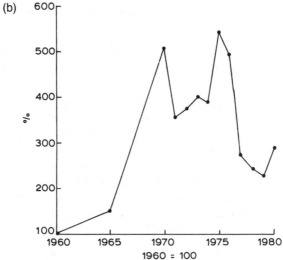

Fig. 16. Variations in annual allocations for agricultural research. (a) Ecuador, (b) Peru. (Based on data from Trigo & Piñeiro (1984), reproduced by permission of ISNAR, The Hague.)

Current decision processes

The systems in general use are mostly hierarchical. ISNAR (1987a) describes a typical case of how budgets are made and approved. The research director, sometimes in consultation with heads of units, prepares the budget of his institution. The budget figures appear in lump sums as salaries, capital and operational expenditures. With the undersecretary of the ministry, he determines where certain cuts should be made. The undersecretary, accompanied by the ministry's budget officer, meets the central budget director of the government. Again cuts are made, either across the board, or in certain areas. The research director may, or may not be present. Further cuts may be made when the cabinet of ministers meet to approve the national budget. Usually research budgets, in particular operational funds, are the first to be cut. Research leaders are not involved in any policy decision concerning research budgets. National budget directors are usually more interested in making budget cuts than in discussing the urgency or quality of services provided by research institutions.

In the past, most government funding for agricultural research in the Latin American countries was allocated to the research institutions leaving the decisions on specific spending to the research organisation itself (Trigo & Piñeiro, 1984). This system had the advantage that it involved fewer administrative costs per unit of resources used for research (Ruttan, 1982), and minimised the time researchers spent in obtaining funds for their projects (Brehdal *et al.*, 1980).

More recently, many countries have incorporated project funding for specific research proposals, thereby increasing government and private sector control over the research programme, and improving ties with the specific users of the research.

INIA (Chile) for example, formally combines two types of financing: overall institutional funding to maintain core personnel and cover certain operational costs, all other operations to be covered by contracts for specific research projects, funded by interested parties.

Some Latin American countries have established a National Agricultural Research Fund, in order to complement the regular funding procedures with funding for special research projects.

A proposed procedure

There is no objective reason why an unsatisfactory procedure in preparing and submitting the budget, as that described, should be acceptable to a NARS. The following is the outline of a desirable procedure to be followed.

The authorities indicate informally to the research management what budgetary increases over the current year, if any, can be expected. If no such indications are forthcoming, the research director should exercise his own judgement on the increase that he considers realistic in the light of the overall financial climate. These indications are transmitted to the heads of departments, indicating the tentative ceilings for planning the budgets in each field of research. From this stage onwards, the budgeting process works from the smallest research units upwards, with the research worker or project leader preparing the detailed estimates of budget requirements for each individual research project. Each head of the division then totals the estimates for all the proposed projects. If these exceed the ceiling indicated by management, a succession of weeding-outs of the less important projects usually makes it possible to bring the division's overall proposal within the limits of the guidelines set by management. If this objective is still not achieved, a review of the estimates of the individual projects is made, paring costs whenever this is possible without affecting the viability of the project. Both the screening and the cutting down to size of the individual projects must be done in consultation with the research workers concerned. They may be overruled but should not be pressured into accepting estimates they feel to be unrealistic. The objective should be 'to maximise the probability of achieving the desired research effort, and yet minimising the cost' (Seiler, 1965).

The proposal then moves upwards to the head of department, and then to the research director's office, with decisions being made at each stage. At each stage changes may also have to be made as a result of new allocations or new restrictions.

Research management then prepares the formal budget request, usually in the form of a book, grouping all the projects in each research area with all the necessary details. To these are then added all the 'overhead' expenditures of the research organisation. Then starts the stage of negotiation, of determining priorities, etc.

Budgeting for research on a yearly basis has serious drawbacks. The continuity and stability of research activities and programmes calls for financial support commitments for a number of years ahead. Without such commitments the research organisation is faced with difficulties in recruiting personnel, in deciding on building plans, acquisition of equipment, etc. The morale of the research worker is adversely affected if he has no assurance that his project will not come to an abrupt end for reasons not directly related to the progress of the research itself.

However, the administration procedure in most countries insist upon annual estimates to be approved by the legislature, within the framework of the annual budget. Private foundations usually make grants available for a number of years, and governments in different countries are also showing a greater flexibility in this respect. Whilst the formal insistence on annual appropriations is still the rule, commitments of an unofficial nature may be given, at least in relation to certain development projects of an institutional character or even for individual projects. Certain countries require a 'forward look' budget for the next five years, in addition to the annual budget; approval can then be construed as tacitly covering the period that has been under discussion. A useful device is the so-called *'three-year rolling grants'*, whereby a grant is approved for three consecutive years, during which period it is not modified unless special circumstances arise. At the end of each year a new grant is negotiated three years ahead of time. This system enables both the Treasury and the research institution to know where they stand for three years ahead, and to plan accordingly.

Another useful device is the 'five-year forward look', whereby each year the institution agrees with the Treasury on a firm estimate for the coming year, a tentative but reasonably firm estimate for the following two years, and very tentative 'forward look' estimates for two or more years.

Many investigators, department heads, and young administrators are faced with the difficulty of preparing their budgets without any previous experience. The organisations of short courses or a seminar on administrative procedures is recommended.

PROGRAMME BUDGETING

Systems of budgeting differ substantially among countries. A Programme Budgeting System (PBS) is being used or in the process of introduction in a number of countries. The system can be operated using standard clerical procedures. However, inexpensive and easy to operate microcomputers and their associated programmes have revolutionised financial management.

A Programme Budgeting System for research 'is the process by which research activities, with well-defined programme objectives, are aligned with available resources'. 'As an organisational tool, PBS is a formal system for collecting information on the programmes, sub-programmes, projects and sub-projects that have been selected and their attendant budget and personnel allocations' (Marcotte, 1987).

PBS is a project-based management tool to facilitate programming, budgeting and accounting. By providing detailed information on costs of the component activities, it enables research administrators to monitor programme versus plan, and evaluate progress towards goals.

The ideal budgeting process includes the following components (Marcotte, 1987):

- designation of the year's provisional budget;
- a clear statement of experimental objectives, justification and methods;
- a breakdown into operations and activities that can be costed on a standard basis, within the constraints imposed by the existing financial and human resources:
- an appropriate presentation for formulation of aggregated programmes and investment;
- receipt of a balanced programme and budget approved with respect to the expressed objectives.

BUDGET REVIEWS

A monitoring process should provide managers with current information on money spent, people employed, materials used, and on progress in each of the research projects. The monitoring should, however, not be more time-consuming than the benefits justify (ISNAR/SPAAR, 1987). Computers now make it possible to supply this information regularly and accurately.

Budget reports for each project should also be made available to the project leader at regular intervals.

By comparing actual with planned expenditure at sufficiently frequent intervals, deviations beyond an allowed tolerance can be kept in check. It should, however, be kept in mind, that the budget report is a measure of financial activity, and it would be highly misleading to use it as a measure of the technical efficiency with which the project is carried out.

When expenditure is substantially less than what was budgeted, a check is necessary to make sure that the research work is not lagging behind what had been originally planned; if it is, the reasons for the delay should be ascertained.

Chapter 7

Planning the Research Programme

Research planning includes the setting of research goals in accordance with national development objectives, the selection of a strategy, and the determination of detailed tactics (Libik, 1969).

Research planning encompasses two distinct phases:

(a) establishing a programme which is based on overall national policy and which defines the areas in which research is required and their relative importance;

(b) determining priorities within areas in accordance with the guidelines of the programme.

National programmes are predominantly policy-making devices; programme structures are mainly research management devices; projects are the devices used by the researcher (Fishel, 1971).

The *objectives* of planning are to:

- determine what specific investigations are required in relation to the national development plan;
- assign responsibility for specific areas of research;
- communicate what is being done;
- determine the size of the budget required.

THE NEED FOR PLANNING

Research planning is essential if limited resources are to be applied as efficiently as possible towards solution of the most important problems. Planning must be realistic, in the sense that the resulting programme can be executed within the capabilities of the research organisation.

The need for planning the research programme is generally recognised. In many developing countries, one may find 'overplanning' or unrealistic planning, for which the means are simply not available; this may be used as 'a substitute for action' (Gresford, 1962).

Many political leaders and development planners take a sceptical, if not negative, view of the potential contributions of agricultural research to development. Some are not negative but simply unaware of these potentials. The result is the same: low government priority to support agricultural research with finances and manpower.

There may be reasons for negative attitudes. There may be poor understanding of the research process, and of what can and cannot be achieved through research. Also, many developing countries have inherited a costly research organisation, one established by colonial powers for purposes that may no longer be relevant to the needs of these countries; some such systems remain isolated from present-day development goals. Modifications occurring in these research organisations since independence have frequently been haphazard, in response to demands from pressure groups or have been copied uncritically from institutions in developed countries where needs and resources are entirely different. A multiplicity of institutions and different ministries may be involved in research, with the likely result that agricultural research is fragmented and inefficient.

When countries do not have a well-established national research strategy, the results are generally found to include such weaknesses as those identified by the International Bank for Reconstruction and Development (World Bank, 1981):

- incomplete coverage of subsectors and regions within a country;
- unnecessary duplication of research efforts;
- generation of research that has no lasting impact in terms of significant technological or institutional development;
- work at research facilities directed to activity important to particular projects but not toward problems and opportunities that may affect large numbers of rural people who live outside the project area.

When research efforts are inefficient or not well understood and making little impact on the economy, it is difficult to convince funding authorities that they should increase allocations to research. A self-defeating circle develops and is perpetuated: inadequate research does not encourage funders to recognise the value from a research contribution; thus funding for research is inadequate, and the research continues to be inadequate.

Increasing awareness of the problems and the importance of planning and programming agricultural research have become apparent in recent years. A special urgency has been lent by the spiralling of research costs, the increasing difficulties in mobilising research personnel for agriculture, and the need for developing countries to capitalise on their meagre human resources.

Funds hinge on planning and probable useful results

For the research organisation to obtain funds depends to a large extent on its ability to show a high probability of useful results. This includes demonstrating that relevant and meaningful criteria have been used in the planning process; especially that the research to be undertaken promises to lead to results that are economically and socially desirable and in line with the goals of the national development plan.

One indication of the importance of planning for national research can be seen in the prerequisites stated by the World Bank as conditions for its funding of agricultural research. (This institution is a major source of funding for the strengthening of national research programmes in developing countries.) They are (World Bank, 1981):

- The development of an overall strategy for the national research system. The strategy should deal with research priorities, organisational structures, physical facilities, staff development, and institutional linkages.

  Such a strategy could be prepared with the assistance of joint national and international 'task forces' and, if necessary, funds could be provided under a first-phase technical assistance component from the World Bank.

- Adoption of a research philosophy that is oriented toward the practical production problems of farmers.

  Such a philosophy should accept the use of multidisciplinary teams conducting research in different commodity and farming systems, backed up by field studies of circumstances of farmers. Research programmes must have strong links with extension work to facilitate the flow of information from research staff to farmers and from farmers to research staff.

- The acceptance by governments of the need to give appropriate priority to the national research effort, to pursue economic policies that are conducive to the generation and adoption of efficient technology, to provide adequate funds for support of the national research system, and to ensure that personnel policies attract and retain qualified staff.

THE RESEARCH PROGRAMME

Who should have authority for determining research policy?

On this question, Elz (1984) writes: 'Research policy, planning and research allocation are not simple technical exercises that can be left in the hands of research scientists and managers. Judgements about the priority of public sector support for agricultural research in relation to other demands for public resources must evolve out of an intricate bargaining process between national legislative bodies, executive agencies, and the research community.'

The dilemma
It is generally agreed that an organisation carrying out agricultural research requires stability and continuity in the implementation of approved policies. It is further generally recognised that a research organisation, which should emphasise creativity, cannot be managed efficiently when subject to the same rules and procedures that govern administrative departments in regard to staff, pay, procurement, etc.

These factors lead to a dilemma: the need to enable effective implementation of research programmes without undue administrative interference must be reconciled with the legitimate demands of the ministries for adequate control of research activities.

The dilemma can be resolved by separating the function of determining research policy from the function of implementing research.

The authority to *define research policies and priorities*, and to allocate the necessary resources, is appropriately vested in bodies responsible for agricultural development. Such a body requires the power and authority to direct agricultural research in accordance with national development needs; that authority is assured by the inclusion in their own budgets of funds for agricultural research plus the right to approve all publicly funded agricultural research projects.

On the other hand, an agricultural research organisation, free from constricting government administrative procedures, should have authority for *implementation* of the research policy.

There are certain drawbacks when responsibility for both the formulation of research policy and implementation of programme is vested exclusively in a ministry. Such a body may be mainly concerned with day-to-day problems, tending to give priority to short-term research and responding to 'brush-fire' or 'pinching-shoe' problems.

It is thus necessary to have a mechanism assuring that the national research programme includes essential long-term research and other components of a balanced research programme.

RESEARCH GOALS

Goal definition

OTA (1981) defines a goal as '*the end* toward which *effort* is directed'. The end must be definable and achievable, at least in principle. A goal that is open-ended is not achievable. For example, to define the objectives of agricultural research in any country as: 'to provide an ample supply of nutritious food for the consumer at reasonable cost with a fair return to the farmer within an agricultural system that is sustainable in perpetuity' (OTA, 1981) is unassailable as a statement of what is desirable, but useless as an explicit definition of a goal to serve as a guide to planners.

Elz (1984) stresses the importance of establishing an easily understood purpose for research through a goal, or a set of goals, that can be monitored for progress through objective measures. 'Without such goal setting, scientific work will reflect either the individual interests of the scientists involved or rapidly become an *ad hoc* pursuit of just about anything that seems connected (or possibly connected) with the general area of the research.

Changing goals

Government policy is not constant and unchangeable; hence, the definition of goals requires constant re-evaluation and adjustment to the changing economic and social climate that characterises a dynamic agriculture.

The specific emphasis in broad national objectives or goals will differ from country to country for many reasons, including stages in their development or certain pressures arising from unique economic or social needs. And the emphasis may change over time within a single country.

If the planning process is to be effective, it must be dynamic, adjusting objectives to the changing economic and social climate.

Research planning objectives have become more and more diverse. Where formerly they were almost uniquely concerned with increasing productivity and efficiency and reducing risks in agricultural production, at present they have been enlarged to include social objectives as well as environmental and societal impacts of any new technologies.

The following are two examples of goals stated for agricultural research in developing countries (Pinstrup-Andersen & Franklin, 1977):

Economic goals: Increased productivity (of land, water, labour, energy inputs); more even cash flow to farmers; higher degree of self-sufficiency in basic foods; increased foreign exchange earnings.

Societal goals: Since the 1970s, many new public issues have become prominent in agricultural research work, such as food safety, environmental protection, nutrition, allocation of water resources (Konkle, 1982).

More equitable distribution of income between sectors or between regions, increased productive employment; increased net incomes to small farmers; improved human nutrition.

In many cases, it is most crucial to increase the productivity of scarcest resources. For example, research aimed at increasing the volume and productivity of usable water is of the greatest importance in the Near East; while land productivity would have priority in Peru.

In a country with an adverse trade balance, such as Israel, increasing agricultural production for export may be a most important goal. In many countries of Africa, where up to two-thirds of the population suffer from undernourishment or malnutrition, the major goal may be to augment the production of food crops.

Underemployment is rampant in most developing countries, and major importance in research there may be given to the search for technologies that increase opportunities for productive employment.

ECONOMIC RETURNS TO AGRICULTURAL RESEARCH

Benefits of agricultural research

Primary benefits of research include improved production, conservation and preservation of food, and improved production of fibres, which usually lead to lower costs to the consumer. Secondary benefits include improved human nutrition, improved food quality, improved international trade balance, release of labour to other sectors of the economy, elimination of drudgery and increased labour productivity (Konkle, 1982).

Rate of return estimates

Though it is difficult to quantify many of the benefits mentioned in the previous paragraph, numerous *ex-post* studies on the economic benefits in agricultural research have been carried out. Most of these studies show consistently, both on a commodity and on a national basis, that the returns are generally two to three times higher than likely returns from most alternative investment opportunities (World Bank, 1981).

The estimated rate of return may be biased upward: the full cost of developing and applying a new technology may not be included in the analysis; cost of unemployment of resources induced by the research may be omitted; inappropriate welfare weights may have been assigned to gains and losses in calculating research returns; the cost of research failures (dry holes) may not have been properly included; and costs of supporting programmes for the development of technology may have been understated (Schuh & Tollini, 1979).

However, the calculated rates of return are consistently so high, that even if the possible bias upward is discounted, they still show that rates of return realised by agricultural research conducted by both national and international research organisations are still very high.

The international rate of return on investment by IRRI in rice research through 1975 is estimated to be about 80 per cent. Roughly similar returns have been estimated on investment in rice research by CIAT, and on national programmes in Colombia (World Bank, 1981).

In the keynote address to a Workshop on Agricultural Research Policy in Small Countries, Ruttan (1985) mentioned that Evenson and co-workers have calculated that the supply of rice in all developing countries was 12 per cent higher than it would have been if the same total resources had been devoted to the production of rice using only the varieties that were available prior to the mid-1960s. A recent study by Joseph Nagy indicates that the gains to Pakistan alone, from the wheat research conducted by CIMMYT, would have been more than enough to cover the cost of the entire CIMMYT wheat programme from its inception to 1980.

The financial value of Mexico's increased wheat and maize production that has been contributed by research is the equivalent of about 400 per cent annual interest on the total money spent for all research on the Mexican co-operative programme with the Rockefeller Foundation from 1943 to 1962. The returns to the wheat programme alone, when comparing wheat research expenditures with benefits, are at least 800 per cent a year (Ardito-Barletta, Jr., 1965).

Assuming that opportunity costs of capital in most developing countries are in the 10–15 per cent range, the World Bank (1981) states that 'underinvestment in agricultural research appears to be substantial, and considerably more money could be invested in this activity with the expectation that returns would exceed both the opportunity costs of capital and the returns from most feasible alternatives in rural development'.

EVALUATING PRIORITIES IN AGRICULTURAL RESEARCH

In many countries, research programmes are established simply by aggregating the research activities already in process (Busch & Lacy, 1983).

A considerable research effort has been devoted to the development of a methodology that can be used in making a choice of research projects, that in the aggregate 'generate the largest total value'. This value can be expressed as an econòmic or a social return.

Assessing benefits according to economic criteria

Mainly during the 1970s, different techniques of procedure and numerous mathematical models, proposed and described in the literature, have attempted to provide a rational basis for making decisions on evaluating priorities in agricultural research.

The principal methods that have been tested for priority setting are:

- use of cost-benefit analysis, to select research areas and commodities;
- mathematical programming to choose an optimal research portfolio, incorporating multiple goals and constraints;
- project rating and scoring systems, based on the establishment and weighting of multiple criteria for ranking research areas, commodities and research projects.

The most sophisticated and most logical of the methods developed—the Minnesota Agricultural Research Allocation Information System (MARAIS)—has been found to be too costly, and it is not being applied widely. The present situation is summed up well by Ruttan (1982), one of the developers of MARAIS: 'None of the methods proposed can be better than the judgment of the persons doing the planning as to likely costs of the research, the probability of success of the research efforts and the potential benefits of the research findings.' It is well to bear in mind that it is practically impossible to quantify political, social, and even most economic factors which need also to be taken into account.

Many of the methods of research evaluation that have been proposed in recent years are mathematical, difficult to interpret, costly to apply, and require data that may be unreliable or unrealistic.

As matters stand at present, the kinds of data required to support a purely economic or mathematical process for selecting research projects and drawing up an optimum portfolio of projects do not appear to exist; or they are subject to such great uncertainty that little confidence can be placed in the conclusions derived from this 'objective' analysis.

Assessing benefits according to social criteria

A large part of agricultural research is conducted and financed by public agencies. It can be regarded as a social investment and thus subject to social criteria for evaluation of its benefits.

An economic analysis approach assumes that the major objective of research is 'profit' in money terms; by contrast, the criterion for planning a considerable proportion of research investment in agriculture should be the 'greatest social return possible for total funds available' (Paulsen and Kaldor, 1968).

When estimating social returns, it is essential to define which groups of people these returns are to benefit: is it the people of a province or region where the research is carried out; is it a special sector; is it the people of the nation who contributed the funds; or is it humanity in general?

A given research project is not always equally useful to such differing groups as individual farmers, the agricultural industry, and the society as a whole. It is difficult to assign relative weights to the demands of such diverse groups, and benefits are seldom restricted to certain groups. Generally, the results of agricultural research are freely available to all farmers (in fact, considerable effort is even expended to persuade farmers to apply the results). Free publication makes research results available on an international scale, which means that the benefits derived from research financed and developed by one country are generally available to others—agricultural research generally differs from industrial research in this respect.

In a study on the 'pervasiveness' of new agricultural knowledge, Latimer and Pearlberg (1965) show that a farmer may gain more from agricultural research as a member of society than as a producer. For example, improved technology may bring a decline in real costs in farm production that benefits all consumers, including the farmers; but farmers may lose more in the process as producers than they gain as consumers (Schultz, 1970).

Certain types of research may lead to clearly negative effects. There are numerous examples of technological innovations introduced through research that led to increased pollution of soil, air, water, and food products. Such adverse effects are sometimes left out of research evaluations, and there is a notable lack of research specifically directed to finding remedies for such potential adverse effects (Heady, 1970). For example, it is now known that there should have been research on the unintended consequences of DDT (and ways to avert them) concurrent with research on actual use of the chemical in agriculture. Technological progress in agriculture has accentuated some important problems of poverty and welfare, as in the case of mechanisation: machines reduced the farm labour force and caused migration to the cities, where the migrants' lack of education made their economic absorption difficult, resulting in a heavy cost to society. Also, many products of research have brought increased profits to one group of producers at the expense of other producers (soybean products replacing butterfat and mechanised agriculture making horse-breeding unprofitable are two among many examples that could be noted.)

It becomes clear that some of the data needed for estimating the social returns of research are not even predictable, let alone quantifiable, before the research is carried out.

In brief, while the use of measures of social value might have been the most fruitful approach for planners of research projects in agriculture, the facts are that social benefits are even more difficult to evaluate than economic benefits, because of the lack of suitable evaluation methodology, subjective value judgments which change over time, benefits no longer judged only by cost–benefit ratios, but increasingly by their effect on distribution and on the environment (Elz, 1984).

Summary

If they are to think constructively on potential impact, research planners need a thorough grasp of principles related to consequences of changes in productivity and returns to the factors of production. Certain farm products have an inelastic demand, which means that a small increase in supply causes a larger drop in price in the market—or vice versa. As a result of this situation, plus the highly competitive nature of agriculture, an increase in agricultural productivity in some cases can actually depress farm prices and reduce the total revenue of farmers (Peterson, 1970).

Research results will be applied under an extremely wide variety of conditions. The contribution to the national economy and consequences on various groups will depend on a number of factors, including: the speed and extent of adoption and the ability of individual farm units to apply them advantageously; their duration of use—from a short to an indefinite period. The results may have unforeseen effects on other practices and on other industries. They may be used internationally, creating more competition among suppliers in export markets.

RATING SYSTEMS

In lieu of the mathematical formulas mentioned above, much thought has been devoted to devising rating systems which can serve as guidelines for the groups that have to determine research priorities. These rating systems are not an alternative to group judgment but rather a device to strengthen objectivity in what is essentially a subjective process.

Prerequisites for the usefulness of a rating system are that:

- the rating should include major criteria of importance in judging the individual research project;
- these criteria should be amenable to quantitative rating;
- the rating method should not be too cumbersome.

If too many criteria are involved, it becomes difficult to combine them into a composite criterion which can usefully serve the group that has to exercise judgement on the priority to assign to a project. Therefore, only the most relevant criteria should be included.

The rating systems have the following *advantages:*

- they openly use guesses, whereas the economic formula use the same guesses in a more sophisticated and time-consuming form;
- they are helpful in partly eliminating personal bias;
- they provide a certain measure of consistency.

The main *weaknesses* of the rating systems are that:

- the ratings are subjective and may be erroneous; however, figures by their very nature tend to give a false sense of accuracy and objectivity;
- all the criteria are given the same range of numerical values, however the relative weight of the different criteria may be markedly different.

Because of the subjective nature of the rating systems, choosing the right people to do the rating is at least as important as the method adopted.

Scoring models is a rating system that is easy to apply and is adaptable to a wide range of conditions and changing circumstances. Some countries are using the 'Delphi Technique'† for the identification of technological requirements and research needs. This technique is a formalised method designed to promote consensus (Daniels & Nestel, 1981). It consists of a series of individual interrogations to a group of experts, interspersed with information and opinion feedback. Enquiries are made regarding the reasons for previously expressed opinions. Each respondent then receives a collection of such reasons and is given the opportunity to reconsider his earlier opinion. The Delphi Technique aims at improving the committee approach by subjecting the views of individual committee members to each other's criticism whilst avoiding face-to-face confrontation (Ruttan, 1982).

Scoring models

A scoring model is a rating system using a questionnaire which includes the major criteria considered important in judging individual research projects. Each criterion receives a score, ranging from 1 to 5 for example, according to the degree to which the project meets requirements (Table 3).

In Table 3 all the criteria are given the same range of numerical values; however, the importance of different criteria is not necessarily the same. This can be taken into account by assigning different weights to the criteria. In this case, the scores are multiplied by the respective weights assigned to each criterion and their sum provides the total score indicating

---

†For more details on group techniques in general and Delphi methodology in particular see Cetron & Bartocha, (1972) and Delbeck *et al.* (1975).

TABLE 3

A scoring model adapted from Mottley & Newton (1959)

Criterion	Question	Range of answers	Score
Technical feasibility	Is scientific manpower available?	No, would have to be trained or imported	1
		No, but can easily be recruited	2–3
		Yes	4–5
	Is technical support available?	No	1–3
		Yes	4–5
	Is additional equipment necessary?	Yes (from expensive to inexpensive)	1–3
		No	4–5
	How much will the research cost until completion?	Unforeseeable, but high	1
		From high to low	2–5
Research direction and balance	To what degree is the research compatible with the directives determined by programme policy?	Implications are not clear	1
		Useful, but not within priorities	2–3
		In accord, according to importance	4–5
	Does the project contribute to a balanced program?	Yes, in different degrees	1–5
Timing of research	What is the probability of the research being completed before the problem becomes obsolete?	From low to high	1–5
Impact of the research	How great is the probability of success?	From low to very high	1–5
	What is the potential effect on the economy?	From low to very high	
	Will it be a long and expensive process to adopt the results into agricultural practice?	From long and very expensive, to short and not costly	1–5

the relative importance of the project.

The main weakness of the scoring systems is that the ratings are subjective and may be erroneous; however, figures by their very nature tend to give a false sense of accuracy and objectiveness.

A graphical presentation of scoring results

A graphical presentation of the scoring results (Table 4) has the advantage of facilitating a rapid appraisal of each project, after the ratings have been painstakingly prepared as shown above. The criteria which are considered of great importance can be emphasised by the use of a different colour, by listing them by order of importance, or any other appropriate technique. Numerical weighting may then be superfluous.

TABLE 4.
Project rating profile

	High	Medium		Low	
	Excellent	Good	Fair	Poor	Undesirable
Availability of scientific manpower	⑤	4	3	2	1
Availability of technical assistance	5	④	3	2	1
Need for additional equipment	5	④	3	2	1
Cost of the research	5	④	3	2	1
Compatibility with policy directives	⑤	4	③	2	1
Appropriateness of timing	⑤	4	3	2	1
Probability of success	5	④	3	2	1
Potential effect on economy	5	④	3	2	1
Ease of adoption	5	4	③	2	1
Cost of adoption	5	4	③	2	1

A circle around the appropriate figure indicates the score for each criterion. The location of the circles in the profile shows at a glance the overall rating of a project, its strong and weak points (as exemplified in the profile).

Advantages of the graphic presentation are its simplicity, and the small amount of paperwork involved. This does not mean that much thought and frequently careful investigation are not required to reach the proper assessments for rating each criterion. Each member of the panel may prepare his own assessment, and these are then pooled into a single graph representing consensus, or a single person may be charged with studying all the background material and preparing a proposal for approval by the group.

Research evaluation based on group judgement

In evaluating research projects, there appears to be no alternative to using the intuition, knowledge, experience and judgement of competent persons brought into interaction with each other in special working groups or panels, and to whom all relevant information is made available. Such persons may include those involved in various aspects of agricultural development and research: administrators, planners, researchers, extension workers, farmers and representatives of funding sources, for example.

The necessary conditions for good judgement include: a clear understanding of the varied conditions under which a commodity is produced, the opportunities that can be exploited, and relevant and timely information provided in an appropriate form to those who have to make the evaluation.

Commodity panels

It has been explained in previous sections why there is little prospect of replacing group judgement as a basic tool in the evaluation of research projects, and that appropriate rating systems could provide a tool for the work of the groups required to evaluate the research projects.

Most research projects are related to specific commodities, therefore decisions regarding the relative importance assigned to problems within each branch of agricultural production are best made by 'commodity panels'.

How these commodity panels are constituted and how they function will be described in the following section.

Composition of the commodity panels

The composition of the panels should reflect the following interests:

(a) the 'clients'—the appropriate department of the ministry concerned; commodity, production and/or marketing boards; the relevant farmers' organisation, other agricultural services;

(b) the 'contractors'—the relevant research unit(s) from the research organisation and the faculty of agriculture.

(c) the specialised extension worker who is a link between client and contractor.

The commodity panels are thus fairly balanced as to the vested interests of their members in relation to their professional backgrounds. This broad membership allows different professional viewpoints to influence the discussion. If a consensus is not reached on certain issues discussed, this may indicate that the research projects concerned are probably not of high priority.

The ministry representatives. In addition to production (which can be represented by the subject matter specialist of the extension service) it is desirable that the ministry team should include a planner and an economist, who could serve on one or more panels (see below). In addition, the regional directors should participate or be represented (see below). It is important that the ministry representatives should have the necessary authority and competence to ensure that the interests of the ministry and the directors of the development plan are given due consideration.

Commodity or regional production and/or marketing boards. In many developing countries, the responsibility for development for a specific commodity or a region is vested in a governmental organisation with practically autonomous status. Frequently these boards set up their own specialised research institutes, which are mainly financed by a tax on the commodity for which the board is responsible (p. 225–6). The only way avoid the resulting fragmentation of research is to give the boards adequate representation in the planning of the research programme, so as to ensure that their interests and requirements are met by the research organisation.

Farmers' representation. Farmers' participation in the research planning process aims at making the research priorities more responsive to the farmers' needs as perceived by them. It also allows farmers' organisations to act as a driving force in obtaining resources for research (Delbeck *et al.*, 1975).

Even if we assume that the specialist members of the research panels have the competence and motivation to ensure that agricultural research reflects the real needs of the farming sector, a major way to convince farmers that this is actually the case, is by involving them in the process of defining research goals, problems and priorities. Paraphrasing a well-known adage on justice—research must not only be appropriate, it must also be perceived to be appropriate. The presence of farmers' representatives has the advantage of not only avoiding a divorce of research from agricultural realities, but also of assuring the farmer that his problems are being brought to the attention of the research worker and that the necessary attempts are being made to provide a solution according to priorities determined in consultation with his own representatives. The psychological importance of this form of contact and reciprocal influence of research worker, farmer, extension worker, and ministry official cannot be overstressed.

Where there are active farmers' organisations, their formal representation on the commodity panels should not pose any special problems. The main difficulty that may arise is that the farmers' representatives, if they are not sufficiently articulate, may feel that they have a lower status in the panel than the specialists, that their presence is more a matter of window-dressing than real participation, and that their influence on decision-making is minimal. Such attitudes will of course be counter-productive, and much depends on intelligent and tactful chairmanship to avoid such a situation.

However, the number of developing countries with active and representative farmers' organisations is still small. Whilst the owners or managers of plantations or large commercial farms generally have no difficulty in communicating their problems and have the political clout to ensure that due attention is paid to their research needs, the vast majority of the farmers, namely—the small, subsistence farmers and share-croppers, have no such possibilities and are effectively barred from the decision-making process by institutional, social and cultural barriers. This is a situation that requires action by the governments concerned, and is not the responsibility of the research organisation. However, the absence of adequate formal representation by farmers should not deter research workers from attempting to establish contact with them, mainly by pre-extension research and on-farm trials. These aspects will be treated elsewhere.

The research workers. The research workers on the panels and committees should be the specialists responsible nationwide for the respective commodities. They should be able to present succinctly and clearly the state-of-the-art in their field, the constraints encountered, and submit research proposals encompassing all the disciplines required.

Each commodity specialist (generally in the fields of plant breeding and agronomy), is usually aware of the constraints encountered in production of 'his' commodity, and should certainly be the best qualified to judge what are the research needs in his field, the potential pathways to be taken in the solution of the problems, the effort required and the probabilities of success. It is true that whilst farmers are generally the major source of information on problems and constraints that they encounter, the ideas or even the demand for specific research proposals do not generally originate in the farm sector, even in developed countries. As a rule, it is the research workers at the experiment stations who realise the potential impact of certain research efforts and explain or demonstrate them to those from whom they expect to secure funds for their research. If funds were unlimited, there would be no major disadvantage in leaving decisions on the research programme as the exclusive domain of the research organisation. However, priorities *have* to be established, and even worthwhile projects cannot always be funded. In decisions on *priorities*, the research worker has no inherent advantage over the other participants in the commodity panels or committees.

If he is open-minded, he will realise his limitations in this respect, and the legitimacy and need of involving the 'users' in the programming process. He will also soon realise that the commodity panels and committees can play a crucial role in ensuring close contact between research, the relevant ministry, extension and the agricultural community.

Communications between these sectors are improved, because the research worker has to make an attempt to show the relevance of his research to the current problems of agriculture, and so in a language that can be understood by laymen.

Number and kinds of commodity panels
The number of commodities for which research projects need to be evaluated is generally quite considerable. It is, however, desirable to avoid excessive fragmentation of the decision-making process, as well as excessive work-loads on the people involved in the process.

Therefore the following device is suggested: a limited number of panels, each concerned with large groups of commodities with common characteristics; each panel establishing a number of committees each of which treats a single commodity or small group of commodities.

Arable crops panel—with committees on cereals, legumes, industrial crops, vegetables, forage and pasture crops.

Plantation crops panel—with committees for coffee, cocoa, oil-palm, rubber, fruit trees, etc.

Animal husbandry panel—with committees for beef and draught cattle, dairy cattle, small ruminants, equines, poultry, animal health.

Cross-commodity panel—research on problems cutting across commodities production, such as irrigation techniques farming systems, soil fertility etc., are referred to a special panel, which must judge the importance of a given research project in relation to its composite effect on a number of branches of production. This panel can also be assisted by an appropriate number of committees.

The composition of the committees is similar to that of the panels, but the members have a more circumscribed area of competence. The committees review the research proposals submitted, identify needs that are not met in the proposals and assign priorities within the framework of available resources. They thus establish a programme for their commodity, encompassing all scientific disciplines essential for the development of new technology.

They report to the panel, which produces a composite report that is submitted to the Director of Research.

Regional committees

The location-specific nature of agricultural research makes a decentralisation of the implementation of the research effort essential. This is achieved by establishing a network of regional stations, each with its substations, pilot farms, and experimental plots in farmers' fields etc. Not only does the relative importance of the various commodities vary in the ecological regions of a country, but the constraints encountered in the production of a given commodity may also be entirely different. In one region, there may be need for breeding a short-season variety that fits into the dominant farming system; in another it may be the need for resistance to a specific disease.

It is at the regional level therefore that grass-level contact between research, extension and farmers can and should be established. The formal framework for this contact should be an advisory committee for the regional experiment station. The composition of this committee would be similar to that of the panels at national level. Important sources of information would be obtained from the local implemented pre-extension research and on-farm research.

The research priorities required by these regional committees must be incorporated in the decisions made at national level. In order to transmit effectively farm-level demands for research, the region has to be represented on the national panels and committees, preferably by the director of the regional experiment station or his representative(s).

Research workers do not, as a rule, readily accept the intervention of panels or committees in determining the priorities to be accorded to their research proposals, especially when the majority of the members are non-scientists.

We have already pointed out why, in mission-oriented research, the involvement of laymen in determining priorities is justified. Researchers who are public servants must accept this situation. However, tensions between panel members and research workers may arise, if the prerogatives and obligations of the commodity panels and committees are not clearly stated. A frequent source of friction is the confusion of programme planning with programme implementation.

All members of a panel should participate in defining the needs and problems of the commodity under consideration, each contributing from his experience and knowledge, and from his point of view. It should be made clear to all involved, that the function of the commodity panels is to give the research institution overall guidance on the needs and problems of the 'users' and to indicate the relative importance of the research proposals submitted. It should be equally clear that the planning, direction, and conduct of the research projects remain the exclusive responsibility of the research institution.

PROCEDURES FOR EVALUATION OF RESEARCH PROPOSALS

In this section, methods for the evaluation of research proposals submitted to group judgement, which are adequate without being too time-consuming, will be described.

Inputs and criteria

Each panel or committee, whose attention is centred on a specific commodity, has at its disposal three main inputs as a starting point in their deliberations (Daniels & Nestel, 1981):

- The technological diagnosis of the production problems of the commodity, which identifies major technological constraints, and makes a preliminary evaluation of the importance of each constraint. This can be presented by a member of the ministry team.
- Brief state-of-the-art reports presented by the responsible research worker, summarising the present research effort and the available technologies for the solution of the technological constraints, mentioned above.
- The knowledge, experience and competence of the participants.

Pinstrup & Franklin (1977), describing the procedures followed by CIAT, indicate that the first step taken is to identify the factors associated with low productivity in a specific crop and then to proceed and identify researchable problems expected to improve productivity and efficiency in production. The committee then evaluates the research projects submitted by the research organisation, using an appropriate rating system based on a number of criteria. Actually, in practice, the problem is not to distinguish between 'good' and 'poor' projects. It is very rare that any of the latter type reach the commodity panel or committee, and they are screened out at an earlier stage (cf. p. 407). The major difficulty is to decide on priorities between projects that are *all* potentially beneficial, when financial and human constraints limit the number of projects that can be carried out. However, the difficulties involved should not be overrated.

The specific *criteria* which are to form the basis of the recommendations of the commodity panels and committees will have to be determined according to the specific conditions of each country, the importance (present or potential) of the commodity, and its possible interaction with other commodities.

Brady (1975) lists four major criteria in setting research priorities at this level and in ascertaining research projects to be initiated:

- *Relative significance of different constraints:* The extent to which removal of a given constraint will contribute to the achievement of important agricultural and social goals, is perhaps the most important long-range criterion.

- *Feasibility of constraint removement.*
- *Cost of research required to remove the constraint.* Even if the cost/ benefit return cannot be calculated the cost of research can be an important consideration.
- *Probability that others* (international institutes for example), *will do the research.*

The weighting of different criteria is possible in order to arrive at a scoring system, but it generally 'gives a specious air of precision to what should be recognised as being fundamentally a subjective process' (Pinstrup-Andersen & Franklin, 1977).

Assigning priorities on the basis of commodities has the disadvantage of ignoring the major differences which exist between types of farms in respect of their ability to adopt new technologies because of differential access to resources. Another disadvantage is that much of agriculture is diversified, and hence the priorities ignore the complex interactions that occur in agricultural systems.

In EMBRAPA (Brazil), when defining research priorities at the commodity level, three basic and closely related factors are taken into consideration: growth, equity, and reduction of risk. In addition to these, 14 additional criteria are considered important in the definition of priorities: the importance of the commodity, its role in nutrition, price elasticity of demand, its role in the balance of payments, the possibility of an immediate response, the industrial demand, price movement, the availability and use of resources, possible beneficiaries, regional equity, risks and uncertainties, the technology employed, the competitive capacity in the production of technology, and the possibility of importing and adapting technology (Lopes Neto, 1975).

A special, and potentially awkward problem is how to incorporate the priorities requested by the regional committees into programme decisions made at the national level. This difficulty can be overcome in the following manner: Each regional station, in addition to its responsibility for research at the regional level, should be a 'lead' station for one or more commodities that have a dominant role in the economy of its region. The researcher who has national responsibility for a commodity should be stationed at the 'lead' station for that commodity. It would be his responsibility to prepare and implement the national research programme to be carried out in all the regions of the country in which the commodity is produced. In establishing the programme, he would take into account the recommendations of the regional committees; because of his nation-wide responsibility he would have every reason to be objective in allocating resources for his programme between the regional stations involved.

In this way, the research work carried out at the regional stations is incorporated into a number of interlocking, cohesive, multidisciplinary and multiregional research projects, which together form the programme submitted for scrutiny to the respective national commodity panels.

The research projects submitted to the panels are generally multidisciplinary and multiregional. The disciplines that will participate in the project and the nature of their contribution must be stated. They must also indicate in which regional stations the project will be implemented (preferably in the form of a sub-project for each region involved). In deciding on how to allocate the research effort among regions, due consideration should be given to the priorities indicated by the regional committees.

Criteria for evaluating basic, long-term, exploratory research, etc.
Because the potential benefits of these categories of research are generally less direct than those of applied research, this does not imply that they are necessarily smaller. They may even be the source of breakthroughs of major consequence.

For basic and for long-term research that is directly concerned with the solution of specific problems of practical or economic importance, the criteria for determining priorities among applied research projects already discussed may be applicable. These include elements of compatability with objectives, risk and uncertainty regarding output, probability of success, and costs and time required. No such criteria can be devised, however, for basic research that is not directly concerned with the solution of a specific practical problem. The knowledge resulting from basic research is generally the starting point for more applied research, which must pass through a number of stages before it is applicable in the field. The actual contribution of basic research to improved technology begins only at this stage.

As to the other categories of research—uncommitted basic, exploratory, educational etc., because of the uncertain return of this output and of their potential impact on agricultural development, it does not appear possible to devise a simple, rational basis for the allocation of resources to this category, nor are there objective criteria for determining priorities in exploratory research or educational research. In exploratory research an original idea is investigated, and in the early stages it is not even possible to guess at the probability of something of practical, economic importance arising from it. A certain effort has to be invested before any evaluation can be made. Yet to neglect this type of research is to stifle initiative and perhaps miss important opportunities. The only solution is to assign a certain proportion of the total funds available to agricultural research for these and similar kinds of research—what Weinberg (1964) calls 'scientific overhead' (10–20 per cent appears to be a reasonable proportion). Whatever the actual proportion decided on, it is bound to be arbitrary.

In assigning priorities among different research proposals that fall in the categories described above, important guidelines can be their scientific promise and feasibility and the reputation of the researcher who submitted the proposal. Specific criteria proposed for this purpose are (Brooks, 1966):

- whether a scientific answer to the problem proposed can be reasonably anticipated;
- the reputation of the investigator, or the promise of a young scientist;
- whether facilities and support needed for the research are available;
- whether the field appears 'ripe' for intensive research, and whether there are real opportunities for major progress;
- whether the results of the proposed research may contribute to the long-term goals of the agency that provides the support;
- the originality of the research, and its technical soundness;
- the scientific 'significance' of the research in terms of affording new understanding of fundamental laws, providing a critical test of current theory, and exploiting new techniques;
- the possibility that it will illuminate work in their fields of agricultural research.

Levels of decision-making

Agricultural research policy has to be formulated at various levels, each concerned with certain aspects and segments of this policy. Basically, there are two distinct phases: (a) decisions on what *needs to be done* and how much *can* be allocated for this purpose (definition of broad objectives, goals to be achieved and budgetary decisions), and (b) what should be done (specifying the components of an optimal research portfolio of projects).

In this process, two categories of persons are involved: (a) individuals capable of broad conceptualisation and strategic thinking, and (b) technical experts who can identify priority projects to be included in the research programme (Delbeck *et al.*, 1975).

Major levels
Decision-making will generally take place at the following levels:

- *national* (cabinet/parliament) where national development goals are determined and policy guidelines are established;
- *ministerial* where sectoral objectives are defined in accordance with national guidelines;
- *institutional* at which policy guidelines are translated into institutional goals and research programmes are established;
- *departmental*—this is the operational-professional level at which institutional goals are translated into specific projects;
- *individual* (researcher) who prepares, proposes, and implements specific projects in his field of competence.

The decisions at each level are made independently; they should, however, be connected by a flow of information, 'one level feeding information to the other sequentially and in both directions until a final decision is reached' (Javier, 1987). At each level, the decision-makers will have to take into account the informed judgement of appropriate panels or committees.

National level
National science policy has been defined as 'the consideration of the interactions of science with policy in all fields' (OECD, 1963). At this level financial support for scientific research has to compete with the needs of defence, health, welfare, agriculture, industry, etc.

National policy generally decides priorities by assigning budget resources directly among competing national development goals on a fixed ratio over a specific planning period; the ratios are reviewed as governments change or as progress is achieved in meeting specified goals (Javier, 1987).

National development plans are generally formulated in five-year plans, and annual government budgets. Javier (1987) writes: 'Since the national goals are invariably broad, noble and ambitious, there have not been constraints in programme priority setting' (for agricultural research).

The national level is not directly concerned with agricultural research planning *per se*, which starts at the ministerial level.

Ministerial level

In allocating resources to agricultural research, one of two pathways is generally followed:

(a) If agricultural research is the responsibility of a special ministry of science, the allocation at ministerial level will be divided among research activities in various fields of health, agriculture, defence, industry, and so on.

(b) If agricultural research is the responsibility of a ministry of agriculture, or is divided among several ministries involved in agricultural development, research will have to compete with other activities of the ministry(ies). In the following outline of levels of decision-making, we will assume that primary responsibility for agricultural research policy and funding is vested in a ministry of agriculture.

After the larger problem of competition between research for limited resources and the other ministerial services, such as development, regulatory functions, etc., has been solved, decisions have to be made on how to divide available allocations for research among horticulture, field crops, livestock, poultry, and other production areas. Export crops compete for research funds with crops for the local market, old important branches of production with new ones that are still struggling to establish themselves, and so on. Various yardsticks may be used in making these decisions, but essentially, they are political decisions.

At the ministerial level, policy-makers must take a longer-term and broader view of priorities than do farmers; however a policy is unlikely to be successful if it is not acceptable to farmers, hence the need for appropriate economic incentives, otherwise farmers will not produce in accordance with government policies (Baxter & Thalwitz, 1985).

Policy decisions regarding the relative emphasis to be given the different branches of agricultural production or the need to establish new branches are usually made by a top committee consisting of the director-general of the ministry and the respective directors of research, extension, and planning. A committee of this type may have attached to it a small advisory team of scientists with economic training, along with competent assessors.

The criteria to be used are (a) *economic*: national value, foreign exchange considerations, fiscal role of the crop, etc. and (b) *social*:

income distribution bias to small farmers, to marginal regions, consumer welfare, environmental conditions. Where there are conflicts in criteria between economic and social considerations, for example, the decisions have to be political, and made at the ministerial level. Decisions at this level will usually distinguish between fields in which all-out research effort is required, those in which research should be limited to a few problems, and those in which no further research effort appears to be justified.

Research organisation level

The main policy concern at the level of the research organisation will be to fulfill its obligations in attaining the goals set at the national and ministerial levels, and at the same time maintaining a high scientific standard and ensuring a balanced programme of basic and applied, short- and long-range research.

Individual scientists and research leaders can and should contribute judgements and data that will be useful in the evaluation process, but they generally lack the breadth of competence and responsibility for final programme decisions. These will be incumbent on the research director, aided by an advisory board who have to attempt to balance the resources required for the high priority projects proposed by the panels in different research fields with the total resources available to the organisation. Project selection in this context will generally mean choosing for each field of research a package of those projects which have received high-priority ratings from the commodity panels, up to the 'cut-off' indicated by the budget available for each problem area, as determined at ministerial level (see above). However, examination of the research proposals may indicate the desirability or even the necessity of increasing the budget allocated to certain research areas. This may involve the search for outside funds or the transfer of funds from one research area to another, or a reappraisal at ministerial level.

The board that is required to support the director of research in establishing the final programme of research before it is submitted to the funding authorities should have as Chairman the personal representative of the Minister of Agriculture (preferably his scientific adviser) and as members—in addition to the Director of Research, the Director-General of the Ministry of Agriculture, and the Chairman of the Commodity Panels (see p. 388). The representative of the Minister of Agriculture will safeguard the priorities in those fields of research most needed for the national agricultural development and production programme; the Chairman of the Commodity Panels (including the Cross-Commodity

Panel) can defend the needs of a balanced programme; the presence of the Dean of the Faculty of Agriculture provides an effective link between government and university research.

In addition to its role in establishing the programme of research of the national research institute, this panel would have the following functions:

1. co-ordinate the research programme with that of other institutions in the country (universities, private, etc.) so as to avoid unnecessary duplication;
2. allocation and training of scientific and technical manpower for research;
3. initiate and encourage new fields of research if needed;
4. follow-up the implementation of policy with regard to research.
5. advise the Director of Research on any other organisational problems with which he may be faced.

Because the national research panel is not directly concerned with the allocation of research *within* each research area (this is the responsibility of the commodity panels) it has to concentrate on integrating the individual proposals into an optimum portfolio, within the constraints imposed by the research budget. The major criterion will, therefore, be the potential contribution to the achievement of the directives of the national development plan.

The final product of the procedure described above is a national programme of agricultural research with the following attributes:

- it conforms to objectives and the priorities set at the national and ministerial levels;
- it remains within the limits of the budget allocated for national agricultural research;
- it maintains a balance between short-term and long-range research, between applied and basic research, between commodities and between regions.

These objectives are achieved in three stages:

(a) determining the principal problems facing each of the commodities;
(b) defining these in terms of research projects;
(c) deciding on priorities based on potential impact, cost and probability of success.

Within the research organisation itself, all levels are involved in the process: research direction, department heads, directors of regional stations, research workers—each level contributing information and judgements. An effective system of feedbacks between these levels is essential.

Research department level

A research department comprises a number of divisions, each responsible for an area of research, and grouping scientists who have certain common interests and similar professional background.

It is the responsibility of the head of the department to ensure that his department makes, within the field of its competence, the full possible contribution to the problems posed by the guidelines transmitted to him, with the resources available to his department.

He must ensure that research proposals are properly evaluated before submission to the Director of Research, from the scientific point of view, their compatibility with the research policy as interpreted by the Director of Research, and in relation to their budgetary demands. He will pay particular attention to whether all aspects of work proposed are within the professional competence of the team, and examine the need for co-opting additional specialists to the research team in order to make the research more fully effective.

An agricultural research organisation usually comprises two types of department: those devoted to specific commodities (or groups of commodities) and those based on disciplines. The research objectives of the 'commodity' departments cannot be achieved without the active support of the 'discipline' departments. The major *raison d'être* of the latter is to provide this support. It is the responsibility of the heads of commodity departments to co-ordinate the interdisciplinary activities necessary to resolve the problems encountered in the production of the respective commodities. The research programme proposed for a specific commodity therefore consists mainly of multidisciplinary projects based on co-operation between commodity and discipline specialists.

In support of this view, Javier (1987) writes:

A closer analysis should reveal that many of the targets are in fact commodities and/or objectives; targets, thrusts and disciplines which can be readily accommodated as the research activities are built-up within commodities . . . The preference for a priority-setting exercise expressed in the main as commodities . . . provides the one-to-one correspondence between research priorities and development goals which our public looks for.

However, assigning priorities exclusively on the basis of commodities does have certain shortcomings. It ignores the major differences which exist between types of farms in respect of their ability to adopt new technologies; because much of agriculture is diversified, priorities based on commodities ignore the complex interactions that occur in agricultural systems.

Problems which cut across commodities, and cannot be accommodated within the commodity projects, have to be treated separately.

Research staff level

Research workers will generally struggle for their right to set the goals of research themselves and thereby determine research policy. They will generally concede, however, that it is legitimate for policy regarding mission-oriented research to be set by the organisation which 'employs' them, in conjunction with the farming community whose interests they are required to serve.

The true interests of the research workers themselves are a powerful argument for their support and even pioneering of rational methods of research planning. Agricultural research is financed largely by public funds; unless research activities are geared to the goals of the public that supports them, it will become more and more difficult—and justifiably so—to mobilise the funds needed for research. Does this signify that the research worker is doomed to be a passive recipient of 'directives' regarding his research programme? Far from it. The individual research worker has a dual share in planning.

However sophisticated the methods developed and adopted for project evaluation and selection, they are fruitless unless good project proposals are available. In order to prepare an effective research programme that can make maximum use of available resources, it is not sufficient to evaluate only projects that have actually been proposed by individual researchers. It is necessary to know what scientific investigations could be conducted and what would likely to be their potential outcome. In this respect the individual research worker can play a major role in shaping the research programme and ensuring its overall value.

Research workers can take the initiative in proposing subjects for research based on their own experience, understanding of the requirements of the farming sector with which their work is concerned, or knowledge of developments that have occurred elsewhere in the world. They are usually in the best position to know which areas have the most potential promise. In order to be in a position to do this effectively, researchers need to be aware of the economic aspects of their work from the national viewpoint. They should attempt to maintain contact throughout the year with the farming community and their colleagues in the advisory and planning services of the ministry. The initiative should not, however, rest exclusively with them; both the ministry and the research management should endeavour to provide them with essential information on overall policy. Researchers should be partners in the actual shaping of policy; the results of their research should be reviewed and taken into account in shaping policy for the future. They must also be alert to new developments and needs, and should be able to recognise problems in their specialised areas before they become obvious to others.

Finally, the researchers play a most important role in the evaluation of research projects. The information which they have on related projects and costs of alternative projects in their field of competence is the primary source of essential data for the different kinds of analysis involved in project evaluation, whatever method is adopted (Schultz, 1970).

In brief, the researchers should:

- identify the constraints to, and opportunities for, improved practices or products;
- collect and interpret what is already known nationally and internationally on the proposed topics;
- design research projects to generate new locationally specific technology.

INTEGRATING RESEARCH PROJECTS FROM AUTONOMOUS BODIES INTO THE NATIONAL RESEARCH PROGRAMME

Even in countries in which overall responsibility for agricultural research has been vested in a National Agricultural Research Institute or Service, there are other bodies engaged in agricultural research, such as marketing boards, private commercial or industrial interests, farmers' associations, universities, etc. Some of these bodies may have their own sources of funding and are therefore generally impervious to policy decisions made outside their institutions. Others are dependent, entirely or partially, on government funding. If all government funds for agricultural research are channelled through the National Research Institute, all research projects proposed by these bodies can be submitted to the procedures of evaluation and priority setting in force, together with proposals emanating from the establishment itself. Projects that are approved can then be included in the national research programme, to be implemented by the bodies that have made the proposals. A case in point, and probably the most important is that of the faculties of agriculture.

It will be noted that in the procedures for the establishment of the national agricultural research programme, we have proposed the participation of agricultural faculty members at all the levels of decision-making. Faculty members are therefore involved in the definition of research policy, and are at the same time in a position to indicate the research areas of projects in which they, or their colleagues, can contribute to the implementation of the research programme.

SEQUENTIAL STEPS IN THE DECISION-MAKING PROCESS

Establishing a research programme, involving as it does many levels, different interests and numerous individuals, is by its very nature a fairly complicated process. Every effort should therefore be made to keep procedures as simple and streamlined as is compatible with the objective to be achieved.

The operational framework

Basically, a framework for research programme approval consists of two components:

(a) The socio-economic requirements of the country as stated in the Development Plan. These should diffuse downwards through all levels.
(b) The proposals of the research workers based on their experience and professional competence, as providing answers to the problems inhibiting production. These undergo a process of evaluation and screenings, passing upwards through the various levels until a decision on the final research portfolio is reached.

A process for programme approval has to be established so as to ensure congruence between the proposals of the research workers and the stated socio-economic requirements.

The general sequence of actions in the proposed decision-making process is as follows:

1. the top level decision-makers determine research policy guidelines and indicate a tentative budget;
2. proposals are prepared in accordance with the guidelines;
3. the programme is finalised at institutional level and submitted for formal approval;
4. a final decision is taken at ministerial level on the research budget and the cut-off point in the proposed research programme.

This sequence involves seven steps, as shown in Fig. 17, and will now be explained in more detail.

Policy guidelines (ministerial level)

At ministerial level policy guidelines are determined; the goals and priority problems are defined; a tentative budget is established, and a tentative allocation among commodities and among regions is outlined; the target groups are indicated. The Director of Research is invited to assist in these deliberations. He has an important role to play, by providing information regarding the development potentials of specific research topics and the feasibility of their goals (Nestel & Franklin, 1975).

Policy interpretation (institutional level)

The Director of Research, assisted by a panel comprising the heads of departments, an economist and the Scientific Adviser to the Minister of Agriculture (cf. p. 399), defines the overall goals of the research organisation in accordance with the guidelines received, and assigns them to the appropriate departments of the organisation, with an indication of the probable allocation of budgetary resources among the departments. The Director informs the heads of departments of his interpretation of the guidelines, and what will probably be available to each of them.

At this stage, it is important that the Director ensures that all participants in the next steps involved in the preparation of the institutional research programme should be fully aware of the requirements of the programme and understand the procedures to be followed.

For this purpose, an annual workshop should be organised, with the participation of research workers, programme co-ordinators, heads of departments, directors of regional stations, and high-level administrative personnel. The Director presents his analysis of the national objectives which he has received from the ministerial level, and guidelines for a research programme congruent with these objectives, indicating priorities in regards to areas of research, regions and farming sectors.

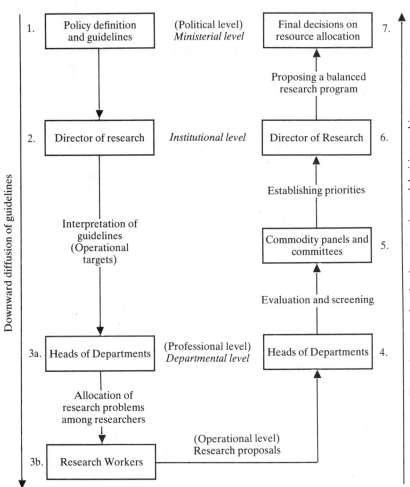

Fig. 17. Sequential programme planning steps.

All the participants should be encouraged to take part in a discussion of the Director's presentation, thereby complementing the upwards and downwards feedback system, which is an essential component of the proposed procedure. The workshop should serve not only to familiarise the research workers with the requirements of the programme, but also to ensure that they feel a personal involvement in the planning process.

Project preparation (professional-operational level)

(a) The heads of departments request from their research workers reasoned research proposals, both for individual and interdisciplinary investigations, in accordance with the specific policy directives made by the Director.

(b) The research workers propose subjects for research, in accordance with the guidelines, based on their experience, understanding of the requirements of the farming sector, and knowledge of developments in their respective fields that have occurred elsewhere in the world.

Before a project can be evaluated, its objectives, methods, resources required, cost, duration, etc. must be expressed clearly in meaningful terms. Therefore, the researcher must prepare a proper "project proposal." In view of the importance of this first step, suggested rules for project presentation are given in Chapter 8.

(c) The projects of each department (as well as cross-commodity projects) undergo a preliminary screening by the respective heads of departments. This screening is desirable before the projects are submitted for formal evaluation, in order to ensure that they contribute to the attainment of the goals that have been set up at the institutional level and to maintain scientific standards.

The following are some of the factors to be considered:

- Is the subject within the general field of assignment and professional competence of the researchers working on the project, or does the team need strengthening with additional specialists?
- Does the proposed subject fit into the overall programme of research of the department and the institute?
- Is the idea novel?
- Are the necessary scientific manpower and technical supporting personnel available?
- Are the technical facilities in the field and laboratories adequate?

- Is the research plan appropriate to the objectives to be attained? Have the experiments been properly designed, and are they amenable to statistical analysis?
- Are the cost estimates realistic and consistent with the possibilities of implementation of the project?

This pre-evaluation of a research project from the point of view of its scientific, technical and organisational merits is usually carried out personally by the head of the department concerned, who may at his discretion invite comments on the proposal from referees. After pre-evaluation the proposal can be transmitted for formal evaluation, rejected or 'recycled'—that is, returned to the authors for improvement, correction, or trimming of the budget in the light of the reviewers' comments or requests.

Project evaluation (commodity panels)
The proposals that have been provisionally approved by the heads of departments are submitted for evaluation to the commodity panels, whose function is to construct, each in their respective fields, an optimum portfolio of research projects matched with available resources. (It is legitimate to exceed by up to 20 per cent the share of the provisional budget allocated to research on a given commodity.)

Establishing the overall research programme
The pooled recommendations of the commodity panels then return to the Director of Research, who with the assistance of his senior staff chooses for each field of research a package of those projects which has received high-priority ratings from the commodity panels, up to the 'cut-off' indicated by the budget available for each problem area.

The result of the project evaluation and selection process should be a feasible and balanced programme of research that is congruous with the funds, personnel, and competences available and makes the most efficient use of these resources for the achievement of well-defined goals of the research organisation.

In all the stages mentioned above, a re-evaluation of projects already being carried out is included, on the basis of a periodical review measuring accomplishment against project implementation processes, schedule, and expenditure. Continuation, changes in plan, or cancellation of the project can be considered in relation to current priorities (see below).

Decision on final research budget and its allocation—ministerial level
The overall research programme is submitted to the top ministerial committee that has initially established the guidelines and the provisional
budget, for formal approval and for a final decision on the size of the
research budget to be included in the Ministry's overall budget proposal
to the Ministry of Finance.

This document should provide a clear statement that:

(a) enables legislators and the general public to judge the merits of the
    programme and the degree to which it corresponds to national
    requirements and justifies the resources invested;
(b) provides clear guidelines for those responsible for the implementa-
    tion of the programme;
(c) provides a feasible and balanced programme that is congruous with
    the funds, personnel, and competences available, and makes the
    most efficient use of these resources for the achievement of well-
    defined goals of the research organisation.

DIFFICULTIES AND CONSTRAINTS IN RESEARCH PLANNING

The administrative load

When adopting a decision-making procedure the amount of administrative cost and effort involved must be a major consideration. Extreme
care must be taken to 'avoid a system that imposes heavy bureaucratic
procedures on the research workers' (Ulbricht, 1977).

Improved planning systems inevitably increase the administrative load
of the research organisation—paper work, attendance at panels and committees, preparation and review of research proposals, etc., all of which
may lead to an 'energy crisis' in the research organisation (Delbeck *et al.*,
1975).

Careful thought must therefore be given to make the procedure as
efficient as possible by eliminating all unessential time-consuming pro-
cessess. For example, the procedure usually followed when planning the
annual research programme is to evaluate all ongoing research projects
together with new projects at all the stages of proposal, screening, evaluation, and final decision. This imposes an enormous work-load at all
levels of decision-making and is clearly counter-productive. A research

project that has been approved cannot possibly show results that justify a re-evaluation after one, or even two years. It is a source of vexation and frustration for the research worker, whether as an individual or a member of a research team, to have the continuation of his work on a given problem questioned year after year. On the other hand, after three years (an admittedly arbitrary time-period) the research project should be sufficiently advanced to justify review, and a decision on continuation, modification or cessation to be taken.

The annual decision-making process should be confined to the evaluation of new proposals and re-evaluation of one-third of the ongoing projects (those that were approved three years previously). The work-load on everybody involved in the process will thereby be considerably reduced, and more time will be available for a thorough appraisal of the reduced number of projects under review.

Another wasteful procedure is to engage in the decision-making process without any pre-determined budgetary framework being made known to the persons involved in the process. As a result, practically all proposals made by research workers pass through all the levels of decision-making, increasing the work-load of all concerned considerably. Finally, when the inflated research programme reaches the ministerial level, the programme is cut down to size in relation to the budget, and the whole procedure of priority rating becomes meaningless. And yet, this is a procedure most frequently followed in countries that have taken the trouble to organise a formal decision-making process.

Lack of balance in the research programme

Up to this point, we have been concerned with the problem of how to ensure that the national agricultural research programme responds to government and farmers' priorities. This exclusive concern carries with it an inherent danger: responding to government and farmers' priorities to the exclusion of other considerations will almost inevitably emphasise immediate needs to the detriment of a balanced research programme.

Agricultural research cannot, and should not be confined to the solution of immediate problems that can be achieved in a relatively short time. Not only must it foresee the problems that may arise as a result of economic or biological developments, but it must also pioneer new developments and open up new horizons. In other words, it must not only work for the future, but also have a hand in shaping the future.

There are additional obligations that a national agricultural research organisation must consider if it is to fulfil all the functions for which it is ultimately responsible. These will be briefly discussed.

Long-term research
While it is not always possible to define clearly the limits between long-range and short-range programmes, there are certain fundamental differences between them. The short-range programme must be dynamic and adaptable to changing conditions and requires a great awareness of agricultural realities on the part of the research worker. This is particularly true and important in developing countries during the period of transition from subsistence to commercial agriculture. Problems that are of great importance at a given stage of this process may lose all significance in a subsequent stage. A dynamic situation requires periodical reappraisal and readjustment of research programmes.

By contrast, long-term research requires stability and continuity; it must also take into account forecasts of scientific, technical and economic developments and their implications.

In principle, there are two types of long-range research:

1. Research on problems of immediately apparent significance in the circumstances prevailing at the time the decision is made, but the execution of which is expected to be of long duration. Typical examples are the breeding of new varieties of fruit trees adapted to specific needs, the studies on farming systems to replace traditional farming (in particular shifting cultivation), prevention of deterioration of natural resources.
2. Research that must be initiated in the present in order to be ready for future needs. The future of agriculture cannot be predicted accurately. It is influenced and shaped by many factors: population growth, technological changes, and economic developments, such as consumer income levels and preferences, changes in foreign markets, and development of industry and services. Projections can be based on trends and knowledge of factors affecting them or on formal statistical models. At best, these can only provide indications of what may reasonably be expected and are usually expressed in terms of various possible alternatives.

It is essential that a balance should be struck between short-term and long-term research, although in practice it is not too easy to assure a true balance. Especially in developing countries, research workers are usually under continuous pressure to solve important pressing problems, of which there is usually no dearth. It is not easy to create a favourable climate for the allocation of funds and workers, both in short supply, for long-range problems, but it would be a most unwise and shortsighted policy not to withstand these pressures. While the demand for the promotion of long-range research is frequently frustrated by lack of understanding of its importance on the part of those who have to approve financial support for this purpose, another and possibly greater danger is that the research worker himself, under the continuous pressure of 'brush-fire' problems, will not find the time or acquire the knowledge needed for planning ahead and visualising future needs and problems.

The need for long-range planning of agricultural research is frequently denied on the premise that the progress and direction of agricultural development cannot be foreseen accurately. There is truth in the allegation that unexpected developments or even setbacks may occur, but this cannot serve as an excuse for neglecting long-range research planning. The conclusion to be drawn instead, is that long-range research must also be periodically reviewed and updated in accordance with changing circumstances. Of course, it sometimes happens that the original judgement was faulty and a wrong decision was taken in embarking on a specific long-term research project.

Such a miscalculation is far less objectionable than facing a crisis which has been envisioned and being unprepared because of a short-sighted research policy. The same applies to developments which may have great economic importance, but cannot be exploited because the necessary knowledge has not been acquired in time.

One example of long-term research, which we have called 'insurance research', in which the importance of a research project cannot be assessed on the basis of its calculated potential economic value, is research aimed at preventing a given situation from occurring. Such research, like insurance, is undertaken in the hope that it will never need to be implemented.

Complementarity relationships between research fields and projects

One of the big problems involved in the planning of agricultural research is the complexity and interdependence of the research fields within the systems. Many problems cannot be solved in a piecemeal fashion, but require that professional competence be developed in a number of specialised fields. Therefore, the demands on resources available for research needed in different problem areas are not necessarily independent and competitive; on the contrary, there exist many important complementary and synergetic relationships which cannot be neglected in planning the research programme. Thus, even when interest is focused on certain fields, a balanced research programme is still essential. For example, in a country where agriculture is entirely dependent on irrigation, research aimed at increasing the productivity and volume of usable water is of the greatest importance. It must be fully realised, however, that irrigation investigations cannot be divorced from other relevant research problems: methods of increasing soil fertility, improved tillage techniques, the creation of improved varieties of crops, effective control of diseases, pests, and weeds—all have a bearing on the efficient use of water which is no less important than optimum irrigation rates or efficient irrigation methods.

The introduction of animal draught will involve simultaneously problems of introduction and breeding of a suitable animal, health problems, forage supply and hence changes in the cropping system, improved equipment, etc.

Hence, even specific problems require a very broad base of investigation and support from many and varied disciplines in order to obtain significant results. It therefore follows that no major field of research can be neglected without affecting the ability of the research organisation to carry out its obligations. This does not contradict the need for selective emphasis on certain problem areas as determined by national policy (cf. p. 378).

Giving a certain project a low rating cannot be justified when by so doing other projects with high ratings are handicapped because they depend on the results of research which in itself would merit low-priority rating. Research results usually generate new projects. Thus the 'output' of one project becomes an 'input' for one or more new projects. It is therefore clear that individual projects cannot be considered in isolation, and their potential effects on other fields of research must be taken into consideration whenever possible.

Investments required for applying the research results

Investment in a specific research project is not a one-time act; if successful, it generally starts a cycle of investments required for the implementation of specific research results, including additional investigations, planning and execution of extension efforts, and investments in infrastructure.

For example, a recent study of the potential effects of the introduction of new high-yield varieties of wheat in the Punjab State of India indicated that the future role of the high-yield varieties is indeed great. It also demonstrated, however, that the fertiliser supply will most likely be a very critical factor influencing the adoption of the high-yield varieties for a number of years. Hence the returns from investments in research on varieties will depend on the solution of fertiliser problems. Once the problem of fertiliser supply is solved, it gives way successively to such problems as credit, marketing, storing, and distribution.

Hence, the importance of a research project will depend not only on its potential benefits, but on the possibility and cost of adoption. This may be delayed due to various factors unrelated to the research proper.

Poor communication between research and development

Political leaders, administrators and staff in planning and finance ministries who either assist in the programming of research, or determine the amounts of resources made available, are frequently handicapped by limited understanding of the production problems faced by the farmers, lack of knowledge of the available and potential solutions, and have preconceived ideas on agricultural research, which they think is unable to make significant contributions to agricultural development.

> Because most of the resources that can be made available for research are often allocated by politicians in positions of power, leaders of agricultural research systems must learn not only how to deal with politicians but also how to influence them. Politics can be a hindrance, but it should and can be regarded, as well as used, as a facilitator and a means of innovation (Jose D. Drilon, Jr. in IADS, 1982).

Effective communication between the policy-makers and their advisers on one hand, and the research leaders on the other is therefore essential. The scientific adviser to the minister of agriculture (preferably a veteran research worker) can be made responsible for liaison between research and development. Also, policy-makers and their advisers have to be provided with all the relevant information in the form of concise and clear reports and during the meetings by verbal presentations explaining the research proposals, their potential impact on development, and how they complement the work already undertaken. Where political decisions are involved, the information will stress economic and social aspects; for professional decisions more attention will be given to production problems and constraints. The participants in the decision-making process are all busy people, and long reports, with masses of data, are counterproductive.

An effective information system can make a considerable contribution to the establishment of an optimal research programme, and can also affect the extent of the resources allocated to agricultural research, as a result of a better understanding of the issues involved by the persons who have influence in this respect.

Shortcomings of group procedures

We have shown that the lack of 'objective methods' on which to base decisions on research priorities, make group judgements an essential component of the decision-making process, and we proposed a number of panels and committees, each functioning at a certain level in the overall process, as vehicles for these group judgements.

However, the value of the work done by these groups can easily be impaired and even made useless by faulty implementation.

When the group is dominated by a few high-status, strong individuals, there will be a feeling that open discussion is impossible (Chung & Ferris, 1971). The value of group-decisions lies in the diversity of competences that participate in the process. If some of these are silenced, decisions become one-sided. We have already pointed out that the farmers' representatives are particularly exposed to this situation, in their confrontation with scientists and development administrators. Only a strong and tactful chairman can avoid such situations, by making all members of the group aware of the need for equality of participation and by giving them the necessary opportunities to do so.

Delbeck *et al.* (1975) point out that 'group processes too often leave participants exhausted and discouraged because of the seemingly endless meanderings into unfruitful by-ways. What group has not experienced the general lack of creativity and absence of a sense of closure which leaves them with a feeling of impotence, boredom and frustration?'

These shortcomings can be overcome by keeping the work-load to a minimum (see above); by having smaller groups with divergent competences but common interests (sub-committees); by ensuring equality of participation (see above); by a clear and succinct presentation of objectives on one hand and of research proposals on the other, and, above all, by a firm and consistent guiding hand of the chairman.

The influence of extraneous factors on decision-making

The guidelines or criteria adopted at the various levels are to some degree objective. However, a number of purely subjective factors may intervene and may have a marked influence on the allocation of shares of the total budget to different areas or categories of research. Examples of subjective factors are: the importance of agricultural research in the eyes of individual policy planners, and at other levels: the personality of the department head, team leader, or research worker involved in a certain area; the predisposition of the director of research toward a certain department or research area; the competence of a department head or of a team leader in solving urgent problems; and the special interests of pressure groups.

A characteristic of agricultural research in the developed countries is the considerable influence that the sector of the electorate most directly concerned (e.g. farmers) has on research policy. Farmers in the developed countries are often able to evaluate the meaning and importance of the problems with which they are faced, and since they are usually well organised, they are able to exert pressure on government agencies responsible for research. Busch & Lacy (1983) mention the influence of some organisations that seek to intervene directly in public agricultural research in the USA, by lobbying for funds for a particular commodity, or by providing partial funding for specific work. This is perfectly legitimate.

However, in extreme cases, some organisations have sought to prohibit research that might lead to results unfavourable to their special interests. 'Through the punishment of people who raise critical or embarrassing issues, or who produce data that undermine established and institutionalised relationships, a normative climate is established . . . that makes clear which topics are controversial, difficult, not-to-be funded, and therefore to be avoided' (Friedland & Kappel, 1979).

In the developing countries, it is generally only the large farmers, the plantation owners, and corporations that are articulate and who can effectively exert pressure, whilst the small farmers, who are the vast majority, have no representation and cannot influence the orientation of the research programme.

Another factor involved in research planning in the LDCs are external donor agencies and international centres, which can play a significant role in setting priorities in national institutions through earmarked funds of technical assistance for specific commodities (Pinstrup-Andersen & Byrnes, 1975). These authors are right in pointing out that 'this situation places great responsibility on those external agencies to ensure that their priorities do in fact correspond to national needs'.

Some scientists and research administrators are more effective than others in arguing their proposals and thereby mobilising support for them. Management's decisions may also be influenced in its appraisal of the proposal by the sponsor's persuasiveness and persistence.

Some of the pressures exerted in the process of assigning priorities are summarised by Elz (1984):

- research workers want to do what interests them and can further their careers;
- politicians want solutions to problems in their specific areas;
- urban dwellers want cheaper food;
- ministry of finance want export crops, new crops to open new markets, import-replacing crops.

Due to conflicting pressures, decisions are not always made on the basis of objective appraisal.

Chapter 8

The Research Project

A research project is a 'self-contained area of investigation, with a specified objective, which terminates when its specified set of objectives has been achieved' (Lawson, 1962).

ESTABLISHING THE RESEARCH PROJECT

Source of ideas for projects

A research idea may originate from various sources; a specific proposal may be initiated by a research worker, or suggested by the head of division, head of department, or the director of research. It can also originate outside the research organisation: it can be requested by the ministry of agriculture, by a farmers' organisation, by the commodity committee. New problems related to existing branches of production are usually raised by the farmers' representative or the extension worker, while proposals for new lines of research are generally presented by the research worker. Information sources that lead to ideas for research projects are the literature on scientific and technical progress in agriculture, basic research findings, market research and population trends.

Screening ideas

There is usually an undefined area between the conception of an idea for research and the preparation of a formal research proposal. It is desirable that before an idea is formally submitted to the research man-

418

agement and the efforts of the researcher are expended on the preparation of a research proposal, it should first be screened mainly on the basis of whether it fits in with the priorities as determined by the current research policy. The screening process may involve informal discussions with colleagues in the field, the immediate superior, and occasionally the research director who will either encourage some exploration or the preparation of a research proposal if he feels the idea has merit, or will discourage further involvement. Considerable discretion must be exercised so that personal preconceived ideas, lack of imagination or understanding should not choke off at its inception an original idea that may have great potentialities. Some exploratory research needs to be carried out before any objective evaluation is possible.

Original ideas are usually presented by a researcher who is deeply convinced of their importance and potentialities. In approving exploratory research on such a proposal, the director is usually more influenced by the degree of confidence he has in the researcher than by the merits of the idea!

Proposal—form and content

A research proposal may be defined as a written document requesting both authorisation and funds to undertake a specific project. Whatever the source of the research proposal, it usually follows the path outlined below:

(1) A project leader is nominated by the head of the division if the project is entirely within the division, by the head of the department if the project is a joint undertaking of two or more divisions of the department, or by the director if it is an interdepartmental project. It is important to make sure that the group is 'sold' on the idea. Lack of enthusiasm or outright scepticism are sufficient to doom the project from the start.
(2) The first, and an extremely important, step is to assemble all the pertinent data and to search through all the pertinent literature.
(3) The project leader, in consultation with his colleagues on the team, then prepares a formal project proposal, which includes:

  (a) the objectives of the research, in which the problem is clearly stated;
  (b) the importance of the work, with all the pertinent data needed to substantiate the significance of the research;

(c) the outline of previous work in the field, indicating what is new in the proposed approach;

(d) the plan of work: this must be clear, specific and appropriate to attaining the objectives proposed; the responsibilities of co-operating departments must be clearly defined;

(e) the cost estimates: these must be realistic—both overestimates and underestimates should be scrupulously avoided; the number of people needed and their skills should be stated, as well as the facilities required; the duration of the project must be estimated.

IMPORTANCE OF RESEARCH PROJECT

The research proposal serves several useful purposes (Seiler, 1965):

(1) It obliges the researcher to study what is already known on the subject, thereby avoiding wasteful duplication and pitfalls, and providing useful information on methods and equipment needed.

(2) The researcher must give careful thought to his proposal, and formulate it in such a way as to convince himself and others of its potential significance.

(3) He must carefully plan his approach to the problem and the strategy of his research.

(4) In the case of teamwork, the responsibilities and duties of each of the team members are clearly defined before the work is started, thereby lessening the possibilities of friction and misunderstanding between team members.

(5) The proposal serves as a formal document for obtaining approval and funds after critical appraisal.

(6) The research project has its own identification symbol, facilitating follow-up and evaluation.

It is usual to provide researchers with standard proposal forms, which have the advantage of ensuring the inclusion of all pertinent information and data required for evaluating the proposal. This need not prevent the researcher who submits the proposal from including any additional information that he deems relevant.

One beneficial effect of programming by projects is that researchers are obliged to take into account the time required for carrying out the research, the number of man-years involved, and the financial and other resources required.

Scheduling

A realistic estimate of the time involved for carrying out the research is essential for the following reasons:

- A research project is usually a team effort; the scheduling of the various activities involved and of personnel assignments is therefore extremely important for reasons of efficiency and economy; it helps to avoid making commitments which it will be difficult to meet and to co-ordinate the work of people involved in several projects.
- An essential element in evaluating the project is the time limit during which it is hoped to conclude the research. A proposal that may be of great economic and practical value at a given time may be obsolete before the research is completed. This is especially true in the dynamic situation which usually characterises a developing agricultural economy.
- Scheduling is an essential element in the monitoring of the project and reviewing its progress; a considerable amount of leeway should be allowed for, however. In estimating the time required for the research, account should be taken of the time needed for designing, assembling and purchasing equipment and for establishing research facilities; for the experimental work itself (in agriculture, field work usually has to be repeated for a number of seasons and the possibilities of a crop failure must be taken into account); and for analysing the results and presenting the conclusions thereof.

For complex projects, involving a large team, complicated procurement problems, an important building programme, etc., it may be desirable, in drawing up a schedule, to use modern techniques, such as PERT (Program Evaluation Review Technique) or CPA (Critical Path Analysis). These techniques may make possible savings in time and expenses by ensuring co-ordination between the different elements of the programme, and providing a sound basis for reviewing programmes according to schedule.

Scientists are usually not over-concerned with time commitments. One of the reasons for the apparent disdain of researchers for scheduling is their drive for perfection, the desire for further improvements and additional data and proof (Hinrichs, 1963). However, it is essential to decide on a cut-off point for every research project, if it is to serve its declared purpose of solving a specific problem at a given time. Scheduling should, however, not be arbitrary or imposed from above but determined in consultation with the people responsible for the execution of the project. It should also be liberal, allowing time for the researcher to read and to attend seminars, for additional investigation outside the project (exploratory research) and also for possible delays.

The reason why a research project should not exceed five years is because it is difficult to plan in advance for a longer period or to foresee the changes that might be required after a longer period. At the end of the three- or five-year period the progress made should be evaluated and a final report written. If it is found necessary to continue the research on the subject, a new project can be prepared, which takes into account the progress achieved, the new problems that have appeared, the new techniques that have become available and the new approaches that appear desirable. Possibly, changes in the composition of the team may be indicated if new or different disciplines are involved.

Costs

In calculating the costs of a project, information on the cost of researchers and the supporting system is essential. For example, Ampuero (1981) gives some figures in use in several countries in Latin America for estimating the annual costs of research: a chief investigator, US$ 20 000–30 000; a research assistant, US$ 10 000–20 000. These figures include salaries, transport costs, equipment and materials, and administrative overhead. Ampuero states that in his opinion, these figures are very low in relation to real needs.

In the international centres, the costs per researcher are estimated at US$ 100 000 to 200 000 annually.

Project evaluation

The allocation of funds for the implementation of the research project is dependent on the priority level it achieves in competition with other projects, when the research programme as a whole is being established. The procedure for establishing priorities has been outlined in Chapter 7; the authority for deciding priorities is in the hands of a public body, in which the research organisation is represented but is in a minority. These priorities are based mainly on the importance of the *problem to be investigated* to the economy; the public body has neither the authority nor the competence to judge the scientific, technical and organisational merits of the project as such. It is obviously the responsibility of the research director to ensure that the project has been properly evaluated on its merits *as a research proposal* before submitting it for priorities rating.

Criteria used for screening project proposals

Project proposals are screened using the following criteria:

A. *Research direction and balance*
  1. compatibility with the short- and long-term objectives deriving from ministerial guidelines;
  2. contribution to a balanced programme of research;
  3. urgency of the research.

B. *Technical factors*
  1. availability and efficient utilisation of the necessary scientific manpower for the research unit (available or not, need to be trained or imported, case of recruitment);
  2. adequacies of research facilities in laboratory or field;
  3. availability of technical support staff;
  4. probability of success.

C. *Economic and social factors*
  1. cost of the research;
  2. impact of the research;
     (a) main target area—all farmers, only commercial farmers, especially small farmers;
     (b) what are the costs of and constraints to adoption? Will it be difficult to convince farmers to adopt eventual results from the research? Will adoption be a long and difficult process? Are infrastructure, institutions and deriving adequate?

D. *Timing factor*
  1. research completion in relation to requirements of the economy;
  2. will the research be completed before the problem has become obsolete?

The evaluation of research projects is usually carried out by the deputy director of research, who, at his discretion, can invite comments on the proposal from referees. After evaluation, the proposal can be rejected, or returned to the authors for correction and adjustment in the light of the reviewers' comments or requests.

DISADVANTAGES OF THE PROJECT SYSTEM

Not all scientists view the project system favourably; some consider that 'projectitis', with its requirement of a precisely specified goal and its detailed outlining of research procedures does considerable damage. The requirement to give precise indications of probable results encourages deception; there is a tendency to buy unnecessary equipment in order not to 'waste' available funds; and any wandering up side-paths (that might lead to significant results) is discouraged (Hinrichs, 1963).

Gamble (1984) attributes serious drawbacks to the project approach, as it generally functions in the developing countries, in relation to projects funded, partially or entirely, by foreign-aid sources. He writes:

> Through it, assistance agencies may exert undue influence on the content of national programmes. Donors, bilateral as well as multilateral, may dominate programmes through their own perceptions of priorities, through efforts to maintain a commitment schedule (e.g., push ahead with facilities even though they cannot be domestically staffed), or through a self-interested desire to market 'tied' inputs (to sustain political support from suppliers, universities or consulting firms at home).

Gamble concludes that: 'selectivity in programme support and donor bias may result in the funding of projects that are inconsistent with national priorities or national programme development. It should however be clear that these comments are not an indictment of the project as a system, but of the way it can be misused.'

REASONS FOR DISAPPROVAL OF PROJECTS

Research projects depend for their financing, as a rule, on the approval of appropriate advisory boards. The approach of these bodies to the proposals submitted to them depends largely on the composition of the body; if laymen predominate, the main concern will be with the problem itself; if scientists predominate, they will be equally concerned with the scientific merits of the problem, with the approach adopted by the researcher to solve the problems and his competence to do so.

Many of the research projects submitted to these bodies fail to obtain approval; these failures reflect serious shortcomings of the researchers who have submitted the proposals. Allen (1960) made a study of the reasons for which research proposals are rejected by scientific committees of peers, on the justified assumption that knowledge of the most frequent shortcomings of research proposals, which lead to their rejection, can help the researcher by enabling him to apply self-criticism before requesting a grant.

In this study, involving over 600 disapproved research proposals, Allen found that in 58 per cent of the cases, the importance or timeliness of the problem was in doubt, in 73 per cent of the cases, the reason for rejection was that the method of attack, as proposed, would not yield sufficiently useful data; in 55 per cent the scientific competence was found to be inadequate, and in 16 per cent of the cases, miscellaneous reasons were involved.

MONITORING AND EVALUATION OF PROJECTS

Monitoring

McLean (1987) defines monitoring as 'the ongoing process of recording, analyzing, reporting, and storing data during the implementation of an activity' (Fig. 18).

The purpose of monitoring is to determine whether: (a) the work is progressing according to schedule; (b) it is being carried out competently; (c) the problem is still relevant; and (d) expenditure is in accordance with budget.

McLean stresses that 'managers must ensure that the monitoring system is not more time-consuming than the benefits justify, that no superfluous data are collected, that data analysis, interpretation and feedback are timely, and that researchers perceive it as useful'.

Evaluation

Whereas monitoring is concerned with whether a project is proceeding as planned, evaluation is designed to examine the effectiveness of the

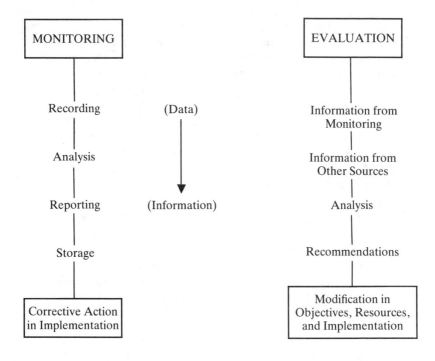

Fig. 18. Relationship of monitoring to evaluation. (From McLean (1987), reproduced by permission of ISNAR, The Hague.)

implementation of the project and achievement of the envisaged objectives (Acharya, 1987). An important aspect of evaluation is to determine
whether the problem investigated is still relevant in relation to the situation in the field.

Monitoring and ongoing evaluation are carried out during the implementation of a project. They enable research management to judge
how efficiently resources are being used, and to make adjustments in
accordance with problems that are identified in the course of implementation.

An *ex-post evaluation* is an assessment of performance after completion of the project; it is used 'to determine whether the project objectives
were attained, and the causes of any discrepancies. The lessons learned
can be incorporated into subsequent planning and implementation'
(McLean, 1987).

Evaluation of a research project after adoption by farmers
An evaluation of the impact of the results of a specific research project
by an experienced economist, after its completion and application has
a certain interest, despite the limitations of economic analysis described
in the previous chapter.

It is also important that the funding authorities, and the public, be
made aware of the benefits that accrue from research; this is also good
politics, because the research director should be able to show, in specific
cases, that research pays its way, both to justify the expenditure already
made on the research and to obtain new allocations.

A potentially misleading assumption is that *'adoption of a technology
by farmers must be a basic measure of research effectiveness'* (World
Bank, 1981). This criterion is justified only when the results of research
are *not applicable*, or are not applied because the cost of application
exceeds the benefits. But *excellent* research results may fail to be adopted
because of factors beyond the control or responsibility of research, such
as government economic and fiscal measures, lack of credit, inadequate
infrastructure, institutional defects, etc. In the latter case, evaluating
research by the rate of adoption would be completely unjustified.

The fallacy of using adoption rates by farmers as a criterion for the
success or otherwise of a research project is demonstrated by the data
on adoption rates of new varieties resulting from a varietal improvement
programme on four crops in Colombia, as reported by Ardila *et al.*
(1975) (Table 5).

The high returns to *soybean* research are mainly due to the rapid and
high levels of adoption (98 per cent) of the improved varieties. This
rapid adoption was due to the strong demand for the improved seeds,
the geographic concentration of producers which facilitated the rapid
diffusion of the new varieties, and the fact that these producers are
among the most progressive of Colombia's farmers.

The high rates of return of the *rice* programme were partly due to the
yield advantage of the new varieties, their adoption on nearly half the
total rice area and the large area sown; however, the high returns were
mainly credited by the authors to the low cost of the programme. This
was due to the fact that information, know-how and plant materials
were supplied by CIAT, IRRI and the USDA. The national programme
therefore involved the relatively low-cost screening of easily available
material.

The *wheat* programme was extremely successful in achieving a significant yield advantage (46 per cent), even more so than the soybean and

TABLE 5

Selected data on the rice, cotton, wheat and soybean varietal improvement
programme

Concept	Unit	Rice	Cotton	Wheat	Soybeans
Estimated net internal rates of return	Per cent	60–82	0	11–12	79–96
Estimated yield advantage, 1971	Per cent	25–39		46	17–36
Land area planted with improved varieties, 1971	Per cent	41	100	35	98
Total research costs/ value production, 1968–1971	Per cent	0.5	0.1	3.0	0.1
Average yields, 1971 Colombia/United States	Ratio	0.68	1.03	0.53	1.01

Source: Ardila *et al.* (1975)

rice programmes, and yet the internal rates of return were relatively
low. This is ascribed by the authors to large and sustained imports of
cheap wheat from the USA under PL.480, which depressed the local
market.

The estimated net internal rate of return for the *cotton* programme
was nil, and yet the entire land area of cotton was planted with the
improved varieties. This apparent paradox is due to the nature of the
research programme, designed to introduce, test locally and distribute
in Colombia the best-yielding US varieties to be revealed. However, no
significant yield differences in yields among these varieties were un-
covered. Hence, the US varieties to be distributed among the Colombian
farmers could have been chosen at random, and, in hindsight, the research project had proven to be unnecessary.

Review by the director of research

Reports are a useful device for keeping the director of research informed
of the progress of research on individual projects. Reports alone, how-

ever, cannot be the sole source on which the director of research bases his evaluation of the competence and contributions of the responsible research workers.

Personal contact with the researchers, by regular visits to their units, affords an opportunity to discuss the problems encountered, to obtain a first-hand impression of the quality of the work done, the urgency with which it is done, and the 'morale' of the research group. It is also an effective means of communicating to the researchers the needs and objectives of the organisation, and occasion for discussion and guidance. As an important 'by-product' it improves morale by giving the researchers the certitude that management takes an interest in their work, that their efforts are properly appreciated and that their difficulties are given sympathetic consideration.

REPORTING TO RESEARCH MANAGEMENT

Researchers have the responsibility of submitting routine reports to research leaders and administrators as one of the tools for monitoring and evaluation of progress in a research project. Most researchers consider this obligation as a time-consuming, unnecessary imposition. If quarterly and annual reports are requested, as is standard procedure in many research institutions, the researchers may even have a certain justification for their antipathy to this requirement.

In agricultural research, dependent as it is on seasons, requiring reports to be submitted on a routine, calendar basis is unrealistic. The research worker, when obliged to do so, will supply reports at regular intervals, but they will be of little intrinsic interest and value, whilst causing a loss of his time.

By contrast, progress reports, written at specific stages of the research project, are of great value to the researcher and to research management alike. In agricultural research, the work encompassing a whole season is usually the suitable unit for a progress report. Such a report should present the relevant data, their interpretation and the tentative conclusions to be drawn therefrom. If changes in the direction of the research appear warranted on the basis of the results achieved, this should be indicated. Such a progress report will make interesting reading, can serve as a basis for verbal discussion, and provides a satisfactory solution to the need for keeping research management informed of the progress being made on a certain problem, in that it presents a detailed report of the work carried out during the current season, and offers tentative conclusions. The preliminary report is also of advantage to the research worker. Agricultural research projects usually require three to five years for their completion; instead of being faced, at the end of this period, with a mountain of undigested data and nebulous memories of the experiments themselves, all that is needed is to co-ordinate a few progress reports, in which all the results have already been tabulated and statistically analysed, and consolidate the tentative conclusions into a more comprehensive and final form. Writing a report at each stage of the experiment has the additional advantage of showing up any weakness in the original plan, at a stage when improvements are still possible, and suggests new or additional approaches. Finally, these reports make possible a periodic review of the progress of the research, thereby ensuring that the resources devoted to the work under consideration are being used effectively.

Coercive measures are frequently used to ensure the presentation of a report without unreasonable delay. Administrators of research grants frequently withhold a fixed proportion of the grant (up to 25 per cent) until receipt of the report. The reports and papers serve for the evaluation of the researcher's performance as a *sine qua non* for promotion; this provides a very powerful stimulus to productive writing!

In addition to exerting pressure, some organisations facilitate the presentation of reports by providing a statistical service that does the statistical analysis and computing of the figures, clerical help for typing, and editorial staff to help the researcher in writing up his report.

Oral reports

Another important form of reporting is oral presentations on research in progress. These presentations are made to the commodity or cross-commodity panels, seminars of research workers, end-of-season meetings with extension workers and farmers. They provide an opportunity for valuable feedback from peers and from 'clients', and for discussion when multidisciplinary work is involved.

Action to be taken after review

The project should be evaluated both in the light of the progress achieved and also the possible changes in the scientific or economic status of the problem investigated.

Seiler (1965) points out that the reviewer is usually influenced in his judgment by extraneous factors. When progress is rapid or the economic prospects in the field are favourable, there is a natural tendency to become overly enthusiastic; if, however, research does not progress as anticipated, or the economic situation in the field is no longer very favourable, there may be a shift to defeatism. 'Pacing' is what Seiler calls 'the calm and systematic appraisal of the many facets of the situation, which should safeguard against extreme fluctuations between enthusiasm and defeatism, during the various stages of a long research effort'.

After review, the following courses of action should be considered (Brooks, 1966):

1. alternating the allocation of funds and resources to the different components of the project;
2. modifying the original time-table;
3. modifying the original goals, in the light of results already obtained or changes in the external situation;
4. changing the priority of the project (for better or worse);
5. termination of the project.

Discontinuance of projects

Worthwhile new projects may be rejected for lack of resources because of commitments to earlier projects, even if the new proposals have greater potential worth than certain of those being implemented. Hence there is a perpetual conflict between the adoption of new, promising projects and the need to maintain stability in the on-going programme.

Because of the very nature of research, a project rarely has a well-defined ending point, where it is possible to state categorically that no further experiments or additional approaches are justified. However, a project must be wound up, preferably at the optimum point, when further investment of funds and effort are no longer justified. The ability to decide on the 'optimum' point is 'the product of intuition, long experience, and an awareness of the strengths and weaknesses on the research staff' (Seiler, 1965).

Discontinuing successful projects
All too frequently, a fairly large proportion of research projects continue longer than necessary. Among the reasons for delay in termination: (a)

the researchers' reluctance to conclude the project and start a new assignment; (b) the unavailability of facilities for a new project; (c) the lack of new projects for the researchers in specific areas of specialisation; and (d) the lack of review by the research management, leading to continuation of projects through inertia.

In agricultural research, with its greatly fluctuating environmental conditions, there is also the fear of the researcher to commit himself, before he feels that his conclusions have been tested for a sufficient number of seasons and are therefore relatively 'foolproof'.

Discontinuing unsuccessful projects
There is usually a greater difficulty in deciding to discontinue an unsuccessful project than a successful one, and hence a still greater delay. Among the reasons for this presented by Seiler (1965) are: 'the hesitancy of the researcher to concede failure and his desire "to try one more time" ("hope fortified by substantial investment is hard to turn off"); bias in favour of the project, by those who have approved the project in the first place; objectives of the project are so vaguely defined that decision is made difficult.'

PROGRAMME REVIEWS

In addition to reviewing individual projects, it is also useful to conduct peer reviews of the entire programme of a department at appropriate intervals, by a team of scientists from other institutions or departments who sit down with the scientists of a department or a discipline and review their research programmes intensively. The programme can be evaluated in terms of the effectiveness of the research carried out, the adequacy of the team-work and of the means available, whether the balance between applied and basic research and between short-term and long-term research is satisfactory, and the relevance of the work to the needs of the farming community. The long-range goals of the department can be discussed. The management of the research institution can then be provided with an unbiased evaluation of the research programme of each unit by scientists who are competent in the field under review.

Preliminary information is supplied in writing to the review panel members before they start work, and the panel then meets with station representatives, who outline the work in progress and its objectives, as well as the organisational methods used. Panel and staff, together, evaluate and criticise the work. As a result of the preliminary preparation, the comprehensive review itself can be concluded in three to five days.

It is felt that although station directors and department heads have a more intimate knowledge of the problems under review, the panellists have a more objective view. They can therefore stimulate new thinking and suggest new approaches. At the conclusion of the review the panel submits a report to the director with recommendations for revising the department's research.

References to Part Two

Acharya, R.M. (1987). Organisation and Implementation of research evaluation and monitoring in the ICAR research system. In International Workshop on *Agricultural Research Management*, ISNAR, The Hague, pp. 184–6.

Aldrich, D.G. (1966). Agricultural research, a key to understanding our environment. In *Symposium on Research in Agriculture*. USDA Research Service, Washington, DC, pp. 50–53.

Allen, E.M. (1960). Why are research grants disapproved? *Science*, **132**, 1532–34.

Alves, E.R. de A. (1984). Major issues in resource allocation. In *The Planning and Management of Agricultural Research*, ed. D. Elz, The World Bank, Washington, DC, pp. 27–39, 122.

Ampuero, E. (1981). *Organización y Administración de la Estación Experimental en los Países en Desarrollo*. Progr. in Intern. Agric. Cornell University, Ithaca, New York.

Anderson, J.R. (1972). Allocation of resources in agricultural research. *J. Austr. Inst. Agric. Sci*, **38**, 7–13.

ARC (1967). *General Principles Governing Forward Policy*. Agricultural Research Council, London.

Ardila, J. & Valderama, M. (1974). El Proceso de la Toma de Decisiones para la Assignación de Recursos en la Investigación en un Instituto Nacional. Paper presented at a Seminar on Methods for Allocating Resources in Applied Agricultural Research in Latin America. CIAT, Cali, Colombia.

Ardila, J., Hertford, R., Rocha, A., & Trujillo, C. (1975). Returns to agricultural research in Colombia. In *Methods of Allocating Resources in Applied Agricultural Research in Latin America*, ed. P. Pinstrup-Andersen and F.C. Byrnes, CIAT, Series CE-11, pp. 53–56.

Ardito-Barletta, N. (1965). Costs and social returns of agricultural research in Mexico. Ph.D. Dissertation, Department of Economics, University of Chicago.

Barnard, C.I. (1957) A national science policy. *Sci. Amer.*, **197**(5), 45–49.

Barre, H.W., Call, I.E., Kendall, J.C., Moore, C.A., Allen, E.W., & Jardine, J.T. (1928). Report of the experiment station committee on organisation and policy. In *Proceedings 42nd Annual Convention of National Association of Land-Grant Colleges and Universities*, Washington, DC, pp. 203–5.

Baxter, M. & Thalwitz, W. (1985) National policies and institutional constraints to linking research with extension in Asia. In *Research–Extension–Farmer: A Two-Way Continuum for Agricultural Development*. World Bank/UNDP, Washington, DC, pp. 42–48.

Boyce, K., & Evenson, R.E. (1975). *National and International Development Programmes*. Agricultural Development Council, New York.

Brady, N.C. (1975). Criteria for establishing research priorities and selecting research projects. *Methods of Allocating Resources for Applied Agricultural Research in Latin America*, ed. P. Pinstrup-Andersen and F.C. Byrnes. CIAT, Cali, Colombia, pp. 33–35.

Bredahl, N.E., Bryant, W.K. & Ruttan, V.W. (1980). Behaviour and productivity implications of institutional and project funding of research. *Amer. J. Agric. Econ.*, **61**, 371–83.

Brooks, N. (1966). Organising research for social and economic objectives. In *Symposium on Research in Agriculture*. Agricultural Research Service, USDA, Washington, DC, pp. 10–13.

Brooks, H., Freeman, C., Gunn, L., Sant-Geours, J., & Spacy, J. (1961). *Government and Allocation of Resources to Science*. OECD, Paris.

Brown, L.R. (1965). *Increasing World Food Output*. Foreign Agricultural Econ. Rep. No. 25. USDA, Washington, DC.

Busch, L. & Lacy, W.B. (1983). *Science, Agriculture, and The Politics of Research*. Westview Press, Boulder, Co.

Carew, J. (1964). As it looks to me. *Amer. Veg. Gr.*, **12**(11), 18–21.

Cetron, M.J. & Bartocha, B. (1972). *The Methodology of Technology Assessment. Gordon and Breach, New York.*

Chambers, R. (1980). *Rapid Rural Appraisal: Rational and Repertoire*. Institute of Development Studies, University of Sussex, Brighton.

Christanssen, R.P. & Yee, (1964). The role of agricultural productivity in economic development. *J. Farm. Econ*, **46**, 1051–61.

Chung, K.H. & Ferris, M.J. (1971). An enquiry of the nominal group process. *Academy of Management* J., **14**(4), 520–24.

Dancy, J.C. (1965). Presidential address to educational section. Meeting of the British Association for Science. *Nature*, **208**, 1025.

Daniels, D. & Nestel, B. (eds) (1981). *Resource Allocation to Agricultural Research*. Proceedings of Workshop, Singapore, IDRC, Ottawa.

Delbeck, A.L., Van de Ven, A.H. & Gustafson, D.H. (eds) (1975). *Group Techniques for Program Planning*. Scott, Fresman & Co. Glenview, Ill.

Dillon, J.L. (1966). The efficiency gap in agriculture. *Farm Policy*, **6**, 62–67.

Elz, D. (ed.) (1984). *The Planning and Management of Agricultural Research*. The World Bank, Washington, DC.

Evenson, R. (1973). *Investment in Agricultural Research*. A survey paper prepared for the Evaluation Group on International Agricultural Research, Yale University.

FAO (1970). *World Plan of Action for the Application of Science and Technology Development*. FAO, Rome.

FAO (1972). Accelerating agricultural research in the developing countries. In *The State of Food and Agriculture 1972*. Rome, pp. 141–54.

FAO (1984) *National Agricultural Research*. Report of an Evaluation Study in Selected Countries. Rome.

Fishel, W.L. (1971). The Minnesota agricultural research allocation information system and experiment. In *Resource Allocation in Agricultural Research*, University of Minnesota Press, Minneapolis, pp. 344–81.

Friedland, W.H. & Kappel, T. (1979). *Production or Perish: Changing the Inequalities of Agricultural Research Priorities*. Project on Social Impact Assessment and Values. University of California.

Gamble, W.K. (1984). *Improving the Global System of Support for National Agricultural Research in Developing Countries*. ISNAR, The Hague.

Gresford, G.B. (1962). *Organisation and Planning of Scientific and Technological Policies*. Paper presented at UN Conference on the Application of Science and Technology for the Benefit of the Less Developed Areas, Geneva.

Griffen, K. (1973). Policy options for rural development. In *Rural Development and Employment*, ed. E.D. Gotsch, Ford Foundation Seminar, Ibadan, pp. 18–79.

Gross, M. (1964). *The Managing of Organisations: the Administrative Struggle*. Free Press, Glencoe, Ill.

Heady, E.C. (1970) Welfare implications of agricultural research. In *Resource Allocation in Agricultural Research*, ed. W.L. Fishel, University of Minnesota Press, Minneapolis, pp. 121–36.

Hertz, D.B. (1957). The creative mentality in industrial research. In *Human Relations in Industrial Research Management*, ed. R.T. Livingston and R.H. Milberg. Columbia University Press, New York, pp. 45–59.

Hildebrand, P.E. & Waugh, R.K. (1983). Farming Systems Research and Development. *Farming Systems Support Project Newsletter*, **1**(1), 4–5.

Hinrichs, J.R. (1963) Creativity in industrial research. In *The Management of Scientific Talent*, ed. J.W. Blood. American Management Association, New York, pp. 155–78.

Horowitz, I. (1960). Regression models for company expenditures on end returns from research and development: where we stand. *IEEE Mgmt*, **EM-7**, 8–13.

IADS (1971). *Preparing Professional Staff for National Agricultural Research Programs*. Report on Workshop, Bellagio, Italy.

IADS (1982). *Report 1981*. International Agricultural Development Service, New York.

ISNAR (1987*a*) *Working to Strengthen National Agricultural Research Systems— ISNAR and its Strategy*. The Hague.

ISNAR (1987*b*). *International Workshop on Agricultural Research Management*. The Hague.

ISNAR/SPAAR (1987). *Guidelines for Strengthening Agricultural Research System*. World Bank, Washington, DC.

Javier, E.Q. (1987). Priority setting and resource allocation in agricultural research at the national level. In *International Workshop on Agricultural Research Management*, ISNAR, The Hague, pp. 63–73.

Johnson III, S.F. & Kellogg, E.D. (1984). Extension's role in adapting and evaluating new technology for farmers. In *Agricultural Extension: A Reference Manual*, ed. B.E. Swanson. FAO, Rome.

Kellogg, E. (1977). Role of social scientists in cropping systems research. In *Proceedings 1st National Conference on Cropping System Research in Thailand*. Chiang Mai University, Chiang Mai, pp. 500–47.

Kislev, Y. & Evenson, R. (1973). *Agricultural Research and Productivity: an International Analysis*. Yale University, New Haven, Conn.

Klopsteg, P.R. (1945). Increasing the productivity of research. *Science*, **101**, 569–75.

Konkle, W.W. (1982). *An Assessment of the United States Food and Agricultural Research System*. Office of Technology Assessment, Washington, DC.

Koontz, H. & O'Donnel, C. (1955) *The Principles and Practices of Management*. McGraw-Hill Book Co, New York.

Kornhauser, W. (1963). *Scientists in Industry–Conflict and Accommodation*. University of California Press, Berkeley.

Latimer, R. & Pearlberg, D. (1965). Geographic distribution of research costs and benefits. *J. Farm. Econ.* **47**, 234–41.

Lawson, E.H. (1962). Must agricultural scientists be illiterate? *J. Aust. Inst. Agric. Science*, **28**, 304–5.

Libik, G. (1969). The economic assessment of research and development. *Mgmt Sci*, **16**, 33–66.

Lopes Neto, A.S. (1975). Mechanisms for allocating resources in applied agricultural research at EMBREPA, Brazil. In *Methods for Allocating Resources in Applied Agricultural Research in Latin America*, ed. P. Pinstrup-Andersen and F.C. Byrnes, CIA, Cali, Colombia, pp. 39–40.

McCalla, A. (1977). Politics of the US Agricultural Research Establishment. Paper prepared for the Agricultural Policy Symposium, Washington, DC.

McLean, Dianna (1987). Monitoring and evaluation in NARS. In *International Workshop on Agricultural Research Management*. ISNAR, The Hague, pp. 173–81.

McNamara, A.S. (1972). *One Hundred Countries—Two Billion People*. Praeger, New York.

Marcotte, P. (1987). Programme formulation and programme budgeting. In *International Workshop on Agricultural Research Management*. ISNAR, The Hague, pp. 157–65.

Mottley, C.M. & Newton, R.D. (1959). The selection of projects for industrial research. *Opl. Res*, **7**, 740–51.

Myrdal, G. (1965). *The 1965 McDougall Memorial Lecture*. 13th Session, Conference of the Food and Agriculture Organisation of the UN, Rome.

NAS Committee (1972). *Report of the Committee Advisory to the US Department of Agriculture*. National Technical Information Service, US Department of Commerce, Springfield, Va.

Nestel, P.L. & Franklin, D.L. (1975). Summary of Workshop decisions. In *Methods for Allocating Resources in Applied Agricultural Research in Latin America*, ed. P. Pinstrup-Andersen & F.C. Byrnes, CIAT, Cali, Colombia, pp. 13–21.

Norman, D. & Hays, H. (1979). Developing a technology suitable for small farmers. *National Development,* **21**(3), 67–75.

NSF (1959). *Science and Engineering in American Industry.* National Science Foundation. Washington, DC.

OECD (1963). *The Measurement of Scientific and Technical Activities: Proposed Standard Practice for Survey of R & D,* Paris.

Oram, P.A. & Bindlish, V. (1981). *Resource Allocations to National Agricultural Research—Trends in the 1970s.* ISNAR and IFPRI, The Hague and Washington, DC.

OTA (1981). *An Assessment of the United States Food and Agricultural Research System.* Office of Technology Assessment, Washington, DC.

Paulsen, A. & Kaldor, D.R. (1968). Evaluation and planning of research in the experiment station. *Am. J. Agric. Econ,* **50**, 1149–61.

Peterson, M.L. (1970). The returns to investment in agricultural research in the United States. In *Resource Allocation in Agricultural Research,* ed. W.L. Fishel, University of Minnesota Press, Minneapolis, pp. 139–62.

Piñeiro, M. (1986). *The Development of the Private Sector in Agricultural Research: Implications for Public Research Institutions.* PROAGRO Paper 10. ISNAR, Rome.

Pinstrup-Andersen, P. (1982). *Agricultural Research and Technology in Economic Development.* Longman, New York.

Pinstrup-Anderson, P. & Byrnes, F.C. (1975). *Methods of Allocating Resources in Applied Agricultural Research in Latin America,* CIAT, Cali, Colombia.

Pinstrup-Andersen, P. & Franklin, D. (1977). A systems approach to agricultural research resource allocation in developing countries. In *Resource Allocation and Productivity in National and International Agricultural Research,* ed. T.M. Arndt, D.G. Dalrymple, & V.W. Ruttan, University of Minnesota Press, Minneapolis, pp. 416–35.

Röling, N. (1984). Agricultural knowledge: its development, transformation, promotion and utilisation. In *Proceedings of 6th European Seminar on Extension Education,* ed. R. Volpi & F.M. Santucci, Centro Studi Agricoli, Borgo a Mazzano, Italy, pp. 124–205.

Ruttan, V.W. (1982). *Agricultural Research Policy.* Univ. of Minnesota Press, Minneapolis.

Ruttan, V.W. (1985). Toward a global agricultural research system. In *Agricultural Research Policy and Organisation in Small Countries.* Report of a Workshop. Agricultural University, Wageningen, pp. 19–35.

Safrenko, A.J. (1984). Introducing technological change: the social setting. In *Agricultural Extension: A Manual,* ed. B.E. Swanson, FAO, Rome, pp. 56–76.

Salmon, S.C. & Hanson, A.A. (1964). *The Principles and Practice of Agricultural Research.* Leonard Hill, London.

Schmookler, J. (1964). Problems of definition and measurement. *The Rate and Direction of Inventive Activity.* National Bureau of Economic Research, Princeton University Press, Princeton, NJ, pp. 43–51.

Schuh, E. & Tollini, H. (1979). *Costs and Benefits of Agricultural Research: the State of the Art.* Staff Working Paper, No. 360. World Bank, Washington, DC.

Schultz, T.W. (1964). *Transforming Traditional Agriculture.* Yale University Press, New Haven, Conn.

Schultz, T.W. (1970). The allocation of resources in research. In *Resource Allocation in Agricultural Research,* ed. W.L. Fishel, University of Minnesota Press, Minneapolis, pp. 90–120.

Schultz, T.W. (1977). Uneven prospects from agricultural research related to economic policy. In *Resource Allocation and Productivity in National and International Agricultural Research,* ed. T.W. Arndt, D.G. Dalrymple & V.W. Ruttan. University of Minnesota Press, Minneapolis, Minn.

Schutyer, W.A. & Coward, E.W. Jr (1971). Planning agricultural development: the matter of priorities. *J. Dev. Areas,* **6,** 38–39.

Seiler, R.E. (1965). *Improving the Effectiveness of Research and Development.* McGraw-Hill, New York.

Simons, H.A. (1960). *The New Science of Management Decision.* Harper & Row, New York.

Tichenor, P.J. & Ruttan, V.W. (1970). Problems and issues in resource allocation for agricultural research. In *Resource Allocation in Agricultural Research,* ed. W.L. Fishel, Minnesota University Press, Minneapolis, pp. 3–24.

Toulmin, S. (1966). Is there a limit to scientific growth? *Science J.,* **2**(8), 80–85.

Trigo, E.J. & Piñeiro, M.E. (1984). Funding agricultural research. In *Selected Issues in Agricultural Research in Latin America.* ISNAR, The Hague, pp. 76–98.

Ulbricht, T.L.V. (1977). Contract agricultural research and its effect on management. In *Resource Allocation and Productivity in National and International Agricultural Research,* ed. T.M. Arndt, D.G. Dalrymple and V.W. Ruttan. University of Minnesota Press, Minneapolis, pp. 381–93.

Weinberg, A.M. (1964). Scientific choice, basic science, and applied mission, *Minerva,* **3,** 515–23.

World Bank (1981). *Agricultural Research. Sector Policy Paper.* Washington, DC.

World Bank (1984). *Toward Sustained Development: A Joint Program of Action for Sub-Saharan Africa.* Washington, DC.

Yudelman, H., Butler, G., & Banerji, R. (1971). *Technological Change in Agriculture and Employment in Developing Countries.* Development Centre, OECD, Paris.

Zuckerman, S. *et al.* (1961). *Report of the Committee on the Management and Control of Research and Development,* HMSO, London.

Part Three

*HUMAN RESOURCES IN AGRICULTURAL
RESEARCH*

Chapter 9

Formation of the Research Worker

THE PROBLEM

In the developed countries, agricultural education has evolved in the course of more than a century, and is capable of forming adequate numbers of agricultural professionals of high calibre. Its major shortcoming is generally a lack of grounding in the practical aspects of farming and over-specialisation of the graduates. It is, however, in the Third World countries that the problems of agricultural education (Fig. 19) are acute, and this chapter is therefore mainly oriented towards these problems.

Prior to independence, practically no attempt was made to train research workers from among the native populations; with the exception of technicians, practically all the agricultural research work was carried out by expatriates.

The need for agricultural development, as a precursor to overall economic development, is now belatedly recognised in the developing countries. Policy-makers are also becoming increasingly aware of the potential role of agricultural research as the spearhead of agricultural progress. However, the ability of agricultural research to fulfil this role is largely dependent on the competence and motivation of its research workers and their supporting staff.

Few of the developing countries have anywhere near adequate numbers of professionals with the necessary education, experience and motivation to develop an effective research programme, or even to apply effectively the existing fund of knowledge already available to their country's specific needs.

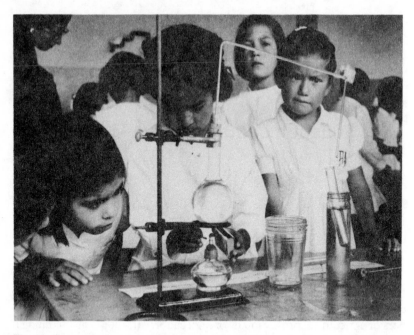

Fig. 19. Research workers of the Future. (From *Give us the Tools*, IDRC Publication 131e, reproduced by permission of International Development Research Centre, Ottawa.)

The formation of personnel for agricultural development in general, and agricultural research in particular, has generally been very slow and inadequate, and as a result, agricultural development has been retarded by an acute shortage of 'well-trained indigenous research workers to assist their countries in both policy guidance and in generating or adapting appropriate science and technology to increase food production' (Eicher, 1982). However, in addition to the inevitable time-lag involved, there are a number of inherent difficulties to overcome.

The number of young people qualified for training at university level is far below the minimum needs of the countries involved; there is, therefore, severe competition for suitable candidates among the various careers open to them for which training is available. In view of the low social standing of agriculture in these countries, and the almost unlimited possibilities for advancement in political and other fields, training in the social sciences, for example, is far more attractive to most young people in the developing countries than is training for agriculture. To this must be added that a career in agricultural research usually involves living far away from the main urban centres and is usually not financially attractive.

In a recent attempt to assess requirements for trained personnel for agricultural research systems in the developing countries, the World Bank estimated that an approximate doubling of the present number of research scientists would be needed. This is probably a conservative estimate: among 65 countries surveyed, it was found that the number of professional staff engaged in agricultural research was hardly more than the number active in Japan alone! (World Bank, 1981).

There were also sharp differences between low-income and higher-income developing countries. Among the former, an average of 16 research workers per million people in the agricultural population were active, as compared to 50–62 per million in the higher-income developing countries. Nearly two-thirds of all agricultural research workers in the 65 countries were found to be located in only 77 countries!

Increasing the number of agricultural scientists also implies substantial additional managerial, technical and support personnel. To build the framework necessary for training the professional manpower requirements of the developing nations is a formidable task, which cannot be accomplished overnight. The magnitude of the task can be illustrated by the fact that only about one thousand agronomists are trained annually in all the countries of Latin America, as compared to Japan which alone trains 7000 agronomists annually! The situation is still far worse in Africa.

Few of the developing countries have strong educational and research institutions capable, in their present context, of providing adequate numbers of properly trained and motivated research workers. The stop-gap measures adopted have been reliance on relatively large numbers of expatriates, training of nationals at foreign universities and at the International Agricultural Research Centres. The expatriates have played an important role in the agricultural research institutions of developing countries, but this role should, by its very nature, be transient. The governments of the LDCs are generally understandably anxious to replace the expatriates by nationals in the shortest possible time. Training at foreign universities is very expensive, but more important still, is generally not appropriate to the needs of the developing countries. The training programmes of the International Agricultural Research Centres have made a very constructive contribution to the formation of research professionals, but their capability to do so is limited, and should be reserved to training the future leaders.

From the foregoing, it is clear that the LDCs themselves will have to undertake the major role in training the research personnel they require. In view of the economic and human constraints faced by these countries, it is imperative to achieve maximum and effective co-operation between their educational and research institutions, a subject that will be treated in some detail later in this chapter.

The term 'formation' in this context has been used advisedly, in order to stress that the education of agricultural research workers requires far more than the provision of scientific knowledge and practical know-how; it also involves developing the will and desire to contribute actively to agricultural progress.

The formation of the agricultural research worker is actually a continuous integral process from kindergarten up to and including university education. The family also has a considerable influence on the abilities, attitudes, motivation and aspirations of the future research worker.

Unfortunately, agriculture is held in very low esteem in most developing countries. Even though there is prestige associated with the ownership of land, the cultivation of land is frequently regarded as a degrading occupation. It is not uncommon in countries in which 70 per cent and more of the people are engaged in rural occupations, for only 5 per cent of the university students to elect agriculture as a field of study—and many of these students enrol in agriculture only after having been turned down for study in more prestigious fields.

The dreams and expectations of most farmers in the developing countries for their children are generally away from farming among those that aspire to a college education. There are exceptions, but these are unfortunately rare.

The need to instil the concept of the dignity of farming throughout the formative years of the future students in agriculture in general, and of trainees for agricultural research in particular, cannot be overstressed. However, a detailed treatment of the whole educational process is beyond the scope of this book.

The aspirant research worker generally comes to the research organisation imbued with the negative attitudes to agriculture and to work described above, and it is incumbent on the research organisation itself to change these attitudes if it wishes to ensure that its professional staff will be concerned with national needs and not exclusively with personal promotion.

It will be futile to expound the 'dignity of honest labour', as long as agriculture remains primitive and consists of back-breaking work producing meagre results. But it is possible to awaken the latent idealism of youth by making aspirant research workers aware of how exciting agricultural research can be in its efforts to change agriculture from ill-rewarded drudgery into interesting, productive work and a rewarding way of life.

The changes in attitudes and motivations of young research workers cannot be achieved by 'morale-building' courses, but mainly by personal influence and example. For this, leaders who are themselves imbued with the necessary values are essential, and their formation is a prerequisite to creating the capability of the research organisation to engage in the formation of their research personnel. The International Agricultural Research Centres can play a major role in helping the LDCs achieve this objective.

In this chapter we will present a concise review of present knowledge of the constraints encountered in developing human resources for agricultural research in the LDCs and proposals on how to overcome these difficulties.

There are still many gaps in our knowledge on how these proposals are best implemented; these gaps will hopefully be filled in the course of 'learning by doing'.

FORMAL EDUCATION OF THE RESEARCH WORKER

Problems faced by higher education in agriculture

Higher education in agriculture in the LDCs is faced with a number of problems:

- entering students are ill-prepared for university education in general, and agricultural formation in particular;
- the curricula courses and textbooks are generally unsuited to the specific requirements of the developing countries;
- lack of human and financial resources impose severe constraints on the scope and level of teaching;
- teaching is attuned to foreign experiences and is routinised.

The students

In developing countries, most students with sufficient education to qualify for university admission come from urban backgrounds, have limited knowledge of plant and animal life, and little understanding or empathy for rural life. Many have entered agricultural college after being denied admission to more prestigious faculties and have hopes for city-based jobs in ministries or commodity production boards.

Though the secondary schools are mainly geared to prepare their students for higher education, the latter arrive at the universities with insufficient general education, in particular in the cultural, humanities and social fields.

For these reasons, the faculties of agriculture have often produced graduates 'who are not only mediocre, but disinterested as well' (Evenson, 1981).

Every effort should be made to encourage rural youth who have the necessary aptitude to enrol in faculties of agriculture. Universities can help by providing scholarships and other incentives, and by setting up special levelling courses to close the educational gap with urban youth.

The curricula

The organisational structures and curricula of most of the faculties of agriculture or agricultural colleges in the LDCs have been patterned on models from European or United States institutions which are discipline-oriented rather than production- and problem-oriented (Gray, 1974). Most of the students in the latter institutions have chosen agriculture as a vocation, either because they have a rural background and farm experience or because they are genuinely eager to serve in agriculture. For students with this background, the specialised, discipline-based curricula are entirely appropriate and give them an adequate preparation for work in a highly specialised agriculture.

In the LDCs, the curriculum content tends to be theoretical with emphasis on rote learning, while most textbooks are based on agriculture in the temperate regions. The graduates may have a fair grounding in agriculturally related basic sciences, but are seriously lacking in many of the qualifications required for agricultural research in developing countries.

Few of the faculties of agriculture provide training beyond the BA or BSc level, but the undergraduate programme itself is generally aimed at graduate work and not at training specifically for the functions that will be undertaken, at least in the first years of their professional life, by the majority of those graduating from college with a BSc degree. For all these reasons, if the novice research worker is to perform effectively, the competence he lacks must be imparted by in-service and on-the-job training by the research organisation staff.

Whilst it is widely recognised that many technologies that are suitable for developed countries cannot be adopted in the LDCs, and the need for an 'appropriate technology' for developing countries is axiomatic, the suitability of these curricula in higher education in agriculture is rarely questioned.

The American and European discipline-oriented educational models need to be replaced by functionally oriented instruction, which provides the future agriculturists with 'knowledge, skills, and experience in agriculture, plus an understanding and appreciation of the interrelatedness of the components of agriculture. Not only should they know their field, they should understand how it relates to, and impacts upon, the society at large. Perhaps most of all, the educational process should produce perceptive people who will respond, through their agricultural occupation and activities, to national aspirations, values and goals. The preparation of such individuals will require substantial rethinking and revising of the existing curricula' (Gray, 1978).

The basic programme of undergraduate studies

Though future research workers require a distinct kind of professional preparation, they share a common basic preparation with all university professionals in general and with other agricultural professionals in particular. The university curriculum for the undergraduate should be designed so as to provide a basic preparation for future professionals. Specialisation in one of the agricultural sciences that will be required for the ultimate high-level careers should be left for the graduate programme.

Initially, most of the natural agricultural sciences were derived from biological science. They are now becoming increasingly interdisciplinary in nature and are also increasingly drawing upon chemistry, physics, and mathematics as tools. Therefore, any professional in the fields of agriculture will need a sound, general grounding in the basic principles of these sciences in order to understand the changes that are transforming techniques and procedures in agronomy.

A solid academic scientific training in these subjects is an essential prerequisite for the future research worker but is not, in itself, sufficient to ensure his professional competence in his future profession.

The technical curriculum outlined above needs to be complemented by appropriate courses in social and economic sciences, and their application to rural development. A certain amount of time must also be allocated to the humanities.

At present, the curricula of many of the faculties of agriculture are often inflexible and not well geared to the realities and needs of national agriculture (Oram, 1977).

The dynamics of development and of science and technology will require a constant up-dating of the courses and the ways they are presented.

The need to make a major effort to develop the communication skills of the students cannot be overstressed. 'Too many agricultural college graduates are academic illiterates' (Myers, 1965). One way of achieving this objective is to allow the students to prepare and present in class selected topics in each course.

The student needs to know at least one foreign language in order to be able to read scientific literature, communicate with his peers in other countries, and to keep abreast of innovations in agricultural science and practice.

The programme outlined above will entail a heavy learning schedule, and will therefore require a certain amount of flexibility. A combination of a basic core of required subjects and a certain freedom of electives selected with the assistance of competent faculty advisers has given good results in universities that have adopted this approach. Even so, it will not be possible to give the students knowledge in depth of all the disciplines he has been taught, nor is this necessary. The objective of the courses should be to awaken the student's permanent interest in scientific and humanistic subjects, to form habits of self-study, reflection and ability to read the professional literature critically.

This is a considerable challenge to the teaching staff because these objectives are far more difficult to achieve than the routine presentation of information, which is the hallmark of the mediocre teacher.

GRADUATE COURSES TO MASTER'S LEVEL

In most developing countries agricultural training in a university, culminating in a first degree has generally been considered as adequate for candidates for agricultural research positions. Experience has shown that such training 'provides at best' only a minor foundation for research and does not prepare personnel adequately in areas such as definition of research programmes, research methodology and analytical research.

Such undergraduates require a relatively long period of supervision in the field by a senior scientist experienced in research methodology (ISNAR, 1982).

There is little doubt that whilst a first degree may provide a grounding in essential subjects, it cannot suffice as training for a research career, for which a Master's level must be considered as essential.

In general, training for a Master's degree in agriculture or its equivalent is provided in the LDCs by the National university, by fellowships at foreign universities and at the International Agricultural Research Centres.

National universities

The capabilities of the faculties of agriculture in LDCs vary greatly. However, few of them are yet able to provide training beyond the Bachelor's level. The undergraduate programme generally aims at providing a fair grounding in agriculturally related sciences and not at training specifically for the functions that will be undertaken by high-level professionals in research development planning and higher administrative positions.

Requisites

Graduate studies should be offered only by faculties of agriculture which have the following requisites:

- a sufficient number of high-calibre, able and dedicated professors, each a specialist in his chosen field;
- adequate facilities for graduate work; specialised laboratories and facilities for field experimentation;
- a consolidated teaching and research programme;
- capability to supply training in farming skills;
- sufficient financial resources for investment and operation of a graduate school; scholarships and boarding facilities for students.

The cost of a graduate course is far higher than those of the lower levels, and it is doubtful whether many of the faculties of agriculture in the LDCs could afford them. Most lack adequate research facilities and experience which would enable the professors to provide advanced teaching and research programmes. A major deficiency is their inability to give training in the practical aspects of farming.

Under the circumstances, it appears unrealistic to expect that many of these faculties will be able to develop a graduate programme on their own.

It would appear more rational for them to apply their generally meagre resources to the improvement of the existing structures and the level of undergraduate education they are providing and to give the necessary academic support to a graduate school of agriculture that could be established together with the agricultural research institution (we will return to this subject later).

Advanced studies at foreign universities

In view of the great difficulties that the majority of the LDCs face in providing graduate education for the qualified personnel they require for their development programmes, they have relied heavily on expatriates and on advanced training for their nationals at foreign universities. In the vast majority of cases these were North American and European institutes of higher learning. Experience has, however, shown that the discipline-orientation of the professors in these universities provides a highly specialised training in a narrow field of basic research, using sophisticated equipment, and on subjects which are usually of little relevance to the developing countries.

Graduate education at these universities concentrates on theoretical principles in a very narrowly defined field of study. It is *not* job specific and it does not directly prepare research personnel to work in technology development. The research fellows' experience is generally limited to the skills involved in carrying out the thesis research project. Seldom do they have the time or inclination to learn the additional research skills needed to apply the knowledge to mission-oriented research (Swanson, 1977).

To these shortcomings, Vinyu (1979) adds: 'the Western type of education that the researchers receive in the developed countries stresses specialisation in a narrow-defined area, is not a desirable preparation for mission-oriented research that is carried out in a wider context and implies interdisciplinary work. The tools and facilities used during the training are not readily available in their home countries'.

The students also become imbued with certain concepts and values that handicap them in their future work, such as the superiority of basic research over applied research, the emphasis on the importance of writing papers for prestige journals, etc. Other undesirable tendencies that are usually already present, such as the desire for life in town and for work that provides prestige, are thereby strengthened.

In addition to these drawbacks, the trainees are removed from their homes for a long period, and often face serious problems of adjustment after having been educated in a foreign environment. Some students may never return if they are attracted to better positions abroad (Harbison & Myers, 1964). And finally, the cost of training at the foreign universities has escalated markedly (IADS, 1979). Recognising these shortcomings, many of the courses provided by the universities in the donor countries include some period of training in the home country or similar conditions.

TRAINING PROGRAMMES AT THE INTERNATIONAL AGRICULTURAL RESEARCH CENTRES

One of the major goals of the IARCs is 'to help national commodity programmes increase their scientific and technical manpower in research and production' (Fernandes, 1977). Therefore, educational and training opportunities have always been an important component of the activities of the centres.

The centres (though not all to the same extent) provide opportunities for non-degree research training, graduate work, post-doctoral fellowships and training for research support services (Vega, 1979).

Non-degree research training programmes

These are 'apprenticeship-type' programmes in which the trainee works under the guidance of a centre specialist on a specific research project.

Short courses are also provided for training production specialists, who will conduct adaptive research.

IRRI's research training centres around the research project. Trainees work in a disciplinary department and are closely associated with an IRRI scientist on one of the latter's research projects. In some cases, his work results in a jointly authored research paper. The trainees obtain solid research experience, learning each step involved in planning, designing, executing and reporting on a research project. Swanson (1975) found that after returning to their home countries, IRRI trainees tended to emphasise experimental research (in field, laboratory and greenhouse) aimed primarily at generating new knowledge about rice production in the tropics rather than in the direct development of location-specific rice technology (in agronomic field trials or genetic crosses).

CIMMYT's training stresses a 'team' or integrated approach to wheat improvement. The emphasis is on interdisciplinary teams that are well integrated in function. Trainees learn all the essential skills and techniques during each stage of the growing season and the varietal development process, with each test first being discussed in the classroom and demonstrated in the field. After he has become reasonably proficient the trainee helps to carry out each research task within the on-going CIMMYT programme. In brief, the CIMMYT training programme concentrates on genetic technology which has wide adaptation and far less on production technology which is by nature location-specific. As a result of this training strategy, the trainees were found to have focused on wheat breeding after their return home, and have continued to work in close collaboration with CIMMYT in its international programme. By contrast, they generally did not specialise in research problems within a particular scientific discipline or engage in developing complementary packages of production technology.

The differences in the type of work in which former trainees from the two international centres engage in is also reflected in their tendency to publish. In the course of two years, only 37 per cent of the former CIMMYT trainees, working in their national wheat programmes, had produced technical papers as compared to 84 per cent of the IRRI trainees working in their national rice programmes (Swanson, 1975).

Academic graduate degree programmes

These are for candidates for the MSc or PhD degree, who conduct their thesis research under the guidance of a centre scientist. This type of programme is implemented in co-operation with universities (cf. p. 456–7). This programme is reportedly most successful in preparing middle to senior level developing country nationals, who, besides achieving a degree, 'gain understanding of the challenges and the opportunities in helping solve food production problems in developing countries' (Vega, 1979).

Post-doctoral fellowships

These are one- to two-year appointments for young PhDs from developing and developed countries, who carry out advanced research in collaboration with centre scientists.

In view of the limited number of candidates that these programmes can handle, it would appear that the interests of the LDCs, as well as of the centres, would best be served if the training opportunities offered by graduate degree programmes and post-doctoral fellowships were reserved for potential research leaders from the LDCs. We will return later to this subject (cf. p. 456).

Training for research support services

The types of training provided include management of experiment stations, documentation and information exchange, and seed production.

The primary reasons why IARC training programmes have been successful is 'their concentration on experience-based training or learning by doing' (Swanson, 1977). As a result, the trainees find it natural to work in the field, conducting demonstrations and field experiments. They achieve self-confidence because they have become skilled in research and production through experience.

Difficulties faced by returning trainees

The considerable investment in training research workers abroad is frequently largely wasted, because after their return graduates encounter obstacles in applying their newly gained confidence.

Byrnes lists the difficulties that are frequently encountered by returning trainees:

- their assignments are not appropriate;
- they do not receive adequate administrative or financial support;
- they lack seniority status;
- others may view their newly acquired capability and knowledge as a threat to the status quo;
- suitable institutional frameworks are missing;
- adequate supervision, stimulation and guidance are lacking;
- they are burdened with other duties and do not have sufficient time to perform as they were trained.

The conclusions to be drawn from the above are self-evident and do not require elaboration.

DEVELOPING A CAPACITY TO PROVIDE ADVANCED TRAINING IN AGRICULTURE

We have reviewed the existing facilities for graduate training available to agricultural students and professionals from LDCs. The faculties of agriculture in the LDCs are generally limited in their resources and only very few of them will be able, in the foreseeable future, to provide an acceptable level of graduate education. It will be a major achievement if they will be able to upgrade the teaching levels of undergraduate education.

The educational process current in most universities of the developed countries has been developed for ecological, social and cultural environments totally alien to the realities of social structure and economic relationships of the LDCs (Fernandes, 1977). Besides being expensive, fellowships at foreign universities have all too often provided training that is not geared to the specific needs of the LDCs.

The graduate and postgraduate training provided by the International Research Centres has proven to be markedly free of most of the drawbacks of foreign training outlined above. However, the numbers of trainees that they can accept without curtailing their research programmes is limited, and these should be reserved for training 'leaders', a subject we will discuss further on.

From the foregoing it is clear that most of the professionals required to staff a developing country's education, research and development institutions for agriculture will have to be trained at home. It is, therefore, *essential* for each country to develop its own national graduate programme in the agricultural sciences, and the question arises whether this is *possible* in view of the limitations imposed by the human, physical and economic constraints faced by these countries.

Castronovo (1984), writing on the situation in Latin American universities, states that:

'notwithstanding all good wishes and good will, our universities, in general, do not educate students to carry out research. But how can they be expected to, contending as they do with reduced means and inadequate facilities? What use are such institutions if their teachers do not conduct research themselves: if, to state the matter bluntly, they lack any essential connection with the medium which might enable them to give a correct orientation to the applied research which students should undertake?

CO-OPERATION BETWEEN THE FACULTY OF AGRICULTURE AND THE RESEARCH ORGANISATION

The assumption that graduate education is too expensive for developing countries is only true if it is intended to copy slavishly the objectives and methods of graduate education practised in the rich countries. Every nation that has a reasonably well-developed research organisation and a faculty of agriculture can establish and maintain a centre for graduate training, based on co-operation between the university and the research organisation.

This 'solution' is realistic because it is based on the systematic and effective use of existing institutions and resources, in particular the facilities of the research organisation that need only to be complemented at a reasonable cost as a means of building high-level scientific manpower.

The aim would be to provide professional training of the same type as that provided by the International Agricultural Research Centres, namely, of 'learning by doing' combined with a number of courses under the aegis of the faculty of agriculture. Most developing countries have a fairly well-developed research organisation staffed mainly by professionals who have been trained at foreign universities and reinforced by expatriate specialists. In some cases the latter form the majority of the staff.

The essence of this proposal is that the faculty of agriculture and the national agricultural research organisation in each country should set up jointly a *National Graduate School of Agriculture* under the aegis of the university, in which the faculty would take responsibility for the necessary courses at graduate level and the research institution would provide the experienced research workers as tutors and laboratory and field facilities for training the students, who would be involved in research and research-related activities. Actually the formal distinction between the respective roles of the faculty and of the scientists of the research institution would be more apparent than real. The faculty staff could be involved, on a personal basis, in the overall national research programmes, with access to the research facilities of the research organisation, whilst senior research workers could be accredited by the university to serve as tutors in the field and use made of their specialised knowledge for teaching assignments under the aegis of the university.

The relationship between faculty and the research organisation would be truly symbiotic and would have the following advantages:

For the faculty
- The faculty of agriculture will be able to upgrade its teaching up to the Master level, and at a later stage, to the postgraduate level, at minimal cost. (Costs for postgraduate studies at foreign institutions have been estimated as being about US\$13 000 a year for an

MS and US$23 000 for a PhD, as compared with US$4000 to US$5000 a year, respectively, for in-country degrees (IADS, 1983).)

- The research organisation can provide the faculty with a pool of experts in various disciplines and commodity research, who at little cost, can serve as experienced teachers on various specialised topics, which the faculty could not hope to provide from their own resources.

For the faculty staff

The inclusion of research specialists in teaching assignments will reduce the teaching load of the individual faculty member, allowing him more time for research and self-learning.

- The faculties can usually develop only a research programme on problems of limited importance. Co-operation with the research organisation will enable the faculty staff to engage in research of national importance, using funds and already available facilities which it might take decades to build up within the faculty itself. This will also indisputedly have a beneficial effect on the quality of the teaching.
- The staff will benefit from interactions with the scientists of the research organisation, and will be made continuously aware of new developments in research and development.

For the research organisation

- The undergraduate from the faculty of agriculture cannot have the precise skills, knowledge and personal motivation required for a successful career in research. The research organisation will be able to participate actively in preparing candidates for its future cadres, and concurrently upgrade the existing research personnel.
- The research organisation will be able to ensure that the students will be formed in a research environment, learning by doing, and will acquire the qualifications and motivations needed for research in a developing country.
- Whilst the major objective for the research organisation is to train their own cadres of research workers, the combined capabilities of faculty and research organisation can handle a larger number of graduate students. Not all the students engaged in the graduate course need to remain in research or will do so. This affords the research organisation a unique opportunity to select research

workers for their permanent staff on the basis of a prolonged contact with the aspirants. This makes it possible to judge their ability and motivation during their period of 'apprenticeship' in the graduate school, and to retain only the most suitable and promising graduates for the permanent research.

- The graduates who move on to functions outside the research organisation as planners, future policy-makers, extension specialists, specialists for agricultural production and marketing boards, will have understanding of research potentials and empathy with the objectives of the research organisation. This will have the long-term advantage of strengthening public support for research.
- Under the direction of competent research workers, the graduate student will be able to make a significant contribution at low cost, to the research programme. Projects which require much manpower, such as soil maps, phyto-ecological surveys, the search for indigenous plants as breeding material, field surveys of pest and disease incidence, participation in pre-extension research teams, and on-farm trials are some examples of the type of work to which the students could make a valid and important contribution to the research effort (Casas, 1977).

For the research worker
- By collaborating in a graduate course under the aegis of the university, the research worker will gain prestige, credit for professional promotion by serving as a part-time teacher at the faculty and tutor for aspirant graduates.
- The research worker will gain collaborators who bring to their tasks youthful energy, fresh ideas and readiness to endure hardships.
- The veteran research workers who are formally and actively involved in the training programme of the graduate school are motivated to keep abreast of developments in their field in order to maintain their effectiveness as leaders and mentors of the junior research staff.

For the graduate students
- The aspirant student has a much greater free choice among disciplines than he could find at the faculty itself, even if it was able to provide graduate training.

- The students will work under the guidance of specialists, who can impart both research methods and up-to-date knowledge in their respective fields.
- The students will gain practical farming experience on the experiment station farms.
- The years spent in 'apprenticeship' in diagnosing problems and in research provide an excellent preparation for responsible work in administration, extension, teaching, policy-making, etc., for those who do not achieve, or do not ask for, permanency in the research organisation. They will have learned what are the problems faced by agriculture, what are the potentials and constraints in achieving solutions; they will have received good grounding in practical agronomy, etc.

The Graduate School of Agriculture

The concept of a graduate school of agriculture based on a formal agreement between university and agricultural research organisation is very similar in principle and operating procedures to the medical postgraduate schools, in which 'learning by doing' is achieved by a partnership between university and a chosen hospital.

Graduate schools for agriculture based on this principle, have been set up by several international agricultural research centres in collaboration with universities, within a developing or a developed country (Vega, 1979). One example is the graduate school in Mexico set up by CYMMIT in co-operation with Cornell University (Contreras *et al.*, 1977). Regional institutions, continuing graduate study and research in agriculture, have been established in different parts of the world. The Southeast Asian Regional Centre for Graduate Study and Research in Agriculture provides a continuing programme of graduate education leading to the MSc and PhD degrees for the SEARC's participating countries at about one-fourth the cost of graduate education in the United States (Drilon, 1975).

The need for developing countries to establish institutions capable of training their own requirements of scientific personnel has been explained above. The idea of graduate schools of agriculture based on co-operation between agricultural research institutions and faculties of agriculture is neither new nor original.

Such graduate schools are already functioning in a number of Latin American countries. In Colombia, for example, the Instituto Colombiano Agropecuario (ICA) promoted a general agreement between the

Ministry of Agriculture and the National University to organise jointly a Graduate School of Agronomy, Veterinary Medicine and Animal Husbandry. The school was to be under the direction of ICA but with the academic collaboration of the National University. Specific contracts were made to ensure the co-ordinated functioning of the university's agronomy colleges and the ICA's experimental centres and stations (Garcés, 1974).

A similar proposal has been made by an ISNAR mission to the Ivory Coast (ISNAR, 1982) as an effective and economic framework for training national research workers who could gradually replace expatriate experts, who at the time accounted for 72 per cent of the agricultural research staff.

In 1980, an accord was signed between the national research institute in Brazil (EMBRAPA) and the Faculty of Agriculture of the University of Sao Paulo, whereby a joint Post-Graduate School of Agriculture was established.

Problems

One should not underestimate the difficulties encountered in endeavouring to establish such a co-operative undertaking between a government research organisation and the faculty of a university. Castronovo (1984) comments in this respect that 'with few commendable exceptions, postgraduate education has not succeeded in becoming integrated into the university system (in Latin America). It is frequently regarded as a nuisance, maintained by non-institutional mechanisms under the pressure of a small number of teachers concentrated in specific academic units. In Argentina, wide-ranging and patient deliberations to institutionalise the Graduate School in Agricultural Sciences floundered when the outside force which sustained the school lost its credibility.'

Marzocca (1983) writing of the failure of viable, well-conceived and academically sound training programmes, states that these disappointing results are due to 'political change, inopportune changes of norms, economic regulations or administrative insensibilities'. However, our own experience indicates that it is the friction between the university and the national research institute that has led to the breakdown of many successful co-operative projects. The demise of the graduate school in Colombia is an example (cf. p. 171–2).

In order to avoid friction with the national university, and at the same time to obtain the maximum benefit at lowest cost, it is desirable that the graduate school project should not be a one-sided undertaking, but based on formal co-operation between the university and the research organisation.

The graduate schools, or research-training centres, should not interfere with any existing universities or programmes. They would have a close tie-in with any university prepared to co-operate in such a programme of national significance. 'Policies and programmes would be developed on a cooperative basis and would therefore supplement and not duplicate, interfere or compete with University activities.' (Cunha, 1967).

Whilst the trainees at these graduate schools would be required to take a number of formal courses, most of the training would be the result of teaching by example, in the course of scientists and trainees working together, informal conversations, open meetings for discussion of research programmes in the light of national and social priorities, etc.

Objectives

The major justification for establishing a graduate school of agriculture as a joint undertaking of the faculty of agriculture and the agricultural research institution is to provide a graduate education in agriculture that is appropriate to the needs of a developing country, in other words, relevance-based training that will produce a competent cadre of researchers, teachers and development specialists. The training would be similar to that provided by the international research centres, namely, in-service, combined theoretical and practical individualised training for undergraduates who wish to pass from the Bachelor to the Master level (or their equivalents). The students would be trained as specialists in specific disciplines, or as multidisciplinary commodity specialists. They would become qualified to engage in a career in research, or as extension specialists, production specialists, agricultural development, administrators, planners, economists, etc.

Selection of trainees

Typically the trainees will be staff members from within the research organisation who are eligible for training towards a higher degree, and undergraduates from the faculty of agriculture who require a higher level of education in order to qualify for research, teaching or higher level work in the administration and development sectors.

Because all the future graduates will be candidates for higher-level work that is vital to the development of agriculture, the selection of candidates should be vigorous. It has been suggested that individuals with university training in sciences could also be considered as a potential source of manpower for agricultural research, and that it would not be too costly to train these people for agricultural specialisation required (ISNAR, 1981*a*).

Priority should be given to members of the research personnel with Bachelor degrees who would be encouraged to work toward a higher degree whilst actively engaged in research. The directors of the research institutes should propose candidates for such advanced training on the basis of demonstrated research ability, job performance, commitment to research and development, etc. The total number enrolled should not exceed the number of qualified professionals available as advisers on the staffs of the faculty and the research organisations.

Procedures
An undergraduate who wishes to progress to a higher degree, will serve in the research organisation for about two years, during which time he will participate in one or more research projects, acquire practical farming skills in the experiment station farms, and attend a number of formal courses, as required by the university, all in accordance with a predetermined training plan.

Each student is attached individually to a senior research worker, accredited by the university, with whom he works in laboratory and field and who provides guidance and supervision. The ideal ratio is one student per adviser but should not exceed two to one.

Besides imparting theoretical and practical knowledge in the disciplines or commodity research in which he is working, the adviser directs the attention of the student to appropriate publications and individual papers which then serve as a basis for seminars and discussions. The student should become accustomed to self-study, to make full use of library facilities, to read critically scientific literature and to keep up with developments in the field of research in which he is involved.

He will be given responsibility for a specific component of a research project, preferably as part of an interdisciplinary team, which will serve as his Master's thesis.

The planned duration of the training would be two years, and successful students would obtain a Master's degree or its equivalent from the university.

Financing

The additional costs involved by the university and the research organisation in setting up and maintaining the graduate school of agriculture would be minimal in relation to the expected benefits to the country in general and to the institution involved in particular.

For the faculty of agriculture, a provision would have to be made for time devoted to teaching by research staff who are co-opted into the teaching programmes and for stipends for students; for the research organisation, funds would have to be invested for student lodgings, lecture rooms, additional laboratory space and equipment. These costs would be largely offset by benefits described above, by the provision of a relatively cheap source of eager, young research workers, and by largely replacing expensive foreign training by training at home. It will also become possible to gradually replace expatriate scientists and save the considerable expense involved in salaries, housing provisions for home leave, and other emoluments, which are far higher than the salaries of nationals.

Semi-public and private organisations that send members of their staff to graduate school, in order to improve their capabilities, should make appropriate contributions to the cost of the training.

IN-SERVICE TRAINING

Because the educational institutions cannot provide graduates with all the skills required by the agricultural research organisations, the latter must develop systems for the in-service training of all levels of research personnel: recently recruited research workers (research assistants), young and mid-career scientists, research managers, directors of experiment stations and supporting staff.

Many aspects of in-service training will have to be provided within the organisation itself, such as, regular staff meetings, lectures, personal guidance, self-study, etc. Other types of training will depend on possibilities provided outside the research institution whilst the research worker is pursuing his work, such as attendance at international and national seminars, workshops, symposia, etc.

Outstanding research workers should be encouraged with help to work towards a higher degree. This will improve their ability to do the work in which they are already engaged or to prepare them for new assignments for which the need has arisen. The need may also arise for conversion courses for personnel who work to move into a new field of research or whose speciality has become redundant.

In-service training is a never-ending process. It cannot consist exclusively of what is provided by the research organisation or by attendance at courses, seminars, colloquia, etc.; the researcher must continue to train himself by personal effort—reading, individual learning, etc. He should be encouraged to do so, by the provision of the necessary facilities, such as a good library, information service, learning aids, etc.

Formation of the junior research workers

The future researcher is faced with severe personal demands. He will have to work in a rural environment and to understand rural people and their problems. He must not only be competent professionally but must look upon his work with a sense of mission, and even of urgency.

The balanced curriculum that has been outlined above will hopefully provide the future professional with the basic scientific and technical requirements for whatever agricultural career that he may finally adopt. However, major determinants of agricultural productivity in a national agricultural research system are, besides professional competence, integrity and motivation of the administrators, research workers and extension officers involved. In the training of these aspects, the agricultural colleges have generally failed, and 'these failures have a pervasive adverse effect throughout an entire national agricultural system' (Gray, 1978).

Hence, the major importance of imbuing the young research worker with the values he has not received during his formal training—the sense of purpose and social responsibility mentioned above.

In view of the background of most candidates for research described above, this objective is not easy to achieve. Ideally, the social motivation of the individual needs to be developed from the earliest age, but for many and diverse reasons very few, if any young people have at any stage of their formal training been submitted to such influence.

The research organisation must provide what is probably the last opportunity to inspire attitudes and aspirations that are compatible with a true spirit of service and a sensitive social conscience (Aranjo, 1979).

There is no doubt that this is difficult to achieve, but it is not impossible. By drawing on the energy and latent idealism of youth, one can give the young research recruits a sense of mission, dedication and values which give their professional lives new content. This is the one great advantage that the developing countries have over developed countries; namely, to offer their youth in general, and the young research worker in agriculture, in particular, numerous challenges.

These objectives cannot be achieved by providing ethical courses, it can be done: (a) by creating an appropriate research environment; (b) providing adequate facilities, resources and technical assistance; and (c) appropriate incentives and reward.

Recently recruited research workers

The typical candidate for a post as research worker in the LDCs, has received formal training to the Bachelor's level, or its equivalent.

The minimum educational requirement for research in general, and agricultural research in particular, is a Master's diploma or its equivalent, and few developing countries have faculties of agriculture that already have the capability to train their students to graduate and postgraduate levels.

It will certainly take time to develop a national capacity to provide graduate training in agriculture, even if a joint NARS–university graduate school of agriculture is established, so that in the interim, the research organisation will have to provide newly recruited research workers with the competence in a number of areas in which he is lacking, and to instil in him 'a sense of purpose and social responsibility' (Moseman, 1965).

Competences to be provided by in-service training

Byrnes (1974) has advanced what he calls a 'competency model for production specialists' that assumes that at least five competencies are required: Technical (knowledge of the subject matter); scientific (how to obtain or validate knowledge); economic (how to make decisions and recommendations in the light of economic and management relationships); farming (skills involved in growing a crop or livestock management); and communication (how to teach or share with others what you know). These requirements are also valid for research workers.

In all these areas, excepting farming skills, the incipient research worker will have received some grounding during his formal education, which will, however, require reinforcing and constant updating by on-the-job training.

In addition, he will need additional knowledge, skills and practices as well as motivation to which his formal education has contributed little or nothing. These, too, are an integral part of his 'formation' and must be developed in the course of his work.

Skills in practical farming
The negative attitude to practical farming that characterises most students of agriculture—the so-called stigma of practicality (Fernandes, 1977) must be changed soon after the incipient research worker begins his career. This includes a willingness to work with his own hands, and even of being proud of having overcome his prejudices in this respect.

A significant deficiency of agricultural education, at university level, is lack of training in the practical skills required in farm management.

Crop production and animal husbandry
The undergraduates may know in theory how to grow a crop, but when they are in the field, they are not able to do it. At first sight, these skills do not appear to be essential for research workers. The need for such training is generally not recognised by the universities or the students, who frequently feel that working at farm level is degrading. Familiarity with practical aspects of farming is, however, of considerable importance, for all researchers engaged in field experimentation. Experimental plots must be models of good management if the results are to be meaningful, and this cannot be achieved without acquiring farming skills. This practical background will also serve the researchers in good stead in their contacts with extension workers and farmers, by their ability to handle farming skills confidently, and identifying practical problems requiring research.

The international centres have compiled some evidence on the knowledge, understanding and abilities of 'agricultural specialists' with respect to the production of one or more crops. Working agronomists from Asia, Africa and Latin America have consistently shown alarmingly low scores on these subjects.

In five production-training courses at IRRI, involving 173 agronomists from 30 countries, the average entrance score was 31·6 per cent.

In a practical test at CIAT of 34 agronomists in diagnostic and management skills on diversified crops, the mean score at the beginning of training was 36·4 per cent. Most animal service graduates enrolled at the CIAT training courses were found to be unable to handle or restrain animals for diagnosis or treatment—a clear demonstration of their lack of opportunity in learning such basic skills. IITA reports that the beginning score for

23 persons from 11 African countries was 22 per cent. About half the group could not identify a rice seedling from an ordinary grass!

The conclusion drawn from the above is that 'all crop research trainees should be capable of raising a crop. They should gain experience in actual production operations, training which includes commercial field experience, and on-station and on-farm trials in which they perform all cultural operations' (Byrnes, 1974).

At CIAT, on completion of 'the Courses, practical experience was by far the dominant choice among all trainees as the most valuable aspect of their training period. Apparently a good programme of practical training also generates an increased appreciation of the value of practical experiences. Even former trainees who have risen to higher positions in their home institutions mention that though they no longer physically used the practical skills they had learned at CIAT, they had gained confidence and a better understanding to evaluate research proposals' (Byrnes, 1974).

The experiment station farms can provide the apprentice research workers with an excellent opportunity to acquire all the practical farming skills associated with advanced forms of agricultural production and they should be provided with the opportunity to do so.

Orientation to problem-solving type of research
Incipient research workers must learn to identify the significant problems in their field that limit production, to understand the social, economic, cultural and institutional environment in which the farmers they intend to serve operate, the problems of communication in a research organisation, the constraints encountered in adopting research results and the social and economic consequences of successful adoption.

Research techniques
In the best of cases, the undergraduate will have learnt at the university the rudiments of experimental design and statistical analysis. Courses and practical field work on the planning, designing and implementation of a research project, analysing the data and how to draw appropriate conclusions, should be organised as part of on-the-job training.

By its very nature, much of agricultural research work consists of field experimentation, whether at the experiment stations or on-the-farm trials. Unfortunately, because of the lack of farming skills of the young researcher, these trials are frequently poorly executed. Results from such trials are bound to be unreliable, and the recommendations that follow may be highly original, but will hardly commend themselves to farmers.

In order to avoid the wasted funds and effort that result from poorly designed and executed field experiments, and worse still, the advancing of unwarranted conclusions, every aspirant research worker should go through a period of apprenticeship in carrying out well-designed experiments in the field, in which the farming skills he has acquired are put to the test.

Ability to work as part of an interdisciplinary team

An aspect of agricultural research that receives little attention when students are trained in the universities is the team approach, where each team member makes his contribution to the achievement of the team's objective. Scientists in general tend to be discipline-oriented rather than problem-oriented, but it is the problem-orientation which is most important in the developing countries.

Working with problem-orientation, a team, such as a plant breeder, plant pathologist, entomologist, agronomist and agricultural economist, analyses the constraints and then takes an integrated team approach to resolve them. This team approach focuses on the problem, the various disciplines interact to overcome the problem. This approach usually results more quickly in a product with more adaptability to farm conditions than does the organisational approach of passing the product from one discipline to another and then expecting the extension officer to put it all together (Gamble *et al.*, 1981).

Many research projects of major importance are essentially multidisciplinary efforts that involve teamwork. In particular, farming systems research is an area in which teamwork is essential, and in which the young scientists can make significant contributions.

Research managers frequently encounter difficulties in organising problem-oriented, interdisciplinary research because by disposition and training, most research workers are individualists. The general tendency is to narrow one's interests and contribution to one's own field, without considering the interrelations with other disciplines.

It is, therefore, essential to overcome these psychological barriers by example, persuasion and appropriate incentives.

Cornell University and the CYMMIT have introduced an innovative programme for graduate training in agriculture in which students are involved in problem or commodity-oriented multidisciplinary research at the graduate level. Their project has demonstrated that 'with proper motivation and research, students can be trained to appreciate the problems and benefits of integrated interdisciplinary research' (Contreras *et al.*, 1977).

In view of the success of the project, and the greater advantages of the integrated approach to graduate training than those arising from a traditional, individual approach, the faculty staff and senior researchers involved in the project became convinced that it should be attempted under almost any institutional arrangement.

Selection criteria of trainees who are able to function together as a successful team are not yet well understood. In the Cornell–CYMMIT experience, diversity in education backgrounds was found to be an asset rather than a liability. However, differences in personalities, politics, religion, taboos and prejudices could become sources of potential frictions. This could only be overcome by selecting team members with sensitivity, patience and understanding towards other team members. The team leaders must provide the trainees with a clear perception of the expectations and goals of the team, the potentialities of team research and the problems involved in its execution. All the team members should gain an awareness of other disciplines involved through courses, lectures and assigned readings on the basis of each discipline.

A major incentive to encourage participation in interdisciplinary teamwork is to assign adequate academic value to these team-associated activities (Contreras *et al.*, 1977).

Knowledge of local farming procedures and social and cultural background of the farming sector
Research workers who are ignorant of the customs and desires of the rural sector that are supposed to benefit from their work, will lack the basic requirements to engage in meaningful, mission-oriented research; namely, an understanding of, and empathy with the farmers. Excellent opportunities for achieving this purpose are provided by involving the students in farming systems research, which requires detailed surveys of existing farming systems and the constraints to adoption of improved practices. On-farm field experiments, in which traditional procedures serve as controls when testing new varieties or management practices provide another opportunity for familiarising the trainees with the farming milieu and ensuring contact between them and farmers. This knowledge will serve them in good stead, whether they continue in research or engage in other high-level agricultural occupations.

Byrnes (1974) mentions that at the beginning of a CIAT training programme trainees served for two months as interviewers in an agricultural economics study of 300 small maize farmers. The trainees were given a week of intensive training in interview techniques. As a result of their work, 'an unexpected by-product was the motivation and insight the trainees developed with respect to the problems of the small farmer. For most, this was the first time they had been on such farms, or talked with farmers'.

Young research workers

The useful knowledge acquired during formal education and stored in the brain at the time of graduation has an estimated half-life of 5 to 10 years. This leads to two conclusions:

(a) the major objective of university education should be to prepare the graduate for a life-time of learning;

(b) whatever the background of the research worker and the quality of his academic training, learning cannot end on completion of his formal studies, hence the need for in-service training.

The objectives of in-service training should be to impart knowledge on:

- subjects essential to research work for which adequate training has not been provided in the formal study programme;
- to keep abreast of advances in technology; and
- topical subjects concerned with government policy and initiatives in rural development.

Organised and informal discussion should be encouraged in order to ensure the involvement of all the research workers in the organisation in its work as a whole. This will lead to a broadening of horizons of the individual research workers and an awareness of the problems of their colleagues, their identification with their institution, a cross-fertilisation of ideas and mutual constructive criticism. A major constraint in interdisciplinary research 'is not lack of capacity in an individual's own discipline, but insufficient understanding of other relevant disciplines' (IADS, 1979).

Regular annual meetings on the work in progress on a given commodity should be held, at which all research workers involved should present their findings and the problems encountered in the course of their work. By inviting other persons, such as marketing economists, agricultural extension officials, national planners, and seed production specialists, and including some of them on the programme, the annual meeting of the research project can serve as a national forum for spreading and discussing the latest information on the commodity (IADS, 1979).

Agricultural research services in all countries find the sources for much of their work in advances made elsewhere in the world; one of the major roles of agricultural research is to interpret and adapt these results to local conditions. It is, therefore, essential to keep the staff in constant contact with world agricultural research, in their own and neighbouring fields, by the provision of appropriate library facilities and documentation services. Regular meetings should be held between research workers in different disciplines, as well as colloquia on specific subjects, study groups, special courses including lectures, case studies and discussions, trips and excursions, sabbatical programmes, educational leaves and other devices. Specialists should be invited to give lectures on subjects, such as the socio-economic problems of the region, farming systems, cultural characteristics of the rural population, governmental policies and development plans.

Mid-career researcher workers

In addition to the need to keep continuously abreast of developments in their own and related fields of research, research workers in the middle of their careers encounter specific problems.

Refresher training
Even the best equipped researcher requires continual training. It only takes a few years after graduation for much of the knowledge accumulated during the formal education of a scientist to become obsolete. Young, recently recruited scientists are frequently more familiar with new techniques and theories than are their older colleagues, who completed their formal studies a few years earlier. As to still older colleagues, they are frequently hopelessly behind the times. Researchers should, therefore, be encouraged to keep abreast of developments in the subject matter of their own and related disciplines. Much of this will be based on self-study, but the research organisation should help by allowing attendance at university courses and symposia, and by organising special courses, especially in new techniques. With the ever-increasing rate of technical and scientific advances, even senior research workers need refresher courses and sabbaticals.

It has been estimated that at least one-third of the time of a working scientist has to be devoted to keeping up with new knowledge in this field, and this study must continue throughout his scientific career.

Work towards an advanced degree
A constructive way of overcoming a typical mid-career feeling of stagnation is to encourage the research worker to work toward a PhD degree; and the research institute should provide the essential facilities to do so. A pre-condition should be that the subject of the thesis chosen is within the research area in which the researcher is working, or in which he will engage after completion of his doctorate, and which can make a significant contribution to agricultural progress.

Normally, the opportunity to work toward the PhD degree should be reserved for researchers who have several years of research experience, whose field of interest is clearly defined and whose permanence in the organisation is assured, and for specialisation in critical areas required by the research organisation.

It has been stated that 'the most expensive item in higher agricultural education was the training in depth of specialists and particularly the provision of facilities for post-graduate study and research. Furthermore, it was also the activity where costs were rising most rapidly because equipment was more sophisticated and as the demands for trained technicians to service and maintain such equipment have increased' (Coombs, 1970).

In consequence of the high cost of post-graduate education, most candidates from developing countries were expected to get their post-graduate training at foreign universities in developed countries, the costs being recovered by foreign aid.

We have already pointed out the major shortcomings of post-graduate training in foreign universities. In particular, PhD degrees have suffered greatly from being too much of a pure research degree by thesis. Whilst in the past there were few, if any, alternatives available to the post-graduate student other than to study towards an advanced degree in a university in a developed country, this need no longer be the case.

The graduate schools of agriculture which can be set up in close cooperation between the faculty of agriculture and the agricultural research service, which we have discussed above, can, in due course, develop post-graduate training programmes. In the meantime, the international centres can fill the breach to a certain extent.

The candidate for a PhD degree, should be given the opportunity to spend a few months at the appropriate university before starting the research in the field, discussing his problem in the light of his former experience, learning techniques, studying the literature, and taking the courses required towards the degree, and do the actual work at his research station.

TRAINING FOR LEADERSHIP

In many agricultural research organisations in the LDCs, inadequately trained national personnel is performing at a moderate level, supported by expatriate specialists.

A major problem that LDCs face is that of developing a local leadership for agricultural research, capable of providing the in-service training we have discussed and of imparting the sense of personal commitment required in order to create a highly competent and motivated national research personnel, who can gradually replace expatriate staff and provide specialists in all the disciplines needed. The most urgent requirement is to form, as quickly as possible, a 'critical mass' of well-trained senior research workers in appropriate disciplines who are imbued with the necessary values and motivations and capable of inspiring and training young research workers.

The international agricultural research centres have the greatest potential to supply the necessary training. The main advantages of these centres are:

(a) their commodity focus which makes them the world's centres of excellence for the commodities involved;

(b) their 'learning-by-doing' approaches to training in intimate connection with the research activities and the technology they generate;

(c) the high calibre and scientific quality of their staffs;

(d) the contagious sense of urgency and seriousness of purpose which they instil; and

(e) the effective utilisation of training by alumni and their national organisations (Fernandes, 1977).

It is at the international centres that trainees from the research personnel of developing countries will probably have their first experience of a research environment that is development-oriented and sensitive to the needs of a developing country.

The training role of the IARCs is high in quality; it is, however, limited in magnitude. The number of professionals that the centres can receive without impairing their research programmes is obviously limited. Since the mandate of the centres is to achieve a significant impact on the capabilities of national agricultural research systems to perform effectively, it is in their interest to be involved primarily in high-level training of those professionals, who, after completing their training, will assume leadership positions in their national research organisations, which will eventually generate their own momentum.

By 'priming the pump' the centres will enable the national systems to eventually undertake or improve their own training programmes, and in due course, strengthen the interactions of the centres and national institutions.

Establishing priorities and the selection of the potential leaders are difficult and critical tasks. Mistakes in the choice of the candidates will be costly both in terms of money and of wasted opportunity. Certain pressures will be experienced, but must be resisted, such as selecting candidates who are not sufficiently qualified, but are the best available. Another form of pressure is to give preference to protegés of political figures. Training selection committees should be strong enough to resist these pressures.

The trainees will need 'to develop sensitivities to questions of morale, leadership techniques, and decision-making criteria' (Peabody & Tully, 1977).

As soon as the research worker returns from training at an international centre, 'it is urgent to capitalise on the enthusiasm he brings with him and provide him with the resources to get his programme started' (Baumgardner *et al.*, 1971).

TRAINING FOR RESEARCH MANAGEMENT

The productivity of agricultural research is greatly influenced by the quality of research management.

'Probably the most needed and least available talent today is the able, experienced scientist who is also a skilled manager—a person who can judge the research needs of his country and accurately evaluate scientific results and, at the same time, manage finances and personnel and interact effectively with his superiors' (IADS, 1979).

Scientists who have been concerned with research on biological, economic or social problems are not well prepared to handle entirely different topics, such as deriving research priorities from national development priorities, support mechanisms, technological requirements, and administrative procedures.

Research workers, when accepting research management responsibilities, are making a decision that will have major consequences not only on their future lives, but also on the research organisation which they are to direct. Most research workers undertaking the duties of research management have training and experience in research, but are not even aware of how abysmally ignorant they are of the basic principles of management, and that learning managerial acumen can help them to solve the innumerable administrative problems with which they will be faced in their 'new career'; budgeting of time, money and effort, building and maintaining effective relationships with other organisations, creating a research climate, etc.

Having always considered competence as an essential for a scientific career, they should realise that they must also acquire competence in research management if they are to be successful in their new and vital role. There are directors of research units who deny the need for their fulfilling a managerial role and who insist that their main responsibility is to guide and direct the scientific work of their organisation and to provide inspiration for the researchers. Such a role is possible at the lower levels of research management, such as research projects, laboratories, and departments. At the higher levels this attitude is possible only if someone else assumes responsibility for the managerial role. If this 'someone else' is a scientist, we are simply begging the question; if he comes from governmental administrative services, nothing remains of the axiom that the man who effectively manages research should himself be a scientist. People trained exclusively in general management, without a research background, do not understand the potentialities of research, the idiosyncrasies of the researcher, or how research has to be carried out.

In order to facilitate the transition from a vocation as a researcher to the managerial duties to be assumed, educational programmes have to be designed to help scientists acquire management capabilities and the specific knowledge needed 'to analyse, make decisions, take action and bear responsibility for creating national agricultural research systems, and attach themselves to problems ranging from national development priorities and strategies to technological requirements, support mechanisms and research priorities' (Rigney & Thomas, 1979).

An excellent scientist may become 'a wretched administrator, an unhappy man and a lost expert' (Gross, 1964), if he does not have the natural attributes needed for his new position as a research manager, and if he does not acquire the necessary basic knowledge of management and administrative procedures.

Training programmes

The object of training programmes for research management is to teach specific management skills and to impart sensitivities which may be relevant to the work of scientist administrators (Peabody & Tully, 1977).

At a workshop organised by the International Agricultural Development Service, a three-month course of highly intensive training for rising mid-career scientists aspiring to managerial posts in agricultural research in developing countries was proposed (IADS, 1979). This appears to be a reasonable duration for such a course: long enough to present a curriculum including the most important topics, but not too long, so as not to disrupt the work of the participants in their home countries.

The subjects should be presented and illustrated through formal lectures, detailed case studies, simulation exercises, and visits to research organisations and institutes. A combination of approaches stimulates different kinds of interests and brings the problems of management and administration to life (Peabody & Tully, 1977).

Case studies should include examples of research management successes, difficulties and problems. Also envisaged are discussions with experienced authorities and on selected readings. The participants should be invited to present and discuss problems of research management that have occurred in their own countries.

Even well-designed organisations cannot be effective without adequate operating resources. There is no doubt that some of the responsibility for inadequate funding is borne by directors of research who do not know how to present their case in interacting with policy-makers and politicians (ISNAR, 1981). Therefore, particular emphasis should be given to the discussion of styles of management, including organisational diplomacy, relations with governing boards and government authorities.

International training facilities

Recognition for the need for managerial training to facilitate the transition from research scientist to research management is increasing in recent years. The newness of the area and a lack of personnel qualified to accept responsibility for preparing and implementing training programmes are obstacles that still have to be overcome.

There are a number of organisations or agencies that offer some type of training and/or experience in agricultural research management, including the following (IADS, 1979):

- Southeast Asian Regional Center for Graduate Study and Research in Agriculture (SEARCA);
- International Agricultural Research Centres (CIMMYT and IRRI) (see below);
- US Department of Agriculture;
- World Bank—Agricultural Management Case Studies and the Economic Development Institute;
- Harvard University—management case studies;
- In-Service Training Awards Program (INSTA), Wageningen;
- International Course for Development-Oriented Research in Agriculture (Europe);
- Indian Institute of Management (IINA), Ahmedabad, India;
- Cornell University has announced a course on the graduate level on 'The Administration of Agricultural and Rural Development' which 'provides multidisciplinary perspectives on the implementation of agricultural and rural development activities in developing countries. It is oriented to students trained in the agricultural and social sciences who are likely to be occupying administrative roles in government action agencies, research organisations and universities during their professional careers'.
- ICAR has set up new centres to train agricultural scientists at all levels and science administrators in agricultural management. To be named the National Academy for Agricultural Management, the centre is located in Hyderabad.

Most of the training programmes for potential research managers are still in the formative stage in most of the institutions. While they are not yet fully formulated, the above organisations and others under consideration (by FAO in Africa, for example) provide the potential for a regional approach to an accelerated training programme for research managers.

Training experiment station managers

Oram (1977) has defined the role of the experiment station manager as:

The day-to-day management of field operations at a major experiment station, or a group of stations under the aegis of a major institute, working across a range of crops and environments. The role of such a manager is to ensure that everything is provided at the right time and the right place to enable scientific staff to work smoothly; he is responsible for land preparation, cultivation, irrigation, equipment, fertiliser, labour and the prevention of avoidable losses to experiments from thieves, predators, birds, etc. He must also be a skilled observer, and as such can be of considerable help to scientists who may not be able to visit the station every day.

In view of the now generally accepted view that the implementation of agricultural research must be decentralised and carried out at a number of regional experiment stations in accord with the ecological conditions of the country, it becomes essential to appoint a superintendent of the national network of regional experiment stations. His duties should include the general supervision of the regional stations and their dependencies; initiation of improvements in efficiency and organisation; liaison between the regional stations; allocation of equipment, etc.

The managers of the regional experiment stations and the superintendent should have a university degree in agriculture and require additional general and farm training in management and administration, and should be fully informed of the requirements of agricultural research in general and field experimentation in particular.

Oram further points out that 'Such people do not grow on trees, and it is a common error to assume that anyone with a good knowledge of research on how to run a field experiment or even a farm can successfully take on such a job. To our knowledge, specific courses designed to impart the research skills and judgment required, including "man-management" do not exist, and this is a gap urgency in need of repair.'

It is our view that many of the subjects included in the courses proposed for research management could also be useful for experiment station managers if a few additional topics are included, such as farm management and field experimentation.

The international agricultural research centres offer practical training for research support services, including the management of experiment stations (Vega, 1979).

Training technicians for research

Many kinds of trained technicians for laboratory, fieldwork and maintenance are needed, who require different kinds and levels of education.

In most developing countries the lack of an adequate cadre of laboratory and field technicians is a serious constraint on the effectiveness of the graduate staff (Oram, 1977); through neglect and lack of maintenance, expensive equipment is not effectively used (Gamble *et al.*, 1981).

Some may be educated adequately in vocational high schools; for some, technical agricultural education will be required. There is, therefore, urgent need for first-class vocational schools.

Field technicians
Obviously, countrywide experimentation on the farms cannot be done by the researchers alone; trained technicians will be needed at all stages of implementation, including the collection of data. Whilst field technicians will have received their basic training at agricultural vocational schools, no formal framework exists in which technicians receive training in experimental techniques. It is, therefore, essential that those who are involved in experimental work should be trained by the research organisation. This requires the active participation of experienced research workers in the in-service training of technical personnel. Seminars and field days should be held regularly in which research and production techniques are taught and demonstrated. An important incentive for research technicians is to establish ladders of promotion leading eventually to middle-level and managerial tasks previously reserved for graduates (ISNAR, 1982). They would require special training to achieve these changes through part-time or night school programmes. Technicians who have proven themselves as interested in the work, and who have the necessary aptitudes, should be encouraged by the research organisation to study towards a degree in agriculture and they should be helped to do so. The number of technicians with necessary qualifications will never be great, but the possibility of moving across the technician/research boundary can serve as a powerful driving force for those who are ambitious and able.

Chapter 10

Manpower Needs, Recruitment and Personnel Policy

DETERMINING MANPOWER NEEDS

The first requirement for an agricultural research system is a well developed manpower plan—'a plan that looks at the problem, the programme, the priorities and the support available, and puts all this together within a realistic time frame for staff development' (Gamble *et al.*, 1981). There are, however, very few agricultural research establishments that have carefully planned the types of specialisation required, the optimum number of research workers in each category, the level of competence required, the ratio between research and support personnel, etc.

The research establishment has generally developed over the years, first in accordance with the special interests of the former colonial powers, and since independence, in order to carry out research projects, frequently initiated and funded by donor agencies. Once these projects are concluded, the funds expire, but the research personnel remain on the payroll of the research organisation. As a result, an increasing proportion of the recurrent budget for agricultural research is spent on personnel emolument, with a concurrent reduction in the funds available for conducting research. Not only does a completely unplanned and an unbalanced establishment come into existence, mainly through inertia, but a progressive decline occurs simultaneously in the funds available to do research.

In their review of the agricultural research manpower in Kenya, an ISNAR mission found that approximately 71 per cent of the recurrent budget for agricultural research was being spent on personnel emoluments. 'On the assumption of a 7·5 per cent annual rise in personnel,

481

[assuming a 4 per cent rise in the budget] all recurrent expenditures in agricultural research will be for personnel by approximately 1992' (ISNAR/NCST, 1982). There is no reason to assume that Kenya is exceptional in this respect.

As a result of the foregoing, the present situation in most national agricultural research establishments in the LDCs as regards professional staff is characterised by:

- Inadequate number of staff in food production research and disciplines of major importance, whilst areas of secondary importance may be overstaffed;
- a preponderance of expatriate staff, many of them on short-term contracts;
- a large proportion of the national staff is inadequately trained for research;
- a concentration of research workers in locations near the seat of government and lack of qualified workers in outlying regions;
- a considerable turnover of research workers;
- a disproportion between scientists and supporting staff;
- a disproportion between the numbers of researchers and funds for research.

These inadequacies indicate how important it is that the LDCs place their research manpower requirements on a sound footing. They need 'to take stock of their existing trained manpower, to relate it to foreseeable national priorities, and to gear their future education and training plans to the realities of the existing situation in respect of those priorities' (Oram, 1977).

Manpower planning is now accepted as an integral component of national plans for agricultural research in some countries, and has achieved priority among other demands on limited resources.

Methodology

There is no generally accepted methodology for estimating future manpower requirements. Before elaborating our own concept for determining the manpower requirements of an agricultural research organisation, it may be useful to examine briefly the more common approaches which are in current use.

Three manpower planning methodologies have been widely employed in both developed and developing countries since the early 1960s (ISNAR/NCST, 1982).

1. International comparisons of the manpower situation in different countries in order to derive common patterns of manpower development. Studies of this kind frequently employ multivariate statistical techniques and are generally undertaken at a high level of occupational/skill aggregation. The assumption underlying this method is that the present situation in different countries is satisfactory or that different countries have the same research requirements. This assumption is, however, entirely unwarranted. At best, broad international comparisons might be useful in gaining a better understanding of the relative position of the agricultural research system in a given country *vis-à-vis* other countries.

2. Manpower projections based on the extrapolation of past trends in the growth of occupations within a country.
   In this case, it is erroneously assumed that:
   (a) the situation in the past is a model worth copying;
   (b) the future will not be different from the past;
   (c) that past trends, such as a rapid increase in research personnel, will continue into the future.

3. The use of sample surveys of employers in each of the major sectors of the economy. Employers (in our case, Directors of agricultural research institutes, heads of departments and laboratories, managers of experiment stations) are requested to estimate their additional occupational requirements over a specified period; the results are then aggregated to develop a national target (Daniels & Nestel, 1981). This is the most popular manpower planning technique, but it is probably also the least logical and most likely to lead to an inflated research establishment. It is based on the subjective opinions of the respondents, whose main concern may be to expand their individual 'fiefs'. It not only perpetuates an existing system which may be unsatisfactory, but actually expands it.

4. More recently, an approach has been suggested independently by FAO and the Rockefeller Foundation in which research staff needs are related to the production of major commodities (Oram, 1977). An arbitrary minimum area has been defined as justifying a national research effort, and assumes that any country having a larger area than this of a specific commodity ought to have a significant research

programme devoted to it, for which a staff requirement could be calculated and compared to the existing manpower resources.

For example, a study on the number of well-trained people needed to support a research and production programme in a given country has been carried out by CIMMYT staff. Using maize as an example, it was estimated that countries cultivating from 30 000 to 100 000 ha would need a national team of 11 agronomists, including breeders, entomologists, pathologists, research production agronomists, extension agronomists and protein evaluation specialists. For countries cultivating 100 000–500 000 ha, this number would increase to 26 and to 55 for countries with 500 000 to 1 000 000 ha (Violic, 1973).

Using this method, it may be possible to 'develop crude estimates of the number of research scientists required globally' (World Bank, 1981), but it cannot provide a reasoned and balanced plan for manpower development for a given country.

In brief, the methodologies in use either give a distorted estimate of scientific manpower requirements, or at best involve a high degree of conjecture in making these estimates. 'Weaknesses in base data on current numbers of research scientists add to the difficulty' (World Bank, 1981). Some other method of assessing manpower needs is, therefore, required.

The target-setting approach

Many factors influence the number and kinds of personnel needed in national agricultural research and related programmes. Among these are changes in the growing population, mounting food demands, the increasing desire of nations to be self-sufficient, and the changes in policies and emphasis of donor and technical assistance agencies (IADS, 1979).

Agricultural research is itself a major agent for inducing change in a developing economy. The question, therefore, arises how can one accommodate a stable structure of an agricultural research establishment with a dynamic situation, in which requirements are bound to change with time. In other words, is it possible to make manpower planning meaningful by using criteria that are not likely to undergo major changes even in a changing situation?

The following criteria are not likely to be affected in the course of time:

- The major commodities produced or potentially important;

- the major research disciplines required;
- the size of the country and the number of different ecological regions with specific research problems.

These criteria can be used to prepare a personnel master plan which:

(a) identifies a core of research specialists that are required 'to form a critical mass of research personnel in sufficient numbers and specialisation to be able to respond to the existing research needs';
(b) determines where they are to be located;
(c) estimates the technical support and administrative staff required (ISNAR, 1981).

The target is to implement a training programme within a predetermined time limit, for the recruitment and training of the personnel required for this master plan.

It is necessary to stress that the master plan for research personnel is planned in accordance with the *fields of research* that will be required by the research organisation to meet research challenges, and is not based on personnel needs for implementing a specific *research programme*. The core of specialists who are considered in the master plan will have to be supplemented by additional personnel, as and when required by the research programme. It is, therefore, necessary to reconcile two apparently conflicting requirements: on one hand, to expand research personnel in order to carry out research projects as and when required, according to a national research plan—projects which by definition have a predetermined life span; and on the other hand to avoid distortions in the research establishment by accretion of personnel that becomes redundant once the project for which they have appointed is concluded.

The obvious solution is to establish an *ad hoc* research team for each project. The researcher or researchers from the core staff, who have the requisite specialised knowledge are called upon to serve as project leaders or specialists on the project team. Additional research workers, if and when required, are hired on a contract basis (or graduate students are used; cf. p. 459).

Upon completion of the assignment, the research team is disbanded and its core of permanent staff becomes available for other assignments. This approach provides stability and continuity in each essential field of research, and at the same time makes it possible to expand or retract the numbers of research staff according to the number and complexity of research projects to be undertaken, whilst retaining a permanent core of specialists, thereby preventing an unplanned increase in the research establishment with each new research project.

Establishing the master plan

In practical terms, the following steps need to be undertaken: (a) Define the essential fields of research for the country as a whole. (b) Assign to each field of research one or more research units in accordance with regional requirements.

A research unit would consist of a senior research specialist, and one or two technicians with permanent status, and a number of assistant or junior research workers (preferably graduate trainees).

The research units would essentially be of two kinds: those concerned with a discipline that is not specifically commodity-oriented, and those that are commodity-oriented. Because of the inherent interactions between plant-breeding and agronomy, the commodity specialists would have to have basic training in plant or livestock breeding, physiology and management practices.

There are no accepted guidelines as to what the balance between scientists and supporting staff should be (Oram, 1977).

At the international agricultural research centres, and in many developed countries the ratios are 1:2 up to 1:4 (World Bank, 1981).

It is recommended that a ratio of 1:1 or 1:2, according to the type of research, should be adopted for the purpose of estimating permanent technical staff requirements.

In determining the number of research units one would have to take into account:

(a) The *regional dispersal of research workers* in accordance with the major ecological regions. A strong, multidisciplinary team of research units will be required at each station, in accordance with the specific crops and problems of the region it serves, but not necessarily in all the disciplines and commodities.

(b) The *relative importance of certain disciplines or commodities.* For example, one of the research units in entomology could be concerned with diphtheria in general, or, if need be a special unit may be devoted to the tsetse fly. Allocation of personnel to commodities would be related to their ranking in importance.

A commodity research unit may be concerned with groups of crops with common characteristics, such as all pulses, or with a single commodity of major importance, such as sugar cane.

(c) In addition to research units, there will be need for scientific personnel for management and scientific supporting services. A major advantage of this method of assessing manpower requirements is that it consists in setting a target based on *present needs* and not on crystal gazing.

A personnel master plan for the research organisation is established, based on the criteria described above, and the number of permanent research manpower required is calculated. Concurrently, the exisiting situation is analysed, deficiencies and redundancies in the research establishment are pin-pointed.

Once this is accomplished, the 'target' consists in recruiting and training the number and kind of specialists that are lacking, or needed to replace expatriates or researchers who will reach retirement age in the next few years. Other redundant personnel may need to be retrained.

The next step is, therefore, to adopt a time plan for achieving a balanced manpower establishment and choosing an appropriate strategy for implementing the plan.

Strategy

The broad outlines of a strategy for training and indigenisation of agricultural research manpower, proposed by an ISNAR team for a specific situation (ISNARNCST, 1982), can serve as a model for the strategy of manpower planning in LDC's in general.

A National Agricultural Research Manpower Development Plan is envisaged, which should consider a programme of yearly allocations of training positions and fellowships which will enable present research staff and the additional staff required, to be systematically trained to the required level in the course of a predetermined number of years. The institutions that should provide the training need to be identified. Priority should be given to training leaders, who will gradually make it possible for the research organisation to create its own momentum for training its own research workers.

Responsibility for the manpower development plan and its implementation should be vested in a training officer, who 'in consultation with the research director, deputy director and the research programme leaders, would identify, place and supervise suitable candidates in the training programmes relevant to the needs of the research service'.

External financing for the training programme can be sought for the staff development project. The World Bank in particular, has expressed its interest in supporting such programmes (World Bank, 1981).

Advantages of a master plan

A well-reasoned master plan serves as a guide in building a research organisation that has the necessary capabilities for coping with the conventional problems of improving agriculture in a dynamic situation and that can adjust itself to change or unusual situations without distorting the balance between its essential activities.

The master plan not only provides guidelines in institution building, but is also a powerful tool for obtaining the necessary backing and funds for implementation. The usual procedure, which consists in attempting every year to wrest from the ministry of finance and from the civil service commissioner approval for a few additional posts for research and/or service workers in the research organisation, is usually time-consuming, ineffective and frustrating.

By contrast, if the director of research presents a detailed, balanced, reasonable blueprint, planned according to the needs and capabilities of the country, every component of which can be shown to be indispensable for fulfilling the goals of the institution as defined by Government, it will be far easier to obtain the necessary means for implementation.

MIGRATION OF RESEARCH WORKERS

Many developing countries have made major efforts to provide training for incipient and practising agricultural research workers. National funds have been invested and considerable technical and financial support has been received from a wide range of international institutions and bilateral assistance from various developed countries.

A high proportion of the researchers trained in these programmes do not return to their native countries. If they are able and industrious, they usually find it possible to obtain employment in the countries in which they were trained, at conditions far more attractive than can be provided by their home countries. If they do return, in view of the shortage of trained personnel, they find it easy to enter into administrative work. Finally, for those who wish to remain in their chosen career, it is easier to accept teaching functions at a university, which also confer higher prestige than does applied research in the field.

Another basic weakness of most training programmes is that trainees are usually individuals who, once the course is concluded, will have no further contact with each other.

If the training is successful, each individual receives excellent preparation in a given field, and is imbued with enthusiasm and a keen desire to apply what he has learnt when he returns home. But when he does return, he finds that alone he simply cannot shoulder the burden of mobilising the means necessary for the innovations which he has in mind, and at the same time train co-operators and imbue them with the same sense of purpose and enthusiasm, convince people to accept the necessary changes and innovations, organise the work that has to be carried out in the field, etc. Alone, he rapidly becomes disillusioned and frustrated and will usually give up his efforts, and search for greener pastures elsewhere.

The high rate of migration of trained research workers (the so-called turnover phenomenon) from the research organisation is a major constraint in developing an effective research organisation. It is not uncommon to find institutions still trying to build a competent staff after 10 years of active training programmes (Baumgardner *et al.*, 1971). The high rates of research staff turnover inevitably cause a decrease in the average work experience of the remaining staff and therefore of their productive potential.

Attrition rates in excess of 7 per cent annually are the norm in Africa, as compared with 3–4 per cent in industrialised countries. Such turnover rates undermine productivity of research programmes and increase the need for training. The situation is further aggravated by the shortage of opportunities for postgraduate training in Africa (ISNAR/SPAAR, 1987).

An instructive example is provided by Kenya, a country that has invested substantially in overseas training for agricultural research personnel in the last two decades, and in which the University of Nairobi provides postgraduate training; and yet, one of the major, and perhaps most important constraints in the agricultural research system, is lack of trained manpower! The estimated average length of employment in the research division of the Ministry of Agriculture for Kenyan personnel is two and one-half years. In the course of three years, the research division lost 58 research scientists, mainly to other units of government, universities, colleges, and the private sector (ISNAR, 1981).

In an investigation on human resources in agricultural research in Latin America, Ardila *et al.* (1981) have even found that the migration process at the country level has produced net losses in the impact of the training programmes on strengthening research capabilities. Much has been written about brain-drain and defection of trainees, and efforts have been made to bind trainees contractually, in order to recuperate the costs of their training. This has generally not been effective (Daniels & Nestel, 1981).

With the development of the private sector, a new and strong competition for research workers has arisen. In many cases, the private sector is willing to pay salaries and benefits far more advantageous than the public sector can offer. 'These payments are possible because they represent not only an investment against incomes to be derived from future research, but also the accumulated knowledge embodied in the researcher' (Piñeiro, 1987).

Whilst there is no doubt that salary levels and conditions for promotion are important factors in the decision to leave the research organisation, it was found in the investigation mentioned above that economic issues serve as a motivating force to leave the institution only when other negative institutional factors are present. The authors ascribe much greater importance to the relationships between the research worker and the institution as well as stability in the working environment than to salary levels (Ardila *et al.*, 1981). The same conclusion was reached a decade earlier by Baumgardner *et al.* (1971) who ascribe the defection of trainees to the poor climate for professional development and the low rewards to be expected. They conclude that 'the ultimate solution lies in relieving the causes of defection by the formation of reasonable and effective employment policies that stimulate staff performance and that elicit institutional loyalty'.

In their study of the causes of migration of researchers in three institutions—INTA (Argentine), ICA (Colombia), and INIA (Peru)—Trigo *et al.* (1982) reached the following conclusions:

- a very high proportion of the scientific personnel of these three institutions have left the posts for which they have been trained; furthermore an important proportion of these migrants left for foreign countries. In Peru, for example, 70 per cent left the country.
- migration of scientific personnel reaches a peak during periods of internal crisis in the respective institutions, when public recognition of agricultural research is at a low level.

- economic factors play a role in the decision to leave, but are not the major factor.

The causes for the loss of personnel are many and complex, among others: the lack of adequate social recognition; inefficiency of the research organisation, leading to frustration; personnel policies; lack of public support.

Political factors may be a major cause of personnel turnover. In many Latin American countries, whenever a new government comes into office, personnel changes follow, almost at all levels. In Chile, for example, the bulk of the losses of scientific personnel were the result of the 1973/74 political crisis (Castronovo, 1984).

Castronova (1984) therefore has good grounds to conclude that a considerable part of the funds, both internal and external, have been, and are still misspent on trainees who do not return to the institutions which have made the training possible, or else leave the organisation after a short time, even if they thereby break a contract.

The huge sums invested in training have not yet provided the agricultural research organisations of Latin America with all the competent personnel required. Castronovo therefore expresses doubts whether 'it is reasonable to expect more funds to be earmarked for training, unless waste can be eliminated, and training benefits for sponsoring organisations improved'.

The problem is acute and of serious proportions. All possible measures need to be taken to eliminate the factors conducive to migration. Organisational stability, institutional efficiency, a good research climate, research with a mission are essential components of a social system conducive to researcher identification with the organisation and its objectives. (This subject is treated in detail in Chapter 11.)

RECRUITMENT OF RESEARCH WORKERS

The identification and selection of candidates who have the aptitude to become research workers are of considerable importance. Taylor (1963) reports that the characteristics stated by scientists themselves as important in their jobs are: drive, mathematical ability, resourcefulness, cognition, integrity, desire for facts, desire for principles, desire for discovery, information ability, skill, flexibility, persistence, independence, discrimination of value, co-operation, intuition and creativity. Whilst there is no gainsaying that these are all important characteristics, each of which is essential to the make-up of a research worker, it is doubtful whether a catalogue of these desirable, or even essential qualities is of great help in assessing the potentialities of candidates for research work.

One of the easiest criteria to use and which is very frequently adopted is the scholastic record of the candidates. In a study carried out to verify the value of using undergraduate grades in the recruitment of young scientists, it was found that there is very little statistical evidence to show that the use of undergraduate grades is justified as a predictor of future success as a research worker. Taylor concluded that 'learning old knowledge at school and mastering what someone else has produced is a different psychological process from producing something new of your own'.

One must regrettably accept the fact that at present there are no objective criteria available that can substitute for on-the-job performance.

Even obvious criteria such as 'the desire to fit into constraints of the problem-oriented research as opposed to academic research' and 'the ability to get on well with colleagues' (Burdekin, 1970) cannot be properly evaluated at the time of recruitment. The only practical possibility is to appoint candidates with good general training (MSc or equivalent) and provide them with the possibility of proving themselves. After a period of one or two years, it is usually possible to decide on the suitability or otherwise of the young worker, and the necessary conclusions should then be drawn without delay.

The monetary rewards for agricultural research novices are usually low and promotion is slower than in other fields. It is therefore fairly safe to assume that only young people attracted to research will apply for vacancies in this field, providing at least for some 'self-selection' in the choice of candidates. For workers engaged in agricultural research, familiarity with farming problems is a great asset. This is no problem when the candidate comes from a farming background; agricultural schools at the secondary level and work for farmers during vacation periods are good substitutes. One of the provisions of many faculties of agriculture is that student candidates must have at least one year's experience in practical farming.

Training in basic research should be an essential part of the make-up of the research worker. However, he must learn, early in his career, to apply this training to definite, well-defined problems of practical importance.

Apprenticeship period

Several NARSs in Latin America are using a system of scholarships for recent university undergraduates, to enable them to participate in research programmes of the departments or the experiment stations, for a one-year period.

The beneficiaries of the scholarships are required to spend half their time on their graduate thesis and the other half on departmental research work. At the end of the year the research institute can decide whether to offer a trainee the opportunity to continue work at the institute. This procedure makes it possible to select among the trainees those who show the most promise as researchers.

A variant of this approach has been adopted in Chile, by INIA. Students from the faculty of agriculture are invited to work during their summer vacation in the harvesting of crops grown on the experimental plots of the regional experiment station. Those that show real interest in the work are provided with a scholarship for one year, during which they can work on their Bachelor thesis, under the guidance of an experienced research worker. The best trainees are offered a 2-year contract. If the work during this period is highly satisfactory, they can apply for a scholarship at a foreign university to work for an advanced degree. On their return, they are offered a permanent position at the institute.

The lack of objective criteria for the recruitment of the most promising candidates is not such a serious handicap as might be assumed at first glance. It should be a basic principle of the research organisation that no researcher should achieve permanent tenure in the organisation before he has achieved the status of junior scientist, which requires several years of service as an assistant.

A research institute, in addition to a few top research workers who are exceptionally gifted, needs a relatively large number of researchers of average ability. As an unnamed Soviet scientist remarked: 'Scientific research must have its infantry; the outstanding individual strikes the spark, but only the mass can exploit the results' (OECD, 1961).

Much necessary and even important research, especially in agriculture, consists in the application of already known principles to specific environments and conditions. It requires careful planning and meticulous execution, painstaking observation and recording, but not necessarily a high level of originality. The qualifications for this work are a good scientific background, a good knowledge of the subject, training in research and plain industriousness as well as perseverance, intellectual honesty and the ability to work with others. These qualities can be easily identified during the 'apprenticeship' period.

PERSONNEL POLICY

Satisfying the legitimate needs of the researchers

If the research organisation does not recognise the importance of defining the essential needs of its researchers, and does not make the necessary efforts to satisfy these legitimate requirements, it is highly improbable that it will obtain the scientific performance from the researchers at the level and intensity required, or reverse the trend of high turnover of researchers.

In his choice of research as a vocation, the major motivation of a scientist is to find work of interest to him, that provides a challenge and an opportunity to prove his ability.

In a study of the motivations of agricultural research workers in the USA, Busch & Lacy (1983) found that 'enjoying doing my kind of research' is the single most important criterion for research problem choice. The second most important criterion was 'importance to society'. These two criteria need not be mutually exclusive, and can even reinforce each other if appropriate motivations are chosen.

Conditions of Service

Conditions of service relate to the following issues (Mook, 1986):

Careers: What kind of a career can a research worker expect, how rapidly will he be promoted and what criteria will be used for promotion?

Income: What level of salary can the researcher expect in the course of his career, and what other monetary benefits will he be eligible for?

Non-salary benefits: Housing, transport, schooling for children, medical facilities, social environment, etc.

Professional development: What kind of on-job training is provided by the research organisation; possibilities of interaction with other scientists, responsibility for research programmes, etc?

Careers

It is essential to provide possibilities for advancement throughout the entire career of the research worker, and these should be linked to significant increases in income.

Promotion along the lines of the public service, based on seniority and administrative pyramids, is definitely not suitable for researchers, whether they work in government institutions or others. This is, however, the system adopted in many NARSs, where promotion is based on length of service rather than on productivity of the researchers. In the countries of Sub-Saharan Africa, many research workers achieve their maximum rank within a relatively short time (5–7 years) and their salaries are increased only slowly, according to the years of service (ISNAR/SPAAR, 1987).

The position of the research worker should be based on his competence and professional achievements. His upward mobility should not be limited by the imposition of management or administrative duties. It is, on the contrary, perfectly admissible, for example, that the research department be headed by a relatively young research worker, with drive and leadership, while one or more members of his department may have reached a higher grade in the promotional ladder as the result of their research achievements. The professional ladder strengthens the scientific incentives of the researcher.

Promotions based on seniority are also detrimental to the research organisation because of the disincentive involved in automatic promotions. A dual ladder of promotion is justified in the research organisation; one hierarchical—for management in which promotion results in increased authority, and the other for researchers based on professional achievement—leading to increased autonomy and scientific standing. This will allow the researchers to move into higher earning and status brackets without changing their careers. Appropriate and fair rewards—titles, remunerations and status—should also be available to those engaged in research, without regard to hierarchy.

Criteria for promotion
It is extremely difficult to evaluate or even to define the research performance of individual scientists or of research groups. A programme of evaluation of research productivity is, however, essential as an objective base of promotion of the individual scientist, and as a means of estimating the progress of the various research groups in the organisation.

Taylor (1963), in a study of performance of scientists at a large physical science research organisation, obtained at least 50 different measures of their performances. On these criteria, factor analysis techniques were used through a computer which reduced overlap between the 50 criteria to the 14 categories shown below:

1. Originality of work and thought
2. Creativity and productivity rating by higher-level supervisor
3. Overall evaluation by supervisor
4. Total work output
5. Productivity in writing
6. Recognition for organisational contributions
7. Quality, independent of originality
8. Likeableness as a research team member
9. Visibility
10. Scientific and professional society memberships
11. Current organisational status
12. Contract monitoring load
13. Status-seeking tendencies
14. Total scientific experience.

There is little doubt that with slight modifications most, if not all of these criteria, with suitable weightings, can be adapted for the evaluation of agricultural research workers' performance.

Each organisation can, of course, adjust the specifications and the relative weightings of each attribute to its own requirements and outlook. Those used for evaluating research accomplishments, by the Eire Agricultural Research Service, for example, are:

1. The inherent difficulty and complexity of the research;
2. the degree of responsibility for selecting, planning, organising, executing and reporting; the amount of supervision received, and the nature of authority to speak for the organisation;
3. originality and initiative in identifying problems for study and in developing methods and techniques;
4. the scientific attainments and qualifications of the individual: his contributions in the field, his publications and professional standing.

An inherent weakness in these evaluations is the danger of subjective appraisal. To overcome this drawback as far as possible, the final evaluation should be made by a group of qualified and responsible research workers.

Stearns (1958) points out the fallacy of evaluating a researcher's work on the basis of success or failure only. This approach would favour the man who sticks to sure things, whilst the imaginative man with initiative and courage would be discriminated against.

The effectiveness of research can usually be evaluated only many years after its inception, and sometimes only after its conclusion. A considerable delay may therefore occur before a final evaluation is possible.

Even if the results of a research project are disappointing, negative results are not in themselves failures; they become failures only if the necessary conclusions from the negative results are not drawn and applied by institutions or individuals (Starkey, 1988).

The technical evaluation of a research project must relate to how and what the researcher has done, and not specifically to the achievement of positive recommendations. A distinction should be made between what ODI (1988) calls informative and uninformative failure. The latter includes, for example, deficiencies in the planning, conduct and monitoring of experiments that make the results unusable; informative failure 'results from well-planned and implemented research from which lessons can be learnt, and the results of which can prevent the exploration of blind alleys being repeated in other work'.

In a wide range of subject areas, 'there is a reluctance to publish (and among editors to accept) reports of the failure of experiments' (ODI, 1988). As a result, 'there has been less learning from each other's experiences, less effective use of human and financial resources, and consequently less overall progress' (Starkey, 1988). The author concludes that it is most important that researchers 'should feel as proud of a well-presented negative lesson as of a positive one', to which I would add that this work should not be overlooked in assessing the justification for promotion of the researcher.

The technical evaluation must refer to *how* and *what* the researcher has done, and not necessarily to the achievement of success. Further, it should be stressed that none of the criteria used is sufficiently precise to make an absolutely objective appraisal possible; whilst these difficulties and inherent weaknesses cannot be overcome, they should not be ignored.

In view of the difficulty of assessing scientific productivity, there is a tendency to rely too much on the number of published papers.

Papers can vary extremely in value, and therefore an analysis of each paper is necessary. Additional drawbacks result from co-authorship of papers making it difficult to assess the individual contribution of each author (the principal author is *not* always the researcher who has made the main contribution); the same subject may be repeated in different terms, in a number of papers; a single subject may be split up so as to provide a number of papers, etc. There is also the danger that over-reliance on 'papers', as a yardstick for the evaluation of the research worker, will affect the choice of research subjects, favouring those that facilitate a steady flow of papers.

On the other hand, it is known that the number of papers written is usually a reliable indication of the productivity of the scientist. Most scientists who publish frequently are considered as high performers by their peers.

It has been proposed that the number of times that a paper is cited in the literature could serve as an index of excellence. In a survey carried out by Price (1965), it was found that only a small fraction of the papers published would qualify for a high rating if number of citations was to be considered a suitable criterion of quality. Frequent citations may also indicate notoriety, as shown by the innumerable references to Lyssenko's writings, before he was discredited in his own country.

A qualitative appraisal of the research paper could, of course be based on:

- the logical approach to the subject;
- the clarity and accuracy of the presentation and the conclusions drawn;
- the importance and timeliness of the subject, its scientific and/or practical contribution;
- the technical excellence of the research and/or its contribution to innovation in experimental procedures.

The adoption for publication in prestige journals as a major (if not exclusive) criterion for promotion can actually be counter-productive. 'It is almost inevitable that professional rewards, gained through academic research and publication in journals, will exert an important influence on the type of research preferred by the scientist' (World Bank, 1981). The social system of science rewards not only solving a problem, but for being the first to do so. This reward system motivates the researcher to choose topics that have a high probability of publication in professional journals (Busch & Lacy, 1983).

The excessive reliance on the number of publications and where they are published is also unjust, and creates a conflict between the goals of the research institution and the researcher. The Director General of the National Research Institute of Venezuela (FONAIAP) writes:

Many of the objectives to achieve are not innovations, but adaptations, with limited value for publication, but of considerable importance for advancing agriculture in the country, and should therefore be encouraged and rewarded. It must be clear, that FONAIAP is not an institution in the service of science but of agriculture. And that should not be a reason for complexes; first, because it is as dignified a charge as any other, and secondly, it requires intelligence and imagination to the same extent as other types of scientific enquiry (Gerente General, 1978).

In brief. If properly evaluated, publications can be an important criterion, but not the only, or even the major, measure of the professional progress of the research worker. Additional criteria are therefore essential, which together reflect as accurately as possible the contributions made by the researcher to progress in the agriculture of his country, such as:

- achievements in basic and applied research, as evidenced by publications, reports, patents or any other document that describes his contributions in the field;
- information on the productivity of the researcher: contribution of original ideas, providing leadership in teamwork, training of young research workers, ability to keep his research projects up to date, ability to draw valid conclusions from his research and apply them in practice;
- any other relevant criteria that are adapted to the special functions of each research institute, such as carrying out research under difficult conditions (remote field stations for example), or on especially difficult problems.

Teamwork

A special problem arises when assessing the work of a researcher who is a member of a research team. Outsiders will find it difficult to ascertain the relative contribution of the team members to the overall performance of the team. Castronovo (1984) proposes assigning a certain number of merit points to the team as a whole, and leaving it to the team members to divide the merit points among themselves.

Horton (1984) points out that the present system of professional incentives inhibits interdisciplinary teamwork, which is becoming more and more essential in agronomic research. 'As a rule, university thesis committees do not look favourably upon interdisciplinary research projects aimed at solving practical problems. Attainment of "high scientific standards" generally requires students to use sophisticated procedures that are often not suitable for use at farm level.'

Unfortunately, academic members of promotion committees for the research staff of the agricultural research institution apply the same unsuitable criteria when evaluating the work of these researchers. Members of a research team generally suffer most from this type of evaluation.

Grading systems

Various grading systems for agricultural research workers have been devised in different countries.

One possibility is to adopt a system that provides grades and emoluments that are equivalent to those of faculty staff with similar qualifications.

The following model is based on this principle, and can be adapted to any situation, as required.

Model grading system for agricultural research workers
The requirements for each grade, the titles of which are presented as examples only, are as follows:

Grade 1 (research assistant). A candidate for this grade must be the holder of an MSc or equivalent degree, with passing grades that make him acceptable as a future candidate for a doctorate thesis. He usually serves as a research assistant to researchers in grade 3 and upward, who closely supervise and guide his work (corresponding university grade: *assistant*).

Grade 2 (junior research worker). Candidates for this grade are either workers in grade 1, who show promise in research, after two years' work in the research institution; or workers from outside the institution with equivalent research experience. Workers in this grade are supposed to carry out experimental and technical work without supervision, but are not yet responsible for defining or planning research projects (corresponding university grade: *instructor*).

Grade 3 (research workers). Candidates for this grade are either workers in grade 2 who, during two years' work in the institution have had consistently good ratings in their performance reviews and have made significant contributions to research; or workers from outside the institution, with equivalent research experience. A worker in this grade has achieved the status of a researcher capable of planning and executing research functions that are assigned to him. He should be competent to give guidance to workers in grades 1 and 2. At this stage, he should be encouraged to work towards a PhD degree (corresponding university grade: *lecturer*).

Grade 4 (senior research worker). Candidates for this grade are workers in grade 3, who have received consistently high ratings in their performance reviews for at least three years; have shown a satisfactory research performance, as evidenced by their publications, internal reports or achievements in the field; and have received favourable appraisals by 2–3 scientists from outside their own institution. Workers in this grade are the backbone of the research organisation. They should show initiative and originality in proposing new research projects, and be able to plan and execute them independently. They will be able to head research teams and apportion part of the research to workers in grade 3. Preferably, they should have completed a PhD thesis, within the framework of their research area (corresponding university grade: *senior lecturer*).

Grade 5 (principal research worker). Workers in this grade have maturity and competence in research and in training researchers. They are responsible for a specific field of research in which they must exercise initiative in proposing new problems and devising the appropriate approach, based on extensive knowledge of existing theory and techniques in their field. They must have already made substantial and original contributions (corresponding university grade: *associate professor*).

Grade 6 (chief research worker). This is the highest grade the research worker can achieve, and is, therefore, reserved for individuals who have made outstanding contributions to agriculture, authored important publications and are recognised authorities in their field. They take full responsibilities for formulating their research plans, and for executing them, provide leadership to research workers in the organisation, and head interdisciplinary teams (corresponding university grade: *professor*).

Incomes

The whole system of grading and promotion described in the previous sections has meaning only if translated into terms of monetary rewards. However, the situation in this respect, in many countries, is far from satisfactory.

Generally, agricultural researchers in the NARSs are government employees, even if the institution in which they work enjoys a semi-autonomous status. Their salaries are frequently linked to those of personnel engaged in the public service; the authorities are understandably anxious not to break this linkage. As a result, scientists employed by the private sector, parastatal research institutions and universities receive much higher salaries and enjoy better working conditions than their colleagues employed by government. This is especially true in developing countries. In Tunisia, for example, senior university staff earn up to twice as much as equivalent research personnel employed by the Ministry of Agriculture (Bennell, 1986).

Promotion in the research establishment needs to be based on achievements by the individual, and not on seniority or hierarchy, as is the case in the civil service. This signifies that professional advancement in the research organisation should not be automatic, and is therefore more difficult to achieve than in the civil service. This alone justifies the adoption of an independent scale of salaries for research workers. The most simple way to do this, without causing perturbations in the civil service, is by linking the salary scales of all research workers in government employ to those of faculty staff in the universities.

Remuneration for special functions

If grading and promotion are not linked to the hierarchical position occupied by a research worker, this needs to be taken into account. Administrative responsibilities, besides being an onerous burden, also reduce the time the researcher can devote to this scientific work and is therefore a constraint on the advancement of his career. He should therefore receive adequate remuneration for his administrative and leadership role, to be withdrawn as soon as he is freed from these functions.

Duties of the research worker

In a previous section of this chapter, we have discussed how the legitimate needs of the research worker should be satisfied. It is therefore appropriate to mention the duties that the research worker is expected to perform.

The research group, to which a specific area of research endeavour is assigned, is the basic unit of the research organisation. The focus of such a group is a scientist whose function it is to solve the problems in his area of competence. In contrast with the higher levels of research management, he is not concerned with the co-ordination of other research units. His duties involve the following:

(a) to show initiative in proposing and preparing research projects in his field;

(b) to undertake the solution of problems suggested to him by his 'superiors', extension officers or farmers;

(c) to co-operate in research teams in which his professional competence is needed;

(d) to train inexperienced research workers (assistants);

(e) to keep abreast of professional literature in his field and maintain interest in scientific subjects of wider interest;

(f) to develop methods and to design or help design the necessary equipment for his research;

(g) to plan and to design the course of the experiments he has to undertake, including collecting background information, identifying and controlling variables; to analyse the results, decide on their significance, draw the necessary conclusions and present the whole without undue delay in the form of a report or a paper fit for publication;

(h) to communicate to the best of his ability, the results of his research to other scientists, to extension workers and to farmers, by the most appropriate media in each case.

He is not expected to work alone in carrying out these responsibilities but will be assisted by the appropriate number of research assistants, technicians and unskilled labour. This, of course, entails a certain minimum of administrative duties. He should be given the necessary authority to direct and control all the resources that have been approved as necessary for the efficient execution of his work and allocated to him in personnel, equipment and funds.

In addition to the research activities described above the researcher may devote part of this time to teaching, lecturing in extension courses, playing host to visiting colleagues, participating in field days, symposia and work-groups, attending special conferences, attending meetings of various committees.

It is axiomatic that teaching at university level must be combined with research. It is less self-evident, but highly desirable, that research workers, whenever possible, be encouraged to participate in teaching in their field, both at the faculty of agriculture and in seminars for research workers, extension workers and progressive farmers. Teaching is a challenge which obliges the research worker to keep continuously abreast of developments in his field and to formulate clearly his own opinions and policies. Time devoted to teaching need not necessarily affect the scope of his research work. Teaching also entails serving as a mentor for graduate and postgraduate students, and this can well be fitted into the research programme of the teacher, thereby extending the scope of his work and his possibilities to cope with the additional responsibilities.

Taylor (1963) found that the higher the level attained by the researcher, the greater the number of activities in which he is engaged, which range from four (for the youngest researcher) to 15 for senior researchers. This wide spectrum of activities—all essential, and to which he is required to contribute, especially if he works in a governmental research organisation—make a consistent, sustained and organised research effort extremely difficult. As Taylor. (1963) correctly points out, it is, paradoxically, the young inexperienced research worker who is left to concentrate on his research, whilst the more experienced, mature and successful scientist is engaged in so many activities that he can no longer devote to his research the time and thought essential for truly creative work.

In addition to his essential activities, serious inroads on the time available for research are made when the researcher does clerical or technical work that could be assigned to others, does not have adequate facilities, does not plan his work properly, devotes part of his working time to activities not relevant to his work, etc.

Chapter 11

The Research Organisation as a Social System

The research organisation is a social system in which a number of groups of people are working: scientists, technicians, administrators, labourers. Each group has its characteristic attitudes, traditions, values and forms of behaviour. It is the relationships between these groups, and those between the groups and the formal organisation itself which transform the research organisation into a social system. They are also the source of many of the strains and stresses which exist within the research organisation, all of which may not be immediately apparent. 'The status and political preoccupations do not always operate overtly, and may give rise to intricate manoeuvres' (Burns & Stalker, 1961).

In addition to the relationships *between* groups, there are also interpersonal relations, mainly *within* groups, giving rise to a feeling of 'belonging' and of collective interests, which express themselves as 'morale'; the degree of interest and pride shown by the individuals in 'their' group and in the organisation as a whole. These relationships can therefore have a considerable influence, for good or for bad, on the general climate of the organisation.

It is highly doubtful whether a research organisation, in which the human element plays a far greater role than in most other organisations, can prove effective in the long run, if it is not able to satisfy the special human needs of the people involved.

Barnard (1938) distinguishes between 'effective' and 'efficient' organisation. The *effective* organisation is concerned exclusively with the achievement of organisational goals; the *efficient* organisation seeks, in addition, to satisfy the needs of its members.

However, in a research organisation, there are certain basic conflicts between the special human needs of the scientists and the goals and objectives of the organisation, which are not easily solved, for it is not admissible that the organisation should be adjusted to serving the needs of its research workers exclusively, instead of pursuing its own goal. For the organisation to be effective, adjustment must be achieved between the many conflicting social requirements and the goals of the organisation.

THE SOCIAL GROUPS INVOLVED IN A RESEARCH ORGANISATION

Five main groups can be discerned in a research organisation: the leadership, the research workers, the administrative personnel, the technicians, and manual labourers.

Each of these groups forms a 'sub-culture' or 'sub-system' (Leavitt & Pondy (1964).

Though the groups are mutually interdependent, they perform basically different functions and require different administrative approaches and practices.

The main social and organisational problems of the research organisation derive from the special traditions and status of the research workers; the social problems of the last three groups would not, generally speaking, be different from those encountered in other organisations. However, the differences in values between the latter groups and those of the researchers create problems that are specific to the research organisation, especially if the people engaged in administration come to consider administration as an end in itself and not as a means of furthering research, which is the basic justification for the work of all the people in the organisation.

THE SCIENTIST

Traditions, values, characteristics

In the developed countries the typical scientist works in an academic atmosphere, on problems of his own choosing. Whilst the conditions of work in a large, problem-oriented research organisation are entirely different from those in an academic environment, the trainees from developing countries become imbued with academic values of the host institutions and oriented towards independent research. Managerial personnel in a research organisation tend to consider research workers as prima donnas, who make life difficult for them and have to be handled with kid gloves.

The statement that scientists are 'different' is based on a number of assumptions: that scientists tend to be opinionated, are attracted by what is bizarre and unusual, are perennial rebels who will not conform to organisational procedures, and resent encroachments on their autonomy. They are primarily governed by professional standards, rather than by organisation objectives; are confirmed individualists with an ingrained propensity to work alone; and have contempt for administrative and non-scientific occupations. They are jealous of their prerogatives and are prone to disparage the professional competence of colleagues.

According to Hinrichs (1963), the results of a survey of 3500 scientists and engineers active in industrial research contradict the popular conception that scientists are 'different' and 'difficult'. This survey found that scientists in industrial research want to work under relatively close administration; have an urgent need to know precisely what management requires of them; exhibit some personal traits generally found in ambitious people in non-scientific fields; resent being characterised as freaks; and have career problems and gripes that are similar to those of other workers: desire recognition, information on what is going on, facilities to do the job, definite assignments and professional status. In as far as it is possible to generalise, it will usually be found that the scientist is well educated, is used to reasoning analytically, and does not accept the *status quo* as a matter of course.

Roles of the researcher

Stein (1963) distinguishes four roles that the industrial research chemist has to fulfil: as a scientist, a professional, an employee and a member of a social group. Stein's analysis is also largely appropriate to the roles of the agricultural research worker:

(a) As a *scientist* he is concerned mainly with adding to the store of human knowledge. He also conforms to an ethical code, which, according to Merton (1949), prescribes:

1. *universalism*: the obligation to subject the data to 'pre-established impersonal criteria';

2. *communism*: accepting the fact that all scientific progress is the result of social collaboration and hence constitutes a common heritage;
3. *disinterestedness*: avoidance of personal, subjective motivations; and
4. *organised scepticism*: the use of logical criteria in appraising of beliefs and avoidance of premature judgements.

(b) As a *professional*, by accepting a position in the agricultural research organisation, the scientist has implicitly and explicitly accepted the obligation of attempting to solve the problems of the farming community according to priorities of which he is not the sole judge. This obligation necessarily limits his freedom as a scientist, in the choice of his research problems.

(c) As an *employee* he has the obligation to be productive to the best of his ability, to show progress in the course of his work, and to be concerned with making effective use of the research funds at his disposal. He must also abide by certain rules and regulations of a disciplinary or formal nature, essential for the normal working of an organisation: such as regularity at work, the need to keep accurate records, to submit reports, to be available for work on teams, etc.

(d) As a *social member* of the group, the scientist is concerned mainly with his informal relationships, with superiors, peers and subordinates. Satisfactory relationships are essential for the establishment of a research 'climate', in which the researcher can be creative.

Requirements

In his choice of a job, the major motivation of the scientist is to find work that is of interest to him, that provides a challenge and an opportunity to prove his ability. Hinrichs (1963) lists 13 basic job factors desired by scientists engaged in industrial research. There is no reason to assume that there are any essential differences in respect to agricultural researchers.

- Freedom to publish results and to discuss the work with other members of the scientific community (this desire is also related to the need for professional recognition);
- opportunity to associate with high-calibre colleagues;
- a scientifically trained management;

- freedom to choose problems or projects on which to work;
- an organisation with a reputation for scientific achievement;
- a research director and staff with first-rate reputations for scientific achievement;
- adequate facilities, resources and technical assistance;
- opportunity for advancement and flexibility in advancement policies;
- equitable financial and status rewards;
- security;
- suitable living conditions;
- individual treatment;
- an opportunity to continue formal education while working.

Of these 13 requirements, seven are typical requirements for scientists and only one—the freedom to choose problems or projects on which to work—may conflict with the goals of the research organisation, though this is not always necessarily so.

In a research project reported by Stein (1963), it was found that the three factors that were rated most important by researchers in industry are: the opportunity to acquire new abilities or knowledge, the possibility to use one's present ability, and salary. The next group of factors included: the opportunity to work with good people, the freedom to carry out one's own ideas and having an important job in the organisation. Of the ten factors considered, those four that received the lowest rating were: the scientific prestige of the organisation, contributing to basic scientific knowledge, a sense of belonging to an organisation with prestige in the lay community, and association with high-calibre persons having important positions.

While it is possible to agree 'that scientists, as a group, show certain attitudes and behaviour patterns which distinguish them from some other groups in society', and that 'recognition of these values is a *sine qua non* for effective administration of scientific personnel' (Shapiro, 1957), it would be patently absurd to assume that all scientists have the same attitudes towards oriented research, or the same need for independence.

Heiman (1965) distinguishes between two groups of researchers, each with his own characteristics, whom he calls, respectively, 'thinkers' and 'workers'. The 'thinker' has a good knowledge of literature, including subjects outside his own field. He is unorthodox in his approach to problems. He may have spurts of activity, followed by periods during which he appears to produce little. He is allergic to administrative control. The 'worker' has an orderly approach to problems which he attacks with well-proven methods and perseverance. He is aware of the latest developments in his field, but has little interest in anything beyond his area of interest. He is best at solving pressing problems, provided the objectives are well-defined. He does not, as a rule, show resistance to administrative control.

Heiman further points out that, between the extremes of 'thinker' and 'worker', there exists a broad spectrum in which many researchers fall. Assignments should be distributed according to the individual's capabilities, and the amount of control should be tailored to his need.

THE CONFLICT BETWEEN THE ORGANISATION'S GOALS AND THE RESEARCH WORKER'S NEEDS

The main sources of strain and conflict in the research organisation are: (1) the organisation's fundamental policies; (2) authority relationships; (3) rules and procedures; (4) pressures from outside sources; (5) rewards and incentives; and (6) conflicts between intra-organisation sub-groups.

The organisation's fundamental policies

The first source of strain between an organisation and its scientists is their divergent aims. The research worker, by training and inclination, wishes to engage in research that is scientifically rewarding, that will increase his standing with his peers, and will promote his own career most effectively. He therefore usually prefers free, uncommitted research to working on the practical problems of oriented research. An agricultural research organisation is, however, entirely committed to the goal of solving the problems with which the farming community, which it serves, is faced. This requires a directed programme, oriented towards the solution of specific problems, a need which is incompatible with the free choice of research subjects by the scientists.

The average research worker is oriented more to his own discipline than to the institution. He feels that he is sole arbiter of what is needed and should be done in his field. This trend has become ever more marked with increased specialisation. He shows little concern with institutional problems and decisions excepting as they may affect his own personal work or that of the small group to which he belongs. This attitude is a source of conflict with a management committed to serve the needs of the organisation as a whole, and with other research units, whose demands clash with his own. As a result of this individualistic approach, the worker resists intrusions in his area by his superiors and colleagues and considers the organisation obligated to supply his needs.

Authority relationships

Both the scientists responsible for directing research and those conducting the research proper have been trained to prefer the academic way of life, with its autonomy and its antipathy for administrative restraints. A problem-oriented research organisation, to be effective, must be based on some form of hierarchy, on co-ordination and on orderly procedure.

In the academic tradition, authority is extremely diffuse. It is based on professional competence and whatever control exists is excercised by the colleague group. This tradition conflicts with the authority problems of the research organisation, in which executive authority is more effective and clearly defined. Control is based on a hierarchical structure, to which scientists are usually allergic, even if the control is exercised by scientists and the hierarchy itself is based, to a large degree, on professional competence. Thus, as Kornhauser (1963) points out, any combination of science and organisation becomes a source of strain, and it is when the organisation attempts to direct science to practical ends, that the conflicts become most severe.

Rules and procedures

The effectiveness and viability of a research organisation require that the individual conforms to certain rules and procedures that are essential if co-operation is to be achieved. Scientists, as a rule, however, show a considerable resistance to conformity, and resent the imposition of personnel practices. They feel they have the right to decide themselves what procedures are appropriate.

Code conflicts are not confined to research organisations. Barnard (1938) points out that code conflicts are a normal part of organisational life, and frequently arise between formal and informal codes. In a governmental research organisation, the more obvious form of conflict is that between the informal codes of the research workers and the formal codes that have been elaborated for the civil service as a whole. For the sake of uniformity of practices in government services, standard procedures, regulations and rules of conduct are imposed; these place limitations on the research administration's ability to establish an environment and practices adapted to the specific requirements of research. As a result, productivity is reduced, morale is adversely affected and it is difficult to retain competent scientists in the organisation. When externally imposed regulations continuously create frustrations for both the administration and the researchers, attempts are made to avoid compliance with the regulations, and a situation occurs that Gross calls 'code inversion' (Gross, 1964).

The forms of code inversion he mentions are: open deviation (which is relatively rare), disguised deviations, exploiting loopholes, over-compliance (*ad absurdum*). The most legitimate form of code inversion is, however, to initiate action aimed at having the code changed. Gross hypothesises 'a universal law of inevitable deviation of imperfect compliance', and concluded, therefore, that a certain amount of '*laissez-faire*' must be tolerated if the social structure of the organisation is to continue to operate without undue strain. He stresses, however, that the deviation from the code must be kept within tolerable limits.

Rewards and incentives

The criteria of success and promotion in management are different from those suited to a scientific career. In the former, the ladder of promotion is based mainly on the administrative duties of the researcher, and only partly on his scientific achievements. Each promotion actually curtails the amount of effort he can apply to his scientific activities. The higher he rises on the ladder of promotion, the less effort he can devote to the work for which he was trained and the greater the effort spent on work for which he was not trained. This subject is discussed in Chapter 10.

Pressures and demands from outside sources

The two principal types of pressure from outside sources with which the agricultural research worker is usually faced are to engage in 'brush fire' research, and to provide a rapid answer to the problems investigated. The need to cope with problems as they crop up, without regard to their intrinsic scientific interest, and the disruptions this causes to orderly, systematic work is a major source of annoyance and frustration to the average research worker.

The demand for immediate results is of course in direct conflict with the researcher's training to carry out research with painstaking attention to detail, not to draw conclusions before sufficient and conclusive data are available, to check and control both his results and his conclusions. Submission to pressure entails a lowering of professional standards; withstanding these pressures may cause alienation from colleagues in the extension service, conflicts with farmers' organisations, etc. The research worker needs latitude in the time and resources at his disposal and his work should not be evaluated prematurely.

Conflicts between intra-organisation 'sub-groups'

The human relations of supporting non-research personnel must also be taken into account; not only the researcher experiences strain in a research organisation. Different sets of rules and procedures for the different sub-groups will cause resentment and conflicts emanating from the non-research personnel.

Many scientists tend to view non-scientific activities as 'low-level'; helping in purely technical work, in the field or laboratory as demeaning; and rationalise this attitude, with its unconscious snobbishness, as opposition to 'uneconomical use of professional time'.

Of course, insufficient technical help, which obliges a trained research worker to spend a considerable portion of his time on chores which could be done by a field or laboratory assistant, is wasteful and will impede the research effort. However, the ability of the scientist to do technical work competently, to be prepared to lend a hand when needed, and to show respect for the work of the technician and appreciation of his contribution, will contribute greatly to reducing tension between the sub-groups.

A factor that contributes to inter-group tension is the low esteem in which scientists hold administrators and administrative activities. Cleveland (1960) writes: 'in faculty folklore—administration is the lowest form of sub-professional endeavour on the campus. The administrators create nothing, initiate nothing, they exist to serve the faculty, and since they keep forgetting that this is their primary role, they must be periodically reminded of it on public occasions by exquisitely worded shafts of faculty wit.' This attitude is too frequently carried over from the faculty to the research organisation.

POSSIBLE EFFECTS OF INTRA-ORGANISATIONAL CONFLICTS

On the research worker

Inability to provide a satisfactory solution to the problems outlined briefly in the preceding paragraphs will result in frustration for the research worker. The degree of frustration depends on the tolerance of the individual scientist who may become emotional, uneasy and antagonistic.

Individuals who cannot adapt to the conditions required by the research organisation can solve the problem by leaving. Others may transfer from research to organisational activities. Others may become apathetic or disinterested. Finally, some use defence mechanisms. One of the defence mechanisms of individuals who feel frustrated in the attainment of their goals is aggressiveness, the source of many of the tensions and difficulties of the research organisation.

Abrahamson (1964) has shown that: '(1) Integration (of the research worker) varies inversely with the amount of academic training. (2) Integration varies inversely with the desire to engage in basic research.' In other words, the better trained the scientist and the greater his scientific curiosity, the more difficult the transition from an 'academic' environment to an agricultural research organisation, in which he is to engage in problem-oriented research.

On the research organisation

The attachment of research workers to the concepts and working methods that characterise the academic environment has an important consequence for organisations concerned with directed research.

Thomson and Bates (1957), state that 'when members of an organisation owe loyalty or allegiance to a profession as well as to the organisation, there is greater opportunity for the demands of the organisation to conflict with those of the profession, and at the same time a greater opportunity for the individual employees to enforce demands on the organisation by invoking sanctions from the profession'.

Their status as specialists, who are frequently indispensible, often makes it possible for the research workers to resist efforts at administrative co-ordination. This situation leads to difficulties even with young scientists, when it is attempted to integrate them into the research organisation. As they become more mature, their status and personal authority increase and their resistance to organisational restraints becomes more effective. This leads to what Steele (1957) calls 'the schizophrenia of the research organisation'.

RESOLVING THE CONFLICTS

It is of course an essential task of the research organisation to find a working solution to the conflicts, stresses and strains outlined above. Attempts to force the research worker into an administrative straitjacket will result in costly failures. Avoiding the issue by ignoring the problems is the most effective way of undermining the research organisation and impairing its effectiveness.

However, the scientist needs the research organisation, as a framework for effective research work and, let it be said, as a source of livelihood. On the other hand, the organisation cannot carry out research without scientists. Hence there has to be mutual accommodation. Research personnel are sufficiently 'different' to make managing them a problem—but they are still enough like other personnel to make it possible to apply certain general managerial techniques, and to adapt others to the differences that indubitably exist. Hence it is necessary to adopt organisational structures and administrative techniques that are appropriate to research.

Experience shows that even when management has an understanding for the special needs of the research worker and strives to be co-ordinative rather than commanding, and even where attempts are made to create an academic atmosphere in as far is compatible with the goals of the research organisation, conflicts, resulting from basic incompatibilities between the desires of the research workers and the requirements of the framework within which they work, do not necessarily disappear entirely.

It is probably unrealistic to expect that the researcher will always feel that his own research interests are identical with those of the organisation for which he works. It is quite possible that one must be satisfied with a situation in which the scientist works willingly towards the objectively defined goals of the organisation.

A certain amount of conflict and dissatisfaction can be tolerated, and may even have a positive effect by stimulating new ideas. The very fact that there is a certain amount of conflict implies an absence of passivity and of conformism that normally characterise a bureaucracy. Hinrichs

(1963) is of the opinion that dissatisfaction with the status quo may even be positively related to research performance.

Establishing a research climate

Much can be done to develop a feeling of 'belonging' and loyalty to the research organisation and thereby minimising turnover, by creating an appropriate research climate and satisfying the legitimate needs of the research workers and eliminating factors which have an adverse effect on productivity.

'Experience has shown that high productivity in a research organisation is promoted and ensured by a favourable research climate and by the norms and values of the research workers' (IESA, 1980).

The academic environment of the university is characterised by an absence of pressure on the researcher as regards choice of research subject, methods chosen to pursue the research and the time needed to achieve an objective. This makes possible a minimum of administrative interventions and procedural constraints. It is therefore not surprising that the academic environment usually serves as the ideal prototype for creating a desirable research climate.

The structure of an agricultural research institution has to be much more elaborate and precise than that of an academic institution. In order to ensure that a predetermined research programme will be efficiently and successfully executed, co-ordination and control are essential. The academic institution, in which freedom of research is an accepted dogma, is therefore not an appropriate model for the agricultural research organisation; but neither is the usual bureaucratic organisation in which controlled hours of work, heavy emphasis on regulations, excessive use of paper forms, reliance on precedents, hierarchical control of expenditure, and many other bureaucratic practices are considered essential for good organisation. Obviously, a balance must be struck between the objectives of the organisation and the aspirations of the scientists who are to carry out the needed research.

The agricultural research organisation must therefore create a climate specifically its own, in which research can flourish and into which as many as possible of the desirable academic procedures, such as evaluation of achievement by peers, 'senior–junior' instead of 'boss–subordinate' relationships, etc., are integrated.

RESEARCH LEADERSHIP

The three basic functions of a leader are to give cohesion and a sense of common purpose to his group; to ensure that the group achieves its aims; and to ensure maximum efficiency of the group on achieving its goal by promoting co-operation, establishing interaction between members of the group and promoting effective communications.

Patterns of leadership

An agricultural research organisation, engaged as it is in problem-oriented research, will tend to appoint research supervisors who are oriented towards achieving the goals of the organisation. The normal tendency of the latter will be to exercise 'executive authority' and to control the work of the researchers. These, on the other hand, will generally resist and resent this type of supervision.

It is interesting to note that in the continental tradition, 'academic freedom' was the prerogative of the professor—the 'Herr Geheimat' i.e. the scientist who had already achieved status. In this case the relationship between the professors and the young research assistant was one of master and disciple, with the former expecting unquestioning obedience from the latter.

Neither the master–apprentice relationship based on scientific superiority, nor the boss–subordinate relationship based on hierarchy, are suitable models for the agricultural research organisation. On the other hand, the view that high research achievement is possible only when research workers are left entirely to their own devices, is also untenable. Too much independence may deprive the young researcher of the stimulation that a competent chief can provide, whilst too close dependence may stifle individual initiative.

In a study carried out by the University of Michigan, which investigated the relationship between supervision and research performance, it was found that frequent stimulation and encouragement from the supervisor contributed to high research performance. At the same time it was concluded that subordinates should be given considerable freedom to make their own decisions and exercise their own initiative. Freedom for the researcher to carry out his work as he sees fit is generally regarded as a primary requirement that should not be curtailed.

Pelz (1956) distinguishes among three patterns of leader–subordinate relationships: a 'directive', autocratic type; a 'participatory', democratic type; and a 'laissez-faire' pattern. In the 'directive' or 'authoritarian' leadership style, the leader dictates policy, techniques and work association, gives detailed instructions and maintains close supervision. All the members of his group are assistants who are doing work for the boss. It is usually leaders with a feeling of insecurity who tend to be authoritarian and to restrict the autonomy of their subordinates as far as possible. This approach is possible only with submissive, ineffectual individuals lacking in initiative and drive. More able research workers react to this type of leadership with hostility and frustration and will usually leave the organisation. Autocratic leadership therefore has no justification in a research organisation that aspires to maintain a high scientific level.

In the participating, democratic type, initiative is encouraged, whilst help and advice are given freely. Glaser (1965) sees in the 'participatory' type, an integrated work relationship between research supervisor and his group, the typical 'colleague authority' based on joint consultation and decision, especially regarding work assignments, problems and implementation. This relationship does not threaten the subordinate's autonomy. He feels that he gets help and advice from his leader whilst retaining a measure of influence on the research activities of his group and sufficient freedom for original thought. This approach emphasizes a relationship of working together without the inequality in status constituting a disturbing factor. The reciprocity in this relationship, the mutual helpfulness and the maintenance of individual autonomy ensure the stability of this relationship. Glaser stresses that supervisors highly competent in research are necessary for this relationship. It is not always possible to find research supervisors with the necessary high scientific competence needed for a satisfactory participatory leadership pattern. Nor is it easy to maintain a balance between taking an active interest in the work of a subordinate, without undue intervention in its implementation. The University of Michigan research mentioned above indicated that in this case it was best for the supervisor to adopt a 'laissez-faire' pattern, in which he practically abdicates his formal authority. Notwithstanding the high value that researchers attach to personal autonomy, they do not really favour 'laissez-faire' leadership, in which the leader devotes most of his time and energies to his own scientific work with little or no contribution to the professional formation and scientific work of his group.

Of the three patterns of leadership in research organisations, 'directive' or autocratic leadership has always been found to be the least effective. But a 'soft' leader also has a bad effect on the morale of his group and of its productivity (Kahr & Katz, 1953). Most leaders cannot be unequivocally classified within one of the three patterns described, but are usually a composite, in different degrees, of all three patterns, with one or another dominating.

It is even desirable for the individual leader to use the different leadership patterns according to circumstances and to adjust them to the individual researchers who work with and under him.

Choice of leaders

Bavelas (1960) makes a distinction between 'leadership as a personal quality', which depends on a combination of personal characteristics (commonly called 'charisma') and 'leadership as an organisational function' which depends on how decision-making authority is distributed through an organisation. Hence, 'the degree to which an individual exhibits leadership, depends not only on his characteristics but also on the characteristics of the situation in which he finds himself'. Consequently, a man who is a successful leader in an authoritarian organisation may not be able to assume the same role in an organisation based on the democratic-participative approach.

Whilst a great number of organisations have sufficient common characteristics to make it possible to generalise in regards to the type of person needed to fill the various leadership roles required by these organisations—the research organisation in general, and the agricultural research organisation in particular, have specific characteristics and problems which in turn determine the personal traits needed by its leaders.

Leadership functions

Whilst the participatory type of leadership is, as we have seen, the most appropriate for a research organisation, there remain certain functions which are the leader's own responsibility. It is his responsibility to ensure that the work carried out by his group is co-ordinated with that of the organisation as a whole, that it is of the highest possible standard and that its contribution to the goals of the organisation is substantial. At times he must make decisions himself without the normal group consultation and still avoid arousing resentment by appearing to have acted in an arbitrary fashion. His is the main responsibility for creating and maintaining in his group an atmosphere conducive to joint effort and a feeling of belonging. He must give scientific guidance without stifling initiative or discouraging original thinking. His criticisms must be constructive; and he must be prepared to assume responsibility for the mistakes and shortcomings of the group. He represents the views and requirements of the group within the organisation, and when appropriate, outside the organisation.

CREATIVITY AND MOTIVATION

Research is primarily creative work. 'Whenever a new idea is developed or an old idea applied to a new situation, then the work is essentially creative and requires new thought' (Dunk, 1964).

Creativity is a process of hypothesis formation, its testing and the communication of the results.

Some people are mainly creative in formulating new ideas, others—who rarely have original ideas—are adept at developing the means to test new ideas and the creativity of others expresses itself in the successful communication of these ideas. Of course, there are individuals who are creative in more than one of these aspects.

Kaplan (1963) points out that most research projects involve a small percentage of creativity and a large percentage of productivity.

The creative researcher is keenly interested in the area of research to which he has decided to devote himself, and will go to great lengths to find problems to work on in his field of interest.

Unlike many kinds of work, research performance cannot be enforced; a number of systematic quantitative studies have shown that there is a positive relation between motivation and the quality of work of scientists (Gollembiewski, 1964).

According to Hertz (1957), the true creative mentality is: (1) receptive to new information; (2) flexible; (3) intellectually curious; (4) able to formulate problems; and (5) able to put information together in many ways as a means of getting to the acceptable solution of a problem.

Conditions favourable to creativity and motivation

Pelz (1956) lists the following conditions as favourable to the creation and maintenance of high motivation in research:

1. Freedom to pursue original ideas and to make contributions to basic scientific knowledge.

2. Frequent contact with scientific colleagues, both those who are working in different fields and those who have similar professional backgrounds.
3. A chief who gives neither complete autonomy nor excessive direction, who frequently interacts with subordinates and gives them the opportunity to make their own decisions.

To these conditions, Glaser (1965) adds differential reward systems, and adequate research facilities and conditions.

Characteristics of the creative researcher

The creative scientist differs from less creative colleagues in a number of characteristics. A single person rarely possesses all the required attributes and there is considerable variation in the degree of creativeness between individuals. However, creativity is not easily measured by psychological tests. The following traits are characteristic of creative researchers:

Originality
This involves active curiosity beyond the confines of his own special field, an ability to toy with ideas, a considerable background of fundamental knowledge, and the ability to discern the fundamentals of a problem (Raudsepp, 1963).

Ability to translate original ideas into practice
This depends on flexibility in thinking and preparedness to abandon preconceived ideas and to attempt new approaches to the solution of problems. Also needed are the patience for tedious and painstaking work, as well as persistence in the face of setbacks. In agricultural research, a considerable fund of practical knowledge and experience in farming are also great assets.

Emotional make-up
Some individuals are, by their personality characteristics, internally motivated for research work. The motivations of the creative worker in agricultural research are not in essence different from those of the 'academic' research worker, but as Cairns (1956) rightly points out, they must be suitably modified to be compatible with his obligations in an

agricultural research organisation—

As a man, he seeks to be productive in terms that he and his associates consider to be important.

As a scientist, the productive urge reveals itself in some form of intellectual curiosity directed toward the study of natural phenomena and the laws that govern them.

In his *role as employee* his primary drives must be effectively related to the particular objectives of agricultural research.

The research worker is sensitive to problems and finds no difficulty in discussing new problem areas with colleagues and the possibilities of arriving at a solution. He is relatively uninterested in small details, is more concerned with the implications and meanings of phenomena rather than the facts themselves.

Whilst he is interested in communicating with others, he is able to maintain what Raudsepp calls 'psychological distance' from his associates and his working environment. He usually prefers to work alone, and is a reluctant team-worker.

Creative researchers are not always likeable individuals. Their impatience of restraints and their non-conformity not only create problems for their superiors, but may also cause disruptions in the work of their colleagues.

Conditions for productive work as perceived by research workers

Of the *unfavourable conditions,* five relate to organisational aspects: instability of the budget; conflict between the demands of management activities and research activities; inadequacies in supportive groups and procedures; inefficiency in selection and placement programmes; and inadequacies of physical facilities. The remainder are factors of motivation and rewards: inadequacies in compensation; lack of professional internship for young scientists; poor communications with scientists working elsewhere on related problems; dearth of suitable recognition; exclusion of scientists from high-level decision-making on scientific problems; insufficient long-range planning in the scientific programme; interference between contract monitoring and regular research activities; and lack of personal counselling for scientists.

The research work itself is the source of the main satisfaction of the researcher, and hence the most important aspect of a favourable research climate is challenging and stimulating work and the encouragement of original, independent, objective and disciplined thinking. From the foregoing it is clear that much can and should be done to meet the legitimate requirements of the research worker.

CONSULTATIVE MANAGEMENT OF THE RESEARCH ORGANISATION

The complexity of the research organisation, the obligation to carry out its commitments whilst at the same time satisfying the legitimate needs of the scientific personnel, the conflicting interests of various groups within the organisation—all these create a complex and difficult situation which requires firm and unequivocal direction to achieve the co-ordination essential to attaining the goals of the organisation.

However, the old-fashioned concept of authority, in the sense of 'possession of an arbitrary, absolute and final prerogative, which assures that the commands of the one in authority are carried out', is certainly not acceptable in a research organisation. People are more and more concerned with the processes by which decisions in which they are involved are taken. Standard forms of democracy—voting, representation systems, etc.—are applicable *only* to organisations in which ultimate authority is vested in the hands of all the members (associations, collectives, co-operatives). This kind of democracy cannot apply to research organisations. With increasing specialisation, horizontal and criss-crossing relations develop which Gross (1964) defines as 'polyarchic'. They add capacity and flexibility to the structure of formal hierarchy.

The right and duty to make decisions is the essential prerogative of the research director. This does not contradict the need for consultative management. Decision-making at various levels should be preceded by consultation with those whom the decisions are bound to affect and who will be closely involved in giving them effect. The degree of 'members' participation in decisions concerning their activities and the activity of the organisation as a whole is a measure of the democracy practised in the organisation' (Gross, 1964).

The advantages of shared deliberation are manifold:

1. A close relationship between the research director and senior research workers is established. The latter have the opportunity to express their opinions and to influence organisation policy, before decisions are taken that involve them.

2. The participants have a feeling of common purpose, shared interest, a sense of involvement in administrative decisions, shared responsibility and, hence, a greater readiness to accept change. Interdepartmental co-ordination is facilitated.
3. The oral interchange of ideas provides stimulation and awareness of the problems with which the organisation is faced.
4. It is an excellent method of communication, with opportunities for emphasis and clarification where required.
5. Collective judgement may be more effective than individual judgement, if decisions have to be made on inadequate information.
6. It serves as a 'check and balance' system, and thereby helps to prevent arbitrary decision-making by individuals.

Against these advantages of involving the researchers in decision-making, there are a number of liabilities (Gross, 1964):

1. Researchers usually do not have sufficient time to study the problems involved and to participate actively in the discussions which necessarily precede decision-making. Frequently they are unwilling to take time off from their research for this purpose.
2. Rarely do researchers have a sufficiently broad understanding of the problems involved, when these transcend their own field of activities.
3. Researchers tend to resist proposals that might encroach on existing prerogatives, without reference to their desirability for the organisation.
4. Researchers are usually more concerned with their individual subject fields than with institutional requirements or policies.

The research director who wishes to encourage participative management should realise that in addition to the advantages listed above, there are difficulties and dangers inherent in this approach. It would be highly unrealistic on his part to imagine that replacing authoritarian direction by group deliberation will magically solve all his dilemmas. On the contrary, he will sooner or later find that the introduction of democratic leadership has opened a veritable Pandora's box of problems. He will become increasingly aware of strains and stresses of which he was previously ignorant.

It is, however, not the democratic leadership process that has given rise to the tensions which become apparent after its adoption. Dissatisfaction that was previously underground has simply been brought to the surface. The establishment of a formal framework that makes possible the discussion of the conflicts latent in the organisation, the dissatisfaction with the way it is managed and other problems of human relationships, contribute to the solution of these problems in a manner befitting a research organisation.

Some of the other difficulties and dangers of participative management will be discussed further. It should, however, be clear already that the two extremes of an authoritarian approach and excessive participative management should both be avoided, the former because it is absolutely inappropriate to a research organisation, the latter because too much participation may lead to organisation breakdown.

Committees

When a problem is of common concern to a number of departments in the research organisation, one of the most useful devices for implementing consultative management is the *committee*, which can take several forms: boards, commissions, task groups, *ad hoc* committees, etc.

Committees as such can undertake managerial functions, make decisions, or simply deliberate and make recommendations. Each type has its place in a research organisation, at various levels of the organisation's structure. A committee may be limited to one hierarchical level (the management committee), or consist of members from different 'levels'. They afford the director the opportunity for formal direct contact with several levels without creating problems of having to 'go through channels'. Committees with permanent functions should therefore be an essential part of the formal organisation.

Committees have many shortcomings, weaknesses and even dangers which will be outlined below, but in many cases they present the sole practical action that can be taken, and the only possible tool for consultative management.

Advisory status
Whilst formally the merely advisory status of the committees must be maintained, leaving the final decision-making as the prerogative of the director, in practice he must exercise considerable restraint and mature judgement before making a decision that runs counter to the consensus of opinion of the committee.

Frequent disregard of the recommendations will lead to disillusionment and to a feeling of futility on the part of the members of the committee. Their reaction will be either a demand to change their advisory status to executive status, or else they may refuse to continue to serve on these bodies.

The director will be well advised, generally, to accept and carry out the recommendations of the committees, even if they run counter to his own personal opinion, reserving the exercise of his prerogative to overrule the committee for those cases in which he feels that the decisions:

(a) are in conflict with the policy of the research organisation and its commitments to the parent body;
(b) are not in accord with those rules and regulations of the civil service which of necessity apply to the organisation;
(c) cannot be accomplished under the existing conditions;
(d) establish conditions which destroy the power of effective leadership; or
(e) are the result of a divided opinion, so that the decision is due to a majority that is not sufficiently representative of the organisation as a whole.

Whatever course of action is finally decided by the director, whether to accept the consensus of opinion or to overrule the committee, it is *his responsibility* and he will remain responsible for any consequences.

In conclusion, it can be stated that the director should exercise his authority, in as far as possible, by attempting a synthesis of group judgement and opinion on the one hand, and the needs and goals of the organisation on the other.

Disadvantages and dangers
The research workers have received their training and indoctrination in universities which are traditionally distrustful of the power of the manager, which they circumscribe by establishing numerous committees. 'Thus in one large university, there are more than three hundred standing committees to share in its administration or advise on policy. Such committees range from the importance of the academic senate and the budget committees to committees on committees, co-ordinating committees, and committees on alumni records, university welfare, and maintenance of order in examinations' (Koontz & O'Donnell, 1955). Whilst this type of set-up may be fully justified in an organisation jealous of its academic freedom, it would be fatal for a research organisation with well-defined obligations in the field of oriented research.

The tendency of the research worker to remember his alma mater as a kind of 'paradise lost' will naturally influence him towards attempting to establish a similar state of affairs in the research organisation. The research director will be under continuous pressure to accord decision-making status to his various committees. Paradoxically, the director who has accepted the principle of consultative management will experience these pressures to a far greater degree than the authoritarian director. There is considerable danger that these exertions to achieve decision-making status will override the need to make constructive contributions to management. It is, unfortunately, common experience that 'factions may exhaust themselves in the struggle to achieve power, leaving little energy for providing benefits to themselves or to others!' (Koontz & O'Donnell, 1955).

Weakness of committees
Many disparaging definitions of committees have been made. Half-humorous sayings such as: 'A committee is made up of the unfit, selected by the unwilling to do the unnecessary' (Koontz & O'Donnell, 1955) or 'A giraffe is a horse designed by a committee', are indicative of the measure of contempt and impatience with which this organisational tool is frequently regarded.

The following are some of the inherent weaknesses of committees:

(a) They are very time-consuming and hence expensive. They may cause delay and even stifle action and initiative.

(b) A committee, in order to be representative, must include members from the different fields of activities of the research organisation. This is especially true for the important committees. In many cases the interests of individual members of the organisation run counter to those of the organisation as a whole. It is, however, a doctrine in public administration that a committee can make collective decisions only in matters that do not contravene the vital interests of any member (Cleveland, 1960).

(c) In cases in which differences of opinion do occur, an attempt will usually be made to find a compromise which is usually weak, and sometimes meaningless. The director, with whom the responsibility for the decision rests in any case, and who will also assume the burden of the consequences, can and should unequivocally act according to what is in the best interests of the organisation as a whole.

(d) It has already been pointed out that in cases in which the general consensus of opinion is in favour of a certain decision, this may conflict with the goal and obligations of the organisation.

(e) Committees may weaken authority and responsibility in the organisation. This may well be one of the main dangers of a decision-making committee. As Koontz and O'Donnell (1955) point out, 'one of the principal social justifications of authority is that it enforces responsibility on its holder to use power effectively and efficiently to accomplish the duties for which it was delegated. Only an individual, however, can feel such an obligation. When authority is dispersed, responsibility is likewise dispersed.'

The committee can also be used as a device to *avoid* making personal decisions for which one must assume entire responsibility. 'Committeeitis' is defined by Gross (1964) as 'the automatic panacea for dodging complex problems'.

(f) An organisation with authoritarian tendencies may use consultative devices as a means to increase control rather than share it, paying only lip-service to the ideas underlying consultative management.

Functioning of committees

With all their weaknesses—some avoidable, some built in—committees can be useful and even indispensable in the research organisation. For committees to be effective:

(a) *The function, authority and responsibility of the committees should be clearly established.* The members of the committee should know whether they are responsible for making decisions, proposing recommendations or simply exploring ideas. Asking advice from a group acting as a committee is all too often interpreted as a delegation of power, and the director may find himself answered with a decision instead of a council. However, the responsibility of the decision rests with him, and not with the committee. If the director allows committees to provide advice in areas in which he has not specifically called for it, he will soon be finding himself subjected to increasing pressure.

(b) *The choice of the members should be appropriate*, being sufficiently representative without making the group too large. Too large a committee will usually be wasteful of time and indecisive; if too small, it will not be sufficiently representative nor encompass sufficient fields of expertise necessary for deliberation. A useful device for overcoming the incompatibility of representatives on the one hand, and the need for restricting the size of the committee on the other, are 'sub-committees'.

(c) *The choice of the subject matter should be appropriate.* Committees should not be used for relatively unimportant decisions or for decisions that can easily be made by the individual, nor for making decisions that are beyond the authority or competance of the individual members.

(d) *They should be well-run.* For this, an effective chairman and good staff work are of great importance. The agenda should be well prepared and all the necessary data should be made available to members well in advance of meetings. Deliberation can be productive only when it is based on well-presented facts and estimates.

(e) *Effective follow-up* of decisions is essential if the morale of the committee is to be maintained. The committee should always be kept aware of the action taken as a result of its deliberation and decisions and the consequences thereof.

Effects of participation in management

We have considered the importance of a consultative form of management from the point of view of improvement of the research climate. Not all scientists are willing or temperamentally suited to participating in administrative or executive functions. They are frequently more concerned with protecting their autonomy than with becoming involved in the organisation.

As a rule, administration and administrators are held in low esteem by most researchers. However, any research function that involves several workers—whether junior researchers, technicians or other forms of the labour staff—requires a certain amount of administrative effort, and the larger the field of interest and the greater the research unit, the more administrative responsibilities devolve on the research leader.

If one accepts Kidd's (1953) definition of administration of research 'as research planning on a broad scale, the development of scientific strategy, the evolution of a consistent philosophy of research, and the difficult task of bringing a sound philosophy to bear upon the conduct of research', surely there is nothing demeaning to the senior scientist if he devotes part of his time to the administration of research units, and even most of his time if the scope of work so requires. In regard to those who are willing to devote time and energy to activities not directly related to their research, the question naturally arises whether this is not achieved at the expense of their research effort. One might, *a priori*, expect that the scientific contributions of researchers working full-time on their research assignments would surpass the contributions of those who spend part of their time on administrative activities.

In a survey of 552 scientists and engineers, in eleven research units—industrial, governmental and university—it was found that only 20 per cent of the PhDs worked full time at research. The rest of their time was allotted to teaching and administration (Andrews, 1964). Scientific performance (as measured by scientific or technical contribution and by general usefulness to the organisation) for PhDs and assistant scientists was *highest for those who spent three-quarters of their time on scientific work!* The higher performance of these 'part-time' scientists was not due to differences in length of experience, status or total time spent on work.

A further surprising finding of this research was that among the PhDs in research the performance of those who spent their 'non-scientific' work time primarily in teaching was lower than that of those engaged in administration.

Andrews (1964) proposes various possible explanations for these findings:

(a) administration has more relevance than teaching for the scientific activities of those concerned;
(b) the more able scientists elected administration;
(c) administration tasks are pushed upon high-performing scientists because they are expected to be more effective administrators;
(d) high-performing scientists accepted non-research tasks as a means of promoting their research work; and
(e) scientists confronted with diversity were exposed to more new ideas, problems or people useful to their work.

It should be pointed out that these explanations are not mutually exclusive, that any one or more in combination may have been effective, depending upon the individuals concerned.

In conclusion, it can therefore be stated that not only is participation by scientists in the management of the research organisation important for improving its research climate, but that a moderate exposure to administrative tasks may actually prove of benefit to the individual scientist. When the researcher is concerned exclusively with conducting his own research he is not very likely to develop a general awareness of problems, or realise his own involvement in the work of others. His isolation will simply reinforce a narrow professional orientation.

The informal organisation

Co-existing with the formal structure are numerous processes of interaction among individuals in the organisation resulting from the need of people to associate with others. These interactions can be analysed according to various theories of group behaviour. As a result of the activities of the informal organisation, the *actual operating* organisation may differ in many respects from the formal organisation.

Kingsbury (1959) points out that research organisations that have essentially the same kind of formal organisation show wide variations in success. He relates these differences to *informal* processes which involve various aspects of the research organisation: personal power, informal communication channels, group performance standards, systems of research, degree of freedom to initiate work and many other factors. A realistic management learns to recognise and live with functional 'informal' organisation which can then be used to implement and strengthen the formal organisation.

The formal organisational structure must define boundaries and divide the organisation into manageable units with specialised responsibilities. When a project has to move from one formal area of responsibility to another, problems and difficulties usually arise. These inevitable gaps between formal units can be bridged by informal relationships. The development of new ideas and the translation of research results into agricultural practice usually involve a major effort and the co-operation of many departments in research as well as extension workers and farmers. Success is therefore frequently dependent on the enthusiasm and emotional involvement of a single individual overcoming the shortcomings of the formal organisation by working through informal patterns of leadership and communication.

Literal application of all the formal rules, regulations and procedures could easily bring work to a standstill and disrupt the organisation; informal relations serve as a safety-valve and help to preserve the organisation from self-destruction (Dubin, 1961).

Informal organisation can also be potentially dangerous. Cliques which exercise undesirable influences, may form in the organisation. These can be of one of three different types (Dalton, 1959):

• *vertical*: involving people at different levels of the hierarchy, providing reciprocal benefits of 'protection' by the higher placed, and 'support' by the lower level;

- *horizontal*: between people of the same status, usually in revolt against a superior;
- *random*: between people attracted to each other on a personal or social basis.

The research workers, for instance, who belong to what has been called a sub-culture with its own norms of conduct, may set standards which may be at odds with the values of the formal organisation to which they belong. Such a group can develop considerable power, which may be directed towards goals that are in conflict with those of the organisation. The modifying informal processes can be considered as an 'overlay' pattern superimposed on the basic formal organisational pattern.

Pfiffner and Sherwood (1960) mention five such overlays:

(a) The *sociometric overlay*, which relates to the purely social relationships among the people of the organisation, both of attraction and rejection.

(b) The *functional overlay*: The influence of specialists or natural leaders upon operations without direct responsibility for the work itself.

(c) The *decision overlay*: In particular in the research organisation, decision-taking does not necessarily follow the structure of formal hierarchy, but is influenced by what the above-mentioned authors call a network of influence.

(d) The *power overlay*: Power and authority are not necessarily synonymous. In the organisation, personal power centres function which do not necessarily coincide with the official structure of authority.

(e) The *communication overlay*: Usually the 'grapevine' which functions in addition to official 'channels' of communication and is frequently more efficient and rapid than the official channels.

References to Part Three

Abrahamson, M. (1964). The integration of industrial scientists. *Adm. Sci. Q.,* **9**, 208–18.

Andrews, F.M. (1964). Scientific performance as related to time spent on technical work, teaching or administration. *Adm. Sci. Q.,* **9**, 182–93.

Aranjo, J.E.C. (1979). Preparing professional staff for national agricultural research and related programmes. In *Preparing Professional Staff for Agricultural Research Programmes.* IADS, New York, pp. 39–41.

Ardila, J., Trigo, E., & Piñeiro, M. (1981). Human resources in agricultural research: three cases in Latin America. In *Resource Allocation to Research,* ed. D. Daniels and B. Nestel. International Development Centre, Ottawa, p. 151.

Barnard, C. (1938). Dilemmas of leadership in the democratic process. In *Human Relations in Administration,* ed. R. Dubin. Prentice-Hall, Englewood Cliffs, New Jersey, pp. 353–57.

Baumgardner, H.L., Ellis, W., Lynton, R.P., Jung, C.W. & Rigney, J.A. (1971). *A Guide to Institution Building.* Office of International Programmes, North Carolina State University.

Bavelas, A. (1960). Man and function. *Adm. Sci. Q.,* **4**, 491–98.

Bennell, P. (1986). Human resources planning and management for agricultural research. In *Improving Agricultural Research Organisation and Management: Implications for the Future.* ISNAR, The Hague, pp. 82–88.

Burdekin, D.A. (1970). Man management of research workers. In *The Organisation and Methods of Agricultural Research,* ed. J.B.D. Robinson, Ministry of Overseas Development, London, pp. 143–47.

Burns, T. & Stalker, G.M. (1961). *The Management of Innovation,* Tavistock Publications, London.

Busch, L. & Lacy, W.B. (1983). *Science, Agriculture and the Politics of Research.* Westview Press, Boulder, Co.

Byrnes, F.C. (1974). Agricultural production training in developing countries. In *Strategies for Agricultural Education in Developing Countries.* The Rockefeller Foundation, New York, pp. 215–32.

Cairns, R.W. (1956). Motivation of the more creative scientists and engineers in industrial research. In *Research is People*. Industrial Research Institute, New York, pp. 22–26.

Casas, J. (1977). *Réflexions sur la Recherche Agronomique dans les Pays Sous Développés*. Série Notes et Documents. Ecole Nat. Sup. Agronomique, Montpellier.

Castronovo, A.C. (1984). The organisation and management of human resources. In *Selected Issues in Agricultural Research in Latin America*, ed. B. Nestel and E.J. Trigo. ISNAR, The Hague, pp. 11–48.

Cleveland, H. (1960). The Dean's dilemma: leadership of equals. *Publ. Adm. Rev*, **20**, 22–27.

Contreras, M.R., Galt, D.L., Muchena, S.C., Nor, K.M., Pears, F.B., Rodriguez, I.M.S. (1977). *An Interdisciplinary Approach to International Agricultural Training*. The Cornell–CIMMYT Graduate Student Report. Cornell University, Ithaca, New York.

Coombs, P. (1970). *What is Educational Planning?* UNESCO, Paris.

Cunha, T.J. (1967). Positive suggestions for the future. In *Rural Development in Tropical Latin America*, ed. K.L. Turk and L.V. Crowder. Cornell University, Ithaca, New York, pp. 413–29.

Dalton, M. (1959). *Men who Manage: Fusion of Feeling and Theory in Administration*. John Wiley, New York.

Daniels, D. & Nestel, B. (eds) (1981). *Resource Allocation to Agricultural Research*. Proceedings of Workshop, Singapore. International Development, Center, Ottawa.

Drilon, J.D. Jr (1975). *High-level Agricultural Manpower in Indonesia and the Philippines*. SEARCA.

Dubin, R. (ed.) (1961). *Human Relations in Administration*. Prentice Hall, Englewood Cliffs, New Jersey.

Dunk, W.P. (1964). The work output of scientists. *J. Austr. Inst. Agric. Sci*, **30**, 181–88.

Eicher, C.K. (1982). Facing up to Africa's food crisis. *Foreign Affairs* **61**(1), 151–74.

Evenson, R.E. (1978). Agricultural research and extension in Asia. In *Rural Asia: Challenge and Opportunity*, Vol. II. Asian Development Bank, Bangkok, pp. 1–48.

Fernandes, F. (1977). Objectives and content of training at the international centres of agricultural research. Paper presented at CGIAR Forum on Training, International Centres, Washington, DC.

Gamble, D.K., Bourke, R.M. & Brookson, C.W. (1981). *South Pacific Agricultural Research Study*. Consultants Report to the Asian Development Bank.

Garcés, C.D. (1974). The experience of the ICA in agricultural development through education and research. In *Strategies for Agricultural Education in Developing Countries*. The Rockefeller Foundations, New York, pp. 141–70.

Gerente General (1978). *Informe para Junta Administradora*, FONAIAP, Caracas.

Glaser, B.G. (1965). Differential association and the institutional motivation of scientists. *Adm. Sci. Q.*, **10**, 82–97.

Gollembiewski, R.T. (1964). Authority as a problem in overlays. *Adm. Sci. Mgmnt,* **9,** 23–49.

Gray, C.C. (1974). Agricultural curricula and instruction for national development: an integrated approach. In *Strategies for Agricultural Education in Developing Countries.* The Rockefeller Foundation, New York, pp. 179–92.

Gray, C.G. III (1978). Suggestions for Land Grant College action—a personal viewpoint. In *Transforming Knowledge into Food in a Worldwide Context,* ed. W.F. Hueg and C.A. Cannon, American Academy of Arts—University of Minnesota, Minneapolis, pp. 72–79.

Gross, B.M. (1964). *The Managing of Organisations: The Administrative Struggle.* Free Press, Glencoe, Ill.

Harbison, F. & Myers, C.A. (1964). *Education, Manpower, and Economic Growth—Strategies of Human Resource Development.* McGraw-Hill, New York, Toronto, London.

Heiman, F.P. (1965). Evaluation of research from the viewpoint of the practising scientist. *Res. Mgmt,* **8,** 139–44.

Hertz, D.B. (1957). The creative mentality in industrial research. *Human Relations in Industrial Research Management,* ed. R.T. Livingston and S.M. Milberg. Columbia University Press, New York, pp. 45–59.

Hinrichs, J.R. (1963). Creativity in industrial research. In *The Management of Scientific Talent,* ed. J.W. Blood. American Management Association, New York, pp. 155–78.

Horton, D.E. (1984). *Social Scientists in Agricultural Research. Lessons from the Montaro Valley Project, Peru.* International Development Research Centre, Ottawa.

IADS (1979). *Preparing Professional Staff for National Agricultural Research Programmes.* Report of Workshop, Bellagio, Italy. IADS, New York.

IADS (1983). Implementing a staff development programme in a national agricultural research system. In *IADS Annual Report, 1982,* pp. 21–29.

IESA (1980). *FONAIAP—Estudio de Organización.* Instituto de Estudios Superiores de Administración. Caracas.

ISNAR (1981). *Kenya's National Agricultural Research System: Report to the Government of Kenya.* The Hague.

ISNAR (1981a). *The Agency for Agricultural Research and Development of Indonesia.* The Hague.

ISNAR (1982). *Rapport de la Mission a la Côte d'Ivoire.* The Hague.

ISNAR/NCST (1982). *A Manpower and Training Plan for the Agricultural Research System in Kenya 1983–1987.* The Hague.

ISNAR/SPAAR (1987). *Guidelines for Strengthening Agricultural Research Systems.* World Bank, Washington, DC.

Kahr, L.R. & Katz, D. (1953). Leadership practices in relation to productivity and morale. In *Group Dynamics Research and Theory,* ed. D. Cartwright and A. Zander, Row-Peterson, Evenston, Ill.

Kaplan, N. (1963). The relation of creativity to sociological variables in research organisations. In *Scientific Creativity,* ed. C.W. Taylor and F. Barron. John Wiley & Sons, Chichester, pp. 195–204.

Kidd, C.V. (1953). Research planning and research policy—scientists and administrators. *Science,* **118,** 147–52.

Kingsbury, S. (1959). Organising for research. In *Handbook of Industrial Research Management,* ed. C. Heyel. Reinhold Publishing Corp, New York, pp. 65–91.

Koontz, H. & O'Donnell, C. (1955). *The Principles and Practice of Management.* McGraw-Hill Book Co, New York.

Kornhauser, W. (1963). *Scientists in Industry—Conflict and Accommodation.* University of California Press, Berkeley.

Leavitt, H.T. & Pondy, L.R. (1964). *Readings in Managerial Psychology.* University of Chicago Press, Chicago.

Marzocca, A. (1983). Sistemas de adriestramiento de personal. In *Convenio IICA—Cono Sur/BID,* pp. 73–88.

Merton, R.K. (1949). *Social Theory and Social Structure.* Free Press, Glencoe, Ill.

Mook, B.T. (1986). Information systems for human resources. In *Improving Agricultural Research Organisation and Management: Implications for the Future.* ISNAR, The Hague, pp. 80–81.

Moseman, A.H. (1965). Education of the agricultural specialist. In *Agriculture and the University,* Institute of International Education, New York, pp. 134–39.

Myers, W.M. (1965). Premises and principles of general studies. In *Agriculture and the University.* Institute of International Education, New York, pp. 117–22.

ODI (1988). Notes on discussion papers. Newsletter 18, 6–28. Agricultural Administration Unit, ODI, London.

OECD (1961). *The Measurement of Scientific and Technical Activities: Proposed Standard Practice for Survey of R & D.* Paris.

Oram, P.A. (1977). Training requirements in research and its application. Paper presented at CGIAR Forum on Training, Washington, DC.

Oyer, E.B. (1982). Implementing a manpower development programme in a national agricultural research system. Paper prepared for IADS Annual Report for 1982.

Peabody, N.C. & Tully, R.C. (1977). *Teaching Agricultural and Rural Development Administration.* Workshop Report. Programme in International Agriculture, New York State College of Agriculture and Life Sciences, Cornell University, Ithaca, New York.

Pelz, D.C. (1956). Some social factors related to performance in a research organisation. *Adm. Sci. Q.,* **1,** 310–26.

Pfiffner, J. & Sherwood, F.P. (1960). *Administrative Organisation.* Prentice-Hall, Englewood Cliffs, N.J.

Piñeiro, M.E. (1987). Agricultural production in Latin America and the Caribbean: International Organisations and Regional Programmes. In *The Impact of Research on National Agricultural Development.* ISNAR, The Hague, pp. 89–95.

Price, D.J. de Solla. (1965). Networks of scientific papers. The pattern of bibliographic references indicates the nature of the scientific research front. *Science,* **149,** 510–15.

Raudsepp, E. (1963). *Managing Creative Scientists and Engineers.* Macmillan, New York.

Rigney, J.A. & Thomas, D.W. (1979). Strengthening of organisations. In *Preparing Professionals for National Agricultural Research Programmes.* Agricultural Development Council, New York, pp. 42–45.

Shapiro, T.R. (1957) The attitudes of scientists toward their jobs. In *Human Relations in Industrial Research Management,* ed. R. Livingston and S. Milberg. Columbia University Press, New York, pp. 157–62.

Starkey, P. (1988). Practical agricultural research: lessons from thirty years of developing wheeled tool carriers. Discussion Paper 25, ODI, London.

Stearns, H.M. (1958). Organising for efficient research and development. In *The Dynamics of Management.* American Management Association, New York, pp 122–78.

Steele, I.W. (1957). Personnel practices in industrial laboratories. In *Human Relations in Industrial Research Management,* ed. R. Livingston and S. Milberg. Columbia University Press, Ithaca, New York, pp. 349–60.

Stein, M.I. (1963). Creativity and the scientist. In *Sociology of Science,* ed. B. Barber and W. Hirsh. Free Press, Glencoe, Ill., pp. 329–43.

Swanson, B.E. (1975). *Organising Agriculture Technology Transfer.* International Development Research Center, Indiana University, Bloomington, Ind.

Swanson, B.E. (1977). Research and production training at International Agricultural Research Centres: View from the outside. Paper presented at the CGIAR Forum on Training, Washington, DC.

Taylor, C.W. (1963). A search for a creative climate. In *17th National Conference on Administration of Research,* University of Denver, Col., pp. 106–13.

Thomson, J.D. & Bates, F.L. (1957). Technology, organisation and administration. *Adm. Sci. Q.,* **2,** 326–43.

Trigo, E., Piñeiro, M. & Ardila, L.J. (1982). *Organización de la Investigación Agropecuaria en America Latina.* IICA, San José.

Vega, R.M. (1979). Post-doctoral and other training programmes of International Agricultural Research Centres. In *Preparing Professional Staff for National Agricultural Research Programmes.* Report of Workshop, Bellagio, Italy. IADS, New York, pp. 55–60.

Vinyu, V.V. (1979). The role of research in solving problems of the developing countries: a Third World view. In *Give Us the Tools,* ed. D. Spurgeon, IDRC, Ottawa, pp. 177–88.

Violic, C. (1973). Maize training. Paper prepared for the International Maize and Wheat Improvement Centre (cited by Byrnes, 1974).

World Bank (1981). *Agricultural Research.* Sector Policy Paper. World Bank, Washington, DC.

Part Four

ORGANISATIONAL STRUCTURE AT THE
INSTITUTIONAL LEVEL

Chapter 12

Structure of a National Agricultural Research Institute

It is at the institutional level that the national agricultural research system (NARS) must reflect each country's needs and characteristics: its size, the kind of agriculture practised, its human and financial resources, its ecological diversity, etc.

These characteristics will determine the kind of departments required, the relative importance of the resources allocated to research on different commodities and natural resources, the number and location of the regional research centres, experiment stations and dependencies, the physical infrastructure and supporting services.

A logical and effective organisational structure is essential to enable assignment of specific duties to individual units, and co-ordination among them. The organisational structure should be as simple and flexible as is compatible with the fulfilment of its functions; it should ensure good communication, efficient use of manpower and other resources, and impose a minimum of routine work on the researchers. 'The organisational form adopted must be a joint function of human characteristics and the nature of the task environment' (Simon, 1960). Agricultural research has a specific organisational requirements different from those of other types of activity.

TYPES OF INSTITUTIONAL STRUCTURE

The complexity of the problems involved in agricultural research makes it necessary to divide the organisation into a number of sub-organisations, which are linked together by a central framework of authority, channels

541

of communications, common interests and joint undertakings. There are different ways in which the research organisation can be divided: into groups of specialists, or into interdisciplinary task groups. Specialist groups, in turn, can be established according to scientific disciplines such as pedology, soil physics, virology, genetics, etc., or the field of production of commodities: field crops, horticultural crops, animal husbandry, etc., each with further subdivisions, as necessary. In addition to the above, there will usually be a number of administrative units based on geographical location (regional experiment stations).

In deciding on the form of institutional structure to adopt, the characteristics of each approach and its relative advantages and disadvantages have to be considered. The criteria to be applied can be grouped as follows: Effectiveness in achieving the goals of the organisations; effectiveness in solving research problems; efficiency in administration; and effect on the research worker.

'Specialist group' form of organisation

The specialist group is a homogeneous unit in which each researcher is responsible for a special well-defined field of work. The group is relatively permanent, providing a feeling of security and belonging to its members. The leader of the group is a scientist highly qualified in his field, who can provide inspiration and guidance to his co-workers. The individual research worker is kept aware of innovations in his field, whether of research techniques or scientific achievements. Though they usually work as individuals, the members of the group have common professional interests and can pool services and equipment. The main disadvantage of this form is that a tendency to narrow interests is fostered, which makes it difficult to cross specialist lines.

When organisational units are based on scientific disciplines, this facilitates scientific leadership and linkages between researchers, as well as pooling of highly specialised equipment. However, scientists are usually concerned with basic problems in their field of specialisation, and many have little or no interest in the specific problems of the crops or livestock production of the country. This has manifest disadvantages, which are most strongly felt in developing countries where new crops and techniques are being introduced and tested; however, the tendency toward this type of organisation is easily understood in countries with a high level of agricultural technology and in which considerable efforts are invested in basic research.

Basing departmental activities on types of production is congruent with the accepted procedure of defining technological problems and needs on commodity lines, and with the needs of farming system research. Commodity departments also facilitate communication with extension workers and farmers, and are more effective in relating research to development plans. However, the commodity department is not without its shortcomings.

Basic research may tend to be neglected, and all efforts devoted exclusively to empirical research, with a resultant reduction in the scientific level of the department. If, however, basic research is adequately covered, as it should be, each department tends to become a self-sufficient research station in itself, with its own physiologists, geneticists, pathologists, soil scientists, etc., with the result that overlapping and duplication in basic research carried out in the organisation become difficult to avoid.

The head of a research unit with such a 'mix' of scientists, having different specialisations, cannot provide scientific leadership, nor can services or equipment be provided economically. The members of the group will have few common scientific interests and each specialist will feel himself isolated from his colleagues with the same professional background.

BASIC UNITS OF THE ORGANISATIONAL STRUCTURE

The functional administrative unit in the agricultural research organisation is the *department*. The departments and their make-up reflect the existence of distinguishable specialities of agricultural production or of disciplines related to agricultural production. These, in turn, reflect the make-up of the agricultural economy of the country. The department has a large measure of autonomy, which derives directly from the difficulty which faces the director of research to direct and evaluate the work of a number of widely different, specialised departments. Hence, the department level is mainly an instrument of communication.

Each department, in turn, is made up of a number of *divisions*. The divisions represent the operating level in the research organisation. These are the cadres for specialisation and for the professional advancement of the research worker.

Each division serves as a pool from which specialists, assigned temporarily to project teams, can be drawn. The lowest organisational breakdown of the research activity is the *research group*, to which a specific problem area is assigned. A research group consists of a senior research worker, one or two research assistants and one or two technicians, and has at its disposal a modest, but stable, working budget, in addition to the funds allocated for specific research projects.

Interdisciplinary task group form of organisation

An interdisciplinary task group (or team, or project group) is an organisational unit brought together for the purpose of solving a particular technical or scientific problem. The main characteristic and principal advantage of the interdisciplinary task group is that the different skills needed for the solution of a problem are grouped together so that the team is able to solve problems requiring the co-operation of several disciplines. In the words of Shepard (1956) 'the technical unit (the problem to be solved) is identical with the social unit that is to solve it'.

It is therefore the problem itself which determines the size and the make-up of the organisational unit. An advantage of the task group is that the diversity of the background, the differences in interests and previous experience of the members of the team stimulate 'cross-fertilisation of ideas' and pooling of knowledge and may therefore enhance the creativity of the team members.

However, the 'team approach' in research has evoked considerable opposition. Whyte (1957) writes that: 'Team research has infiltrated the academic world to the point where individualism has largely disappeared and originality has lost out to uniformity of thought and action.' Raudsepp (1963) sees 'in the excessive contemporary trend towards team-work, throughout the entire spectrum of research, a major deterrent to real creativity.'

The specific objections to the team approach are:

1. Creative ideas originate with the individual and not in groups;
2. without full and spontaneous co-operation, the team effort cannot be successful; however the members of the group may not be compatible;
3. teamwork tends to stifle initiative and originality and can therefore produce only superficial ideas, suitable for solving only routine problems;
4. the team members are responsible to two different leaders: permanently to the head of a department, temporarily to the team leader.

However, in practice, it is generally possible to avoid problems by clearly defining the spheres of authority of each of the leaders;

5. the contribution of the individual working in a team may often remain unrecognised, with most of the credit going to the team leader, even if his contribution has been mainly administrative.

Whatever the short-comings of the task group, much of the research carried out in agriculture is too complex for a single researcher to cope with, working alone. The need for knowledge in many fields, the use of expensive equipment and facilities, and other requirements of modern research make teamwork not a matter of choice, but an essential fact of research life. Kelley (1964) writes: 'Research is now a profession, not an avocation, work is largely in teams, not individual, equipment is now complex and expensive—not simple, environments are now organisational, not monastic.'

A common misconception is that freedom and individual initiative have to be stifled in the team effort. This is not necessarily so. The project, as a joint effort, must be jointly planned, each researcher indicating what his own contribution will be and what he needs to complement his own limitations. Each researcher then has a considerable amount of freedom in planning and executing his own contribution to the common goal. Frequent, periodic meetings for 'brainstorming', exchange of information and mutual advice are needed, but these should not be a hindrance to the researchers' individuality, rather the contrary. Nor do these preclude informal meetings and consultations between individuals of the groups. Finally, the interpretation and presentation of results should be a joint endeavour. This last phase may be a limitation of the individual's freedom, but its negative impact should not be exaggerated.

Whilst the team, as a functional unit for the implementation of a specific project, is an essential element of modern research, it is rarely used as the *basic* component of the organisational framework of research.

According to Shepard (1957), the 'project' type is the more primitive form of organisation. When a research organisation is first established, with a small number of scientists, usually one in each field, there is no possibility of establishing groups according to specialities. The whole group works as a team or a number of teams. As the organisation grows, the number of researchers in each field increases, and they tend to be grouped into specialist units. The next stage in the development of the research organisation is the establishment of common services, whose efficiency has a considerable impact on the effectiveness of the departments and the research teams.

Actually, the different types of organisation described above are not mutually exclusive or incompatible.

Combining different organisational forms

An efficient agricultural research organisation can comprise research units based on scientific disciplines and others based on lines of production. These form the permanent framework of the organisation to which research workers are assigned on a permanent basis. In addition, *ad hoc* teams are established to work on formally authorised projects. These are based on close co-operation between individuals from different departments, ensuring adequate coverage of the basic and applicative aspects of a problem, as needed. Upon completion of the project, the team is broken up. In certain cases, when a 'discipline' specialist's entire time is devoted to a specific problem in a 'production' department, he is 'seconded' on a more or less permanent basis to that department, but continues to 'belong' to the 'discipline' department as one of a group of research workers with common scientific interests. He continues to receive scientific guidance from the head of the 'discipline' department, joins in its scientific group activities and returns to his group when a specific assignment or project is concluded. In the case of certain complex problems covering many fields, the project group may achieve semipermanent status.

The combination of permanent specialist groups in the organisation, with *ad hoc*, temporary project groups, makes it possible to retain the project form with its considerable advantages, whilst at the same time mitigating its undesirable aspects.

The weft and warp pattern of organisation

Most agricultural research organisations comprise departments based on disciplines (the horizontal approach) and others based on fields of production (the vertical approach). The shortcomings of each of these approaches, when adopted separately, have already been pointed out.

When the research organisation includes departments based both on 'disciplines' and on commodities, as is frequently the case, it is practically impossible to eliminate all duplication and overlapping. The department based on a discipline is concerned with research aspects that may be common to a wide field of products, whilst the research work of departments based on the production of a certain commodity relates only to those aspects which are of direct concern to that commodity. 'Soil chemistry' will study the fate of fertilisers applied to different soil types, in different forms or in different manners, whilst 'field crops' will experiment with different rates and combinations of fertilisers to ensure the optimum crop yield, high quality, or other aspects of economic significance. However, the 'chemistry' people usually require a plant for their research and there is no reason that they should not use plants of economic importance; the 'crop' people cannot work without using 'chemistry' unless their work is to be entirely empirical. Hence, the numerous occasions for overlapping.

As Gross (1964) points out, a certain amount of overlapping is not always entirely a disadvantage; 'Men would be pretty badly off if their shirt tails were not long enough to overlap with the top of their trousers!' However, overlapping is not the main disadvantage of the 'mixed' approach. The main weakness of this organisational structure is that full effectiveness in achieving the goals of the organisation is not attainable, unless special measures are taken.

We have within the same framework two basically different research units, whose goals tend to be divergent: those based on disciplines, whose major and probably overriding interest is to further knowledge, each in his own field; and those based on fields of production, whose aim and responsibility is to further knowledge which is directly related to improving production, in their respective fields, whether by increasing yields, improving quality, using inputs more efficiently, or creating new uses, etc. Basically, these are also the goals of the agricultural research organisation as a whole, which is not concerned with advancing knowledge *per se*, but cannot achieve its goals without advancing knowledge in a wide number of scientific fields.

It is clear that the goals of the 'production' departments cannot be achieved without harnessing the full potentialities provided by the 'discipline' departments.

In the conventional set-up, both types of research units have equal status in defining their research interests, and, as a result, heads of the production departments have responsibilities which they cannot fully discharge, and the onus for co-ordinating these frequently conflicting interests of 'productions' and 'disciplines' falls on the research management. In a large organisation this may become an unwieldy problem to handle.

In other words, as long as the research organisation consists of separate and distinct 'vertical' and 'horizontal' strands, it does not function as a harmonious whole. What is needed is a 'weft and warp' system, in which the strands are interwoven in a way that ensures maximum effectiveness. To achieve this, it is necessary to define the responsibilities of the two types of research units in accordance with the differences in the nature of their work, providing for the maximum integration of their efforts. The production units should be responsible for furthering the interests of their field of production in all its aspects, and the discipline units should be responsible for the high scientific level of their contribution to this common objective.

The *head of a 'commodity' department* would have the direct responsibility for representing all aspects of research of relevance to his field in the course of the formulation of research policy and priorities; he would then have direct responsibility for the co-ordination required for implementing these policies. In practice, this would mean that he can call on all the various disciplines for providing team-members to work on projects in his field. He would be responsible for establishing the teams and for administering the research programme in his field.

The *head of a 'discipline' department* would have responsibility for scientific leadership in his field. Research projects would be assigned to the individual members of the department in consultation with the heads of the commodity departments. The evaluation of the detailed research proposals prepared by individuals of his department, advice and supervision during the execution of the work, and critical appraisal of research results would be his responsibility. He would have to create the appropriate research climate in his department, all of whose members, though working on problems connected with different fields of production, would belong to a small 'community' with common scientific interests. He would be the judge of the merits of the individuals in his department in regard to promotion.

It would be convenient to give the two types of research units different names, in order to stress the differences in function, e.g. institutes and stations would be suitable for commodity research units, and laboratories for the disciplines.

Recapitulating, the function of the head of the production department would be primarily organisational, concerned with the orientation of research and the head of the discipline would provide scientific leadership concerned with the execution of the research.

Service departments

A service department is established whenever it is decided, for reasons of economy, efficiency and control, to group certain auxiliary activities into a single unit which provides service to several or all of the departments of the research organisation.

A specialised service department serving the entire organisation is more economical and efficient because it makes possible the full-time employment of skilled personnel, the use of specialised equipment, and the supply of necessary services to the departments according to objectively determined priorities. Amongst these services can be mentioned several that are fairly routine, such as a precision equipment workshop, blueprinting and photography laboratory, a supply service, library, accounting, etc. In addition, there may be special types of service departments that are staffed by highly specialised people with academic training, who provide measurements and analyses, which may require expensive and complicated equipment and techniques (such as mass spectrophotometers, electron microscopes, ultra-centrifuges, etc.). A similar kind of service is provided by trained statisticians, who advise on the planning and layout of the experiments, process data for computers, and carry out the routine statistical analysis of the results. These persons have been termed 'professional technicians' and are not considered researchers, who make research decisions.

The service departments of the research organisation play an extremely important role, in that their effectiveness determines, to no small extent, the effectiveness of the research departments. It is, however, their very importance that may be the main cause for conflicts between service and research departments. It is a general rule that service functions become scarce resources. This is not limited to research institutions, but can be observed in industrial organisations (maintenance activities), in universities (typing pools), in hospitals (pathology laboratory). Heads of research units have to exert pressure or invoke authority in order to obtain priorities. This gives the service head power in the organisation, and as a result arrogance is frequently observed. This is not in accord with the basic obligation of the service department to serve the researcher. A possible, if partial solution, is to make the use of the services facultative, with the alternative of using outside services, if they are more economical, rapid or effective.

Another type of service department is found in units that provide specialised services to the research departments, but also carry out research of their own. Chemistry and biochemistry laboratories, for instance, are frequently set up to provide routine analyses for other research units in the department. The first problem to be encountered is that when the chemist or biochemist takes a personal interest in the work, and becomes involved in the planning of the research and the evaluation of the results, he becomes a partner in the research and is no longer a provider of routine services. In the spectrum between routine work and active participation in research, the exact point of take-off from service to research is never clear-cut. What is certain is that there are frequent differences of opinion between the parties concerned on whether the biochemist, for instance, is providing service or is to be accorded the privileges of a research partner. Usually, a liberal attitude should be adopted, by which the 'service' man should be given the benefit of any doubt that may arise as to his personal involvement in research.

A further difficulty is when the service department becomes so involved with research of its own, that it no longer provides sufficient services to the other departments. This then makes it necessary to set up an additional service laboratory, or oblige each research worker to carry out the routine analyses he requires, a procedure that is wasteful of space, equipment and manpower. The only appropriate solution in this case is for the head of the department to use his judgement in maintaining a balance between research and routine work provided as an essential service to others.

Plant introduction service

A special type of service unit is a plant introduction service. The introduction of new varieties, of germplasm with special characteristics, of new crops from other countries, is probably one of the most effective and economical activities in which a research organisation can engage. If there is no organised framework for this purpose, each research worker will take the initiative and contact introduction services or individual scientists in other countries; the result is anarchy. A researcher who has published a paper in which he describes a new variety will be inundated with requests for seeds, generally by a number of people from a single country. In the absence of centralised records the same variety may be introduced, tested and rejected a number of times in succession, without the researcher being any the wiser.

Another point that cannot be overstressed is that with all its potential benefits, the introduction of seeds and especially vegetative material can also be a source of potential danger, that in certain cases may assume the proportions of disaster by introducing new diseases, pests or weeds.

For all these reasons, a central plant introduction service, as part of the agricultural research organisation, is an absolute necessity. The service introduces new varieties on its own initiative, or the initiative of any individual, whether he belongs to the research organisation or not. The introductions are first grown under strict control, under the supervision of a plant pathologist and, sometimes, an entomologist. Screening is carried out in co-operation with the scientists interested in the new material, to whom the promising varieties or strains are handed over as soon as the quarantine period is over. The introduction service maintains a card index of all introductions and their eventual fate. It may maintain a 'seedbank' under suitable cold storage conditions. The service also serves as a clearing station for seeds requested by foreign researchers, and is responsible for providing the sanitary certificates required, etc.

Seed-testing and diagnosis laboratories
Another type of service department is found in units which provide services to the farmer, either directly or via the extension service. A few examples are seed testing; soil and plant analysis for diagnostic techniques in nutrition requirements; identification of diseases and pests. Much soul-searching goes on as to whether research institutes should also concern themselves with services of this kind. The main justification for providing these services is that there is a close connection between the research work carried out by the institution providing the service and the level of the service itself: diagnosis of nutrition requirements cannot be deduced mechanically from the data resulting from soil or plant analyses, but requires interpretation on the basis of plant nutrition research.

ESTABLISHING THE BLUEPRINT OF THE AGRICULTURAL RESEARCH ORGANISATION

It is extremely unusual for an agricultural research organisation to have been planned according to the long-term requirements of the economy, with a blueprint, and then to build up the organisation according to a preconceived plan. In the vast majority of cases, individual research units have been established by professors at agricultural faculties, by various ministerial departments, by private enterprise, or by farmers' organisations. Even in the individual research institutions, research activities have usually developed following the hiring of promising individuals, irrespective of their specialities, and it is their interests which then determine the research programme and organisation and not, as would be logical to assume, the objective needs and requirements of the national economy.

In time, the research activities have grown to an extent that some form of organisation becomes essential in order to ensure: (a) a complete and balanced coverage of all the essential fields of research; (b) reduction of duplication as far as possible; and (c) a mechanism for determining priorities and allocating funds.

It is rare indeed that the planner can start with a clean sheet; on the contrary, he is usually faced with so many vested interests and so many individual personality problems that it becomes very difficult to devise an objective and balanced plan. The unplanned 'growth' of the organisation has usually resulted in unequal development of various essential fields of agricultural research. However, an organisational blueprint, based on what Urwick (1965) calls 'a logical drawing-office approach in which the organiser should draw an ideal structure in a cold-blooded, detached spirit', is an essential starting point, either for the setting up of a new research organisation, or for reorganising an existing set-up, even if this is to be done gradually over a number of years.

In constructing the blueprint of the organisation, the following aspects have to be considered:

1. The organisation must be built on the basis of the functions it has to fulfil. In an agricultural research organisation this is to solve problems of importance to agriculture. The complexity of these problems, as relating even to a single, small country, is considerable.
2. It must maintain a balance, with a minimum of overlapping, between different essential fields of research. This should be shown in an organisation chart, indicating the departments and divisions with their specialised functions, and which provides a framework for a permanent establishment that adequately covers the essential fields of research in which the organisation may be called on to make its contribution.
3. It should allow for flexibility to cope with emergencies or with problems of temporary importance. It is here that the 'project' approach makes a considerable contribution. The permanent establishment of the organisation has to be based on *permanent* fields of activity. This permanent framework must, however, be able to expand rapidly when required, to cope with new or urgent problems. When the need occurs to handle problems that are beyond the capacities of the permanent staff, an *ad hoc* team is established (see Chapter 10).
4. An ideal organisational structure for research should clarify and for-

malise authority relationships, channels of communication, and provide an environment in which research can function efficiently. However, even a logical and perfectly planned blueprint will usually need adjustment to the personality problems that will be encountered sooner or later. Whenever 'deviations from the pattern, in order to deal with idiosyncrasy of person' (Urwick, 1965) are deemed necessary, so that a creative researcher should be able to fit into the organisation with a minimum of strain and tension, the necessary alteration should be made.

It is very tempting to guard the perfection of the blueprint and to make the research worker fit into it. But it is a basic duty of the administrator of research to depart from rigid patterns whenever necessary and, within limits, attempt to find individual solutions whenever indicated. If the plan is well made, the number of deviations will be small, and, hopefully, temporary. In any case, they should be kept to a strict minimum.

REORGANISATION

Need for reorganisation

In every research organisation, continuous changes are needed to adjust to shifts in interests as a result of changing economic conditions or technological innovations. Gross (1964) formulates what he calls 'the principle of permanent reorganisation', and sees in the constant changes that take place in the formal and informal structure 'a contribution to immortality'. Koontz and O'Donnell (1955) confirm this view; they consider a moderate and continuing reorganisation as necessary, merely 'to keep the structure from developing inertia'.

Symptoms of obsolescence

The following are considered symptoms of obsolescence in a research organisation: the organisation becomes slow to adopt new tools and techniques; it becomes increasingly difficult to recruit new talent; and a higher than usual turnover of the more competent staff occurs.

Reasons for reorganisation

Reorganisation may be made necessary by changes in production practices or in the emphasis on certain fields of production. A rigid departmental structure that cannot be adapted to the changing pattern of agriculture will reduce the effectiveness of the research organisation, and contribute to its becoming obsolete.

Sometimes special circumstances favour a certain type of research activity, resulting in a considerable increase in permanent personnel of one of the units. The special circumstances may cease to exist, but the resulting imbalance in relation to other units will remain. A strong personality at the head of a laboratory may 'achieve' similar imbalance.

Gross (1964) states that the tendency for different units of an organisation to grow or decline at disproportionate rates is universal. He calls this the 'law of disproportionality'.

When a research unit grows beyond a certain size, or a researcher attains professional status, so that he becomes a recognised leader in his own right, 'fission' of a research unit, or the establishment of a new group, may be indicated. Fission may also become unavoidable as a result of personality conflicts within the group. Staff-line conflicts may require a re-allocation of authority relationships.

Koontz and O'Donnell (1955) list the following organisational weaknesses which justify reorganisation: excessive spans of management; too many levels; poor interdepartmental co-ordination; excessive numbers of committees; lack of uniformity in decision-making; failure to accomplish objectives; and excessive costs, ineffective financial control.

Implementation of reorganisation

After the need for reorganisation has been recognised and the decision taken, the following necessary steps are enumerated by Ginzberg and Reilley (1957): developing the reorganisation plan; announcing of plan to organisation; detailing new functions and responsibilities; alignment of various operating systems and initiatives to reinforce the plan; instructing the personnel in new methods to facilitate change in their behaviour; monitoring the plan; and adjusting it in the light of experience.

An essential preliminary to a change in the organisation is to make the people concerned understand the need for the reorganisation, and the advantages that will accrue therefrom. Every effort should be made to ensure the support and active co-operation of the majority.

Ginzberg and Reilley (1957) point out that reorganisation will usually involve many unpleasant and difficult actions which have to be taken if it is to succeed, such as earlier retirement; or the transfer of personnel unwilling or unable to co-operate in the new set-up; restrictions or changes in the area of responsibility of individuals, etc. The more painful and unpleasant the decision, the less is it possible for the director to delegate his responsibility to others.

An excellent opportunity for limited reorientation of programmes, and increasing or decreasing the size of individual units obtains when the need arises to replace research workers that leave the organisation on attaining retirement age. Each department will consider the vacancy that occurs as its 'own', a tendency the research director must control by safeguarding his prerogative to appraise the situation on its merits. He will have to decide whether to search for a candidate with the same type of training as the one that he is replacing, or whether circumstances warrant the initiation of a new programme or the strengthening of another area.

The process of reorganisation, under all circumstances, is 'long, uncertain, difficult, and often-disappointing' (IESA, 1980). It requires a constant effort to overcome inertia and conservatism. These are not only due to psychological factors, but may also result from the possible undesirable implications for individuals in the organisation, who may fear a loss of authority or of influence.

The success of any major process of reorganisation will depend largely on the sustained attention and dedication with which it is implemented. However, the leadership of the organisation is largely preoccupied with its routine obligations, and cannot, as a rule, devote more than a fraction of the attention required by the reorganisation process. Hence the need to appoint a competent personality to be responsible for implementing the process and to devote full time to this undertaking. He can be assisted in his work by an *ad hoc* committee.

Reorganisation of agricultural research structures

National agricultural research systems in many countries are characterised by almost continuous bouts of reorganisation (see case histories, Chapter 1).

Generally, two distinct phases can be discerned: as long as agricultural research is incorporated in a ministry, reorganisation is generally a function of politics, such as power conflicts between ministries, or personal 'fief' building within a ministry.

As awareness of the importance of developing agriculture increases, and of the major role of agricultural research in this process, governments begin to realise that agricultural research must be insulated from political vicissitudes if stability, continuity and effectiveness are to be ensured.

Therefore, the solution that has generally been adopted is to assign the responsibility for agricultural research to a semi-autonomous national entity, with adequate safeguards to ensure that the ministry responsible for agricultural development retains the power to direct research policy in accordance with national priorities.

During the second phase, reorganisation is mainly concerned with solving problems of research planning, regionalisation, increasing institutional efficiency, improving channels of communication, resolving conflicts within the organisation, etc. Too frequent reorganisation can, however, be counterproductive for obvious reasons.

Chapter 13

Regionalisation of Agricultural Research

THE JUSTIFICATION FOR DECENTRALISING RESEARCH

The growing importance of the agro-ecological region as a planning unit for development planning, as well as the increasing orientation of the agricultural research effort towards the small farmer, have emphasised the need for decentralisation of the national research service. As a result, regional research centres and their dependencies, have assumed major importance in the NARSs.

Centralisation of agricultural research does not conflict with decentralisation; on the contrary, the two concepts complement each other. Centralisation is essential for overall direction of the national research system, for planning and administration. The work in the field must be decentralised for a number of reasons. The obvious reason for establishing several regional experiment stations, and thereby decentralising the actual field work, is the largely location-specific nature of agricultural research.

Surprisingly, experience has shown that this obvious reason is less critical than is generally accepted. We have found, for instance, in the course of many years of research, that the search for, or the breeding of varieties adapted to the relatively narrow ecological requirements of each region of a country may be futile (Arnon, 1972). In practically every major crop the best results are frequently obtained with varieties of wide adaptability—which usually transcend not only the narrow ecological boundaries of a region, but even those of a country. Outstanding varieties of wheat, barley, sorghum, maize, cotton, sugar beets, etc., are those which have proved outstanding in many countries. The same is true for a great many varieties of fruit trees and breeds of farm animals. It is also by and large true for basic agronomic practices, such as fertiliser application, rotations, irrigation practice, etc. Frequently, greater divergence in fertiliser response is found between adjacent fields submitted to differential treatments, than between fields with similar management in different regions of the country.

Even though certain varieties and management practices are not specific to narrow ecological niches, a substantial part of agricultural research is very location-specific, and even in the case of varieties with wide adaptability, none of these can be recommended to farmers without having been tested locally.

There are a number of additional reasons for the need to decentralise the implementation of research in the field. Applied research in agriculture requires many years before definitive conclusions can be drawn and valid recommendations made. This period can be considerably shortened if the research plan is carried out simultaneously in a number of regional stations, thereby submitting the treatments tested to a variety of environmental conditions.

It is extremely important, for political and psychological reasons, that the research should have a grass-root orientation. The dispersal of the researchers in the network of regional stations, with all the problems it entails, ensures that their work remains closely related to the needs of the regions; this relationship is further strengthened by their active participation in trials on farmers' fields. From the regional stations, farming systems research can be originated, and a network of adaptive trials on-farm can be organised. Information on farmers' needs and contraints is improved, thereby ensuring that the research programme addresses the real needs of the farmers (Cernea *et al.*, 1985).

A pattern of regional stations can dovetail with the regional organisation of extension, thereby improving co-operation between researchers and extension workers, facilitating the transfer of research findings to the farmers and shortening the time-lag in their adoption.

For all these reasons, 'a state system that includes a strong network of branch stations gets more for its research dollar, for a given level of expenditure, than a state system that is more concentrated. What decentralisation gives up in lower costs seems to be compensated for by the relevance of the research and the more rapid diffusion of results' (Ruttan, 1985).

Interlocking the regional stations into a single system

The regional stations should not work as separate, independent entities, but should form an integral part of a national research organisation. Experience has shown that centrifugal forces can easily lead to a fragmentation of the research effort, lack of co-ordination, and wasteful duplication. To counteract these tendencies, a system is needed that can ensure the interlocking of all the regional stations into a single, co-ordinated national research organisation, whereby each regional station would have dual functions: to contribute to the implementation of an overall national programme and to be responsible for research on specific regional problems.

In order to achieve a high level of productivity and efficiency, in spite of the decentralisation of the research organisation, a certain hierarchy of the research units forming the national system is essential. This hierarchy consists of headquarters, a central research station, a number of regional stations, each with its field stations and on-farm experiments locations.

Research headquarters

These should preferably be located near the seat of government, to enable the director of research and his administration to maintain close contact with government institutions, in particular the ministry of agriculture, and national political leaders, to keep the latter aware of the contributions of research to development programmes.

A number of service units, such as a research project office, an econometric unit, a small, but adequate staff of scientists capable of planning and co-ordinating research programmes and evaluating the results of research, central administrative offices, are located at headquarters.

If the central research station is located near the capital city, as is frequently the case, this provides the ideal location for the headquarters of the national agricultural research service.

The central research station

With this there are problems of a national character, which require costly and sophisticated equipment, multi-disciplinary teams of research workers and co-ordinated experimentation in a number of regions. It is therefore impractical and unnecessary to replicate, in each regional station, the facilities, equipment and staff of the central research station.

The scientific level of the regional stations is dependent on that of the centre; therefore, no effort should be spared in developing a centre on the highest possible level, commensurate with the needs and means of the country.

For these reasons, one of the regional stations should have the status of a central station, with a major concentration of scientific staff and supporting services. It should have three parallel functions:

- to engage in research of national significance;
- to support the regional stations, by providing leadership, co-ordination and central services;
- to function as a regional station for the area in which it is located.

The regional stations
The disadvantage of according a special status to one of the regional stations, as proposed above, is that research work conducted at the national centre will enjoy more prestige and will tend to attract the best workers away from the regional stations.

It is possible to mitigate this disadvantage if each regional station, in addition to its purely regional research work, would serve as a 'lead station' for one or more fields of research on a multiregional or national scale.

In almost every region of a country, one commodity or farming system is dominant, and it may also be of importance in other regions of the country. Assigning responsibility to each regional station for planning and co-ordinating a nationwide programme, on an important commodity, to be implemented in all, or most regional stations, will broaden the field of influence of each regional station and raise its status to one similar to that of the central station. This system of co-ordinated, national, multidisciplinary commodity programmes, assigned to different regional stations, in accordance with their agro-ecological locations, is a powerful means for welding these stations into an interlocking and cohesive national research system.

Because of their nationwide scope, these programmes are also very effective in achieving, rapidly and efficiently, well-defined objectives in high-priority projects. The system also dovetails well with the crop campaigns undertaken by the development authorities. Wright (1982) mentions four essential elements for an effective national commodity programme:

- research is carried out in all areas of the country in which the crop is important;
- all scientific disciplines important to the development of new technology for the commodity are involved;
- production problems are the focus of the programme;
- a nationally respected scientist is nominated co-ordinator of the programme.

The co-ordinator's office should be located at the 'lead' regional station for the commodity programme. The co-ordinator, who is also the main troubleshooter for the programme, will need to visit all the co-operating regional stations quite frequently, in order to evaluate progress, give advice and encouragement, and convey information from group to group. He should be 'the country's best informed individual on the commodity he represents, providing information to administrators, planners, politicians and other policy-makers, as well as scientists' (Wright, 1982).

Single-commodity or polyvalent regional stations

An important question is whether the regional structure of research should consist of single-commodity or polyvalent experiment stations.

In many countries, regional stations are frequently devoted to a single crop. Even when several stations are established in one vicinity, they are frequently in different locations and constitute separate administrative entities.

The single commodity experiment station has the advantage of a clearly defined limited objective: research that identifies with the interests of a specific group of producers, who generally participate, directly or indirectly, in the orientation of the research and its financing.

These advantages are most manifest for the commodities produced by the high-income farming sector (plantations, marketing boards, etc.). Focused as they are on the narrow interests of minority groups of farmers, their objectives are not necessarily those of high national priority, and may even be in conflict with national interests.

A station devoted to a single commodity usually reflects the type of monoculture prevalent in the region. A station of this type will tend to perpetuate the shortcomings and risks involved in monocultural systems, and will certainly not be in a position to suggest diversification.

For example, in many countries, research on crops for human consumption has been carried out in research stations different from those devoted to research on crops for animals and on animal husbandry. Sometimes these artificial divisions of research are run under different departments or ministries. 'Research on animal production based on cultivated crops is an integral part of research on the farming system and general research must embrace all aspects of the general farming systems: it must be planned in the context of the farming systems for which it is relevant' (Russell, 1970).

A multipurpose station, by contrast, would be in a position to investigate possibilities of diversification and integration leading to new, improved farming systems. Only such stations can improve the technology of the majority of the farmers of the region.

As a rule, the only single-commodity experiment stations that are justified are those that require special conditions, or for commodities that cannot be integrated into a farming system. Forestry, fisheries and rice research are examples. Such stations should, however, maintain close links with the regional station in their area.

The regional experiment stations should be interdisciplinary

While there is no intention of duplicating the central research station in each of the regions, the maintenance of an adequate scientific level requires a minimum number of research workers in the basic disciplines: plant pathology, entomology, soil science, etc. They can co-operate with a number of production specialists, which would obviously be impossible if the latter were located at different stations, each devoted to a single commodity.

It is essential to avoid isolating research workers. Contact with colleagues, even in different fields, is thought-provoking and stimulating. Grouping several types of production in a single regional station makes it possible to maintain larger groups of scientists than in the mono-type regional station.

Finally, regional stations should not work as separate, independent units, but should form an integral part of a national research organisation. Regional stations have dual functions: to contribute to the implementation of an overall national programme; and to concern themselves with specific regional problems.

Field stations

Each regional station should have a number of field stations, strategically located for adaptative research in specific agro-ecological conditions not represented at the regional station. A few capable resident technicians can be responsible for the implementation and monitoring of the field experiments, under the direction of research personnel from the regional station.

PLANNING THE REGIONAL RESEARCH SYSTEM

Generally, because of financial constraints and shortage of human resources, a network of regional stations and their dependencies has to be established in stages. It is, however, essential to plan the system in its entirety, and to implement the plan in stages, in accordance with national priorities.

Critical elements in the planning of a regional structure of research are: (a) the number of regional stations and the dependencies required; (b) the location of the stations and (c) a development plan for the establishment of the regional structure.

Number of regional stations

The number of regional experiment stations will perforce be a compromise between the need to carry out research in all the ecological regions of the country and the constraints imposed by limited human and financial resources. The number of regional stations must be kept to the bare minimum required for establishing a network capable of effective adaptive research in the main ecological regions of the country.

Regional stations are, by their very nature, isolated from the main stream of scientific effort of the country. It is essential to minimise this defect by affording the scientists in the regional stations the best possible in scientific equipment and library facilities. This is feasible only if the number of regional stations is kept at the absolute minimum necessary, and this is in turn possible only if a regional station groups several types of production instead of allocating these fields to separate stations.

The need for a minimum number of scientists and technicians and mix of disciplines required to conduct a research programme effectively has already been stressed.

The concept of critical mass is often violated by the dispersal of scientists among many stations, even though this arises from the legitimate desire to serve each of the agro-ecological regions or administrative units. Continuing limitations in resources will impose hard choices between efficiency of the research system through concentration, thus ignoring many important needs, and attempting to provide uniform coverage leading to ineffective research. A few well-staffed and well-supported stations have a far better chance of producing useful technology than a widely dispersed system with all the attendant problems of difficult communication, intellectual and social isolation, and lack of support services and social amenities (ISNAAR/SPAAR, 1987).

It is rarely realised how costly is the establishment and maintenance of an effective regional experiment station. Ampuero (1981) gives as an example a well-equipped regional station in Ecuador—Santa Catalina—which required an initial investment of US$ 6 000 000 and an annual operating budget of US$ 1 500 000 (1979 figures).

A variety of pressures—political, sectoral, regional—to which a Ministry of Agriculture may be subjected, can easily result in a proliferation of regional stations. Ampuero (1981) writes in this respect:

Agricultural research programs, particularly in Latin America, have had a tendency to create a large number of experiment stations. This probably resulted from the necessity of promoting regional development or in response to political pressures from ministries or governments, petitions from provincial representatives of the Congress, and, in many cases, demands from organized groups of commercial farmers.

One of the most insidious pressures, and the most difficult to resist, is that resulting from offers of financial aid by foreign donors to establish one or more experiment stations—generally single commodity stations. These pressures should be resisted if an effective and economically viable research organisation that the country can afford is to be maintained.

An important consideration is sustainability, over time, from national resources. Another concern, expressed by Ruttan (1982), is the excessive investment in research facilities development relative to development of scientific staff. 'Premature facilities investment represents a burden on the research system rather than of productivity.'

The specificity in the response of crops and their varieties to environmental conditions is the basic justification for the dispersion of the research effort throughout a country. However, many developing countries, in an excess of zeal, have attempted 'to adapt and tailor technology to relatively minor gradations in soil, climate, and economic conditions' (Boyce & Evenson, 1975). Because of the need to avoid excessive dispersion and fragmentation of the research effort (see above), this proliferation of poorly staffed, inadequately equipped and financially hamstrung research units is counterproductive, and represents a considerable drain on scarce human and material resources.

The number of regional stations should be sufficient to cover adequately the major ecological regions of the country. In each region, strategically located sub-stations and on-farm experiments can take care of deviant ecological niches of economic or social significance.

'Establishment' of a regional station

The smooth functioning of a field station, with its complex of fields (experimental plots and main fields) and laboratories, requires a permanent 'establishment' of technical, administrative and scientific personnel. It must be stressed that the scientific personnel at the regional station, while forming part, administratively, of the region station, belong to the 'establishment' of their respective research departments, and their scientific work is an integral part of the work of each department.

Allocation of research workers to the regional stations
Each regional station requires a minimum number of researchers with different professional expertise in order to:

- provide complete coverage by multidisciplinary teams of all the problems encountered by the regional station;
- enable cross-fertilisation of ideas;
- avoid creating a feeling of isolation;
- make it possible to provide social amenities for the staff and their families.

From 10 to 20 researchers at each regional station is the minimum required to achieve these purposes.

Whilst much of the work at the regional station is interdisciplinary, for example a commodity specialist, soil scientist, plant pathologist and an economist working together on a common project, each specialist may feel isolated from his colleagues in the same or related disciplines. This isolation may have a negative effect on the professional competence and advancement of the individual. Therefore, each researcher, wherever located, remains a member of a national research department, to which all the researchers with common professional competence and interests belong. The individual researcher receives scientific guidance from the head of his department, who is also responsible for the career advancement of all the members of the department, wherever they are located.

The department head must pay particular attention to those researchers of his unit who have been seconded on a permanent or temporary basis to one of the regional stations. He must organise, at regular intervals, meetings and seminars of the entire department, so that all the researchers in a given discipline, including those working in isolation from their peers, can be kept aware of the work carried out by their colleagues, remain abreast of progress in their scientific field, have the opportunity to discuss problems with their colleagues, learn new techniques, etc. Attendance at workshops, refresher courses, congresses, reciprocal visits among regional stations, enrolment in postgraduate courses, sabbaticals, are all means that help to avoid the professional isolation and stagnation of specialists assigned to regional stations.

Where a certain type of production predominates in a region, the regional station can serve as the national centre of research for the particular production, and concentrate the principal efforts in research in that particular field for the whole country. All the other types of production are usually represented by a single research worker, who is the representative of his department or division in the region, and who supervises the regional field work of his colleagues from the central station. In addition, he has his own defined field of research, which is his own responsibility. A worker in vegetable crops in one of the regional stations, whose own field is onion breeding, for instance, will also supervise the work in the field carried out for his colleagues in the vegetable division, such as fertiliser work, irrigation problems, varietal trails, in all the other vegetable crops.

In the arrangement as proposed above, the research worker is responsible to two different leaders: professionally to the head of his department, and administratively to the director of the regional station to which he is assigned. Any problems that might arise from such a situation can be largely avoided by clearly defining the spheres of authority of each of the leaders.

Technical and administrative staff

The research personnel cannot be effective if the supporting staff, tech- ·nical and administrative, is not adequate. Good field technicians (generally graduates from secondary agricultural schools) who have received adequate on-farm training, are absolutely essential if the experimental field work is to be carried out with the necessary care and expertise on which reliable experimentation depends. One field technician at the least should be assigned to each research worker.

The regional station also requires a full complement of administrative personnel, but excessive paperwork should be shunned.

Location of the regional stations

After deciding on the number of regional stations to be established, suitable locations must be found. The major dilemma in this respect is that purely agro-ecological factors may clash with other considerations, such as the social requirements of the research workers, the availability of essential services, communications and others.

Very early in the history of agricultural research stations, there arose controversy as to the appropriate setting and location for these institutions. The German plant scientists were the first to contend that farms were not needed for experimental work; that agricultural research work was carried out most effectively if the stations were located in an urban environment, and, whenever possible, in the vicinity of a large university. A farming environment was actually considered harmful and antagonistic to a scientific orientation of the research programme. It was further contended that laboratories, glasshouses with controlled environment and small test plots were all that were needed, so that the agricultural research institutes could be established on very small holdings.

Considerable importance is generally attached to the proximity of a university or agricultural college to the research insitution. The advantages are: the proximity of scientists available for consultation, and possibly co-operation on research projects; facilities for junior staff to receive postgraduate training: and access to additional equipment and library facilities. The university or college will also benefit because of the possibility of assigning part-time teaching to the senior staff members of the research institute, who will be able to contribute from their experience in up-to-date research to the training of future agronomists.

This consideration is as important today as it was in the past, especially if it is decided to establish a symbiotic relationship between the university and the research institute for postgraduate training (cf. p. 460).

The proximity of an urban centre is also deemed of great importance: it is easier to attract top-rate scientists, since family matters, such as social contacts and activities, education of the children, cultural life, etc., are much easier to solve.

A modern research centre requires a large area for building its laboratories, administration and services, glasshouses and experimental fields, animal husbandry units, etc. These requirements usually cannot be met in close vicinity to an urban area, and certainly not within the precincts of a town or city. There arises a conflict between opposing desires, and the alternative solutions are: to establish the headquarters and laboratories of the central research institute in an urban centre, and, at a distance, the field and farm buildings required for experimental purposes; or to set up the agricultural research centre in its entirety in a rural environment, but as near to an urban centre as possible.

The first solution has manifest advantages, outlined above. Its protagonists will also tend to justify this choice by stating 'the future of agricultural research lies in the laboratory, and not in the field. . .', an argument that, as we have seen, is at least a century old. And yet, fruitful agricultural research has always been a combination of laboratory and field work, the two complementing each other, and there is no sound reason to assume that this will be different in the future. Easy access to experimental fields should be given greater weight than easy access to the cultural facilities of the town. An urban setting for an agricultural research centre, far from its experimental fields, will strengthen natural tendencies to carry out one-sided, unbalanced research programmes in the laboratory and to neglect the requirements of applied research or delegate this type of work to other services.

On the other hand, an experiment station that is ideally located from the purely agro-ecological aspect, but isolated, difficult to access, without social amenities for the staff and their families, will not be viable; excessive commuting of personnel, lack of communication with headquarters and other components of the national research system, will make the normal functioning of such a station extremely difficult, if not impossible.

Hence, the inescapable conclusion that a major consideration, of equal weight (but not greater) to the agro-ecological factor, is the proximity of a township that can provide social services and amenities to the staff and their families (education, health, recreational facilities, etc.) as well as facilities for purchasing equipment and supplies, technical servicing, electricity, etc. A reasonable distance between township and experiment station would be about 20–30 km, or less.

Other essential requirements are: easy access to the station, a sufficiently large area of suitable land (see below) and a potential source of water for irrigation and other purposes.

An important decision that has to be made is whether personnel should commute to work, or residential quarters be provided on the station. Easy access to the station is of major importance, not only for the convenience of the staff, but also to encourage visits by farmers and extension workers, either individually for advice, or in groups for field days, demonstrations, lectures, etc. If, as is highly advisable, the local extension agency is also located at the regional station, easy access assumes even greater importance.

A special case is that of the central research station. In many countries it is located at, or near the seat of government, and not in an agricultural setting. Such a location, which ensures effective communication between research leaders and the political–administrative higher echelons concerned with national planning and budgeting, is an advantage for the headquarters of a national agricultural research service. However, for an institute that is itself engaged in agricultural research, to be located on the basis of political instead of ecological and social considerations, will sooner or later cause an alienation of the research centre from the essential problems of the farming sector.

Physical infrastructure

Appropriate physical facilities at the regional station are essential for effective research work.

The physical infrastructure of a typical regional station includes: administrative and laboratory buildings, residential quarters, greenhouses, library, farm buildings and maintenance facilities, a system of roads, transport facilities, electricity and water supply, and an experimental farm with adequate land and equipment.

Buildings
Adequately planned buildings are a great asset in facilitating research. The laboratories should be well designed, providing sufficient space, proper facilities and with sufficient flexibility to adjust to changes in routine and personnel. For agricultural research, it is important that each departmental building be situated on a fairly large plot of land, so that there is plenty of space available for glasshouses, screen houses and

small plots of plant material, such as breeding plots, requiring constant supervision.

Very often, heavy investments are made in buildings mainly for prestige value. The result is that not only are funds exhausted which might have served for the purchase of equipment or other essential purposes, but also that high upkeep costs swallow a considerable proportion of the annual budget.

Buildings should be kept as simple as is compatible with their functional objectives; the facilities should be as flexible as possible, to allow for changing functions and moving people in response to frequent changes in research needs and priorities that are characteristic of a dynamic research programme. Laboratories can be modularised, as well as offices, conference rooms and standard facilities.

Equipment

The provision of adequate equipment is becoming a more and more essential requirement for modern research, and agricultural research is no exception. Equipment has become more complex and expensive; it also becomes obsolete far more rapidly. For these reasons, it is essential to give considerable thought to analysing the need and justification for each piece of equipment, and once purchased, to assure its equitable use by as large a number of researchers as possible. This will frequently clash with the wish of each department to have its 'own' equipment.

A concomitant of the large-scale use of expensive and complicated equipment is the need for a trained staff for servicing and maintenance. This also requires workshops that are equipped for maintenance work, and for developing apparatus which is not commercially available, or whose purchase might take a long time (as in developing countries).

A special requirement of agricultural research is specialised equipment for experimental field plots. In developed countries, the cost of farm labour is becoming an important factor in research budgets; in developing countries, a shortage of literate personnel, who have an understanding of the need for precision and accuracy in the work, is also a limiting factor.

In recent years, great progress has been made in manufacturing specialised equipment for experimental plots: for tillage, seeding, spraying, cultivation and harvesting of a variety of crops. Similarly, modern equipment is available for measuring and weighing, determining soil moisture, diagnostic techniques, physiological processes measured in the field, etc. Adequate equipment in the field is therefore no less important than in the laboratory. With the advent of relatively cheap compact computers and word processors, a regional station can afford to computerise all the work these wonderful tools can handle.

THE EXPERIMENT STATION FARM

Functions

The basic function of the experiment station farm is to provide suitable conditions for the layout and execution of field experiments. This includes land, equipment and labour for all routine farming operations required. Even in the research station, field experimentation should always be carried out in the context of normal farming practice. There are, however, a number of additional functions that the farm can perform.

Practical training of research workers
Research workers frequently have no rural background and insufficient training in the practical skills required in farm management, crop production and animal husbandry. The opportunity to learn these skills most conveniently on the farm should not be neglected.

Involvement of research workers in farm management
Concomitantly, experienced research workers should be involved in the practical management of the farm, as advisors to the farm manager, each in his own field of expertise, on the crops and varieties to be grown, technological innovations to be adopted, the monitoring of diseases and pests etc. Problems that are encountered in the areas devoted to normal farming practice can become topics for research.

Field-scale testing of innovations
The experiment station farm is normally the first place in which technological innovations are tested under farming conditions. These tests can provide valuable insights on the practical problems that may be encountered when applying production techniques based on experimental results. New prototypes of field equipment can also be tested before deciding on their import on a commercial scale.

Demonstration
The farm can also serve as a demonstration area for extension workers

and farmers on good husbandry in general, and specific practices in particular. Field days should be organised for this purpose.

Services to farmers
Wherever possible, the commercial production on the farm should have additional objectives besides providing income. For example, fields should be used to produce certified seed of varieties recommended by the researchers, and which have resulted from the breeding work or varietal testing on the station. Ampuero (1981) comments on this subject. Pedigreed seed of new varieties, and in the case of livestock improved sires for breeding, can make an important contribution to farming and provide a useful source of income to the research organisation. Nurseries for the production of forest, plantation crop and fruit-tree seedlings; virus-free vegetable propagation stock, etc. is another legitimate and useful activity for the farm to engage in.

Animals and chicks required for experimental work could also be supplied. In this way even the commercial production functions of the farm could be tied in fully with the research work of the regional station.

In no case, however, should these additional farm activities be allowed to conflict with the basic aim of supplying services to the research workers in carrying out their field experimentation.

Choice of the site for the experimental farm

The site chosen for the experiment station, including its farmlands, should have a large tract of uniform soil, representative of one of the major soil types of the region. Good surface and sub-surface drainage is important; a slight but uniform slope of up to 5 per cent may be advantageous in this respect. A slope of up to 15 per cent may be suitable for perennial horticultural crops and permanent pastures, provided the slope is fairly uniform and the area terraced.

It is extremely rare to find a sufficiently large area that is ideal in all these respects, hence the need for a good topographical map, to be able to pinpoint all the areas within the proposed site which are not suitable for experimentation: excessive or irregular slopes, depressions, etc.

It has been pointed out that fields with very steep or irregular slopes are not suitable for experiments; in many areas, however, this may be the typical topography of the region, and a large area of flat land, even if it could be found, would therefore be entirely unrepresentative. The solution in this case is to acquire for the experiment station a large area, which includes sufficient usable land for the conventional layout of experiments, and devote the remainder of the land for studying the problems of large-scale farming and similar activities.

The arable land that is not suitable for experimental plots, can be organised into a number of farming units designed to test different farming systems, in which emphasis is placed, for example, on the maintenance of soil fertility and on methods of soil conservation in relation to mechanical cultivation. Other areas can be devoted to a seed-multiplication unit and a livestock husbandry experiment unit respectively (Arnold, 1976).

One question that is frequently asked is whether an experiment station located on excellent, if not ideal conditions of soil, topography, and infrastructure can be representative of typical farming conditions. The average farmer certainly does not always have uniform, graded, fertile soil at his disposal, and certainly does not benefit from the same kind of infrastructure and equipment available to the manager of the experiment station farm, and the researchers who do their experimental work under these privileged conditions. How then can the results obtained on the experiment station benefit the farmer? The answer to this question should be unequivocal: the objective of the work on the experiment station proper is to evaluate the production potential under research conditions and *not* under farmers' conditions, which in any case are extremely heterogeneous. In experimental work, all conditions, excepting the variable(s) under test, must be kept as uniform as possible, and this evidently requires conditions and actions not available on the average farm.

Before the farmer can benefit from the work at the experiment station, it is necessary, as we have already insisted, to test the results of this work under farmers' conditions. Hence the stress on pre-extension research and on-farm trials to validate or adapt the results obtained on the experiment station.

Size of the area required

The major factor determining the size of the area required for an experimental farm is that almost all field experiments have after-effects which may influence the crops grown subsequently for a number of years (the study of these after-effects may in itself be a worthwhile subject of investigation). Because of these after-effects, no new experiments can be laid down on the same site for a number of successive years. In the interim period, crops receiving uniform treatment have to be grown on the site, for at least three years but preferably more. For this reason alone, the area of the experimental farm must be several times greater than the area required for all the experiments carried out in a given year.

There are also problems for which scale of operations is an essential element in the research, and for which conventional field experiment layouts cannot be used. Much can, however, be learnt from careful observation of non-replicated treatments, even if a statistical analysis of the results cannot be performed.

Therefore, an estimate needs to be made of the net area that will be required for annual experiments. The figure arrived at should be multiplied by factor of 4 or 5. To this must be added;

(a) the area required for permanent experiments (on fruit trees, permanent pastures, crop rotation experiments, long-term fertility trials, etc.);
(b) the area required for the layout of roads, irrigation and drainage, etc.;
(c) the area required for buildings, gardening and recreational purposes;
(d) areas not suitable for experimental purposes (see above) within the proposed site.

Every effort should be made to reach a realistic figure; if the area planned and then allocated for the experiment station is too small it will generally be very difficult and expensive to expand at a later date, and the experimental programme will have to be restricted. If too large an area is acquired, this may become too great a financial and managerial problem.

Farm equipment

The farm should be properly equipped with tractors and equipment for tillage, seeding, fertiliser application and crop protection, as well as for harvesting and processing. Under certain conditions, draught animals and animal-drawn equipment may also be required. Every experiment station should have the instrumentation required for obtaining basic meteorological data.

Maintenance shop

A properly equipped maintenance shop is an absolute necessity. Timely preparation of equipment and machinery and the possibility of emergency repairs can decide the fate of an experiment.

Farm management

Poor management of the experiment station farms has frequently and seriously hampered research programmes in developing countries.

A farm manager with proper qualifications can be a major factor in the successful operation of the regional station.

Oram (1977) has defined the role of the experiment station manager as:

The day-to-day management of field operations at a major experiment station, or a group of stations under the aegis of a major institute, working across a range of crops and environments. The role of such a manager is to ensure that everything is provided at the right time and the right place to enable scientific staff to work smoothly; he is responsible for land preparation, cultivation, irrigation, equipment, fertiliser, labour and prevention of avoidable losses to experiments from thieves, predators, birds, etc. He must also be a skilled observer, and as such can be of considerable help to scientists who may not be able to visit the farm every day.

Generally, the most qualified candidates available for the post of farm manager are agricultural diploma holders with a few years' experience in agricultural production, but little or no knowledge of research requirements. They therefore tend to manage an experimental station farm as if it was an ordinary commercial farm, showing little understanding of, and sympathy for, the requirements of the researchers. As a result, relations between manager and researchers are frequently strained.

The managers of the regional experiment stations should have a university degree in agriculture and require additional general and farm training in management and administration, and should be fully informed of the requirements of agricultural research in general and field experimentation in particular. 'Such people do not grow on trees, and it is a common error to assume that anyone with a good knowledge on how to run a field experiment or even a farm can successfully take on such a job' (Oram, 1977).

It is our view that many of the subjects included in the courses for research management could also be useful for experiment station managers if a few additional topics are included, such as farm management and field experimentation.

The international research centres offer practical training for research support services, including the management of experiment stations (Vega, 1979).

Linkages between farm manager and researchers

In order to achieve mutual understanding between the farm manager and the researchers, and co-ordination of farm management with research requirements, it is useful to appoint a committee of representatives of the departments whose work depends on farm facilities, and who should meet with the farm manager at regular intervals, in order to present their needs and requests ahead of the farming seasons, so as to plan in good time the layouts of the experiments, establish priorities and provide information on their needs for equipment, materials and services.

The farm manager should, however, be entitled to make spot decisions on the use of personnel and equipment. Under circumstances in which weather largely dictates the timing of agricultural operations, a delay of a day, or even a few hours, in supplying equipment for tillage or sowing, may be disastrous for one experiment, while of less import for another. It is therefore essential that someone be in a position to make decisions reflecting these specific and unpredictable needs, in view of the often conflicting demands of the researchers, and when timing is critical.

Young researchers are frequently lacking in actual farming experience. As a result, the preparation of their experimental plots may be substandard, sowing and management faulty and plant protection inadequate. If the crops on the experimental plots are substandard, this not only reflects adversely on the research organisation, but the results of these experiments will be worse than valueless they will be misleading.

Hence the need to establish a small unit for the field work involved in experimentation, consisting of well-trained farm workers, equipped as far as possible with specialised equipment for field trials.

Allocation of land for experimental plots

There are in principle, two ways of allocating land to the researchers for their experimental plots: (a) the allocation of a certain area on a permanent basis to each department; (b) seasonal *ad-hoc* allocation of plots to each department, with appropriate siting of the experiments in accordance with a planned crop rotation.

The first solution has the grave defect that it makes it almost impossible to avoid laying out experiments on sites which have been used for previous experiments, without the benefit of a sufficiently long intervening period of uniform cropping. Nor are the crops tested under proper rotation conditions.

The second procedure requires that the entire farm be planned as a single functional unit, in which the land under annual cropping is divided into a number of fields in accordance with the crop rotation adopted; part of each field is reserved for experiments on the appropriate crop, and the remainder is grown with the same crop under uniform conditions.

This arrangement does not preclude: (a) the setting aside of an area devoted to perennial crops or long-term experiments; (b) the allocation on a permanent basis of small areas in the vicinity of the respective departments for breeding plots, for example, that require almost daily supervision by the researcher.

Farm records

A system of recording must be adopted and strictly adhered to, in which all farm operations on each field are recorded. Of particular importance for future experiments on these sites are records of the crops grown, tillage operations, fertiliser and herbicide applications. Reference to these records will frequently help in the choice of an experimental site and explain unusual responses to experimental treatments. Records on man/days of labour, use of machinery, the amounts of other inputs, can serve as a basis for calculating cost/benefit ratios for innovative treatments carried out on a farm scale.

An archive of annual cropping plans, showing the exact location of all experimental plots in a given year, should be available to the researchers. Otherwise, an unsuspecting researcher may lay out his experiment on an area that has recently served one of his colleagues as an experimental plot. The results obtained under these conditions may well be original and surprising, but will hardly be conducive to producing worthwhile information.

Income

Income derived from farm produce and services may be an important source of research funds. It would, however, be wrong to judge the effectiveness of farm management by the amount of income generated. The services required for the experimental work may often clash with the interests of commercial production and if the manager is judged by the profit he makes from the farm he will be under considerable pressure to be biased in favour of commercial production.

Most of the management difficulties encountered by a farm manager derive from government procedures, mainly those relating to the formalities of purchasing and selling goods, and the handling of funds. The farm manager should at least be allowed to maintain an imprest account, with the authority to allow expenditure for urgent items. The level of the imprest account can be adjusted to the variations in seasonal agricultural activities.

The services provided by the farm to the researchers in the management of their plots and expenditure for materials should be charged to the research project involved. Conversely, income derived from the plots should be credited to the same project. This will encourage the researchers to avoid unessential operations and expenditures and motivate them to ensure good husbandry practices to their experimental plots.

ON-FARM RESEARCH

Researchers in developing countries are becoming increasingly aware of the need to transfer part of the research activities from the experiment stations to farmers' fields.

The rationale of this approach is that work at the experiment stations aims at establishing the potential production under the agro-ecological conditions characteristic of the region, whilst the objective of on-farm research is to determine what the farmer can achieve by applying an improved technology.

The advantages to be derived from on-farm research can be summarised as follows:

- On-farm testing is more 'realistic' than work at the research station and the results will be more readily accepted by the farmers.
- The research can be carried out in a larger number of physical

environments than those provided by the experiment station and its dependencies.

- The feasibility of adoption is determined early in the testing process.
- The involvement of the farmer in the experimental work has considerable psychological value in the modernisation process. It also shortens the time for the flow of information from research to farmer and vice-versa.
- The awareness of the researcher of the real needs of the farmer, the constraints he faces and the techniques he uses, is enhanced.
- The researcher can learn from the farmer's experience and closeness to the land.
- On-farm trials provide a unique opportunity for co-operation between research and extension workers. The procedures involved in on-farm trials, and the resultant linkages with extension workers will be treated in more detail in Chapter 22.

References to Part Four

Ampuero, E. (1981). *Organización y Administración de la Estación Experimental en los Paises en Desarrollo*. Programme in International Agriculture, Cornell University, Ithaca, NY.

Arnold, M.W. (1976). *Agricultural Research for Development*. Cambridge University Press, London.

Arnon, I, (1972). *Crop Production in Dry Regions*. Leonard Hill, London.

Boyce, K. & Evenson, R.E. (1975). *National and International Agricultural and Extension Programmes*. Agricultural Development Center, New York.

Cernea, M.M., Coulter, J.K. & Russell. F. (1985). *Research–Extension–Farmer*. A World Bank and UNDP Symposium. The World Bank, Washington DC.

Ginzberg, E. & Reilley, E. (1957). *Effecting Change in Large Organisations*. Columbia University Press, New York.

Gross, B.M. (1964). *The Managing of Organisations: the Administrative Struggle*. Free Press, Glincoe, 11.

IESA (1980). *FONAIAP—Estudio de Organización*. Instituto de Estudios Superiores de Administración, Caracas.

ISNAR/SPAAR (1987). *Guidelines for Strengthening Agricultural Research Systems*. World Bank, Washington DC.

Kelley, M.J. (1964). Basic research. In *Handbook of Industrial Research Management*, ed. G. Heyel. Reinhold, New York, pp. 136–56.

Koontz, H. & O'Donnell, C. (1955). *The Principles and Practices of Management*. McGraw-Hill, New York.

Oram, P.A. (1977). Training requirements in research and its application—An overview. Paper presented at CGIAR Forum on Training, Washington, DC.

Raudsepp, E. (1963). *Managing Creative Scientists and Engineers*. Macmillan, New York.

Russell, E.W. (1970). National research planning at the national level. In *Organisation and Methods of Agricultural Research*, ed. J.B.D. Robinson. Ministry of Overseas Development, London, pp. 91–96.

Ruttan, V.W. (1982). *Agricultural Research Policy*. University of Minnesota Press, Minneapolis, Mn.

Ruttan, V.W. (1985). Toward a global agricultural research system. In *Agricultural Research Policy and Organisation in Small Countries*. Agricultural University, Wageningen, pp. 19–35.

Shepard, H. (1956). Patterns of organisation for applied research and development. *J. of Business*, **24**, 52–58.

Shepard, H. (1957). Organisation and social structure in the laboratory. In *Human Relations in Industrial Research Management*, ed. R.T. Livingston and S.H. Milberg. Columbia University Press, New York, pp. 185–96.

Simon, H.A. (1960). *The New Science of Management Decision*. Harper & Row, New York.

Urwick, L.F. (1965). *The Pattern of Management*. University of Minnesota Press, Minneapolis.

Vega, R.M. (1979). Post-doctoral and other training programmes of International Agricultural Research Centres. In *Preparing Professional Staff for National Agricultural Research Programmes*. Report of Workshop, Bellagio. IADS, New York, pp. 55–60.

Whyte, W.H. (1957). *The Organisation Man*. Simon & Schuster, New York.

Wright, B.C. (1982). Elements of successful crop research projects. In *Annual Report, 1981*. IADS, New York, pp. 15–25.

Part Five

ADMINISTRATION IN THE SERVICE OF RESEARCH

Chapter 14

Principles of Research Management

'Effective research management is the product of a unique combination of experience, insight, will, and personality' (Ruttan, 1982).

Though research organisations have their own specific structure, objectives, and needs, many tenets of general management of organisations are applicable to them; other tenets require adaptation to the special needs of research organisations and still others are not applicable at all.

In this chapter we present a brief review of management principles in their application to an agricultural research organisation, which in many cases complement the topics presented in Chapter 11.

BASIC ELEMENTS OF FORMAL ORGANISATION

The basic elements of formal organisation proposed by Mooney (1949) are:

(1) The *co-ordination principle:* Division of labour, essential in modern enterprise, will lead to chaos unless co-ordination is practised. This requires authority and leadership.

(2) The *scalar process* is defined as 'the grading of duties, not according to different functions, but according to degrees of authority and corresponding responsibility'. It refers to the need for a clear line of authority to run from the supreme authority of the organisation to every individual in the organisation, by a series of steps giving an uninterrupted scale. This is the *hierarchical* division of authority and responsibility, based on chain

585

of command, delegation of authority, unity of command and the obligation to report. Every individual knows who has the authority and responsibility for each field of activities in the organisation. Individuals should be able to communicate directly across the organisation, provided they keep their direct superior informed of any significant action deriving from these contacts.

(3) The *functional process* deals with the horizontal growth of the organisation and refers to the concept of specialisation. The difference between a head of department and a head of laboratory is one of degree of authority (and hence part of the scalar process), whilst the difference between a geneticist and an entomologist is a question of speciality and hence functional.

It is considered desirable that the work of each individual should be confined to a single function, and that related functions be grouped together under a single head.

(4) *Structure* relates to the logical relationships of functions in the organisation. Traditional organisation usually works with two basic structures: the *line* and the *staff*.

(5) *Span of control* is concerned with the number of subordinates a manager can supervise effectively. According to the traditional theory, this should not exceed six immediate subordinates if administrative efficiency is to be maintained. On the other hand, it should be as wide as possible, to keep the number of levels of management at a minimum.

(6) *Balance*, in a well-structured organisation, has to be maintained in the relative size of its subdivisions; sufficient flexibility should be ensured in the face of set procedures, and overcentralisation avoided.

The authoritarian approach of classical organisation theory is clearly reflected in the nomenclature used, borrowed from the military; hierarchy, chain of command, span of control, communication through channels, etc. Whilst the military nomenclature may be useful, the situations in which it has to be applied are entirely different in a military unit and a civilian organisation, in particular a research organisation.

In the research organisation the carrying out of instructions depends on a desire to co-operate and not on discipline and fear of sanctions.

The authoritarian tradition that permeates classical organisation theory is also no longer in accord with the democratic spirit of our times. A hierarchy, at the top of which stands a 'boss', in whose hands all authority and responsibility are concentrated, is considered an anachronism.

As a reaction against the 'mechanical' or 'engineering' approach to organisation has followed a period in which the human relationship approach has been stressed and the principles of social psychology applied to management science.

In the 1920s, following the famous Hawthorne experiments, Mayo (1960) showed that employees cannot be considered as passive and malleable individuals to be treated as machines, but are human beings whose productivity depends on inner motivation and that a 'sense of belonging' to a stable and cohesive group is essential to their well-being. The reaction of employees is not always rational; output does not necessarily increase if work is made easier or better incentives are provided. Hence the need to seek a better understanding of how people behave in an organisation.

The concept of 'participative management' was developed, whereby subordinates are encouraged to share in the responsibility for setting and achieving the goals of the organisation. The new approach did not envisage the destruction of authority, but advocated exercising authority in such a way as to 'achieve a high level of motivation, enthusiasm and loyalty and was therefore primarily concerned with the human side of enterprise' (McGregor, 1964).

The effective co-ordination of individuals working within an organisation requires that they understand the goals of the organisation and are imbued with a common sense of mission in achieving these goals. The newer concepts of organisation, which are usually identified with the human relations group, still accept many of the traditional tenets of the classical organisation theory. They stipulate, however, that these are not immutable principles, but that they can and should be modified and adapted to the needs of the people working within the organisation.

Since the 1960s, it has become apparent that there is a certain disillusionment with the concept of management as applied psychology. The belief has grown that there has been too much emphasis 'on helping people to grow and mature', and too little attention to productivity (Sayles, 1964). The behavioural sciences were criticised for not providing sufficiently realistic appraisals of managerial activities and of paying insufficient attention to the effect of changing technological systems. It was stressed that executives in a modern organisation are not concerned only with motivation; and that they deal with many people, within and outside the organisation, besides their subordinates.

Whatever the shortcomings of the human relations school in regards to industrial organisations, there is little doubt that their basic approaches are highly appropriate to research organisations and that many of their conclusions can be usefully applied to the latter.

The *scalar* and *functional* processes, do not always stand the test of reality of modern organisations and in particular research organisations. The relationships between the research director, department heads and project leaders have a certain similarity to the scalar process; they are, however, dissimilar in essence. The whole concept of the superior–subordinate relationship, as it exists in governmental or industrial organisations, is uncharacteristic of the relationships between the different levels of research leadership. The need for decentralisation, delegation, participation and consultative management, as stressed by the 'human relations' approach, is applicable to research organisations.

Organisation does not necessarily imply regimentation. Organisation can be adapted to the needs of research, so as to promote creativity and take into account the special mentality of research workers. *Decentralisation* and *delegation* make it possible to give researchers a great degree of freedom and responsibility in the direction of their own activities whilst *participation* and *consultative management*, under proper conditions, give the staff an opportunity to influence decisions that affect their work.

There are, of course, still many scientists who feel that the less management there is in a research organisation, the better. Management 'should arrange for funds, facilities and buildings; engage outstanding scientists and drop by occasionally to ask "What's new?"' (Van Tassel, 1965).

HIERARCHY IN THE RESEARCH ORGANISATION

In an organisation, individuals have responsibility for various major functions. They are assisted by executives who each have well-defined responsibilities, and so on down the line. Thus is established a hierarchy of responsibility and authority, in which the higher positions in the hierarchy provide a unifying symbolism of tremendous significance (Gross, 1964). No organisation can be held together without some pattern of hierarchy. Simon (1960) writes that 'an organisation will tend to assume hierarchical form whenever the task environment is complex relative to the problem-solving and communicating powers of the organisation members and their tools'.

Even in a research organisation, by its very nature averse to the concept of hierarchy, the legitimate framework is provided for making important decisions and for settling internal conflicts, thus making it possible for the organisation to act effectively. As Koontz and O'Donnell (1955) point out, 'without definite lines of authority, the way is prepared for politics, intrigue, frustration, lack of co-ordination, duplication of effort, vagueness of policy, uncertainty of decision making and other evidences of organisational inefficiency'.

Multiple hierarchy

The traditional hierarchic concept assumes that each individual in the organisation has only one direct superior. This is incompatible with specialisation, an essential element of research organisations.

Single lines of authority have never existed in research organisations, in which three different groups of people work together: scientists, technicians and administrators. One line of authority extends from the director of research, through the vice-director to the heads of research projects, researchers and assistants. A second line extends through the assistant director for management. Hence a researcher may be subject to the authority of his head of department, to the administrative assistant, and to a project or team leader. If engaged on a number of projects, he may even have several team leaders to whom he is responsible. Authority may be functional (project leader), professional (head of department) or administrative (budget or auditing personnel). A technician may be responsible to the researcher with whom he is working and to the head of the service department of which he is a member. Instructions are received not only from superiors but also from various specialists. If all these instructions were to go 'through channels', they would rapidly bog down. This necessitates, in the research organisation, what Gross (1964) calls 'multiple hierarchy'.

In general, multiple hierarchy is based on the dualism of 'administrative authority' and professional authority, which incidently are the source of many of the intra-organisational disputes resulting from divided loyalties, or ill-defined boundaries between the different forms of authority.

A special form of multiple hierarchy is found in appeal, grievance and suggestion systems, whose very objective is to provide additional lines of communication and authority. An essential ingredient of an appeals system is to make it possible for subordinates to by-pass their immediate superiors.

LINE AND STAFF

Both traditional organisation theory and the newer concepts are agreed on the need for two types of executives in any large organisation in which specialisation is necessary: line and staff. The line is made up of the people whose work contributes *directly* to the achievement of the fundamental goals of the organisation, and the staff, of those who, by supporting the line, contribute indirectly.

Whilst in an industrial enterprise the research laboratory is staff, in the research organisation it is those who carry out or supervise research who make up the line. The line man has authority to give orders to those carrying out research whilst a staff man, at least theoretically, has no authority over line and can give orders only to his own subordinates; his duties are to provide advice, help in co-ordination and control, and supply services to the line personnel at all levels of the organisation. However, certain key staff personnel, such as those in charge of finance or personnel, may by enforcing rules and regulations, achieve indirect, though powerful authority over the line. The relations between staff and line executives have been the subject of much theoretical discussion.

A potential source of conflict between staff and line derives from those situations in which the line has the formal authority, whilst the specialist, whose duty it is to advise, has the technical competence to make the decisions, which his 'line' superior lacks. As a result, there is frequently a running struggle for influence on higher management, between ambitious staff employees and line personnel (Brown, 1954).

Generally speaking, the expert is no longer considered as a man merely giving advice, his opinion carrying much more weight than mere advice. A relationship frequently evolves which is neither orders nor advice.

The special functions and responsibilities of staff groups in the research organisation are to provide the specialised knowledge that line leaders lack, or to devote themselves to certain specific functions for which the line executive does not have the necessary time. Pfiffner and Sherwood (1960) propose three main categories of staff:

(1) *General staff*, who work on overall plans and policies. Any decision that cuts across departmental lines must, in principle, be made by the director of research. General staff men are those to whom some of the functions that are the direct responsibility of the director can be delegated. The deputy director is the prototype, who handles parts of some or all of the director's principal functions, but no one function in its entirety.

The executive committee, though its members are both line and staff, itself has a staff function. In principle, its function is to advise the director of research, and even if the latter usually accepts the committee's recommendations, he has the responsibility for the decisions, and orders are issued in his name.

Heads of line departments may have staff units of their own.

The director may also appoint an 'assistant to the director', who serves as his 'eyes and ears'. His principal duties are to keep track of what is going on in the organisation.

The Research Project Office is also an important staff unit, which devotes its time to the continuous evaluation of research projects, keeps watch on whether the project leaders are maintaining their schedules and submitting reports, and transmitting their findings to the director, who can only intermittently concern himself with this subject.

(2) The *specialised staff* consists of specialist groups, such as a division for experimental planning and statistical analysis, who have unique skills or knowledge.

(3) *Co-ordinating and auxiliary staff* are those who are concerned with budgeting, accounting, personnel, public relations, maintenance, supply, transport, etc.

SPAN OF CONTROL

The classic concept of 'span of control' remains valid. There is a limit to the amount of information that an individual can receive, evaluate, memorise and communicate. Therefore, grouping remains an essential feature of organisation, with its concomitant span of control.

The idea that there exists a universally applicable span of control is, however, no longer generally accepted. The effective span is considered to depend on the type of organisation and on the functions of the people concerned. A tall structure with short span characterises the tight control of an autocratic organisation and favours passivity, dependence and submissiveness of subordinates (Argyris, 1962). It also makes necessary many levels of management, which slows down communication between levels, increases red tape and reduces administrative efficiency (Karger & Murdock, 1963).

A wider span, which requires a larger measure of delegation of power, is certainly more suited to research organisations. With competent scientists as heads of departments, much less contact between the director of research and the heads of departments is needed than in other types of organisation.

AUTHORITY

The standard definition of authority in management parlance is 'legal or rightful power; a right to command or to act' (Koontz & O'Donnell, 1955). Authority, according to Malinowski (1960), is 'the force that unifies the social group, without which chaos results'. As with many other terms coined in management science, the word 'authority' with its connotations of arbitrariness and despotism, does not accord well with the newer concepts of the 'human relations' approach, and certainly not with the spirit of research organisations. Hence, the tendency to avoid using the word 'authority' and to replace it by others, such as 'responsibility'. However, the terms responsibility and authority are not interchangeable and the indiscriminate use of the one in lieu of the other leads to confused thinking on this basic subject.

Sources of authority

Formal authority
Originally, the authority of the hierarchial chief sprang from his charismatic qualities. Charismatic qualities derive from the personal leadership qualities of the individual rather than from the position he holds. In classical management theory the ultimate source of authority is *institutional*. Government is usually the source of authority for the agricultural research organisation. Through its specialised agencies (treasury, commissioner of manpower, etc.) it enacts rules and regulations, engages and discharges research, technical and administrative personnel, decides on salaries and promotions. Authority to determine research policy, and the appointment of the research director is delegated to the ministry of agriculture, who may in turn delegate this authority to a board of trustees. Authority to execute research policy is invariably delegated to the director of research.

Authority confers the formal right to make decisions and to enforce them as well as to resolve disputes; it is an essential element of institutional organisation (Malinowski, 1960).

Authority and power
Formal authority is the legitimate source of power; there is, however, no

guarantee that every executive invested with formal authority is able to wield the power deriving from his authority. Authority consists of the right to engage in certain actions needed for the guidance of the organisation or of certain units thereof. These rights to exercise power in a specified area include the right to request certain kinds of information; make decisions; initiate action; and allot rewards and punishments.

Authority does not necessarily ensure influence or real power. Figureheads have authority but no power. Individuals with influence within the organisation may exercise power far in excess of their authority. However, authority is necessary to legitimise power, both for the individual administrator as well as for the organisation as a whole. Even if the sole authority to engage in agricultural research is vested in a single central organisation, if the organisation is ineffective, either in its performance or in establishing a close relationship with the ministry, with the extension officers and the farming population, it will find itself powerless against 'bootleg' research carried out under various names and pretexts by individuals acting within organisations that do not have the 'authority' to do so.

Authority will usually ensure real power in the following circumstances (Gollembiewski, 1964):

1. if the way authority is exercised is acceptable to the subordinates, e.g. the research director adopts a participative approach;
2. if the degree of pressure he exercises on his subordinates 'remains above a threshold considered reasonable' by them;
3. a substantial amount of autonomy in planning work, making decisions, etc., is allowed to the individual.

The acceptance theory of authority
Much credence has been given to the theory that the real source of authority lies in the readiness of subordinates to accept the decisions of the superior. The individual, as Tannenbaum (1962) points out, 'always has an opportunity with respect to a decision made by another, directly to affect his behaviour, to accept or reject that decision'. Without this acceptance, authority is illusory. The readiness to accept authority depends on whether the advantages resulting from acceptance outweigh the disadvantages. The supervisor has legal and economic means to enforce his 'power': he can affect the promotion, the working facilities and other important aspects of his subordinates' professional life. He is, however, dependent on the subordinates for his own effectiveness in the execution of his duties.

In addition to reason, technical know-how and special authority, the supervisor needs persuasion, inducement and tactics to win power. Scientists, as a group, tend to resist authority, which they consider as an unwarranted intrusion in their activities. They may, both consciously and unconsciously, attempt to assert their independence on all possible occasions and resist any encroachment of authority on their freedom of action. Whilst one may have reservations as to the validity of the acceptance theory of authority, there is little doubt that for an order to be effective, it has to be understood, be compatible with the recipient's ability and preparedness to carry it out to the best of his ability.

Personal authority
Personal authority is an essential complement to official authority. It is based on intelligence, experience, leadership ability, and professional standing, itself based on past services.

Functional authority
Specialisation, which characterises modern research, provides the specialist with the *technical competence* to make decisions, whilst the hierarchical chief, who retains the formal authority, becomes less and less professionally able to encompass all the aspects of his role. This causes a dispersal of authority in research organisations and an ever widening gap between formal authority and competence. The trend in research organisations is therefore to accord more authority to functional status and less to hierarchy (Gross, 1964). In many cases, the lack of specialised knowledge, or the inability to supervise adequately certain functions of subordinates, causes the partial transfer of authority from the line superior to the specialist—hence, 'functional authority' or 'pluralistic' authority. Functional authority may be exercised by line, staff or service heads of departments (Koontz & O'Donnell, 1955).

Authority in research organisations

It is doubtful whether any of the sources of authority outlined above is, in itself, adequate to ensure the exercise of authority in a research organisation. As Koontz and O'Donnell (1955) point out, different social groups react differently to the exercise of authority. In particular, research workers are allergic to executive authority yet they are still sufficiently aware that without adequate authority, co-ordination of effort and the achievement of a common goal are not possible. Formal authority makes for efficiency, economy and convenience in communications. It must, however, be mitigated by reason and adapted to reality. Whilst in an industrial organisation the power wielded by the manager (based on disciplinary and economic sanctions) may be sufficient to enforce acceptance of authority by all subordinates, in the research organisation, authority is a compound of the formal power vested in the individual by the institution, his acceptance as a leader by his peers and subordinates, and the acknowledgement by them of his scientific and technical competence.

In the research organisation the conflict between formal authority and technical competence is more apparent than real. The needs of specialisation are taken care of in establishing the organisational pattern, the division of work and the delegation of authority. The hierarchical positions are filled by people with the requisite scientific competence. Potential discrepancies between formal authority and competence become most apparent at the top of the hierarchical ladder. This is, however, a built-in defect which cannot be overcome by authority giving way to competence.

At these levels a formal authority is required to co-ordinate between specialist groups. Key hierarchical posts have to be occupied by generalists, 'without whose mediation the warfare among experts might perhaps be fiercer and produce less rational behaviour' (Kaufman, 1962).

An analysis of the administration of research clearly indicates that there are distinct differences in the nature of authority exercised by the 'research line' at each level.

At the top level, the director and his associates do not usually conduct research themselves; their responsibility is mainly administrative: formulation of research policy, selection of personnel, allocation of resources, etc. As a rule this does not involve any direct interventions in the actual conduct of research by the individual scientist. Gross (1964) mentions the fallacy, rooted in 'organisation chart psychology', of thinking of the director of research as the man standing on top of a pyramid, who 'dominates' the organisation.

First, the 'man on top' is responsible to higher authority—whether the minister of agriculture, a national council for agricultural research or a board of governors. He is no less subject to directives than to giving them to others. Secondly, he is continuously submitted to frequently conflicting pressures from all sides: colleagues within the organisation, on whose co-operation he depends, and outsiders who present demands that must be met or resisted. Thirdly, 'he is subjected to a constant bombardment' from below. The people over whom he has authority are constantly submitting requests, which frequently conflict with each other, exerting pressures in view of obtaining action favourable to themselves, and in many and sundry ways attempting to force his hand. The director is not at the top of triangle, but, as Gross defines his position, he is 'at the centre of a confused and whirling vortex of centripetal forces. He buffets and is buffeted.' Whilst the director retains the formal authority to approve a decision, in the *formulation* of the decision he will usually depend to a large extent on consultation and persuasion.

At the level of head of department, the responsibility is mainly one of co-ordination between divisions and ensuring the appropriate and effective framework for departmental research. It is only at the level of heads of divisions or laboratories that direct intervention in the conduct of research may be necessary and legitimate. This is the level at which there is practically no discrepancy between technical competence and authority and yet this is the type of authority that is most likely to be resented by the research worker.

In general, it can be stated that authority and responsibility in the research organisation are diffuse; authority and dependence have an entirely different connotation than in other types of organisation.

Delegation of authority

Authority is centred in a chief executive. However, the manifold and complex duties of the director make it impossible for him to accomplish all his duties single-handed. It is therefore imperative for him to *delegate authority* to a number of associates. By delegating authority, the director will find time and freedom to handle adequately the functions which devolve on him alone.

A further need for delegation of authority arises from the requirement to divide the research organisation into a number of departments, each responsible for a certain field of activities, and subdivide these into divisions, laboratories, etc. This obviously also entails delegation of authority without which decentralisation, which characterises the research organisation, would be impossible.

Substantial authority for making various decisions independently is therefore delegated to department and division heads, project leaders, administrative executives, etc. This results in a situation which grants a considerable degree of autonomy to the individual units of the organisation, with their specialised areas. Once granted, delegation of authority should not be partial or subject to continuous control. The incumbent should be given wide latitude in performing his duties independently, on the tacit understanding that he is carrying out, in good faith, agreed policies, and is not acting in opposition to the views and wishes of his director. In case of doubt, it is obligatory to consult his director before making a decision.

Delegation of authority, and the resulting autonomy, of course create certain difficulties in the attainment of co-ordination and operational efficiency, but this is a price that has to be paid to prevent bottlenecks in the organisation.

Delegation of authority can be specific or general, but it should always be clearly defined. Authority should be delegated as close to the point of action as it is possible, in order to obtain sound decisions. The administrative authority for specific research projects, for instance, should reach right down to the project leader; each researcher should also have authority to expend the funds allocated to him, without reference to higher authority, provided he stays within the broad lines of the approved allocation. He also takes the decisions in relation to technical assistance, the use of research facilities, etc., related to his daily work.

Delegated authority can always be retrieved when changes in policy or programme are contemplated. Every reorganisation involves the recovery and redistribution of authority.

Splintered authority exists when a decision can be made only by two or more executives, the alternative being to shift the onus for the decision to a higher echelon. If the head of the organisation is not to be swamped by problems requiring decision, cases of splintered authority can hardly be avoided. However, too many cases of splintered authority may indicate the need for reorganising authority delegation (Koontz & O'Donnell, 1955).

RESPONSIBILITY

Responsibility can be defined as the obligation of an individual to whom a duty has been assigned to perform the service required. One cannot hold an individual responsible for duties for which he does not have the necessary authority. Whilst authority can be delegated and duties assigned at each level, responsibility cannot be delegated (Koontz & O'Donnell, 1955). No individual can abdicate or reduce his responsibility towards his superior by assigning a duty to someone else. The head of each organisational unit is fully responsible for the acts, mistakes or shortcomings of his unit, even if the direct blame is due to a subordinate.

It is the head who assigns the duties of his subordinates and hence is held responsible for their actions.

CONTROL

In the research organisation, control refers to two entirely different types of activity: performance of the research by teams and individuals; and performance of all the administrative functions of the research organisation.

Control of performance of research

Controlling the performance of the research teams and individuals can be defined as the endeavour to achieve co-ordination of efforts and ensure that the implementation of the programme is according to plan. In view of the special characteristics of research and researchers even the word 'control' has unpleasant connotations. The forms of control used in most other types of organisations are not applicable and special approaches are needed, in order to safeguard the freedom of individual planning by the researcher. The problems of how to achieve essential control and at the same time not to encroach on the initiative and freedom of action of the researcher have been treated when discussing the subjects of evaluation of the results of research, of the work of researchers, budgeting as an instrument of control, etc.

Control of performance of administrative function

Since World War II, many institutes of higher learning and of research have been making efforts to increase their organisational efficiency by making use of new techniques of management. These include the operation of professional offices on institutional research, the use of new mathematical methods of quantitative analysis as a basis for making decisions regarding the internal allocation of resources, and a growing reliance upon automatic data equipment capable of rapidly processing large amounts of information relating to the everyday processes of administration.

Chapter 15

The Scientist as Research Administrator

SEPARATION OF RESEARCH DIRECTION AND BUSINESS ADMINISTRATION

It has been suggested that the separation of the major functional roles within the organisation among scientists, 'managers' and administrators, will contribute considerably to a reduction of the strains and conflicts within the organisation. The scientists would be concerned with the achievement of the research goals of the organisation, the 'manager' (who should be a scientist himself) with the direction and co-ordination of research and, at each level, administrators with no formal decision-making prerogatives, would ensure the business administration of the organisation.

It is a generally accepted principle that the direction of the research organisation should be in the hands of a senior research worker, capable of commanding the respect of scientists and administrative personnel alike. In view of the similarity in background and thinking of the director and the research workers, the latter will be confident that administrative considerations will be subordinate to the needs of research.

By separating the research direction and business administration, the research director will be freed from administrative details. Whilst it is well known that scientists will resist direction from non-scientific administrators in matters directly pertaining to their research work, it is also true that they are more readily prepared to accept direction on administrative procedures from the administrator than from the research director, who usually is not fully competent and certainly not as well qualified in this area as his administrative assistant.

TRANSITION FROM RESEARCH WORK TO RESEARCH ADMINISTRATION

In the research organisation 'leadership' consists of a combination of scientific and administrative work that may be, and usually is, a source of conflicts, that increase as the administrative responsibility increases in relation to the scientific work. The role conflict of the 'researcher-as-administrator' has been described by Gross (1964).

At the lower levels—as head of a research project, a laboratory or even a research division—the administrative responsibility is not a serious burden on the researcher. It is mainly a tool by which he can further his scientific work. The furtherance of his career is almost entirely dependent on his scientific contributions; any irritations and apparent time-loss involved in his administrative activities are more than offset by the authority and prestige that are concomitant with his administrative standing. He is the centre of the research effort in his field, fully conversant with every detail of the work carried out by his unit, personally and professionally able to give guidance and supervision.

With the next step upward in the administrative ladder, as head of a department of a large interdisciplinary team or of a regional experiment station, the role conflict may assume serious proportions. The time and energy the researcher must devote to administrative functions increase— much of his effort is devoted to furthering the scientific work of his colleagues, whilst his own becomes more and more curtailed. He becomes more concerned with co-ordinating and furthering the scientific activities of other specialists, with which he may be familiar but of which he is *not* the scientific leader—a dichotomy results between his role as scientist and his role as administrator. He directs scientists who are the experts in their particular fields. He still attempts, tenaciously, to continue his own research, but he must delegate more and more of his scientific activities to his juniors; he finds it more and more difficult to keep abreast of developments in his own field and a serious professional and personal conflict arises when he realises that he is no longer really 'needed' by his erstwhile associates. He may feel that the status, prestige and possibly monetary rewards concomitant with his administrative duties are not a compensation for his reduced ability to devote himself to his scientific career, for which he has spent many years of preparation, in which he has carved himself a name and which is of far greater interest to him. Conversely, the department is a relatively small administrative unit; it does not involve any complicated problems of organisation or administration. It does not offer any special challenge; the head of the department does not usually have sufficient administrative responsibilities to keep him fully busy, but they are sufficient to interfere with his research work. It is therefore not surprising that these are the functions in the research organisation that are the most difficult to fill. The more successful the researcher is in his field, the less he will be inclined to undertake the onerous duties of head of department, and even after he has taken the plunge, he may reconsider 'que le jeu ne vaut pas la chandelle'. As he has not burnt his bridges behind him, his full-time return to scientific activities is not fraught with insuperable difficulties.

It is the third level which really constitutes the point of no return, when an able scientist is called upon to head a large research institute or the agricultural research organisation, as a whole. He usually has little understanding of the new role awaiting him, and even has negative feelings about it. Milberg (1957) calls attention to a certain inability on the part of scientists to define the research management task. He states that the 'vagueness of terminology and conceptualisation with which scientists describe the job of scientists that move to a managerial function, is highly uncharacteristic of the attitude of scientists to other matters'.

On accepting a managerial appointment, the scientist is not immediately aware of a loss of professional activity, and may even delude himself into believing that he will be able to continue, albeit on a limited scale, his professional work; his delusion is frequently one of the factors in deciding on accepting his new role. He soon realises that adequate attention to research, to writing and to keeping up with professional literature can be maintained only at the expense of his administrative responsibilities. This realisation does not come immediately; the erosion of his professional activities is gradual, he becomes more and more involved with the workings of the organisation, personnel and with relations outside the organisation. By the time he realises that he has actually made a choice between scientific endeavour and administration, he will find that the bridges behind him have already been burnt. It depends on how he faces this conflict on 'whether a good scientist has been sacrificed to obtain a bad, mediocre or an unhappy administrator' (Gross, 1964) or whether he is able to overcome the emotional shock and to find his new occupation a challenge and a source of satisfaction. The most insidious danger is that he may try to 'keep a foot in both camps', attempting to maintain full involvement in the direct conduct of research but reluctant to give up the prestige of his administrative post. This can be done only at the expense of the organisation as a whole, as a crucial position will be filled by a person unwilling and unable to devote himself entirely to the executive responsibilities it entails.

Barrows (1963) summarises the major problems which arise when the director of a research organisation is not fully committed to his managerial duties and/or is not conversant with management science:

- a breakdown in functional organisation planning and consequently in internal communications;
- a lack of balance between scientific personnel and supporting personnel, in order to preserve a 'basic capability for research';
- a tendency to consider cost accounting procedures, budget reports, job evaluations and other administrative controls as needed only to keep government watch-dogs at bay;
- the confusion of organisational objectives with research objectives;
- a vacuum in executive decision-making, resulting from failure to define management responsibilities.

It is clear that a scientist, when accepting administrative responsibilities in research, is making a decision that will have major consequences not only on his future life, but also on the research organisation as a whole.

Steele (1962) writes of the 'homelessness' of research managers: by their former colleagues they are considered as deserters from research, by administrators as scientists unequipped for managerial duties. The first and basic innate quality required of the research administrator is leadership and the ability to deal with a wide range of people. Scientists, as a group, have certain characteristics, some of which are assets and others liabilities, when managing them. They are intelligent, have analytical ability and a certain amount of objectivity, and their work is of paramount importance to them. They are, however, allergic to administrative procedures, dislike controls, and owe allegiance to their profession rather than to the organisation. The research director must find a way of resolving the conflicting aims of the organisation and the individualist drives of the researcher.

Whilst his training as a specialist may have served him well, he must, in his new position, become a generalist, able to co-ordinate the work of specialists and experts in a wide range of fields. In order to keep the research programme attuned to the need of the farming community, he must be well informed on the realities of the agricultural economy of the country.

THE DIRECTOR OF RESEARCH

Attributes required of the director

It is almost axiomatic to state that the director of a research organisation must himself be a scientist, who has achieved a reputation as a capable research worker. This requirement is based more on psychological grounds than on logic; it derives from the natural desire of scientists to have as their superior an individual who fully understands them and their problems.

The prestige that he has achieved by having reached an acknowledged status in his field is a prerequisite for the director's new post, a phenomenon that Gross (1964) calls 'prestige transference'. His scientific competence is also necessary, not so much for the actual direction of the research effort, as for ensuring his acceptance by his fellow-scientists as a leader; a worthwhile and desired source of approval.

The scientific background is also a prerequisite in the making of many of the decisions with which he is faced in the course of his work. If the role of director of research is assigned to a mediocre scientist who is a competent administrator, the research workers will not have confidence in his professional judgements, and will resent his status as a superior. However, a scientific background does not automatically ensure harmonious relationships with his scientific staff or his suitability as administrator of a research organisation. As Mandell (1961) trenchantly points out, 'scientists can be vicious in attacking men with backgrounds similar to their own, who, they feel, subordinate professional objectives to administrative considerations'. The head of a research organisation must be both a scientist and an administrator. The larger the organisation, the greater will be the demands for management skills and organisation ability.

An excellent scientist may become 'a wretched administrator, an unhappy man and a lost expert' (Gross, 1964) if he does not have the natural attributes needed for his new position, and if he does not acquire the necessary basic knowledge of administrative procedures.

The role of the director of research

The research organisation is a complex and dynamic structure, and therefore the job of director is bound to be complex and difficult. The basic objective of the director is to achieve an organisation that is efficient in attaining its objectives. However, it should be realised that the research director does not exercise independent and unchallenged authority in the management of the organisation that he heads.

The functions of the research director are: (a) planning, based on decision-making; (b) organising for the most efficient implementation of the plan; (c) ensuring effective communication within the organisation; (d) co-ordinating the research effort; (e) directing and administering; and (f) representing the organisation.

Planning

'Planning is the executive function which involves the selection, from among alternatives, of enterprise objectives, policies, procedures and programmes.' Planning cannot be separated from managerial performance (Koontz & O'Donnell, 1955).

The ultimate approval for the research programme is in the hands of the minister of agriculture. It is, however, the duty of the research director to screen the individual draft proposals of this programme, to submit them to the board of governors, research committee or other formal agency to obtain approval and support for his proposals. Subsequently, the director of research has to solve the problems involved in carrying out the plan, and at the same time maintain the general scientific level of his organisation and the job satisfaction of his researchers.

The need to guide research along lines of interest to the agricultural economy of the country, and at the same time to protect the researchers from excessive outside pressures, is an inescapable source of tension and strain to which the director is continuously submitted.

In a dynamic agricultural economy, which should be the main characteristic of the agriculture of developing countries, it is difficult to predict the long-term objectives of the research programme. It is the director's responsibility to plot a general course and then to adjust, whenever necessary, its direction according to the changes in the general agricultural policies with which he should be familiar. He should have authority and funds to initiate exploratory or even speculative research projects without reference to any outside authorities in order to encourage initiative and original thinking.

Organising

According to Koontz and O'Donnell (1955), organising 'involves determination and enumeration of the activities required to achieve enterprise purposes, the grouping of these activities, assignment to subordinates and delegation of authority'.

The principal function of the director is holding the research organisation together as a functional working unit and initiating reappraisal of the organisation in the light of changing programmes and policies. He is responsible for the organisation of the extremely diverse activities of a heterogeneous group of researchers.

Co-ordination

Co-ordinating and keeping in proper balance the various units of the research organisation is one of the main responsibilities of the research director. The objective is to ensure that the different parts of the organisation complement each other so that maximum productivity is achieved. This requires, amongst other things, defining the limits of responsibility for different fields of research so as to avoid, as far as possible, overlapping and duplication of effort. The director must therefore reappraise, at regular intervals, the research programme of the organisation in the light of results, and of new situations as they develop.

Directing and administering

It is the director's responsibility to ensure that the researchers have the assistance and equipment that are needed to carry out adequately their research obligations. He makes the physical plant decisions, e.g. what buildings and facilities should be built, and according to what priorities.

He must establish basic procedures and regulations, to ensure an orderly functioning of the organisation, whilst at the same time holding in check excessive demands of the administrative personnel for paperwork by researchers.

He will review the estimates for research work, and will make the final assessment of budgetary proposals.

He will submit the formal report on achievements of the organisation to the ministry of agriculture.

It is his duty and prerogative to supervise the recruitment of scientific staff and the selection of senior administrators. He must activate the machinery to handle general administrative tasks, the provision of supplies and services, etc.

Representation

There are many occasions on which the director must represent his organisation, and this may take up a considerable part of his time. He will have to maintain frequent contact with the highest echelons in the ministry of agriculture, the national scientific council and farmers' organisations. He will have to attend conferences, both national and international, be a member of various boards, co-ordinating committees, *ad hoc* committees, scientific societies, etc.

Delegation of authority

The brief outline of the activities of the director of research given above presents only a fragmented picture of his duties. If we add the many hours he has to devote to reading the literature and reports in order to keep up with developments in the general field of agricultural research and in his own organisation, and that he must devote a considerable proportion of his time to public relations, committee meetings outside the organisation, one may well ask: how it is possible for a single person to shoulder such a wide and heavy responsibility? The answer is, of course, that this can be done only by delegating authority.

It is a basic tenet of administration theory that the head of a large organisation should not be *directly* in charge of any of the major activities of the organisation, so that he is not handicapped in his ability to act as co-ordinator, to determine policy and to plan for the future.

In the agricultural research organisation, the head must have two principal deputies: one for research, the other for administration. The *deputy or vice-director for research* comes immediately under the director in the organisational hierarchy. He assists the director in the shaping of research policy, prepares the data and material on which decisions affecting the research programme are based. He has primary authority for the conduct and co-ordination of research, helping in the selection of research staff, determining the relative needs in personnel and equipment of individual research projects. He advises on the possibilities of collaboration between units within and outside the organisation. He reviews the scientific papers based on the research findings of the organisation. He, in turn, delegates authority for a number of these activities to an assistant for research projects, and to a number of committees.

The assistant director for administration assists the director with budgetary control and personnel policies. He is in charge of the development of plant facilities, the running of the agricultural experiment stations, discipline and labour relations.

He, in turn, is assisted by administrative personnel, each in charge of one of the many fields of administrative activity: head of personnel, head of services, treasurer, superintendent of regional stations, etc.

The next question which then arises is: if every specialised function in the organisation is delegated to someone else, what is there left for the head of the organisation to do? As Braybrooke (1964) expressed facetiously, it would appear that, in a smoothly running organisation, the head would have practically nothing to do but approve the decisions of those who are in charge of the different activities of the organisation, and in all probability would very rarely have occasion to disapprove.

However, this apparent conclusion overlooks one major point and that is that the director may delegate authority, but cannot abdicate from his responsibility. He remains a focal point of responsibility in the organisation and this is his genuine and indispensable contribution.

Bavelas (1960) is of the opinion that the main responsibility of the head of the organisation is essentially the making of decisions taken in the face of uncertainty, or what he calls 'uncertainty reduction'. Whilst authority for decision-making in fields which are within the competence of various units of the organisation can be delegated to the proper levels, it is the head who must make the decisions that affect the organisation as a whole. The amount of knowledge available, on which a decision can be based, is not always sufficient 'to be able to predict the consequences that can follow a particular choice and the probabilities of given consequences' (Dill, 1964). When decisions have to be taken in situations in which the amount of information available is not sufficient for certainty of choice, it is up to the director to decide and act after having listened to, and evaluated the opinions of his associates, and then have the courage to accept full responsibility for the consequences of his decision.

THE HEAD OF DEPARTMENT

As already indicated, the department is, by and large, the functional administrative unit of the research organisation, consisting of a number of divisions each responsible for a certain area of research endeavour, but all together having certain common interests, services and professional background.

Ideally, the head of department should be both a leader in the field of research with which his department is concerned and an administrator. In view of the complexity of the research problems handled by a department, normally consisting of 3–7 divisions, each with a number of areas of research activities, it is rare indeed to find an individual fully qualified to provide scientific leadership in such a variety of fields. In general, the head will not be as well versed in many of the research areas with which his department is concerned as the scientists who work under him. However, his maturity and experience, his scientific ability and broad knowledge of the field should enable him to discuss intelligently the problems of his researchers in whose work he should take an active interest, providing them with the maximum help and encouragement and at the same time avoiding interference as far as possible. His professional duties will be mainly concerned with co-ordinating and ensuring a general environment favourable to research activity. In particular, he will be concerned with overcoming the natural tendencies of the individual divisions to isolate themselves, by improving communications between them and encouraging teamwork. He nominates the team leaders for research carried out by members of two or more divisions of his department. He will initiate new research activities, and reappraise the organisation of his department in the light of agricultural, economical or scientific developments, proposing the merging of divisions, their abolition or the establishment of new units.

An important aspect of his work is to ensure communication between his department and other departments in the organisation, other scientific institutions, the extension service and the farmers themselves.

His functions as head of department do *not* preclude his devoting part of his time to direct participation and personal involvement with research activities in his own special field. It is even important that he should continue to maintain a relatively high degree of research activity and should attend professional meetings, present papers to meetings and journals etc.

Administrative duties

In order to ensure that the department head should not be overburdened with paperwork he should be assisted by an administrator who will handle the routine administration of the department. He himself will be concerned with the maintenance of morale and with personnel requirements. In consultation with the senior scientists of the department (departmental research committee) the head determines policies of operation, departmental budgets, allocation of personnel to research projects, priorities. He assures satisfactory services for the department and is involved in the construction of research facilities and purchase of equipment.

It is the responsibility of the department head to ensure that research proposals are properly evaluated before submission to the director of research, both from the scientific point of view and in relation to their budgetary demands. He will pay particular attention to whether all aspects of work proposed are within the professional competence of the team, and examine the need for co-opting additional specialists to the research team in order to make the research more fully effective.

He recommends promotions, training grants, attendance at congresses for the members of his department.

As a member of the executive committee of the research organisation he participates actively in the formulation of the policy of the organisation as a whole. He should have the authority to make decisions related to matters specifically of concern to the department, without need to refer first to higher authority.

Professional requirements

He should be the holder of a PhD degree, with at least 10 years' research experience of which five at least were as head of a division with appropriate professional achievements to his credit, and command respect from his colleagues as a scientist. He should have a comprehensive knowledge of the agricultural economy of his country, and of the basic tenets of the management of scientific and administrative staff. He should have a broad knowledge of the research activities carried out in the general field of his department, both in the country and abroad. He should have good judgement in evaluating research proposals, their conclusions and implications. He should be able to express himself clearly and succinctly, both orally and in writing. He should have ability in negotiating, both with staff and with management. He needs organising ability.

THE HEAD OF DIVISION

At the level of the division, the head has the closest contact of all research management with the researchers themselves. Hence his involvement with his subordinates is mainly professional and only to a small extent administrative.

His principal duties are to initiate and plan the research programme of his division and to provide leadership and scientific guidance to his researchers, reviewing their methods of work and discussing with them their results and the conclusions to be drawn from them. In establishing his research programme and proposing the individual research projects for approval, he will be guided by the overall research policy of the organisation. He will initiate new research activities within the scope of his division's competence and will encourage initiative and original thinking by his researchers.

Whilst he is the 'head' of his unit, he has to realise that research activity cannot be directed in the usual sense, and that the individual researcher 'is largely his own labourer and his own manager' (Hertz & Rubinstein, 1953). At this level, at which contact between the head and the researchers is extremely close, the conflict between his responsibilities to management on one hand, and the ingrained antipathy to direction of his subordinates—the researchers—on the other, might give rise to unbearable strains and stresses; it is only when the division head is also the recognised leader in his field, that a workable relationship can be established and maintained. To maintain this position he must also be actively and personally involved in the pursuance of research and simultaneously devote a considerable amount of his time to contacts with his researchers, individually and in working groups.

Administrative duties

These include the provision and maintenance of working conditions conducive to the efficient, high level scientific work of his group; the evaluation and judgement of his scientific personnel as a basis for recommendations for their promotion; checking of the estimates required for each individual research project and evaluation of its overall feasibility within the physical and personnel limits available; the submission of reports on the progress of the work on hand; participation in the negotiation of research grants for his group; the supervision of the use of equipment and other resources allocated to his group. He is responsible for the 'work discipline' of his division, and assists in the choice of new personnel.

Professional requirements

He should be a holder of the PhD degree, with many years of experience in research, of which at least four years were as a senior scientist. Should be well acquainted with the problems of the agricultural economy of his country. Should be a recognised authority in the professional field in which the division is engaged. Should have experience and ability in directing research workers. Should have mature judgement in scientific matters in the field of his competence. Should be able to express himself well verbally and in writing; and to present scientific papers, to professional meetings and journals.

THE PROJECT LEADER

There are several essential differences between the project leader and the heads of other research units in the organisation:

(a) his function as project leader is by definition transient; it lapses when the research project is concluded;

(b) his status as research leader is ambivalent: he may be the leader on one project, an ordinary team-member in another;

(c) the main difference, however, derives from the fact that in the team are assembled a variety of specialisations and skills. The project leader cannot possibly be a scientific leader in all these specialisations. Hence his main duties are administrative.

The project leader prepares the detailed plan for the project in co-operation with his team members. He is responsible for the overall implementation of the project and co-ordination between team members, with whom he maintains close and constant contact. He carries out part of the work himself, within the area of his competence.

He is the focal point of the project, to whom team members apply for the solution of day-to-day problems and assures smooth communication between the members.

His basic responsibility is to complete the project in accordance to the proposed plan of work, within the limitations of his budget and within the specified time schedule.

Chapter 16

Administrative Personnel and Service Units

PERSONNEL

Head of administrative services

There are two basically different approaches to the functions of the administrator in research institutes, the one more or less typical for American institutes, the other for European institutes and, in particular, in the USSR.

In the first approach, the administrator has overall responsibility for the administrative activities of the research institution; this includes personnel, discipline, labour relations, implementation of the budget and development of the physical plant. A university education is required, usually in economics, public administration or agriculture. In any case, he must have administrative knowledge and experience. He is very near the top of the hierarchy. In the administration of the organisation he has the authority to make decisions without the need to consult the director unless he feels that policy matters are involved. This approach derives from the director's tendency to delegate authority in all matters non-scientific, so as to be free, as far as possible, from preoccupation with routine administration. The administrator will also usually be consulted on general policy determination. He has direct contacts with scientists, mainly heads of departments, and exercises a certain amount of control over their activities—especially indirectly, through budgetary considerations. This is frequently a source of conflict between the administrator and the scientists. The longer he is with the organisation the more powerful he usually becomes.

To counteract this tendency it is necessary to insist that in all adminis-trative decisions that have direct implication for the research work of the organisation (such as allocation of various resources to different units), the administrator is in the position of a staff officer to the scientist-director and acts only after due consultation with him.

In the second approach, the administrator still has responsibility for the same kind of activities described above, but he occupies a relatively subordinate position in the organisation. He has no scientific background and is typically trained in business accounting and business procedures. He executes policy with no part in decision-making, not even in the field of administration, and hence has no control, direct or indirect, over scientists. His role is therefore never a source of conflict with the re-searchers. This concept of the role of the administrator makes it necessary for the director of research to be more directly involved with administra-tive duties than is the case with the former concept, in particular with those that have a more or less direct bearing and influence on research activities. When necessary, the director will delegate certain administra-tive responsibilities to other scientists and not to the professional adminis-trators. Kaplan (1961) reports that the shouldering of administrative responsibilities for activities bearing directly on research by the scientists does not necessarily infringe on their research time. This apparent paradox is a result of the tendency of scientists to reduce administrative procedures to a bare minimum and the avoidance of conflicts between researchers and administrative staff, resulting from decisions made by administrators in areas in which intimate knowledge of scientific require-ments is necessary. The conflicts are also time-consuming.

A somewhat different approach, described by Bunker (1961), was applied by the National Institute for Medical Research in the UK, after it was found that notwithstanding the appointment of a head of adminis-trative services, the director was overly involved in administrative mat-ters, and especially on problems arising from the effect on scientific work of decisions made by the administrator. The senior post was abolished and a number of lower grade staff officers appointed, who had no authority in their own right, but were the first line of approach for heads of research units who required administrative services.

The director delegated to them the right to make decisions on many matters which they did in the light of their knowledge of the director's mind, of current policy, of the conflicting needs of different units, etc. Only in cases of disagreement between the head of the research unit and the staff officer, would the matter be brought to the director for his decision. Bunker reports that after this procedure was initiated, the load of work on the director dropped considerably and the friction which existed previously between administrative and scientific interests was noticeably reduced. Bunker ascribes at least part of the success of this approach to the fact that the scientific director had an aptitude for administration, but possibly success generally depends mainly on the size of the organisation concerned.

An administrator must have knowledge of the principles of organisation, an understanding of human behaviour and motivation, and of personnel administration techniques. An administrator of an agricultural research organisation must know that the concepts of administration developed for other types of organisation are not always suitable for this type of organisation and need adaptation and adjustment. The knowledge required cannot be based on experience alone, but must be learned.

Supporting personnel

The administrative services required to run an agricultural research institution—financial, purchasing, inventory, upkeep, etc.—are essentially similar to those needed in any other organisation of similar size. However, the personnel employed on the administrative and technical supporting services of a research organisation should not only be highly competent in their profession, but should also have an understanding for the scientific and technical needs of the researchers.

Applying standard business methods in a research institution, without due regard for the specific characteristics of research activities and the researchers themselves, can become a constant source of friction and of frustration, both for the administrative personnel, and for the researchers.

Personnel director
The personnel director is responsible for all problems relating to hiring, firing and discipline, and advises the director on all personnel problems. He must be conversant with the rules and regulations pertaining to personnel in government service, both scientific and in the supporting services, but must at the same time have sufficient authority and understanding to solve specific problems and to treat special urgencies.

Recruiting. The personnel director is responsible for all the formalities and practical problems concerned with recruiting. He must prepare a position description which gives a clear definition of the responsibilities and duties involved and a statement of the qualifications required.

Personnel records. The maintenance of a master file for each employee is also the responsibility of the personnel director. This file should include the personal history of the individual, letter of recommendations received before hiring, information on any special aptitudes or specialities he may possess, on-the-job training and other forms of professional advancement, the annual evaluation form of his performance; grades and promotions; and for research workers, also a list of publications and other achievements.

Accountant

The chief accountant is responsible for all the financial activities of the institution. He supplies all the necessary data for the preparation of the annual budget and is responsible that expenditure is in accordance with the approved budget and accepted procedure. He supplies up-to-date information to researchers and heads of units at all levels on the state of expenditure in their relative fields of responsibility. He must check and approve all expenditures over a certain sum, and assure that they are made in accordance with government and institutional regulations.

Research budget director

In most agricultural research organisations, in addition to the regular annual budgets, there are usually many other sources of financing research projects: farmers' organisations, private firms, foundations and trusts, foreign grants, etc. The research worker is generally neither equipped nor particularly suited to negotiating these grants; this would also interfere with his work, and is, in certain cases, ethically undesirable. This important task should be handled by a special officer with intimate knowledge of the nature of the research work carried out in his organisation, and of research requirements. It may be desirable to combine this function with that of head of the research project office, if the incumbent is suitable and if the scope of his activities makes this possible.

Management analyst

The function of the management analyst is to conduct a constant examination and appraisal of organisational forms and methods. This involves management studies aimed at improving management practice and the performance of individual units, feasibility studies, formulation of codified instructions and job descriptions; proposals for efficient keeping of records, the designing of filing systems and operations, etc.

A well-trained and competent management analyst can make considerable contributions to simplifying procedures, improving performance and communications in the research organisation. He must be free of any direct involvement in routine operations, but should be conversant with all the details of the organisation in which he works and have intimate knowledge of its people.

Superintendent of regional stations
His duties include general supervision of the regional stations: initiation of improvements in efficiency and organisation; liaison between station superintendents, responsibility for the implementation of development schemes in the regional stations; giving advice on the crop planning of the individual stations.

The superintendent should have a university degree in agriculture and have additional training in management and administration; must have considerable experience in farm management and agricultural practice, an understanding of the requirements of agricultural research in general and of field experimentation in particular.

SERVICE UNITS

Public relations

A good public relations unit has the important function of creating a favourable public image of the agricultural research organisation. This is achieved mainly by keeping the public informed of the efforts and achievements of research. The unit is also responsible for organising visits to the institution and its dependencies, by groups and individuals.

Audit unit

The basic functions of the audit units are: (a) to ensure that all administrative operations in the research organisation's units are carried out in accordance with accepted norms, the laws of the country, and the institution's regulations; and (b) that funds are used for the purposes for which they were assigned (IESA, 1980).

The audit unit carries out audits in the different units of the institution, in particular in the regional experiment stations and their dependencies; checks inventories, inspects construction works in progress and checks on compliance to contractual terms; and proposes improvements in administrative procedures in use in the institution.

Legal office

The office represents the research institute in all legal matters, and provides legal advice to the director of the institute.

Management Information System (MIS) Units†

King (1986) defines a MIS as a support system for storing, routing and processing information needed by an organisation to achieve its objectives. Computerisation is not essential in a MIS, but it greatly increases the possibilities for effective research. The availability of powerful, inexpensive microcomputer hardware and software makes computer-based MIS a realistic possibility for national agricultural research systems (NARSs).

King further points out that NARSs in Africa, Asia, Latin America and the Middle East are responding to these opportunities.

Valmayor & Mamon (1987) describe the establishment of a MIS unit by the Philippine Council for Agricultural Research, which comprises the following information systems:

(a) *Research Management Information System* (RMIS) is a computer-assisted system to create files of proposed, new, and ongoing research projects. It provides support for the planning of the national research programme, for the consolidation of budgeting plans, and maintenance files of approved research projects for monitoring and evaluation. It should be stressed that the maintenance of a register of all research projects carried out in a country is an essential component of a rational system of planning and resource allocation.

(b) *Research Information Storage and Retrieval System* (RETRES) is a system designed to establish a databank of terminal reports of completed research projects, publications, monographs, and other library

†A detailed discussion of MIS is beyond the scope of this book. The interested reader is referred to: Davis, G.B. & Olson, M.H. (1985). *Management Information Systems: Conceptual Foundations*, 2nd edition. McGraw-Hill, New York.

materials and develop a query system that facilitates literature searchers. This complements the RMIS which deals with ongoing research.

(c) *Equipment Infrastructure Management System* (EIMS) which provides a computer-based inventory of all equipment and infrastructure resources, providing information on the location, distribution and utilisation of all research equipment in the country.

(d) *Manpower Management System* (MMS) provides complete information on all research workers involved in agricultural and natural resources research; the location and specialisation of the researchers.

(e) *Financial Management System* (FMS) which provides financial indicators to formulate operating plans and institute financial control. It also provides management reports showing a comparative analysis of approved, recommended budget *vis-à-vis* actual releases of individual research projects.

King (1986) stresses that 'the effectiveness of an organisation's MIS is determined not by the complexity and sophistication of its components, but by the quality of the support it provides'.

Library

Inadequate library and information facilities are a serious problem in many developing countries. The costs in foreign exchange of providing adequate facilities and the necessary periodicals and books are extremely high.

The number of libraries in each country is usually very small and they are generally located in the big cities; the books available are limited in number and frequently outdated. Journals are also few in number and generally reach their intended audience after a long delay.

It has been estimated that the average researcher spends about 25 per cent of his time searching for information and keeping up to date professionally (Lufkin, 1966). Hence, a good library service is essential for effective research work.

The effectiveness of a library service depends largely on the professional competence and motivation of the librarian, who can save researchers a considerable amount of time by proper organisation of library facilities, provision of guidance in the search for information, and in particular by providing information on the availability, function and use of basic reference tools.

What is needed is what Monge (1979) calls a documentalist, whose first function is to collect the 'fugitive' material produced in the country, i.e. internal reports, mimeographed papers, etc., which contain valuable and up-to-date information. Second, he/she should screen the vast amount of information produced in the world and make it available to the users in the country so that they are not bogged down with non-pertinent information. He/she should also process, group and disseminate information so that it reaches the user with the least possible delay.

A special challenge for the librarian is to ensure that researchers in the distant regional stations should not be neglected, but should be provided with an efficient circulating library service, in addition to the basic literature that must be available at each regional station.

The librarian should be assisted by a library committee, in which each department is represented, meeting at regular intervals to define policy, prepare budget estimates, decide on priorities in the purchase of books, journals and equipment, and an equitable allocation to regional stations.

Statistics department

The statistics department should be staffed with one or more consulting statisticians, who have experience in biological research.

The statisticians advise the research workers on the design of individual experiments, on the collection of data and on the interpretation of results.

By consulting the statistician *before* starting an experiment, the research worker avoids carrying out his experiment in a way that will make a valid statistical analysis impossible, or will not provide the data needed to achieve significant differences. Working together, with the researcher's knowledge in his own field and the statistician contributing his specialised knowledge, efficient designs can be prepared for each individual experiment.

With the help of trained clerical staff, the department can also carry out the routine statistical analysis, and thereby affect a considerable saving in the research workers' time. The statistician can then advise the research worker in the statistical interpretation of the results and how to obtain the maximum information from the data available.

Maintenance and technical services

A relatively large research institution will usually require its own maintenance personnel: a small building unit, electricians, refrigeration technicians, electronic technicians, carpenters, gardeners, etc. The need for

a fine-mechanics workshop has already been mentioned. The maintenance and technical services should be headed by a civil engineer, who
would also be responsible for planning, cost estimates, supervising and
control of all building activities of the institution.

Careful planning will be necessary in order to decide which services
should be decentralised, and available in the regional stations, and which
can be centralised, such as the repair of electronic or laboratory equipment.

Regarding the situation in most Latin American countries, Ampuero
(1981) writes: 'In most research stations a service department exists, but
this department is too often understaffed, undersupplied in terms of
equipment and spare parts, has a very low financial allocation and is
bound by financial and administrative regulations in the procurement
of supplies'. In view of the importance of this unit for the efficient
functioning of the research organisation as a whole, the situation described by Ampuero cannot be justified.

The central supply unit

The purchasing department should free the research worker from all
but a minimum of care in matters related to purchasing, both of the
routine and the exceptional items he may require for his work.

A centralised stockroom for the research organisation has a number
of advantages: many standard items can be purchased far more economically in bulk, and can be supplied to the researcher as soon as requested.
This may be doubly important for an agricultural research institution in
a rural setting, situated at a distance from the sources of supply. A
central stockroom also obviates the need for individual 'hoarding' by
researchers of standard items, which in the aggregate, may add up to
considerably more than the reserves required by a central stockroom.
To be effective, the stockroom must always maintain a reserve of frequently used items above a certain minimum, and place its orders in
good time to replenish its stock. Much care and judgement must be
exercised in planning reserves of more specialised pieces of equipment
and materials.

The ordering of scientific equipment should always be carried out in
close consultation with the research worker involved, and also with other
specialists who may be in a position to give advice on the subject. The
purchasing officer should have a reference library of catalogues of scien-

tific equipment, agricultural machinery, materials, etc., available for consultation, which he should keep as up to date as possible.

A central supply unit is of particular importance in developing countries. These are usually dependent on imports of many essential supplies required for research work. The inability to obtain materials, equipment or spare parts at short notice can cause not only disruptions in routine work, but also the loss of entire experiments, and in the field, the loss of an entire season.

Part of the stock will have to be kept in the stores of the regional stations, but overall responsibility for purchasing, maintaining the inventory up to date, and distributions should be vested in the central supply unit.

Research service departments providing services to the farmer

Another type of service department is that which provides services to the farmer, either directly or via the extension service. A few examples are seed testing; soil and plant analyses for diagnostic techniques in nutrition requirements; identification of diseases and pests. Much soul-searching goes on as to whether research institutes should also concern themselves with services of this kind. The main justification for providing these services is that there is a close connection between the research work carried out by the institution providing the service and the level of the service itself: diagnosis of nutrition requirements, for example, cannot be deduced mechanically from the data resulting from soil or plant analyses, but requires interpretation on the basis of plant nutrition research.

Buildings

Adequately planned buildings are a great asset in facilitating research. The laboratories should be well designed, providing sufficient space, proper facilities and with sufficient flexibility to adjust to changes in routine and personnel. For agricultural research, it is important that each departmental building be situated on a fairly large plot of land, so that there is plenty of space available for glasshouses, screen houses and small plots of plant material, such as breeding plots, requiring constant supervision.

Very often, heavy investments are made in buildings mainly for prestige value. The result is that not only are funds exhausted which might have served for the purchase of equipment or other essential purposes, but also that high upkeep costs swallow a considerable proportion of the annual budget.

References to Part Five

Ampuero, E. (1981). *Organización y Administración de la Estación Experimental en los Paises en Desarrollo*. Progress in International Agriculture. Cornell University, Ithaca, New York.

Argyris, C. (1962). *Interpersonal Competence and Organisational Effectiveness*. Dorsey Press, Homewood, Ill.

Barrows, B.F. (1963). Scientific management in science. In *The Management of Scientific Talent*, ed. J.W. Blood. American Management Association, New York, pp.189–94.

Bavelas, A. (1960). Leadership: man and function. *Admin. Sci. Q.*, **4**, 491–98.

Braybrooke, D. (1964). The mystery of executive success reexamined. *Admin. Sci. Q.*, **8**, 533–60.

Brown, P. (1954) Bureaucracy in a government laboratory. In *Human Relations in Industrial Research Management*, ed. R.T. Livingston and S.H. Milberg, Columbia University Press, New York, pp. 241–62.

Bunker, L.H. (1961). Administration in Research. *D & M Bulletin*, **16**, 119–206.

Dill, W.R. (1964). Varieties of administrative decisions. In *Readings in Management Psychology*, ed. H.J. Leavitt and L.R. Pondy, University of Chicago Press, Chicago, Ill, pp. 457–73.

Gollembiewski, R.T. (1964). Authority as a problem in overlays. *Admin. Sci. Q.*, **9**, 23–49.

Gross, B.M. (1964). *The Managing of Organisations: the Administrative Struggle*. Free Press, Glencoe, Ill.

Hertz, D.B. & Rubinstein, A.H. (1953) *Team Research*. Eastern Technical Publishers, New York.

IESA (1980). *FONAIAP—Estudio de Organización*. Instituto de Estudios Superiores de Administración, Caracas.

Kaplan, N. (1961). Research administration and the administrator. USSR and US. *Admin. Sci. Q.*, **6**, 51–72.

Karger, D.W. & Murdock, R.G. (1963). *Managing Engineering and Research*. Industrial Press, New York.

Kaufman, H. (1962) Book review. *Admin. Sci. Q.*, **6**, 503.

King, P.P. (1986). Management information systems planning in national agricultural research systems. In *Improving Agricultural Research Organisation and Management: Implications for the Future.* ISNAR, The Hague, pp. 55–58.

Koontz, H. & O'Donnell, C. (1955). *The Principles and Practices of Management.* McGraw-Hill, New York.

Lufkin, J.M. (1966). Reading habits of engineers. *IEEE Trans. Educ.*, **E–9,** 179.

McGregor, D.M. (1964). The human side of enterprise. In *Readings in Management Psychology,* ed. H. J. Leavitt & L.R. Pondy. University of Chicago Press, Chicago, Ill.

Malinowski, B. (1960). *A Scientific Theory of Culture.* Oxford University Press, New York.

Mandell, M.M. (1961). Research management: some clues for selection. *Personnel,* **38**(1), 71–74.

Mayo, E. (1960). *The Human Problems of an Industrial Civilization.* Viking Press, New York.

Milberg, S.H. (1957). Selection and development for research management. In *Human Relations in Industrial Research Management,* ed. R. Livingston and S. Milberg, Columbia University Press, New York, pp. 80–91.

Monge, F. (1979). Making information accessible. In *Give Us The Tools,* ed. D. Spurgeon. International Development Centre, Ottawa, pp. 161–74.

Mooney, J.D. (1949) *The Principles of Organisation.* Harper & Bros, New York.

Pfiffner, J.M. & Sherwood, F.P. (1960). *Administrative Organisation.* Prentice-Hall, Englewood Cliffs, NJ.

Ruttan, V.W. (1982). *Agricultural Research Policy.* University of Minnesota Press, Minneapolis.

Sayles, L.R. (1964). *Managerial Behaviour—Administration in Complex Situations.* McGraw-Hill, New York.

Simon, H.A. (1960). *The New Science of Management Decision.* Harper & Row, New York.

Steele, L.W. (1962) *The Role of the Research Manager.* General Electric Research Lab. Bull.

Tannenbaum, A.S. (1962). Control in organizations: individual adjustment and organizational performance. *Admin. Sci. Q.,* **5,** 458–522.

Valmayor, R.V. & Mamon, Cynthia R. (1987). Research information systems for agriculture and natural resources in the Philippines. In *International Workshop on Agricultural Research Management.* ISNAR, The Hague, pp. 153–55.

Van Tassel, K.R. (1965). Managing research and development. *Research Mgmt,* **8,** 145–57.

Part Six

THE TRANSFER OF TECHNOLOGY

Chapter 17

National Extension Services: Selected Case Histories and Overview

THE UNITED KINGDOM

England and Wales

Historical Background

Technical advice to farmers began in 1890, following an Act of Parliament providing for this service.

The national advisory service was established in 1946, and operated in its original form until 1971, when it underwent a major reorganisation and was combined with a number of other services, to become the Agricultural Development and Advisory Service (ADAS) (Fletcher, 1986).

Structure

ADAS is part of the Ministry of Agriculture, Fisheries and Food (MAFF). The service is under the control of a director general, who is responsible to the permanent secretary of the ministry.

ADAS is divided into four major services: Veterinary, agriculture, agricultural science, and land and water. The Agricultural Science Service has been described in Chapter 1.

ADAS operates at three levels: Headquarters, regional, and divisional. There are six regions in England and Wales, with a sub-centre in each region.

Each region is subdivided into a number of divisions, of which there are 24 in all, each division comprises a number of districts.

Central Laboratories for plant pathology, pest infestation, and veterinary services, respectively, provide logistic support to the other units.

Plant Clinics, 15 in number, provide a fast diagnostic service with appropriate recommendations. All aspects of crop growth are dealt with. The clinics are used by ADAS advisers, advisers from commercial companies and private consultants. The service is no longer free (Fletcher, 1986).

The Extension Development Unit provides advice on extension methods and investigates communication methods.

The Computer Development and Operation Research Unit is concerned with developing and testing new equipment, techniques and systems in operational research and computers.

The Aerial Photography Unit.

The Farm Waste Unit provides advice on farm waste disposal problems.

The Feed Evaluation Unit provides information on the composition and nutritional value of feeds for livestock production (OECD, 1981).

Departmental Farm Projects are carried out on commercial farms where ADAS applies modern techniques which are compared with the existing techniques.

Activities

The major functions of ADAS advisory and related work are: (a) to provide technical advice, practical and scientific, on agricultural matters, (b) analytical and diagnostic services, including a wide range of activities.

There are several broad categories of activities: regulatory and surveillance work; advisory work and servicing; promotional work; research and development.

Advice is provided to farmers, private consultants, trade advisers, outside professional agriculturists, private veterinary surgeons, to administrative divisions in MAFF, other government departments, local authorities, other public bodies and overseas visitors.

Servicing Work includes land use consultations with local authorities, the restoration of coal workings; advice to farmers on estate management; design and construction of farm buildings.

Surveillance work involves information on changes in land prices and rents, on pest or disease infestations, etc. (OECD, 1981).

Research and development have been treated in Chapter 1.

Aid to developing countries provides occasional training for extension workers from the LDCs. Staff is seconded to LDCs for periods of 2–5 years, or occasionally for shorter periods to undertake specific tasks.

Methods and procedures

The educational level of the farmers, most of whom are farm-born, is variable, but most have received a basic level of education up to age 16. Some farmers continue their professional education at diploma and certificate courses at local agricultural high schools. A minority of farmers have obtained university degrees (Fletcher, 1986).

These characteristics of the farmer clientele evidently have relevance to the advisory methods adopted.

The first point of contact for all farmers is the district adviser, who is also the key organiser of meetings, small discussion groups or symposia. In most districts there is one agricultural and one horticultural adviser, and occasionally more than one of each in offices covering several districts.

Though the district adviser is a generalist, he tends to specialise in arable crops, horticultural crops, or in animal husbandry. He is generally a graduate in his subject, and able to deal with most of the routine enquiries in his field. When he needs specialist advice, he consults his regional team. For diagnostic services, he sends material to the plant clinic.

Visits to farmers by specialists are never made on a routine basis, but only in response to a request to investigate a specific problem.

The farmer may also consult directly various other government sources of information, including the regional specialists and research workers in the Agricultural Food and Research Service. This approach is, however, not considered desirable by ADAS, because it is not thought to be the most effective way of using its staff resources.

Often, small groups of farmers, specialising in certain crops, meet on each other's farms to look at techniques or discuss specific farm problems.

Farmers are increasingly making use of telephone information services available to meet local situations in most parts of England and Wales.

Because of decreasing resources and increasing commitments, the written word has become the major means of communication with the farmers: leaflets, bulletins, reference books, news sheets, articles in the farming journals, ADAS divisional publications, and occasionally scientific papers are used.

Extension planning

Advisory programmes are prepared at national, regional and local levels. At national level, a survey is undertaken by a working group, comprising ADAS staff, experts from the universities, the agricultural industry and farmers, to determine staff requirements, co-operation within ADAS and with the agricultural sector, and methods of communication.

Regional and local advisory programmes are designed in a similar manner.

Implementation of programmes is evaluated by special working parties or by headquarters (OECD, 1981).

Extension personnel
Recruitment is based on appropriate qualifications and competitive interview. All advisers are required to complete a probationary period before final appointment.

Grade structure and promotion. Initially, there were advisory service grades and scientific grades reflecting the two major activities of ADAS. The grade structure has, however, evolved over the years and has become a mixture of both grade systems:

(a) the scientific system in which are graded: the staff of the central laboratories, the regional agricultural scientists, and the heads of disciplines (Senior Principal Scientific Officers); the support staff in the regions (in the grades Higher Scientific Officer, Scientific Officer, and Assistant Scientific Officer).

(b) the advisory system in which staff can be promoted from Advisory Grade IV (a training grade) to Grade I.

Promotion from one grade to the next is based on a system of annual reporting through heads of departments. Appointments to higher post vacancies depend on these recommendations and competitive interviews. The main working grade is Grade III, and promotion to Grade II is based on technical and communication abilities and management potential (Fletcher, 1986).

A typical career profile of an adviser: recruitment at age 20–24 in Grade IV; promotion to Grade III after two years; to Grade II at ages 30–40, and to Grade I at 40–50. Most officers do not advance beyond Grade II.

Pre-service and in-service training. There are no specialised opportunities for pre-service adviser training. By contrast, in-service training continues throughout the career of the adviser. Courses vary in length from a few days to several weeks. Financial help, and time-off—from one day a week to full-time absence of 3–5 years to attend outside educational establishments—are available to ADAS employees (OECD, 1981).

Trends and possible development of ADAS

The new demands on ADAS for knowledge and skill transfers, and motivating farmers in a 'post-surplus era' are likely to be very different from the past (Gabriel, 1987). Specialisation will increase, there will be more emphasis on marketing; an increased awareness of environmental problems, pollution and conservation; the integration of agriculture with other countryside matters, etc.

University involvement in agriculture advisory work

There are many contacts between faculty staff with advisers and farmers (Darling, 1970):

- messages are conveyed to advisers and farmers through the graduates and postgraduates they have trained;
- they frequently report research findings in a form understood by farmers, at professional meetings, radio and television talks.
- many advisers and progressive farmers keep in touch with faculty staff engaged on work of interest to the former;
- papers are published in the farmers' periodicals, bulletins on behalf of ADAS, etc.

Private sector

There are commercial firms and private consultants who provide information and advice to farmers. Private sector advisers are more numerous than ADAS advisers, and provide a great deal of the on-farm advice given to farmers (Fletcher, 1986). External sources already far surpass the advisory capability of ADAS (Barker, 1986). This trend will probably accelerate now that ADAS is charging the farmers for most of its services.

Scotland

In Scotland, the provision of agricultural advice is the responsibility of the three agricultural colleges, which are independent bodies financed by government through the Research and Education Branch of the Department of Agriculture and Fisheries.

The general advisory officers and some specialists are based in area offices situated strategically throughout the country.

Objectives, methods, recruitment, in-service training and promotion are similar to those adopted in England and Wales.

THE NETHERLANDS

Historical background

Agricultural advisory work began in the Netherlands in 1876, with the establishment of the first agricultural college attached to an experiment station; the teachers served also as agricultural advisers. In the 1890s, the first full-time advisers were appointed; by the beginning of the 20th century, there were 30 advisers, each specialising in one of the main branches of agricultural production.

The advisory service expanded rapidly after the agricultural crisis of the 1930s, and again after World War II, when it became necessary to repair the considerable war damage and to adapt to changing production conditions (OECD, 1981).

Participants in the extension system

Agricultural extension in the Netherlands has expanded considerably in the course of the years; the amount of information available for diffusion has also grown considerably, and has resulted in a marked differentiation and specialisation within the service.

In addition to the official extension service, other institutions have become increasingly involved in the diffusion of agricultural technology. These include the social-economic extension service, the private sector, the extension input of the research institutions. The media have also made significant contributions to the diffusion of innovations, and last, but not least, the farmers have provided themselves with a sound basis of technological and economic capacity for adopting new technologies (Zuurbier, 1983).

Present structure of the government extension service

The Ministry of Agriculture and Fisheries provides advice on agriculture through the *Farm Development Service* which is organised into 11 provincial directorates, each with 2–7 specialised regional advisory centres.

The provincial directorates have about 20 general advisers, and 80 specialised advisers (including experts on organisation, advisory methods and financing (OECD, 1981)).

The regional advisory centres. Thirty-eight in number, the centres are fairly autonomous and organised according to branch specialisation:

there are regional services for arable cropping, dairy farming, mixed farming, intensive animal husbandry, horticulture, etc.

Advisory services for each commodity are provided by a group of 20–30 general field extension workers who have specialised in the commodity served by their respective centres, each of which provides the administrative backstopping for their group of extension workers. The extension workers focus on technological and economic aspects of farm management and 'are discouraged from providing recipe extension, i.e., being at the beck-and-call for every technical problem that the farmer faces' (Röling, 1987).

The general field workers are supervised by *branch specialists* who have access to information from the experiment station serving their commodity. Each regional centre also has a number of subject matter specialists (SMSs) in various disciplines. These are linked with the research institutions responsible for their respective disciplines (Röling, 1987).

Each of the farm management advisers supervises 10–15 local farm advisers.

General advisory officers. In addition to the provincial directorates, there are 23 general advisory offices, strategically located throughout the country, whose main function is to serve as a link between research and extension and thereby supporting the work of the regional advisory centres, by providing them with specialised information. One group of offices specialises in various disciplines (plant protection, mechanisation, etc.) and another group in the production of various commodities (arable crops, flower bulbs, etc.) (OECD, 1981).

Personnel

The provincial directorates have a staff of 240 people who provide support to the extension personnel and are also responsible for the implementation of government regulations; the regional advisory centres have a staff of 930 and the general advisory offices another 150 people (Zuurbier, 1983).

Clientele

In principle, the Dutch advisory service 'makes no distinction between self-employed farmers and employees, old and young, men or women, or between particular types of farms' (OECD, 1981). In practice, however, 'extension workers and progressive farmers attract each other like magnets' (Röling, 1987), so that extension contact is highly skewed in favour of progressive farmers. A study by Haverkoort (1976) showed that extension workers had direct contact with at most 30 per cent of the farmers and that 65 per cent of the visits took place on 10 per cent of the farms.

Extension strategy

The basic principle of the Dutch advisory service is 'to teach the farmer to help himself'. Therefore, the advisers are not restricted to disseminating knowledge and making recommendations, but help the farmer to make decisions—amongst other things by explaining government measures which affect farm income, and by helping farmers to understand how general social changes will affect them.

The Dutch government consults regularly with farmers' representatives regarding their requests from the advisory services, in order to ensure that the advice and information provided by the service are practical and useful. The National Council for Farm Development at the national level, and the provincial councils for farm development at the regional level, are the formal consultative bodies (OECD, 1981).

Programming

The staff of the regional advisory centres prepare every year an advisory programme based on an analysis of the agricultural problems in their respective regions, setting out priorities and defining a time schedule. Requests from farmers' organisations, trade organisations, study groups, etc., are also taken into consideration (OECD, 1981). Work plans are drawn up for implementation of the programme.

The programmes of the regional advisory centres are co-ordinated by the provincial councils and finalised by the National Council for Farm Development (OECD, 1981).

Pre-service and in-service training of extension workers

The Agricultural University, and some of the agricultural colleges, give courses on extension methods, but there are no special institutions specifically providing training for future advisers.

Advisers are appointed on the basis of an interview and psychological test. Those accepted are employed on probation for two years, during which they must prepare for examinations by self-tuition, according to a given syllabus. They must also attend an induction course on advisory work and methods. The SMSs attend special training courses. At the end of the probationary period, the candidates are assessed and those judged satisfactory receive permanent employment.

Regular in-service training is provided through monthly meetings at national level, and fortnightly meetings at regional level. Short courses on special subjects are organised.

Linkages with agricultural research

Regional advisory officers also serve as directors of the regional research centres, where they can have adaptive research carried out at the experimental farms and market gardens on a limited scale, for specific local problems. More general problems, requiring a more intensive investigation, are sent by the regional advisory centre to the subject matter specialists (SMSs) at national level, who either direct, or are attached to, an experiment station appropriate for the commodity or farm type concerned. Problems which require more fundamental research are transmitted by the SMSs to the appropriate agricultural research institute.

Research results are conveyed back to the farmers through the same channels. Research findings which need to be adapted to local conditions are first tested at the experiment stations or the regional research centres.

The functional link between the advisory service and the research system is provided by the SMSs at national level. They act as advisers to the governing bodies or advisory committees of the institutes relevant to their respective specialisations. These SMSs are attached to appropriate experiment stations or research institutes. They are also responsible for co-ordinating the research carried out on their subject at the various regional research centres. Within the framework of the farm development service, they maintain close contact with the regional SMSs.

Research workers are frequently invited to attend meetings between national and regional SMSs, thereby enabling a free and rapid feedback of research results, and ensuring that the research workers are aware of the actual needs in their respective fields of activity (OECD, 1981).

An important linkage mechanism between the regional extension services, the experiment stations, and the research institutes are roving liaison officers. These general service consultants are highly specialised extension officers, housed at the research institutes and experiment stations, but serving the regional extension services. 'They carry research results down, and field problems up' (Röling, 1987).

Non-governmental advisory services

Besides the Ministry of Agriculture and Fisheries, a number of non-governmental bodies provide extension advice. These comprise: voluntary organisations (e.g. organic farming); animal feed mills (which have separate departments for sales and extension); private consultants; co-operatives and farmers' organisations.

The non-government sector is substantial and becoming increasingly more important. One example of extension by a co-operative, is the employment of extension workers by the Co-operative Flower Auction in Westland for the sole purpose of informing flower growers about the preferences of different categories of cutsomers (Röling, 1987).

The farmers' organisations constitute a four-tier structure with considerable lobbying power at the national level. They are subsidised (up to 90 per cent) by government to provide socio-economic advice in bookkeeping, insurance, taxation, social legislation, tenancy laws, etc., as well as matters concerned with transfer of farmsteads, advice on change of profession, financial aspects of production and household (Zuurbier, 1983).

Private sector extension
A large number of organisations are involved in agricultural commerce and industry. Many of these operate a sales service that also provides advice on the most efficient use of the products they sell (Zuurbier, 1983).

The media fulfil an extremely important role in the diffusion of agricultural information. About 70 per cent of farmers are subscribers to professional agricultural journals, more than 50 per cent listen to the daily agricultural bulletins on the radio, and about 35 per cent see the television series 'Harvest in Images'. A telephone answering service provides information on specific problems. In a single year this service was used 22 000 times by farmers seeking advice on plant diseases control (Zuurbier, 1983).

Study clubs
One very interesting development since the 1960s, is the development of study clubs. These are usually initiated by the extension worker, but

have grown into mutual interest networks. They consist of relatively small groups of farmers who come together regularly, and study each other's farm management. Röling (1987) writes that 'there is an amazing openness with respect to the financial results and the exchange of technological information'. The computer has allowed these networks to become interconnected and to exchange information from the book-keeping office with which they are connected. These study groups play an important role in information exchange and in social pressure to innovate (Röling, 1987).

Present and future trends

The Dutch Government has become increasingly concerned about the effects on the environment of intensive agricultural practices, such as the threat to drinking water supplies as the result of overuse of fertilisers and pollution by manures, and extension will have to address itself increasingly to these problems.

The economic situation and the need for budget cuts will inevitably lead to privatisation of the extension services. With more and more farms going out of business, the remaining farms will need more and better information which will probably be supplied by a non-government information system (Röling, 1987).

ITALY

Historical background

In 1839, following a meeting of agriculturists, in Pisa, the first experiments in non-formal education and technical assistance to farmers were undertaken in a few areas. These led to the creation of *Cattedre Ambulanti di Agricoltura*, which were formally established in the Province of Rovigo in 1890 and gradually initiated in every province of the country. Initially, they were privately financed (Fraser, 1987).

The act establishing the *Cattedre* stipulated that the agronomists appointed to this work had to undertake the following obligations (Volpi, 1982):

- the diffusion of rational agricultural practices;
- provision of technical assistance to *all* farmers who requested it;

- training for teachers at elementary schools in agricultural areas;
- to establish demonstration plots;
- to publish a magazine concerned with agricultural technology.

The *Cattedre* were an enormous success and by 1914, there were 278 such units. Their staff were competent and appreciated by the farmers. Without the 'benefit' of diffusion theories, they worked mainly with large and middle-sized farming units. They developed intensive consultancy work which contributed to the increased use of fertilisers and new crop varieties; they organised joint efforts at controlling crop diseases and pests, and achieved a major success in the control of phylloxera, a disease of vines which had spread through Europe, causing enormous economic damage (Volpi, 1982).

The *Cattedre* were discontinued in 1935, when the responsibility for the training of agricultural workers was transferred to the Ministry for National Education. A further change occurred in 1939, when the Ministry of Agriculture and Forestry created the Provincial Inspectorates for Agriculture. Technical assistance to farmers thereby became a State responsibility.

The reasons for discontinuing a highly successful extension programme are attributed by Volpi (1982) to the increasing intervention of the State in agriculture, with a spate of regulations and development schemes. The agronomists of the *Cattedre* were the only technical experts available; they were mobilised to operate the controls and the schemes. From advisers and teachers to the farmers, they became government officials responsible for regulatory services.

The official view was that the provision of incentives (easy loans, attractive prices for agricultural commodities) was more important for technical progress than educational means, and that diffusion of improved technology would occur spontaneously from the progressive farmers to the others. Finally, it was assumed that the propagation of innovations could be left to the commercial promotion undertaken by the industries involved in the production of agricultural inputs and provision of services to the rural areas.

This official downgrading of the extension services operating in Italy occurred at a time when many other countries in Europe were strengthening their advisory services to farmers.

After World War II, the Marshall Plan began to operate in Italy and emphasised the need to achieve economic and social progress of the rural sector. Centres for agricultural technical assistance were set up in the less developed rural areas (OECD, 1981). However, the official indifference towards agricultural extension and vocational training remained unchanged (Volpi, 1982).

The provincial inspectorates did continue to do extension work and to organise farmer training courses, but these activities slowly withered. By the 1960s, they had come to a virtual halt. With the advent of the European Economic Community, 'a plethora of paperwork was added to that already existing in the Italian administration, and it soon became impossible for the Inspectorates to do more than process administrative matters' (Fraser, 1987).

The void created by the virtual abdication of government from the provision of extension was filled by the major industries producing fertilisers, plant protection chemicals, seeds, agricultural machinery, etc. They organised promotional services, employing agronomists who worked in close contact with farmers. These activities were of course concentrated in the areas of relatively advanced agriculture. The private sector made a considerable contribution to technical progress in agriculture but increased the disparities between and within regions (Volpi, 1982).

The Borgo a Mozzano Project

One remarkable exception to the general lack of interest in rural development in Italy, was the initiative undertaken by a small, but dedicated group of agriculturists and economists in Tuscany, who in the mid-1950s, started a debate on the role of national investment in agriculture. This led to a number of practical experiments, one of which was begun in 1954 in Borgo a Mozzano, a commune in Tuscany. This project has made a considerable contribution to the study of extension work in Italy and abroad, and subsequently led to the establishment of the Centre for Agricultural Studies at Borgo a Mozzano, for the training of Italian and foreign extension workers and research on extension.

The area chosen was representative of peasant farming, with relatively low levels of technology compared to Italian agriculture as a whole; the target population was geographically isolated, and reputed to be extremely averse to changing customs and production methods. These were largely out of date in a market economy.

For full details of this historic project, the reader is referred to Volpi (1982), who also provides a detailed list of references. Suffice it to state here, that the approach and methods used in the Borgo a Mozzano project anticipated, in the early 1950s, concepts which became fully accepted only in the 1970s:

- a preliminary study, which led to the choice of the location;
- a systematic survey of the area and its agricultural economy, including the physical, ecological, infrastructural and demographic environment; an inventory of the farms and their production organisation; trends and variations in income levels, agricultural structures and population;
- on the basis of this study a 'package was constructed, for improving of crop and animal production, labour-intensive production systems, irrigation, improved infrastructure, especially access roads.

Extension work proper was only started after this initial phase. It was based on the following precepts:

- backstopping for the resident agronomist was to be provided by a number of specialists (agricultural chemist, livestock specialist, machinery specialist, etc.);
- introduction of new practices was to be preceded by on-farm trials and demonstrations;
- all aspects of agricultural production, commercialisation and transformation were to be covered;
- extension was to be available to all categories of farmers and labourers; special attention was to be given to avoiding bias in favour of larger farms.
- the individual approach to farm issues was to be combined with group collective actions;
- after a few years, the work was enlarged to dealing with family problems.

Since 1955, the extension work based on these sound principles has been expanded to the region of which Borgo a Mozzano is a part (Volpi, 1982).

Decentralisation

In 1972, technical assistance and advisory work were assigned to the region and the question of recreating an extension service arose. A 12-year programme was adopted, whereby one extension worker for 400 farms would be provided. This objective, funded by the EEC, required the fielding of about 6000 extension personnel (Fraser, 1987).

The organisation of the advisory service in Italy is presently (1987) as follows:

At the national level, following the transfer of responsibility for the advisory services to the regions, the technical assistance and advisory service of the Ministry of Agriculture became a technical division for information, extension, technical and scientific documentation in agriculture, responsible for the systematic dissemination of research findings, thereby serving as a linkage between agricultural research and extension.

At the regional and local levels each region regulates the organisation of its advisory and technical assistance programmes, taking into consideration its specific political circumstances, local needs, and the human and financial resources available.

Many of the regions have maintained the old structures and continued the work previously done by the ministry. Some regions have been very slow to develop an extension policy; this is 'partly because Italy is highly politicised, and extensionists could be a natural channel for political influence among the farming community. There are therefore interests at stake under which auspices the extensionists will be fielded, and to whom they will be responsible on a day-to-day basis' (Fraser, 1987).

Other regions, such as Piedmont Region, have set up Centres for Technical Assistance (CATA) whose programme is controlled by the region but who rely on the professional organisations for the implementation of extension to their members. Piedmont has about 150 CATAs, each concerned with at least 80 farms (OECD, 1981). A technical scientific regional council has been established to orient the regional technical assistance programme (and to counteract political influences). The council consists of representatives of the farmers' organisations, officials of the regional administration, the development agency, and university professors.

By 1987, only three of Italy's 20 regions had an extension service that was rated 'excellent' or 'good' by one of the major farmers' unions. As a result, the private sector does much of the extension work. Most corporations that sell farm inputs have at least one technician/salesman per province. The *Consorzi Agrari,* a co-operative linked to the Small Farmers' Union and to a large political party, have 3000 technician/salesmen operating throughout the country. Company representatives and salesmen are usually considered to be a reliable source of information (Fraser, 1987).

No academic institutions in Italy are engaged in research on extension methodology (excepting Borgo a Mozzano), nor do any of the 23 research institutes engage in any advisory work (OECD, 1981).

FRANCE

France has been a forerunner in the trend to privatisation of the agricultural extension services. Until 1959, agricultural advisory work had been conducted in the traditional way by an extension service of the Ministry of Agriculture. Responsibility for advisory work was then assigned to the farming sector, through the Association Nationale pour le Développement Agricole (ANDA).

The state reserved for itself the functions of control, co-ordination and assistance, participating in the preparation of programmes and monitoring of activities.

The state has the same number of representatives on the Administrative Board of ANDA as the farmers' associations. The board is responsible for decision-making and financial management (OECD, 1981).

USA

Historical background

Non-formal dissemination of information on agricultural innovations began in the USA long before an official extension agency was established.

Prominent colonial landowners exchanged letters about new crops, breeds and methods. Washington and Jefferson envisaged the need for public and private diffusion of improved agricultural practices. The number of agricultural journals increased from 30 before 1840 to 450 by 1913 (INTERPAKS, 1985).

After 1887, the agricultural experiment stations began issuing hundreds of bulletins of research results. Farmers' institutes, travelling lecturers, short courses, all became popular.

In 1882, the Federal Government founded the US Department of Agriculture (USDA). The first known demonstration farm was set up in Texas to show farmers how following USDA directions could help combat boll-weevil damage to their cotton fields (OECD, 1981).

At the beginning of the present century, the recently founded state colleges, in order to improve acceptance among farmers, began to appoint special personnel to reach the grassroots people; this soon included resident field agents.

The American Association of Agricultural Colleges and Experiment Stations saw in extension an opportunity to greatly extend the scope of the land-grant colleges (LGC) while freeing researchers from the time-consuming tasks of dealing directly with farmers. Although the LGCs wanted both extension funds and complete autonomy in their use, they had to compromise. The extension agents' salaries were paid in part by federal, state and county governments, and even in some cases by private organisations. To this day, the question of how the extension agents are to divide their loyalties has not been clarified (Lacy *et al.*, 1980). The extension agents were instructed to work with groups; to facilitate this, they took the initiative in organising Farm Bureaux (Busch & Lacy, 1983).

An Extension Committee on Organisation and Policy (ECOP) was established in 1905, to provide a sense of common mission and purpose. It comprised state extension directors and the deputy director for extension of the Science and Education Administration (SEA) (OECD, 1981).

In 1914, the extension service was formally established by the Smith-Lever Act, which provided for 'co-operative' agricultural extension work between the LGCs and the USDA. 'Ironically, it was called co-operative because both the LGCs and the federal agency coveted the function, and a compromise had to be struck' (INTERPAKS, 1985).

The cutting edge was to be the LGC, but the federal department had to approve work programmes. Federal funding was to be matched by state and local levels (tripartite funding system).

Structure of the Federal Extension Service

In 1978, the Science and Education Administration (SEA) was established, integrating the USDA's ARS, the Co-operative State Research Service, the extension service, and the National Agricultural Library. However, the desired welding of the four disparate entities was never quite realised, so the concept was abandoned in 1982 and the agencies resumed their autonomy.

The *federal extension service* has six programme categories: agriculture, natural resources, food and nutrition, family education, rural development, and 4-H youth.

The state co-operative extension system: from its earliest beginnings, extension was not only seen as a way to deliver new applied knowledge to farmers and other rural people, but also to transmit the interests of the latter to the LGCs. To promote sensitivity of extension workers to rural people, all were required to have a farm background. Farmers also had considerable control over the whole system (Röling, 1987).

The three functions of research, extension, and education are administered separately, and a different set of power relationships marks each of the three activities. 'Popular writers who speak of *the* relationship between the LGCs and the USDA, gloss over what is a complex, multifaceted *set* of relationships' (Busch & Lacy, 1983).

State co-operative extension services work co-operatively with the USDA, and in close liaison with public and private interests at national, state and local levels. Each state extension service has considerable autonomy in determining its own policies and programmes at the local level.

'In many respects, the Co-operative Extension system may be considered as a federation of states working together on a co-operative basis with the Federal Extension Service to provide educational assistance on problems of a regional and national nature which cannot be adequately resolved on a state-by-state or local basis' (OECD, 1981).

4-H programmes
Major objectives for the 4-H (head, heart, hands and health) programme include: involvement of youth in environmental improvement and conservation of natural resources; in health education programmes, in consumer education; in leadership roles and citizenship activities.

A major thrust is aimed at doubling the number of volunteers working with the 4-H programmes.

Structure of the State Co-operative Extension (Fig. 20)

The structure of the State Co-operative Extension varies somewhat from state to state. Thomason and Toscano (1986) describe the set-up in California:

> The overall supervisions of extension work is the responsibility of the Director of Co-operative Extension who delegates the actual running of the extension work to two parallel chains of command: one that deals with county-based advisors and one that deals with campus-based specialists. In the first chain of command, the county advisors answer to county directors, and through them to one of four regional directors, who are administratively responsible to the Director of Co-operative Extension. In the other chain of command, the campus-based

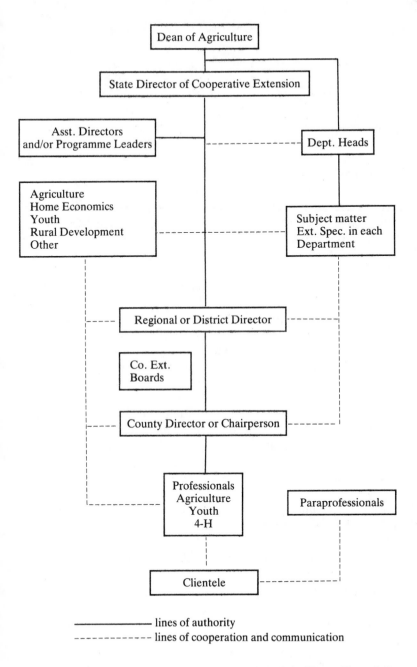

Fig. 20. Typical extension structure in a state of the US. (From Claar & Bentz (1984), reproduced by permission of FAO, Rome.)

specialists are organised in eight extension programmes, each under a Programme Director, who answers to an Associate Director-Programme, and through him, to the Director of Co-operative Extension.

Clear lines are established for administrative and programmatic co-ordination between the two groups.

The federal, state and county partnership

The federal, state, and county levels of extension operate 'without domination by either one and with maximum freedom for independent action' (OECD, 1981). Programmes must be mutually agreed upon between the USDA and the LGCs.

Relationships between the three components of LGC—university system

There are marked differences in the relationships between the three components of the LGC system: Teaching (Office of the Dean), research (Experiment Station) and extension (Director of Extension).

In Texas, for example, the three bodies are separate, and there are very few joint appointments. In Arkansas, the extension service is located off-campus and has no formal ties to federal or state research; in the Mid-Western States, there is full integration of extension specialists in research-oriented departments and many joint extension/teaching and extension/research appointments.

Sources of extension funding

Extension programmes are funded co-operatively from three major sources: federal (about 40 per cent), state (about 40 per cent), and county (about 20 per cent). Percentage of funds from each of the three sources vary considerably from state to state (OECD, 1981).

Extension personnel

Extension agents are federal civil service employees responsible, in principle, to the Federal Extension Service (Lacy et al., 1980). This is in contrast to teachers and researchers, who are state employees.

A single individual's salary may be paid from three different sources (Busch & Lacy, 1983).

Qualification requirements
All county extension agents must have a Bachelor or Master of Science degree before they are hired and most specialists must have a PhD degree. Most agents with a BSc degree are expected to work for a Master's degree after they are employed. Professionals are recruited primarily on the basis of their qualifications, rather than by competitive examination. Probationary periods vary from state to state, but are usually for six months to two years (OECD, 1981).

Pre-service and in-service training
There is no specialised pre-service training for extension workers. A few LGCs have three-week training programmes for future extension workers. Many workshops are provided at national and state levels to provide updating and in-service training in methodology and subject-matter (OECD, 1981).

Number of extension workers
The extension service has a highly competent staff of about 3000 and more than 3000 specialists and specialised multi-county area staff members (OECD, 1981).

Clientele of the extension service

Extension agents and SMSs conducting adult extension programmes in agriculture and natural resources devote approximately 70 per cent of their time to commercial farmers, 25 per cent to small and low-income operators with a gross income of $ 20 000 annually, or less, and about 5 per cent to home gardeners (OECD, 1981).

A national programme that is being accelerated is to help small, limited-resource farmers. This programme is supported by national funds but implemented by state and county level extension staff in accordance with national guidelines.

Extension programmes are designed and priorities established through a high degree of involvement of local people. Many programmes are carried out with agricultural producers through their commodity committees.

Links with agricultural research

At the federal level, extension staff maintain close contacts with their research counterparts in the ARS and the LGCs. At the state level, close working relationships exist between the LGC researchers and the SMSs.

Technology transfer by ARS
ARS is organising its activities for improving the dissemination of ideas and discoveries to the extension and the public by the *Technology Transfer Plan*. Technology transfer will be planned, formally structured, co-ordinated and tracked. Effectiveness and efficiency will be enhanced by increased use of electronic communication and automation systems. Co-ordination of these activities will ensure that different sets of research findings that bear on a particular problem will be presented to users as a package of technology (ARS, 1984).

Programme planning

Co-operative extension has developed a co-ordinated process of programme planning described by Thomason & Toscano (1986). Priorities are set by work groups comprising discipline and commodity-oriented specialists, advisers and university researchers. These assess clientele needs and problems; identify applicable research or research gaps; establish priorities; identify, design and carry out major projects for the development of relevant research; plan the application of research, adoption strategies and diffusion of information; and finally, evaluate the impact of these projects.

Private sector extension

Private consultants have become a major force in the delivery of crop management advice, with pest control information an important component. This trend is expected to continue. Educational efforts by consultants include meetings for their customers, newsletters, scheduled farm visits, and supply of educational material by mail, phone and computer.

Many programmes are the co-operative effort of public and private sector advisors (Blair, 1986).

Future trends

Some specific programme adjustments and redirections currently undertaken include (OECD, 1981):

- emphasis on crop production systems using computer models to integrate information from various disciplines;
- rapid expansion of integrated pest management programmes;
- a greater effort to helping farmers cope with government regulations in regards to improving water quality, involving work on food safety, pesticides, control of plant pests and diseases;
- more emphasis on limited tillage, soil testing and plant tissue analysis, increased efficiency in the use of fertilisers, etc.;
- more emphasis on animal health, and reduction of residue problems related to the use of antibiotics, and sulphonamides;
- more help for farmers in coping with financial management problems;
- with more than 50 per cent of the families now gardening, more help will be given to urban people with gardening information.

AUSTRALIA

In Australia (OECD, 1981), the State Departments of Agriculture are responsible for agricultural extension, and are completely independent of the Federal Ministry of Agriculture in this respect.

From the late 1960s, Divisions of Agricultural Extension were established in each of the State Departments of Agriculture, to co-ordinate extension work throughout their respective states. In the 1970s the tendency to regionalise the advisory services was intensified, especially in the larger states.

Structure

The basic purposes and the broad structures of the advisory services in the different states are similar, though there are some differences in organisation between them.

Taking New South Wales as an example, the state is divided into eight regions. The staff of the Division of Extension Service consists of 7 regional extension officers, 8 publicity officers, 36 field assistants, and 120 clerical staff.

Advisers are commodity specialists assigned to each of the regional offices. They are supported by research officers in such areas as biological and chemical research, marketing and farm economics.

Advisers provide advice on request by the farmer by telephone, office discussion or on farm visits. There are also planned programmes, using group and mass media.

Each region has its publicity officer, responsible for supplying information through the mass media, and has a number of rural youth officers to assist rural youth organisations.

Clientele

The major clientele of the advisory service are commercial farmers and their families. The service is mainly utilised by the more progressive and successful farmers, who also tend to be younger, better educated and in a developing business. The work of the adviser includes technical and managerial information, training and supply of market data.

Liaison is maintained with farmers' organisations and local government. The Division of Extension Service also supplies information to policy-makers and other government departments, as well as other organisations, such as banks, stock agencies, chemical companies, etc.

Methods

Whilst the methods adopted in the different states may differ, most have adopted the following procedures:

- programme planning and evaluation of the advisory work carried out at district and regional level;
- interdisciplinary advice by the staff;
- team-work in the definition of advisory problems;
- liaison with agricultural research, farmers' organisations and agribusiness.

Until recently, advice was being given in most states by highly trained subject matter specialists (SMSs). This approach has been intensively debated since the 1870s, and two states (Western Australia and Southern Australia) have adopted the 'generalist' approach, whereby local extension officers advise on whole farm management, and when necessary, call in a SMS to provide technical back-up. In the larger eastern states, there is a growing interest to follow the same generalist approach, but this tendency is constrained by the vested interests of the specialists who have devoted years of training for their respective competences.

Programming

Programmes which require state-wide planning and co-ordination are generally initiated by the head offices of the state departments of agriculture. Most programmes are, however, decided at regional or district level. Farmers, financial agencies, agribusiness, etc., are usually consulted in the formulation of the programmes.

Queensland has formalised these procedures by establishing district extension committees in each of its 33 advisory districts, comprising representatives of local department research units, and from each extension branch operating in the district. Programmes proposed by these committees are subject to approval by the extension services board, a state-level body comprising the department's deputy director general or his deputy, the director of each technical division and the director of the information and extension training branch.

Victoria has developed a formalised computer-assisted planning procedure for all its activities: the objective-based management system which groups all the departments' regulatory, advisory, research and administrative activities, into about 200 programme areas within which projects are conducted.

Links between extension and research

In *New South Wales* the Department of Agriculture established in 1972 a three-person research–extension evaluation unit, the first of its kind in Australia. Its activities have included surveys of farmers on changes in their farms, contacts with the advisory service, attitudes towards government programmes, adoption of the results of research, analysis of advisory officers' work time, etc. *South Australia* subsequently established a similar unit. *Western Australia* has two senior officers who are responsible for ensuring good liaison between research and extension officers, and in *Tasmania* all information obtained from research work is circulated to extension officers through technical information sheets, for inclusion in a storage and retrieval system called AGDEX.

Training of extension staff

There are specialised schools for pre-training of staff; these are recruited directly from the universities or colleges of advanced education. Recruitment is on the basis of qualification, experience and aptitude for advisory work. There are no entrance examinations. Newly recruited workers undergo a probationary period of 6–12 months before final appointment. There is no formal testing or examinations for promotions, but knowledge and skills in economic, social and technical areas, as well as management of staff are progressively registered and taken into account.

In-service training is possible both as extramural postgraduate training, or in organised in-service course work. During the first 5–10 years of service, an advisory officer will normally receive induction training and training in extension methods and business management. He will also be involved in a continuous process of technical training in his specialised field by means of on-job training, attendance at seminars, conferences, etc.

Recruitment is generally at the age of 21–22; promotion during the first ten years is semi-automatic—subject to satisfactory performance of duties. Promotion to senior positions is generally through competition.

ISRAEL

Historical background

During the period of the Mandate (1918–48), the department of agriculture was organised on the same lines as in the British colonies. Each major group of commodities was the responsibility of a division headed by a chief agricultural officer, who was responsible for development, research and extension in his field.

The workers at field level were agricultural inspectors, who reported to their respective chief officers through a regional officer. The governmental extension system was mainly concerned with Arab farmers, and its major functions were regulatory and supervisory.

The Jewish Agency maintained an agricultural experiment station based in principle on the land-grant college model (though there was still no faculty of agriculture at the time), and therefore operated its own minuscule extension service. Though few in numbers, the extension agents were very active in the Jewish farms, and published technical information bulletins, the first of their kind in the country.

With the establishment of the State of Israel (1948), the new ministry of agriculture had to build an extension service practically from scratch, capable of coping not only with the existing disparate Jewish and Arab sectors, with their completely different types of farming systems and levels of sophistication, but also with the flood of new immigrants—about half of whom came from the concentration camps in Europe, and the other half were refugees from the Arab countries. Though the two groups came from entirely different backgrounds, they had a few charac- teristics in common: they were completely destitute; they came from an urban environment and had absolutely no agricultural knowledge or experience; they were physically so debilitated as to be unable to make a significant physical effort.

It was with this disparate rural population—sophisticated modern far- mers from the old-established communal and co-operative settlements, traditional subsistence farmers in the Arab sector, and a flood of new immigrant 'non-farmers' placed without preparation on some new settle- ments—that the young extension service had to cope.

In addition to the few extension workers from the mandatory period, a large number of veteran farmers were mobilised to live in the new settlements with the immigrants and help them to establish themselves and to become farmers.

A dual system of extension was established: one for the new settlers and the other for veteran farmers. In 1965, a major reorganisation of the ministry of agriculture led to a fusion of the two systems and the regionalisation of the extension service, in a form which has remained largely unchanged to this day.

Present structure

The main structural characteristic of the extension service in Israel is its status as a major Department in the Ministry of Agriculture, to which the Commodity Divisions are subordinate.

In addition to these Commodity Divisions (Livestock, Field Crops and Horticulture), there are Service Divisions providing support to the former: an Irrigation and Soils Service, Mechanisation and Technology, Extension Methodology and Planning, General Administration.

The field service is decentralised, and is the responsibility of ten Re- gional Agricultural Offices of the Ministry.

The service functions both on a national and on a district level. At service headquarters, each main crop, commodity or branch of produc- tion is represented by a department, staffed by subject matter specialists (SMSs). The departments, in turn, comprise several divisions.

The country is divided into ten extension districts, each of which is staffed by 25–70 extension workers, according to the size of the farming population of each district. They carry out their advisory work in their specialities, corresponding to the divisions at headquarters, and receive guidance and support from the SMSs at the national centre.

The district staff is headed by a director, who is assisted by an extension co-ordinator, who provides educational methodological support to the field workers (Elkana & Sagiv, 1981).

Staff

The veteran farmers of the 1950s have all returned to their farms and have been replaced by younger extension workers, many of whom at the time, were graduates from secondary level agricultural schools. With the reorganisation of the service, an intensive upgrading of the staff was initiated. Extension workers were encouraged to study at the faculty of agriculture, in part-time release. At present, almost all extension workers hold academic degrees, and university training is a prerequisite for those who are newly appointed (Blum, 1987).

The regional advisers are all specialists, sometimes in a very narrow field; they are supported by SMSs at headquarters.

The extension staff comprises 450 people, of whom 370 are professionals, for a country with about 75 000 farmers.

Clientele and concepts

Extension workers are not involved in any supervisory or regulatory functions of the ministry of agriculture. Their primary responsibility is towards the farmers.

The service is open to all, but is mainly given to those who demand it. Farmers in a village are informed of the visit of an extension agent, but only those who sign up are visited. Where there are several farmers producing the same commodity in a village, no extension is provided unless they appoint a co-ordinator (Elkana, 1986).

The key concept of the service is that 'it is more effective to work through key people than to reach all' and that 'by spreading themselves too thinly, extension staff find themselves running around a lot but achieving little' (Elkana, 1986).

Planning

Initially, one of the key elements of the 1965 reorganisation, was a very elaborate and detailed system of programme planning. It proved to be extremely time-consuming, both for the extension workers and for the management who were supposed to use it for evaluation and monitoring. It gradually fell into disuse (Elkana, 1986).

Linkages

Notwithstanding all the measures to ensure close links between the extension workers and the researchers, centrifugal tendencies are still evident, such as the gradual development of separate on-farm trials (cf. p. 797).

On the whole, however, the institutionalised linkages—joint commodity panels for programme planning of research, in which extension, research and faculty staff are joined by farmers' representatives; on-farm trials and common services, as well as many personal contacts—have proven themselves, and the overall extension/research linkage is satisfactory, if not ideal.

Extension–Education linkages
In 1979, the faculty of agriculture established on extension centre, which offers a one-year diploma course, with an option to combine it with a Master's degree in agricultural science. Most students take the course part-time (Blum, 1987).

KENYA

Structure

The extension and manpower development division is the largest of the five divisions of the Department of Agriculture. It comprises a network of extension staff. The responsibility for extension is exercised by the director of agriculture, through his senior deputy (national level), the provincial directors of agriculture, the district agricultural officers (DAOs), the divisional extension officers (DEOs), and the locational extension officers (LEOs). The LEOs are the frontline extension workers with the farmers.

Communicating innovations to the farmers

The Ministry of Agriculture has established an *Agricultural Information Centre* which is equipped to produce literature and films depicting agricultural development activities and technological innovations. Research results are transferred to extension workers and farmers through publications, radio, television and the press.

Field days are organised by the experiment stations, to enable the farmers to see and discuss innovations, materials and recommended practices. Simple information sheets and pamphlets, in English, Kiswahili, or major local languages, are made available to the participants and to the extension staff for wide distribution. Farmers who cannot read or cannot attend the field days, are served through radio interviews with research staff.

Extension–Research linkages

Linkages between research workers and extension workers are generally deficient, with each service blaming the other for the indisputable shortcomings.

Since the 1970s, a major effort has been made to improve the relations between the two services. Joint meetings are held regularly at the provincial level to enable exchange of ideas between researchers and extension workers; joint field tours are undertaken, making it possible to identify problems and constraints requiring a research effort.

Linkages with universities

The universities rely mainly on the extension service to convey the results of their research to the farmers. They also organise open days, seminars, demonstrations, short courses for farmers and frontline extension workers (ISNAR, 1981*a*).

Voluntary organisations

Various voluntary organisations provide support to the government effort in various aspects of development, including agricultural extension.

In Kenya, for example, church groups have actively supported farmers' training programmes; some have even established their own training centres for farmers. They have also supported youth training by initiating the development of village polytechnics.

Other voluntary bodies, such as Action Aid, World Vision, the National Greenbelt Movement, also support extension activities in Kenya. For example, Action Aid encourages youth in schools located in disadvantaged areas to engage in vegetable growing, poultry and rabbit keeping, etc., by providing material aid and enlisting school teachers for advisory work (Onyango, 1987).

MALAWI

Structure

Agricultural extension for smallholders is the responsibility of the Department of Agriculture of the Ministry of Agriculture. Advisory services to the commercial plantations are provided by the private sector.

The government extension service is decentralised into eight Agricultural Development Divisions (ADDs). Each ADD has crop, livestock husbandry, land, credit and marketing, women's programmes and visual aids divisions.

The ADDs are divided into Rural Development Projects (RDPs), which in turn are sub-divided into Extension Planning Areas (EPAs) of which there are 180; these EPAs are a compromise between agroecological zones and administrative boundaries.

Provision of extension

The extension service has adopted a modified version of the training and visit system of extension (cf. pp. 712–20). Technological recommendations come from the Department of Agricultural Research (DAR). Each extension field assistant is responsible for a 'block' of villages, with about 750 smallholder families per assistant. The subject matter specialists are stationed at the ADDs.

Farm clubs, run by the villages, have facilitated extension work, the effective supply of credit, and the recovery of credit. Special women's programmes have been provided, both in homecraft and agriculture.

Research–extension linkages

Linkages between the researchers of DAR and the extension staff are mainly informal and based on personal contacts. However, a research liaison officer transmits queries from the extension workers to the researchers, and requests the response of the latter in the form of summarised technical recommendations.

On the whole, research–extension liaison is generally deficient; extension workers are rarely aware of recent research findings, and many extension recommendations remain unchanged for many years. The Agricultural Handbook of Malawi, includes all the basic extension recommendations for the country, but has not been updated since its publication in 1971.

Farming system research

In 1983, CIMMYT and USAID undertook a series of seminars to prepare staff for farming systems research. It is expected that this research will not only prove to be an effective link between research and extension workers, but will also supply valuable information on the farmers' major social, economic and technical problems.

Performance

The quality of the extension services is variable across the ADDS. Not all the blame for lack of impact on the agriculture practised by the majority of the smallholders can be attributed to the extension service; lack of focus of research on the major problems of the farming sector, delayed reporting of results of research and lack of formal linkages between the two services have been factors in the lack of stimulus to agricultural development.

INDIA

History

Extension services in India developed much later than the research system (cf. pp. 114–22). Following independence in 1947, the Government of India initiated a single-line extension service which was discontinued with the introduction of the community development programme (Rivera, 1987).

Village level workers of the community development programme were supposed to provide virtually the entire range of services required for the development of the village community. In theory they were expected to devote about 80 per cent of their time to agriculture, but they had so many regulatory functions that actually only a small fraction of it remained available for agricultural extension (Verkatesan, 1985).

In the absence of a professional extension system, and hence of essential feedback from the field to the research system, the Indian Council for Agricultural Research (ICAR) initiated its own extension-related activities (Jain, 1985).

The State Agricultural Universities (SAUs) undertake four extension-oriented activities:

(a) The lab-to-lab programme has 139 centres throughout the country and supports about 75 000 small and marginal farmers and agricultural labourers. Its objectives are to increase income of these groups by low-cost technology and provide opportunities for self-employment (poultry, dairy).

(b) Farmers schools or farm science centres are supported by ICAR, institutes, the SAUs, and voluntary organisations. They train farmers, farm women, school dropouts, rural youth, and village level extension workers.

(c) National demonstrations, about 2500 every year, are conducted by research workers on farmers' fields to show the production potential of important crops. Usually the poorest men in the village are selected so that the success of the technology cannot be attributed to the effects of affluence but to the technology itself.

(d) Operational research projects are concerned with solving community-based agricultural problems through appropriate technology and identifying the constraints preventing adoption.

The SAUs are also involved in an intensive training programme for subject matter specialists (SMSs). Annually, about 500 SMSs receive in service training in their respective disciplines, and in extension methodology.

While the official research system started extension-oriented activities, the state departments of agriculture (DOAs) established their own research-oriented activities. In a few states, adaptive research centres, (also called field trial stations), were established as part of the departmental activities. These activities naturally created overlapping of functions between the research and extension systems (Verkatesan, 1985). Whether the research or the extension establishment should be responsible for adaptive research is still being debated.

In the late 1950s, the steep population increase and food shortages due to unfavourable weather conditions accentuated the urgency to improve food production. In response to this situation, the Indian government initiated the intensive agricultural district programme (IADP). Eventually,this programme covered 28 districts, and the need for intensive training of extension staff was recognised. In the early 1960s, three extension education institutes were created (Rivera, 1987). In the mid-1960s, short-term training courses in agriculture were organised, in collaboration with the SAUs, research institutes and colleges.

In the 1970s, extension was still the weak link in the services available to the farmer. Otherwise, India was well equipped for the adoption of new technology: a dramatic breakthrough in high-yielding varieties (the 'green revolution') provided the focus for economically advantageous technological 'packages'; widely available inputs and the credit to purchase them; an acceptable level of efficiency of marketing, favourable producer prices, and a sound road network. Extension efforts were focused on village level community development, and technical advice was only one of the many functions of the village level worker (Roberts, 1987).

The training and visit system (T & V)

In 1978–79, after preliminary pilot-testing, the T & V system was intensively developed by pooling staff from existing schemes. By 1985, it was functioning in 13 states. The T & V provided an administrative framework, not present previously, that facilitated the flow of information from farmers' fields to researchers.

Pickering (1987) ascribes the success of the T & V in India to the following factors: the existence of a strong agricultural research system and the availability of attractive research results (the 'green revolution') which had not yet been disseminated; inputs could be purchased freely; credit was available freely; grain markets worked with tolerable efficiency; and producer prices were remunerative enough to motivate farmers to invest in improved varieties and the necessary inputs. The road network and other essential infrastructure were sound.

The weak link in the services available at the time to farmers was extension. Therefore, when T & V arrived in India, extension reforms had become a high-priority issue, and the preconditions for successful extension efforts were in place.

To the factors mentioned above, Rivera (1987) adds that of the human resources available. Beginning in the early 1960s, the Department of Extension in the Ministry of Agriculture and Development had over a decade prior to the introduction of the new system, prepared a large pool of trained manpower for extension at different levels. In Andra Pradesh, for example, a state with 80 million people, the introduction of T & V required an increase in extension staff from 1500 to 4000. These were mostly provided by transfers within the state civil service of over 100 000 staff (Hayward, 1987).

Following the introduction of T & V, extension management training was intensified for the grassroot and medium levels, and—at a later date—for senior level management.

Research–extension linkages

Though there is strong central financial support for extension, in 1985 a formal research–extension linkage was still lacking at the national level (Verkatesan, 1985), and links between research and extension are generally weak in India. Jain (1985) gives three reasons for this situation: (a) scientists at the SAUs and officials in the state DOAs are reluctant to accept changes in concepts and procedures; (b) although the programmes of the SAUs and DOAs bring scientists and officials together, they have no real involvement with, or appreciation for each other; (c) although the physical infrastructure may be built up at regional level, decentralisation of the administration of research is a slow process because of insufficient institutional freedom.

However, a continuous effort is being made to link the NARP and the T & V extension system in each state, with the key link between the two being the SMSs (Jain, 1985).

Agencies involved in agricultural extension

In India, agricultural extension activities are undertaken by the national and state public sectors and the agricultural universities; commodity extension by the private sector; and numerous agricultural and rural development activities by non-governmental organisations (Rivera, 1987).

National level

The Union Ministry of Agriculture and Rural Development is organised into four main departments: Agriculture (DOA); Agricultural Research and Education (DARE); Rural Development (RD); Food and Civil Supplies (FCS).

The first three departments, DOA (through the Directorate of Extension (DOE)), DARE and RD are all involved in agricultural extension. The extension activities of DARE have already been mentioned. RD, which is an outgrowth of the community development department, started some of the first extension activities in India. Like DOA, it supports a number of training and development activities at national and state level, and operates several rural development projects which employ village level workers.

Other directorates of the ministry of agriculture are also concerned with extension activities, but have no field staff.

State level

The overriding leadership and the major field-based, production-oriented extension service are provided by the state DOAs, most of which have adopted the T & V system.

At the state level, formal education in the specialised disciplines of agriculture, agricultural research and extension education, including in-service training, are the responsibility of the agricultural universities.

State departments of agriculture, horticulture, animal husbandry, etc. undertake village-level extension, supported by the specialist and training activities from the AUs.

The supply of material inputs, credit, and various agro-services is handled by various public, co-operative and private institutions. (RAPA/FAO, 1985).

INDONESIA

Present structure (RAPA/FAO, 1985)

At the *national level*, the ministry of agriculture (MOA) consists of four directorates-general (DGs) for food crops, estate crops, animal husbandry and fisheries, respectively; and two agencies: for agricultural education, training and extension (AAETE) and for agricultural research and development (AARD). A national BIMAS secretariat is also active, as well as a national commission for co-ordination of agricultural extension. Each DG has a directorate of extension and the AAETE has its bureau of extension.

At the *provincial level*, each technical agricultural service has its own extension division. Other institutions that are closely related to extension are: a BIMAS secretariat, an agricultural information centre, research stations, agricultural in-service training centres, a forum for co-ordination of agricultural extension. The rural extension centres serve as headquarters of agricultural extension, and also for farmers' meetings, training, demonstrations, etc.

At *district level*, most of the elements of the provincial level are repeated at district level.

Linkages

Extension–research linkages are maintained through participation in agricultural extension co-ordination forums, subject matter meetings, BIMAS meetings, chaired by the ministers involved, a consultation forum attended by officials and contact farmers, field days at experiment stations and universities, and on-farm trials.

Another linkage between research and extension is provided by the agricultural development boards. These are administratively responsible to the provincial governor, but technically responsible to the appropriate director-general in the ministry of agriculture. The boards carry out field trials in which AARD is *not* involved, either in the implementation or the evaluation stages.

In each province, an officer of the ministry of agriculture, the *kanwil*, has an advisory and co-ordinating role with respect to all technical activities undertaken by the ministry in the province. In some provinces, the *kanwil* is also responsible for *rural information centres* (ISNAR, 1981*b*).

Extension–education linkages do not exist between extension and the universities, but diploma and graduate courses for extension personnel are organised by the universities. The secondary schools are under AAETE, making it possible to ensure co-ordination of extension activities and formal secondary agricultural education.

Extension–support services are assured through the mass guidance intensification programme, which operates a co-ordinating mechanism at each administrative level, from the national level to the village level. The linkage between extension and the support services (input suppliers, credit and marketing services) is the responsibility of the chief administrator at each level, who chairs the co-ordinating committee.

Provision of extension to the farmers

In contrast to AARD, whose spatial organisation is based on agro-climatic zones, the AAETE's organisation is based on provincial administrative boundaries.

The extension services, as indicated above, are the responsibility of the ministry of agriculture. Management of the service has, however, been delegated to the efforts of three groups within the ministry: AARD supplies the technology based on its research; the four directorates-general service the individual units of extension services; AAETE provides advice on methodology. At farmers' level, matters become still more complicated because implementation is carried out in co-ordination with provincial governors, who answer to the Ministry of Home Affairs for their administration of certain extension services. The extent to which these agencies co-ordinate their efforts evidently has a strong influence on the quality of services provided to the farmers (ISNAR, 1981*b*).

Most farmers in Indonesia have mixed farms (except in the wet-rice areas), with annual field crops, perennial tree crops, livestock and often fish ponds. The existing extension system obliges the farmer who wishes to obtain information to consult at least four extension agents in different offices, each representing a specific subsector (Baharsjah, 1985). Conversely, a number of different field extension workers converge at different times on one individual's farm.

The role of the directorates-general in the Indonesian extension services is largely an historical one. It is the direct result of a national programme started in the mid-1960s to increase farm production and income by supplying credit and inputs at subsidised prices.

The function of the directorates-general within this programme was to provide the extension workers for implementation. The directorates continue to provide the major share of the extension services in the provinces (ISNAR, 1981*b*).

The DG food crops extension department oversees extension activities, and its crop production department plans and supervises adaptive trials, without these being co-ordinated with AARD, which of course has its own network of adaptive and on-farm trials (Padmanagara, 1985).

The regional agricultural services transmit information on new technologies to the farmers and manage demonstration plots at farm level. Since the services are organised by commodity subsectors (food crops, estate crops, fisheries, livestock), four regional agricultural services give advice to farmers in each subdistrict (Baharsjah, 1985).

AAETE does not have the same authority or status as AARD has for research, and therefore has difficulties in attempting to oversee and co-ordinate extension work in food crops.

The DG for estate crops has an extension department, but several semi-autonomous agencies implement much of the extension work on these crops with their own extension staff. DG animal husbandry and DG fisheries have smaller, but parallel services, and the co-ordination between them is weak.

Deficiencies of the extension services

Padmanagara (1985) lists the major weaknesses of the Indonesian extension services as follows:

- a paternalistic approach to extension, with messages conceived and delivered in a top-down fashion, typical of the colonial and feudal approach;
- extension advice is not generally adapted to differences in agro-climatical zones, socio-economic factors or various levels of access to resources of the farmers;
- lack of feedback from farmers, and inadequate analysis of rates of adoption of recommended practices;
- poor farmer participation, notwithstanding the long-time existence of active farmers' groups;
- a complex structure of the extension service, resulting in a fragmented system, lacking co-ordination;
- difficulties in accommodating national agricultural targets with farmers needs;
- weak linkages with research.

Improving the extension service

Whilst a unified extension system is not likely in the near future, several efforts are being made to improve co-ordination at all levels (Padmanagara, 1985).

- AAETE is to provide co-ordination at national level through a 'project implementation unit' of the national agricultural extension programme.

- The provincial and district offices are to ensure co-ordination at their respective levels. Each provincial head of agriculture has appointed a deputy for training, extension and technical development who is responsible for co-ordinating the activities of the four DGs and the AAETE's provincial agricultural information centre; the district heads have deputies with similar functions.
- At the rural extension centres, the four DGs started pooling their staff, and the most qualified extension officer has been put in charge of a co-ordinated programme for all four services.

Agricultural information centres

AAETE has established a number of agricultural information centres, and the intention is to have one in each province.

At the district level, AAETE maintains institutes for agricultural extension, which are home bases for the extension field workers.

The role of the agricultural information centres is to gather information from the research institutes, transform it into a form that can be understood by farmers, and then distribute it to the institutes for agricultural extension. The latter collect information from the farmers on their problems, and pass it on to the information centres. These also provide a meeting place for AAETE staff and the field extension workers (Baharsjah, 1985).

TAIWAN

The extension system in Taiwan can serve as a model for multi-agency extension services.

The Provincial Department of Agriculture and Forests directs and co-ordinates agricultural research and farm information services in the Province. It directs a dual system of extension: one through government offices and the other through farmers' associations.

The public office extension is operated by the department at provincial, hsien (regional) and township levels. It maintains grassroots contacts with farmers through elected village chiefs and neighbourhood headmen; it is mainly concerned with implementing government food production programmes; and plans and supervises the educational extension work of the farmers' associations. Every year the department contracts with the provincial farmers' association to implement an extension programme through the local township associations. It provides financial, technical, and informational assistance, but no control other than financial. Public office extension advisers are sometimes loaned at township level. By law, 70 per cent of the annual profit of the farmers' associations from credit, supply and marketing is to be used for farmer and housemaker education (Lionberger & Chang, 1981).

Certain cultural and social characteristics that have played an important role in modernising agriculture in Taiwan are unique to that country. Nevertheless, much can be learnt by other developing countries from the way the agricultural institutional system operates in Taiwan. This system has created an agriculture whose productivity has increased rapidly in a few years and where farms of an average size of 1·1 hectare, using mainly animal and human power, produce enough food for a large family, with a surplus for the urban population and for export (Hsieh & Lee, 1966).

Lionberger and Chang (1981) have made a study in depth of the transfer of improved technology to the farmer in Taiwan, and the role of research, extension and the farmer in this process. The following is a summary of their findings explaining the success of the procedures adopted:

An effective linkage system between all the institutions involved in agricultural development

In Taiwan, the three basic functions, agricultural research, extension and education, are carried out in separate institutions: research by the department of agriculture and the faculties of agriculture; extension by the department of agriculture and farmers' organisations, and education by the department of education.

Each of these institutions depends on the others as a source of information on improved farm practices to transmit to the farmers.

Research is responsive to farmers' problems through its links with extension and direct contact with farmers. Although extension is not part of the official duties of the researchers, they are actively involved, mainly by assisting the extension advisers, and most of them feel that they gain in status thereby.

Communication between the researchers in the research institutes and those working in the district agricultural 'improvement' stations is intensive. Researchers also maintain good, though informal, relations with the agricultural faculty staff.

Each year, the Department of Agriculture and Forests contracts with the farmers' associations to execute an extension programme for which the department provides financial, technical and informational assistance. The governmental extension service also maintains direct, grassroots contacts with farmers, through small, local extension units.

The extension workers of both services are in close contact with the research units; are fully aware of research results, and are personally involved in on-farm validation of research results.

Bi-monthly meetings are held by the farmers' associations to consider extension problems and exchange ideas. Similar meetings are held in the government offices to inform the extension staff about government programmes and new developments in agriculture.

Overall co-ordination of research and extension activities

The linkages described in the previous paragraphs are strengthened by careful co-ordination of all research and extension activities.

The Department of Agriculture and Forests directs and co-ordinates all research programmes as well as all farm informational services in the districts, including contract arrangements with the mass media. It also provides informational services for all the research institutes and 'improvement' stations.

The agricultural faculties, though they belong to the Department of Education, form a vital part of the research system.

The extension office of the Department of Agriculture and Forests is not only directly involved in extension through its local offices, but also plans and supervises the extension work carried out by the farmers' associations.

Research oriented to the needs of the farmers

Most available resources are invested in applied and adaptive research, using a largely borrowed scientific base for this work.

Research results are carefully tested for adaptability to local conditions, in the 'improvement' stations and by on-farm trials.

Extension does not confine itself to transmitting information

The extension workers help farmers solve problems that are not of a technological nature, such as planning the farm, carrying out government directives, improving rural living, assistance in obtaining needed supplies and services.

Communication of scientific farm information is mainly through group activities: demonstrations, tours and training meetings; but extension workers are highly accessible to individual farmers. The extension offices are within bicycle distance from the farms within their jurisdiction. On-farm visits by extension workers are frequent, and they also attend farm discussion groups sponsored by local units of the farmers' association.

Satisfactory working conditions for researchers and extension workers

More than half the researchers and extension workers come from a rural background. They are attracted to the agricultural services and remain in them, with good chances for promotion. They feel well rewarded financially, and have personal satisfaction in their jobs. They empathise with the farmers, and enjoy their trust and esteem. Both extension agencies provide periodic in-service training.

Farmers are responsive to innovations

The majority of the farmers are literate. They are eager to learn about new technologies, and adopt them rapidly. Farmers who are considered 'innovators' are highly valued sources of information.

Farmers' representatives are involved at all levels of decision-making

Overall control of development-related activities may be dominated by government, but the farmers have a major role in the decision-making process. An official review committee composed of farmers has to approve all research proposals submitted for funding by the research institutes and the improvement stations. The committee also decides whether research findings are sufficiently validated for release to farmers; no researcher is allowed to by-pass the committee in communicating research results to farmers through the official channels, but they are allowed to publish independently in professional journals and farm magazines.

Conclusions

At first sight, the institutional system in Taiwan appears to be a potential source of conflicts, overlapping and duplication. A complete separation of extension, research and education functions, a triple system of research (departmental, university and by extension), a dual system for extension (departmental and private) appear to be a recipe for confusion. And yet, experience has shown that the system works remarkably well.

The major lesson to be learnt from the Taiwanese experience, as shown convincingly by Lionberger & Chang's study, is that even a diffuse system, in which research and extension are fragmented, can achieve excellent results, provided co-ordination is effective and strong linkages are created between the various agencies involved in agricultural development, research, extension and education. Of particular interest are the exceptionally strong links between researchers and extension staff.

In contrast to the dissatisfaction and frustration that characterise extension workers in many LDCs, the vast majority of Taiwanese extension workers (in both agencies) expressed satisfaction with their work and status. They are also held in high esteem by the farmers (Lionberger & Chang, 1981).

Taiwan has one of the most successful small-farm operations in the world; on an average 1·1 ha farm, the Taiwan farmer, using mainly human and animal labour, produces enough food for his own large family, for the urban population and an excess for processing and export. A major contribution to their success has been made by the farmers' associations that have provided credit, supplies, marketing and extension advice. They have also effectively represented the farmers' interests at township, regional and provincial levels (Lionberger & Chang, 1981).

However, about 40 per cent of the farm families in Taiwan have less than 0·5 ha of land, and many of them are low-income, technologically backward and burdened with debts (Chen, 1981). Though special efforts are made by the extension agencies to help this marginal sector, it is accepted that farmers who cultivate less than 0·5 ha cannot depend exclusively on agriculture, and an improvement in their standard of living can only come from off-farm work, for at least some of the family members (Hong, 1981).

BRAZIL

The evolution of the extension service

A system of technical assistance was developed early in the century in Sao Paulo. Extension activities in Brazil were first performed by a development agency that enjoyed strong support by federal and national governments. The agency was replaced in 1948 by the rural extension service. Though financed by the federal government, the service was based on state institutions.

In 1956, the federal government delegated responsibility for technical assistance to rural areas to a private organisation which was to serve as a national co-ordinating agency named ABSCAR, and supplied the necessary funds for this purpose. By 1971, ABSCAR was providing technical assistance to almost half the municipalities in Brazil, with 1270 field workers; the ratio being one extension agent per 1800 rural families.

In order to provide liaison with research, extension liaison officers were stationed at the research institutions. This system did not prove to be very effective; one reason was probably the paucity of research information made available to the extension workers.

On the assumption, prevalent in Brazil in the 1950s and 1960s, that the fund of technological knowledge already available to the research institutions was in itself sufficient to enable a doubling of yields, it was decided to strengthen the extension service at the expense of resources formerly assigned to research (Alves, 1983).

The Brazilian Corporation for Technical Assistance and Rural Extension

In 1974, the Brazilian Corporation for Technical Assistance and Rural Extension (AMBRAPTER) was established to replace ABSCAR. AM-BRAPTER was created as a public corporation, on the same legal model as EMBRAPA (cf. pp. 142–3). Funding continued to be mainly from federal government sources.

The target population

Until 1964, the Brazilian rural extension service dedicated the greatest part of its efforts to the small and medium farmers. From that date on, there was a gradual reorientation towards medium and large farmers (a process opposite to the one adopted by most developing countries at the time). The reason for the change was the desire to increase agricultural supplies to the rapidly expanding urban sector and for export. The assumption was that the larger producers, with the resources available to them and their higher educational level, could respond better to improved technologies than small, backward farmers (Alves, 1984*a*).

In spite of the considerable effort invested in extension, especially in the period 1974 to 1986, 'the level of Brazilian agricultural technology is still, on the average, very low. One of the basic reasons for this is in great part the fact that innumerable results of the research carried out still do not reach the extensionists and much less the farmers' (Rivaldo, 1987).

COLOMBIA

Evolution of the extension service

In Colombia, the early extension efforts were based on the concepts and methodology of the USA extension service (the so-called USA extension doctrine), and did not reach beyond the more progressive, better educated and well-to-do farmers.

In 1972, in a special report from the Rockefeller Foundation, it was stated that:

> Many peasants do not yet know that there exists a new technology, and those who have heard of it, generally mistrust it. Usually, they have never been visited by an extension agent; they are also too timid and too aware of their own ignorance to visit the experiment stations and assist at farm days, or at demonstration plots in their own vicinity. And so, they continue to farm the same way as their ancestors, hardly surviving and hardly any produce to sell (Streeter, 1972).

In 1954, the Servicio Tecnico Colombiano—Norte Americano (STACA) was initiated, within the framework of Point 4 Aid. For the first time, the hitherto sporadic attempts at extension were replaced by a formal service, which four years later became the Division of Extension, within the Ministry of Agriculture, with responsibility for 17 'agricultural zones'.

Though extension and research were both within the Ministry of Agriculture, co-ordination between the two services was deficient, as evidenced by a statement of the Minister of Agriculture to Congress in 1959: 'in the Ministry we have wished to foster joint efforts by the extension service, agricultural research and socio-economic research, but the relationship between these divisions is practically non-existent, so that the results are generally deficient'.

In consequence of this situation, a director-general was appointed in the ministry charged with achieving a better integration between research and the rural development activities of other divisions of the ministry (Piñeiro *et al.*, 1982).

Up to 1967, the extension service continued to be a division within the Ministry of Agriculture, 'where it had become a routine bureaucracy' (Streeter, 1972). Most of the staff were of urban background, their professional training had been largely academic and theoretical, and they had little or no experience of farming practice or farming people.

Transfer of extension to the Instituto Colombiano Agropecuario (ICA)

In 1967, it was decided to transfer extension from the Ministry of Agriculture to ICA. This organisation, which had previously concerned itself mainly with improving its research effort, now attempted to strengthen the extension service: 60 extension units were activated over the country, each consisting of an agronomist, an animal scientist or veterinarian, a home economist and 5–6 extension workers who lived in the villages. A considerable sum was invested in appropriate means of transport to facilitate visits to farmers, and a number of mobile units of specialists, who stay for several weeks in a community, were established. Two projects, on the lines of the Pueblo Project in Mexico, were launched— one in the State of Antioquia and the other in the State of Cundinamarca.

The most important change that occurred during this period was in the concepts regarding the objectives and work of the extension service and their adaptation to the conditions and needs of a developing country. The original objective 'of attempting to induce the majority of the farmers to adopt the whole fund of technology developed by ICA, was replaced by the more modest objective of improving life for the rural family (Dirección Planeación, 1983). This was expressed by a change in the name of the service—from 'extension agricola' to 'extension rural'.

A research programme on rural sociology, mass communication and extension was initiated in 1970, and the constraints encountered in increasing productivity and improving welfare of the rural population were identified. Fourteen projects of rural development were undertaken.

In order to achieve the aims of extension rural, ICA felt it necessary to co-ordinate its activities with those of other institutions active in the field, and signed a number of bilateral accords with the Caja Agraria, the Division de Organización Campesiña del Ministerio de Agricultura, Acción Comunal, and several others.

The so-called extension activities carried out by the Divison of Research and Technology Transfer are: field days, demonstration plots, conferences, seminars, short courses, professional literature. These are legitimate activities for a research organisation, but hardly justify adding 'Technology Transfer' to the title of this division.

The units of ICA involved in actual extension activities, as they are generally understood, are: rural development, agricultural production and livestock. Certain technical assistance activities are linked to credit that is made available to the modern sector as well as to small farmers.

In addition to ICA, more than 35 other institutions—statal, parastatal, private and international—participate in extension activities of one sort or another. Amongst these, are universities and agricultural schools, growers' associations, departments of the Ministry of Agriculture, the international research centres, IICA, etc., each with its proper fields of action, criteria, objectives, and different sources of information. The large commercial farmers use the services of private consultants who receive their licences as certified extension workers from ICA.

Because of the complex situation arising from the multiplicity of organisms engaged in extension, the need is felt for an appropriate coordinating body with the necessary authority and power, to rationalise the transfer of technology to farmers and to improve the efficiency of the process (Dirección de Planeación, 1983).

ICA has also developed a National Plan of Technology Transfer (PLANTRA) with the object of rationalising all extension activities of the institute. PLANTRA is based on an exhaustive diagnosis of the socio-economic situation of the various farming sectors, their technological requirements, the technology available in ICA, the constraints to adoption, and the existing extension infrastructure. On the basis of these findings, extension projects were formulated, and priorities determined at regional level, within commodities and for each farming sector (Alarcon *et al.*, 1984).

USSR

Adoption of new practices

All recommendations made by the research institutes concerning the adoption of new practices must undergo pre-testing on pilot areas in the state and collective farms.

These recommendations are then transmitted to the Scientific and

Technological Council, either of the Agricultural Directorate of the Province (oblat) or of the Union Ministry of Agriculture.

These councils comprise representatives of the research institutes, officials from the agricultural administration (but the producers are not represented), who evaluate the potential interest for agricultural development of the research recommendations.

A first analysis may result in (a) rejection—a very rare occurrence in view of the rigorous pre-testing carried out by the research institution; (b) a request for additional research; or (c) provisional acceptance.

The acceptance becomes final only after verification trials have been made under the direct auspices of the councils, each of which maintains its own experiment stations for this purpose.

The scientific and technological councils are completely autonomous and there is no hierarchical relationship between the provincial, union state or all-union councils, and no mutual obligations excepting the requirement for exchange of information. Hence, every council is completely free to choose the innovations it considers useful for the region under its jurisdiction.

Exchange of information between the councils at different levels is considered very important in order to evaluate the scope of application of an innovation and finalise recommendations on its application in practice (Casas *et al*, 1979).

Constraints to the adoption of new practices

Notwithstanding the rational procedures for technology transfer described above, there is still a considerable gap between research results and actual practice. The major reasons for this situation, described in detail by Casas *et al.* (1979) are briefly:

- The almost exclusive concern of the research workers with single practices; the lack of liaison between commodity and discipline research institutes, even when the nature of the work required is complementary.
- The lack of interest of the research workers in the field application of the results of their work; many consider their research concluded as soon as it is published.
- Deficient supply, in quantity and quality, of essential inputs—in particular of plant protection chemicals and machinery. Due to

difficulties in production, storage and transport, serious delays in the supply of inputs are experienced.

These deficiencies have been widely denounced in the Soviet press, and as a result, investments in the agricultural sector were substantially increased in the 1968–72 Plan.

Science–production unions

A relatively new instrument for linkage between research and agricultural production is the commodity-oriented science–production unions. Their objective is to accelerate the pre-extension research process and the diffusion of new technologies.

Each union organises groups of state farms that investigate, under large-scale farming conditions, the innovations proposed by the research institute with which a permanent linkage has been established. The farms undertake to increase rapidly the seed or vegetative propagation material of improved varieties, as well as improved animal stock, and to demonstrate to other farms the advantages of the innovations. In brief, a Soviet version of the 'innovative farmer' concept.

The unions establish experimental workshops for mechanical innovations or improvements, and centres for the professional formation of specialists.

The first unions for agriculture were established in 1973, following the example of similar institutions in industry. Their number increased rapidly, as one of the measures adopted by the Council of Ministers of Agriculture in 1976, 'for increasing the efficacy of agricultural science and strengthening its links with production'.

The science–production unions have already demonstrated a beneficial effect on production in those branches in which they have been active; not only in the farms participating directly in the unions, but also in the entire region in which they were established.

The unions have also had a marked effect on the research institutes with which they are associated: the latter have become more effective; they concentrate more on objects that can have a rapid and marked impact on production; more of the research is multidisciplinary, and produces packages of recommendations. The number of research projects has decreased, with a concurrent increase in collaboration between researchers. Large-scale adoption of new practices has accelerated (Casas *et al.,* 1979).

THE PEOPLE'S REPUBLIC OF CHINA

The Agro–Technical Extension Centre

In 1983, the Ministry of Agriculture, Livestock and Fisheries (MALF) established the Agro-Technical Extension Centre, as a change from the formerly commune-centred system.

The centre has four divisions: one for organisation and management, and three for different groups of commodities.

Provincial level

At the provincial level, a division of the provincial bureau of agriculture co-ordinates the work of, and provides support for, the county services. The provincial apparatus seeks to link its work with the Provincial Academy of Agriculture and its research stations at the prefectural level.

County level

The centre of gravity of the extension system is at county level. Each county has, or will have, an extension network consisting of: (a) a 'three-in-one' agro-technological extension centre (experimentation, training and extension); (b) technological service stations in townships and villages; and (c) demonstration stations.

The experimentation component includes adaptive research and on-farm testing.

By 1983, 300 county centres had been established, with 4500 technological service stations and specialised demonstration stations (Yen, 1985).

Linkages

Until 1960, the linkage research–extension–farmer was taken to its ultimate conclusion: the Chinese simply eliminated the extension network as a separate system and merged it with the research system. Each unit at each level was to be a scientific research unit, with no special personnel responsible for extension. Information was to be passed directly and personally between people and between units. In addition, in order to exchange information about experience with specific commodities, conferences were frequently called at national, provincial and county levels.

Provinces also printed manuals summarizing their experience in planting particular crops.

In 1961, the state-run extension system was re-established. Several stations were set up in each county, each one serving a few communes (Stavis, 1978). The independent extension system described above gradually evolved (Zhou, 1987). The three-in-one concept is a carry-over to the new system.

Several forms of research–education–extension linkages have been institutionalised. At national level (RAPA/FAO, 1985):

- a science and technology committee has been instituted within MALF;
- the China Association of Agriculture has been set up, with branches at the provincial level, as well as branches in specialised fields;
- a programme named 'Key National Technological Development Projects' has been launched by MALF.

All these bodies comprise specialists, professors, researchers and extension workers.

However, the existing linkages between extension, research and education are still considered unsatisfactory, and it is felt that 'a comprehensive, systematic, and constant relationship needs to be established' (RAPA/FAO, 1985).

The agricultural extension service also has many links with the non-agricultural sectors, such as commercial, chemical, industrial, and foreign trade sectors.

China has sought financial and technical assistance from outside sources, such as FAO, the World Bank and UNDP, in order to strengthen its extension service (Yen, 1985).

POLAND (Sakson, 1986)

Structure

Since the 1920s, a number of extension organisations have been established in Poland and have undergone several changes following the attempts to collectivise agriculture and changes in the administrative divisions of the country.

Extension units have been established since 1975, as a state service administered at provincial level, through the provincial agricultural progress centres.

The state extension service propagates the official agricultural policy, and carries out the tasks designated by the provincial administration. The clientele includes all farmers, male and female.

Centres for education and progress in agriculture

Research on organisation and methods of extension is carried out by the Centres for Education and Progress in Agriculture.

There are 49 such centres, employing over 10 000 people, and comprising provincial specialists (16 per cent), local specialists (20 per cent), and general agricultural and rural household advisers (64 per cent).

All the centres do not have the same organisation; however, in general, the specialists are organised into sections of plant production, animal husbandry, economics and organisation, home economics, scientific and technological information (including publications).

Each province is divided into regions, comprising several communes. The regions are directed by managers.

Clientele

The extension service works on the 'progressive farmers' approach, and concentrates on the innovative and large farmers, each adviser working with about 20–30 farmers.

Technology transfer starts at the model farms, which are generally owned by the best farmers. Advice is also provided for groups of young farmers and of impoverished farmers. Individual advice is extremely limited because of communication constraints (inadequate telephone services, mobility problems).

Objectives

The major objectives of the extension service are: technology transfer, farmers' training and modernisation of rural households.

In the future, the intention is to concentrate the advisory services on farm management and economic organisation.

Private sector extension

There are *voluntary organisations of producers* for various commodities, whose aim is to protect group interests in relations with the administration and with the customers. They also act as agents in the provision of inputs and the sale of produce.

These organisations employ full-time personnel, but their contribution to extension is secondary and they are far less active in this respect than the official extension service.

Horticultural and dairy co-operatives, as well as state food and agriculture industries, provide extension advice to contract farmers in addition to assistance in obtaining modern inputs. The stress is on production, and the extension activities are marginal.

Mass media

Forty-nine professional periodicals for farmers are published in Poland. A state publishing house specialises in agriculture. About 1000 hours of radio and 200 hours of television are devoted yearly to agricultural programmes.

Problems

The relative lack of effectiveness of the extension service is ascribed by Sakson (1986) to shortages of new and of traditional inputs (fertilisers, feed concentrates, machinery, etc.) and to the frequent reorganisations.

Co-ordination of the state services and those of the private and semi-public sectors is a problem that has not been solved satisfactorily.

THE EXTENSION SERVICES—AN OVERVIEW

Extension services in the developed market economies

The beginnings
The forerunners of agricultural extension in both Europe and North America were the agricultural societies (True, 1928). The first was the Society of Improvers in Agricultural Knowledge of Agriculture in Scotland, established in 1723. In the USA, the American Philosophical Society was founded in 1744, under the leadership of Benjamin Franklin,

followed in 1785 by the establishment of the Philadelphia Society for Promoting Agriculture.

In France, an early Society of Agriculturists became the present-day Académie d'Agriculture, whose proceedings were first published in 1761. The first agricultural society in Germany was established in 1764, and in 1765, the Free Economical Society was established in Russia, with a large experimental farm near St Petersburg.

All these societies had the common aim of keeping their members informed of improvements in agriculture through their publications, newspaper articles and lectures. In the USA, the agricultural societies also organised fairs, which, besides facilitating the sale of animals, farm products and machinery, also served as a venue for lectures.

Agricultural extension proper started with the system of itinerant teachers of agriculture, a system pioneered in 1843 by the New York Assembly employing experienced and progressive farmers to give public lectures throughout the state on practical and scientific knowledge in agriculture. Similar systems began operating in Europe: in Italy, for example, the Cattedre Ambulanti di Agricultura were initiated in 1890.

The first modern agricultural advisory service was established in Ireland during the great potato famine, and operated from 1847 to 1851. Itinerant instructors were appointed to work among the small peasant farmers in the areas most affected by the famine (Jones, 1982).

The agricultural situation and its implications for the extension services
Two major developments are taking place in the developed market economies: (a) the smaller and poorer farmers are being pushed out of agriculture, and (b) increased productivity has led to overproduction in many countries, whilst competition on the world market is becoming fiercer. At the same time, private sector agencies are increasing their activities considerably. There is also a marked tendency to reduce government-provided free extension. In the USA, for example, the federal share for extension has been drastically reduced, with resultant heavier loads on local sources.

As a result of these developments, the national extension services 'are under increasing pressure to reassess their work within what is now a less sympathetic environment' (Stevenson, 1984). This reassessment has to be based on criteria that the extension services were not required to apply in the past, such as cost-effectiveness, investment appraisal. Questions must be asked such as whether the service can be supplied more cheaply by the private sector; is there a significant general social or economic need; do clients value the service, etc. (Stevenson, 1984)? In brief, extension must become more accountable and provide evidence of efficiency and effectiveness.

Since the 1970s a trend in the change of emphasis on various aspects of extension work is apparent (OECD, 1981): (a) a greater interest in the human factor; the farmer is not considered only as a producer and a user of natural resources, but also as head of a family, and member of organised groups, (b) advisory work is no longer uniquely concerned with the individual farm, but also with the rural community and even the region as a whole.

These changes imply concern with trade, food, industry, tourism and environmental problems.

The changed emphasis on objectives and the more integral extension approach have resulted in organisational changes that are reflected in the change of names of many of the national services. For example, in England and Wales, advisory and development work, which were previously in separate services, have been grouped together in 1971 into the Agricultural Development and Advisory Service (ADAS). In the Netherlands, as well as in France, the national agricultural advisory services were reorganised in 1965 and renamed Agricultural Development Services in order to emphasize the orientation towards social development of rural society.

Röling (1987) sums up the situation by stating that: 'the idea that extension will work itself out of a job has so far proved inaccurate. Instead, the ever smaller absolute number of farmers is being supported by an ever larger army of information providers.'

Private sector extension

As agricultural technology has progressed, the scope of agribusiness has widened considerably, and the number of non-governmental extension agencies has increased. These include commercial firms that sell inputs and supply free advice on their effective use, as well as private consultancy firms that provide services for a fee.

In many countries, the private sector sources of advice already surpass by far the advisory capacity of the national extension services. The fact that this development has occurred at a time that the national advisory services provided advice free of charge, 'reflects the demand from farmers for detailed services on technology and management of a kind the official extension services do not have the resources to supply' (Barker, 1986).

Barker then raises the question of how the national extension services should respond to the new situation—by competing, by sharing the work and co-operating, or by withdrawing from areas in which the private sector is active?

The natural tendency is to compete, but the rational and probable adjustment will be for the public resources to be concentrated where there is the greatest need and are most effective; leaving other tasks, in particular in the routine services, to the private sector (Barker, 1986).

Extension services in the developing market economies

Historical background
Colonial policy in the 18th and 19th centuries aimed at producing as much food and raw material as possible for export to the metropolis. In some colonies, expatriate farmers were encouraged to settle and foreign companies were given concessions to set up plantations employing local or indentured labour.

Until World War I, government programmes in the rural sector of the colonies consisted mainly in the enforcement of regulations. It was only in the late 1920s that extension services were initiated for the native farmers. In the British colonies, an agricultural officer was appointed to each district, who relied on local extension staff to reach the farmers. The AO was closely identified with the district civil administration which kept law and order.

In some colonies a few residential courses for farmers and extension workers were organised. In practice, it was the larger farmers in the more favoured regions, growing export crops, who received most of the attention of the extension staff.

Similar situations pertained in the French, Dutch, and Belgian colonies.

After independence, the extension services retained many of the characteristics of the colonial period (Adams, 1982).

Even in the long independent Latin American countries, the majority of the national extension systems were only started during the decade of the 1950s.

Most LDCs were assisted in the establishment of their extension services by bilateral agreements and international aid (FAO). A major role was played by the USA.

The USAID model of extension was introduced into most Latin American countries and part of SE Asia in the two decades following

World War II. The model was based on the American land-grant college system, but few of the LDCs had well-established universities capable of linking research with a dynamic extension service.

The services were normally organised outside the regular government bureaucracies so as to ensure freedom of action for the extension staff. However, on withdrawal of foreign support in the 1960s, the formerly semi-independent services were placed under direct government control, and lost their influential position in the budget.

By 1980, most LDCs already had a national extension service. However, research on food crops was generally at its beginnings, and in the absence of an appropriate technology there was little that an extension service could do to promote productivity of these crops.

The bias towards export crops, grown mainly on plantations and by large commercial farmers, remained in force. Much of the time of extension field workers was taken up with administrative and regulatory tasks required by the increasing bureaucratisation of the government services (Adams, 1982).

Investment policies in extension

In 1959, low and middle income LDCs spent twice as much on extension as on research, relative to the domestic value of agricultural production. In the following two decades they moved to spending approximately equal amounts on both services. In other words, extension during this period showed little or no growth, whilst research was considerably expanded (Evenson, 1985).

Counting salaries and related costs, like those for laboratories and technicians, the ratio of research costs to extension costs is as much as 20:1 in the low-income countries, as compared to 3:1 in the industrialised countries. This difference is mainly due to the cost of training the extension workers: little training in the LDCs and expensive advanced training in the industralised countries. The low cost of extension workers induces governments in the LDCs to hire more extension workers than research workers. They therefore use extension systems based on untrained field workers (Evenson, 1985).

The most common problem reported by extension leaders in Africa and Latin America is the inadequate number and low quality of the extension staff. In Africa, except for Malawi and Liberia, where the ratio is about 1:600, the ratio of extension workers to farmers is as low as 1:2500 (RAPA/FAO, 1985).

Objectives and strategies
Even if the basic objectives of the extension services are very similar in most LDCs, the strategies adopted are different in accordance with the farmers' potentials and the constraints they face in their respective countries.

The Committee on African Development Strategies (1985) describes the major ills of African countries as neglect of agriculture as compared to concentration on 'unfeasible industrial programmes'; to this FAO (1986a) adds: uneven resource distribution; misdirected development strategies such as urban preoccupation, disincentives to farmers, neglect of rural women; political instability; external economic factors such as weakening of competitiveness on international markets for key agricultural commodities and debt problems.

The remedies for recovery, proposed by FAO, include: improved training of manpower, better veterinary services, decentralisation of agricultural development with more responsibility assumed by local communities. Many large and ineffective state marketing systems, and in particular extension services, need to be reorganised.

Emphasis on equity. In most countries, one of the major changes in the objectives of extension in recent years is a new awareness of the need to focus a major effort to improve the productivity of the small farmers, who have limited access to information and resources.

The 'green revolution' has highlighted the fact that the adoption of new technologies, even if they are apparently scale-neutral, can increase the disparity between different sectors of rural society of each country—making the poor poorer, and the rich, richer and more powerful' (Arnon, 1987).

Alcantara (1976) describes how, for example, as a result of modernisation of Mexican agriculture, wealth was increasingly concentrated in the hands of a small fraction of the rural population, whilst an absolute worsening occurred in the standard of living of the lowest-income sectors of the rural population, and the persistence of hunger in the birthplace of the 'green revolution'. Notwithstanding the investment of millions of dollars of the national budget into agriculture, 83 per cent of all farmers could only maintain their families at, or near, subsistence levels.

A new approach to rural development was formalised at the World Conference on Agrarian Reform and Rural Development, in 1979, when the international community adopted a programme of action in which 'national objectives and strategies would be governed by policies for attaining growth with equity, people's participation, and redistribution of economic and political power'.

The redefinition of rural development policies in favour of the disadvantaged sectors of the rural population has coincided with a reappraisal of the concepts, methods, strategies and structures of the extension services. Programmes have been undertaken which focus specifically on the small farmer (Garforth, 1982).

Whilst all countries have adopted considerations of equity and participation of small farmers in principle, ministries of agriculture in many countries still follow a policy oriented towards increasing agricultural output without social considerations. Their number is, however, decreasing (RAPA/FAO, 1985).

Structure

In most countries of the world, extension services are provided by several sources, but the backbone of the system is a government service, located generally in the ministry of agriculture.

Co-ordination between the several agencies is the exception, rather than the rule, with resultant overlapping, duplication, and conflicting interests. Even within a ministry, extension services may be fragmented between various departments.

In Africa, there are still some countries that do not have a nationally organised extension system. This is most common in the francophone countries (RAPA/FAO, 1985).

A radical change in the concept of agricultural extension has occurred in a large number of Asian countries (and more recently some African countries) with the introduction of the training and visit system of extension. It has required a substantial reorganisation of the extension system, involving a redeployment or manpower, new management systems, increased training demands, increased mobility of the extension workers and their supervisors and improved linkages between research and extension (Denning, 1985). In India it has been adopted on a national scale: in some countries, on a pilot project scale.

In *Latin America*, most of the countries of the hemisphere have established extension services, the majority relatively recently: for example: Brazil, 1948; Bolivia, 1948; Paraguay, 1951; Argentine, 1952; Colombia, 1954; Ecuador, 1954.

Most of these services were established as the result of bilateral agreements between the respective countries and the USA, which helped to finance and organise these services during the first years of their existence

(Herzberg, 1975). Without the USA aid their emergence would have been delayed, and probably their development would have been different.

Rice (1974) identifies four phases in the USA extension role in Latin America: the period for creating new institutions; the period for strengthening the institutions through improved programming and training activities; the period for transfer of operational responsibility to the local staff, and then to the host government; and finally the period as advisers to the local leadership.

By 1958, permanent US extension personnel were active in every continental Latin American country, excepting Mexico, Venezuela, Argentine and Uruguay, from which US government missions had been withdrawn.

After 1958, the influence of the extension philosophy of US assistance strategy lessened, 'Whatever the conceptual merits of the extension program, it has not demonstrably altered the level of productivity or the structure of the rural economy, and it has not persuaded the host governments to give more attention to the traditionally impoverished rural sector than they had in the past' (Rice, 1974).

To a certain degree, FAO and the World Bank replaced the USA in supporting extension activities in Latin America. FAO placed top-level extension advisers in a number of countries, and the World Bank provided financial resources for expanding and equipping extension services throughout the hemisphere.

In the late 1960s, USAID re-established its concern for extension, in forms that varied from country to country. It was generally involved in special projects, rather than direct aid to the central extension service. Examples of this form of aid are the setting up of a mobile extension school to train farmers in the Eastern Province of Guatamala; a national crop campaign organised by the extension service in El Salvador; a maize campaign in Costa Rica, also run by the extension service.

Period of transfer of responsibility to local staff. With the demise of the 'Servicios'—the joint development institutions of the national governments and USAID—the extension services were transferred from the 'Servicios' to the respective ministries of agriculture, and became subject to the direct control and administrative procedures of government. The American experts who previously had operational roles in extension activities, stepped back into advisory capacity, whilst the former counterparts took over the principal directory positions.

The situation since the 1970s

Changes in the organisation of the national extension systems are frequent, and there are marked differences in the systems adopted by the different countries.

In some countries, such as Peru, Argentine and Colombia, research and extension are the responsibility of a semi-autonomous institution; in most others, extension is a department in the ministry of agriculture.

Characteristic of most Latin American countries is the fragmentation of extension among numerous institutions, and the lack of co-ordination between the various agencies (Herzberg, 1975).

A change in orientation of extension efforts has also occurred in most countries of the hemisphere, which 'found themselves subjected to a social awareness and a drive for a humanistic approach, supported for political reasons by Government and rural organisations' (Ansorena, 1972).

Institutional linkages

A major problem facing the extension services in most Third World countries is the lack of appropriate research results, compounded by the deficient linkages between research and extension.

Only in a few developing countries have formal linkages been created between research and extension. In Argentine, Jordan, and the People's Republic of Yemen a department or directorate-general of research and extension have been established. Several countries have established extension liaison units in the research institutions (Nigeria, Kenya, Zambia, Malaysia). Incidentally, this is an approach adopted in some industrialised countries such as the Netherlands, which has its roving liaison officers (cf. pp. 635–6).

An innovative institutional linkage between research, extension and training is the three-in-one Country Agro-Technological Extension Centre in China (RAPA/FAO, 1985).

In Asia, in those countries in which agricultural research and extension are under separate organisations, there are generally technical co-ordinating committees from national to field levels. However, there is generally no procedure or staff to follow up the recommendations of these committees, and implementation is left to individual members.

In Sri Lanka, Malaysia and Nepal, extension and research are the responsibility of a department within the Ministry of Agriculture. The director of the department has considerable authority in matters of staff management, programming and co-ordination of the two divisions, of research and extension, respectively. 'However, bureaucracies within the system weaken the linkages between the two neighbouring Divisions' (RAPA/FAO, 1985).

Recapitulation and conclusions

The relative lack of success of many of the extension services in the LDCs cannot be imputed only to deficiencies in the services themselves (FAO, 1971): *politicians* starved the extension services of resources, and never gave them a chance to succeed: *sociologists* focused on extension methods and principles, and neglected the total social and economic environment; *economists* failed to help planners prepare balanced programmes in the light of available economic and social resources; *foreign extension experts* limited themselves to applying an extension doctrine that was inappropriate for a developing country; *national extension leaders* attempted to implement activities independently of other agents of developmental change.

Extension has survived in the LDCs in spite of its general lack of impact because a complete programme for development *must* include extension activities as one of the essential elements for agricultural and rural development.

However, if the extension services are to fulfil their task properly in the future, much must be done: organisational structures and operations must be improved; proper training at all levels must be provided; more realistic planning will be essential; regulatory functions must be avoided; but above all, full integration and involvement of extension activities in the development process must be achieved.

Chapter 18

The Role and Objectives of Agricultural Extension

'Technology transfer' is a broad concept which includes the efficient transfer of agricultural innovations to the farmer *and* the provision of prerequisites needed to make adoption possible.

THE NATURE OF AGRICULTURAL EXTENSION

The nature of agricultural extension changes with the roles that are assigned to it by the authorities. The differences in emphasis on varying aspects of these roles are reflected in the different names given to extension activities at different times and in different countries.

The term 'extension' originated in England to describe a system of university extension education, taken up first by Cambridge and Oxford Universities, and subsequently by other educational institutions in England and elsewhere. The objective was to bring the educational advantages of universities to ordinary people (Swanson, 1984). This approach also influenced the concept of the land-grant colleges. In the American tradition, the term 'extension education' is used to emphasise that extension is an educational activity which seeks to teach people to solve their problems by 'exchanging' information.

In the United Kingdom, Germany, the Scandinavian countries and others, the extension function is called 'advisory work', because its major role is perceived to be solving specific problems.

In the Netherlands, 'voorlichting' means to keep a light in front of somebody so as to allow him/her to find the way. In France and Latin

America 'vulgarisation' means simplifying information so that the *'vulgus'* (ordinary people) can understand the message brought to them. Reactions against the implications of these top-down definitions has led to terms such as 'animation', 'mobilisation', and 'conscientisation' (Röling, 1987). The different aspects of extension have been defined by Röling (1987) as follows:

Information extension: Helping the individual to make optimal decisions among alternatives provided by the extension agents.

Emancipatory extension: Helping the emancipation of the underprivileged sectors of society; correcting structural problems.

Human resource development: Enhancing the capacity of the individual to make decisions, to learn, to manage, to communicate with others, to analyse the environment, to be a leader, to organise.

Persuasive extension: Inducing preventive behaviour with respect to environmental pollution, soil erosion, health hazards, etc.

Extension as a policy instrument: In most countries government uses agricultural extension as one of its instruments for developing agriculture as a whole, and not for helping individual farmers.

American extension doctrine

The major points of this doctrine are (USAID, 1968):

- The extension service should be designated as the chief government organisation through which technical agricultural information should be disseminated.
- The Director of the Extension Service should be attached directly to the Minister's Agricultural Planning and Co-ordinating Group.
- Extension must be strictly educational in function and should not become involved in 'crash' programmes of current interest to the Government. Such programmes may be necessary, but they should be handled by the group involved solely with the promotion of special projects. Too much involvement of extension personnel in such programmes may limit their usefulness for other areas of work.
- Small farmers should not be among the clientele of extension unless social goals are considered more immediate than production goals, since limited staff and budget will make it very difficult to attain the social goal of working with small farmers who will have little impact on agricultural production. This will limit the ability of extension to

provide extension assistance to the larger farmers who are in a position to respond quickly to technical production and marketing information.

Rice (1974) comments that these principles may be appropriate to a modern agriculture in an industrial nation, but ignore 'the economic, institutional and other situational constraints that greatly limit the utility of a strictly informational or educational program for the vast majority of (Latin American) farmers'.

EXTENSION EDUCATION

Extension education is generally the main, if not the only, agent for farmer education in developing countries, and is a specialised form of the broader concept of adult education.

Andersen (1964) defined extension work as:

An educational service for advising, training, and informing the farmer and grower concerning practical and scientific matters relating to his business, and influencing him to use improved techniques in his farming operations which, for this purpose, includes livestock and crop production, farm management, conservation and marketing. The task of the adviser is two-fold: to advise and to influence—to be a source of information and at the same time to inspire and lead his client to be receptive to means of improving his farming operations.

In order to provide a continuous supply of updated farm information that is essential for modernising agriculture, three functions must be performed (Lionberger & Chang, 1981): the *integrative function* —fitting new technologies into on-farm (FSR) situations; the *innovative function* —provided by agricultural research; and the *dissemination function* (farmer education) provided by the extension service.

Farmers require information in the following major areas (Wharton, 1965):

- *New inputs:* this involves up-to-date information on new varieties of crops, breeds of animals, agrochemicals, equipment, etc.
- *Techniques of production:* land use management; rates and techniques of fertilisation and irrigation; effective crop protection, etc.

- *Economic factors of production:* choice of commodities that can be produced with a profit; information on marketing conditions and prices, techniques for preparing produce for the market, etc.

Much of the technological information required by the farmer must come from the research organisation. Therefore an essential function of extension is to select information derived from research or other sources, that can be beneficial to the farmers they serve.

ROLES NOT DIRECTLY RELATED TO AGRICULTURAL PRODUCTION

Extension agencies frequently engage in activities that are not directly related to extension education; policymakers may even view these roles as of primary importance. These activities may comprise the provision of inputs, supervising credit repayment, enforcing government regulations, providing statistical information, carrying out pest control operations, organising co-operatives, participating in community development programmes, supporting political patronage systems, etc.

Already in 1964, Andersen warned against making the extension service perform regulatory, fiscal or enforcement functions. Even if for reasons of economy the temptation to use extension agents for these functions is great, this will destroy the confidence the farmer will have in the extension worker as a counsellor and friend, and thereby seriously damage his effectiveness.

More recently, the training and visit system of extension has adopted as a basic tenet of its approach the precept that extension should confine itself strictly to educational activities.

In the eyes of the villagers, extension agents engaged in regulatory roles are often seen 'as emissaries of government, enforcing mysterious and even senseless regulations, who must be bribed or evaded, in order that the ordinary needs of village life can be met' (Hunter, 1970).

In view of the severe shortage of personnel with which most LDCs are faced, it would be unrealistic, and possibly harmful, to insist that extension agents should devote themselves exclusively to providing professional advice to farmers.

Common sense will usually suffice in deciding what regulatory activities can be entrusted to extension workers—activities which will find general acceptance in farming circles as proper and in the interests of agriculture in general. Supervisory and regulatory functions that are directly linked to the adoption of improved practices, are not only legitimate activities for the extension worker, but may actually strengthen his hand. For example, the provision of credit for essential inputs, to be effective, would require the extension worker to supervise use of these inputs in the field and to certify that they were applied properly; regulations requiring the adherence to certain soil conservation practices, collective pest control, fire prevention measures, etc. can properly be monitored by extension workers, provided they endeavour to explain the need for these measures to the farmer and make him aware that they are intended for his benefit. On the other hand, the extension worker should have no part in regulating or other activities that have no direct relation to promoting production, even if savings in government manpower can be achieved thereby. These activities will undermine the trust in the objectivity of the field worker, a prerequisite to his effectiveness. Similarly, all activities that involve much paperwork must be avoided, as they inevitably cut down the time the extension worker can spend in the field. In certain extreme cases, practically *all* the time of the extension workers may be taken up with paperwork.

Promotion of government development policies
It is generally agreed that the objectives of extension education are not only to bring about an improvement in farming through the application of science and technology but also to promote the social, cultural, recreational, intellectual and spiritual life of the rural people. There is also general agreement, especially in developing countries, that whilst the advisory service has to adapt itself to the existing social framework of the farmers, it must also be active in promoting change towards a more progressive social framework as a prerequisite for technological change.

A question that is frequently raised is whether the extension service should promote central agricultural policies of government. Views range from considering this to be a prime responsibility of extension (the Eastern Bloc, many LDCs) to the conviction that such an approach is to be completely avoided (farmer-sponsored advisory work in Western Europe) (Barker, 1986).

Whilst extension has to be objective in its handling of information and supportive of the interests of the farming community it serves, its activities must be compatible with the political policies of government.

What might be acceptable to all is Barker's proposal (1986): (a) the extension service has a role to interpret and explain national policy to its clients; (b) the extension service should advise policymakers on the probable impact of decisions they intend to apply.

TRENDS IN THE FUTURE TASKS OF THE EXTENSION SERVICES

The following trends described by Röling (1986) are short-term in the developed countries and long-term in the LDCs.

As farmers become better educated, better informed and capable of obtaining information without the help of extension agents, there will be less need for communicating technological innovations to them. There will be a demand for more sophisticated services provided by advisers with high technical competence and an overall view of the farming situation.

Farmers will want direct access to information from data banks, to a reference system adapted to local conditions for production and market needs.

The extension service already has, and will continue to have, the task of building and managing mutual interest networks, based on computer services such as data bases, decision simulation, etc., and educating farmers in the interpretation, adaptation, and use of the information made available. The extension service will have to develop client-oriented software, and to maintain information systems which need to be updated, adapted and improved continuously, based on changes in the field, new needs of producers and consumers, etc.

With the exacerbating problem of overproduction of many basic agricultural commodities, extension will devote less effort to increasing production and devote major efforts to adjusting the supply of farm products to demand; planning optimum systems in the face of production constraints, and marketing in all its aspects.

With the accelerating exodus from farming, extension will endeavour to ensure the viability of rural communities, promote the generation of off-farm income, and devote itself to socio-economic community work.

Where the development of agriculture has caused undesirable side-effects such as overproduction, environmental pollution, erosion and the destruction of habitats and landscapes, extension will have to be used to influence farmers' behaviour in directions which are congruent with public interest (Röling, 1987):

(a) by promoting government policies which require compliance with rules and regulations that aim to reduce environmental and societal damage due to technical innovation and intensification: pollution of the aquifer through wastes and fertilisers, acid rain, loss of ecosystems, loss of employment, etc. This will involve explaining government policies and the consequences of non-compliance, as well as how to comply with these rules. It will therefore no longer be possible for the extension service to remain exclusively a farmers' support system, and it will increasingly become a policy instrument to effect change.

(b) the extension service will have to assist in the transformation of land taken out of agricultural production into new ecosystems and landscapes for recreation, green belts, wood production, etc. This will require a considerable extension effort, especially during the transition period when farms are being taken out of production and redundant farmers need to be retained for non-agricultural rural occupations.

Chapter 19

The Structure of National Agricultural Extension Services

INSTITUTIONAL SETTINGS

In most countries of the world, the extension service is the responsibility of government and is generally located in the Ministry of Agriculture.

However, there can be significant differences in the access of extension administrators to the policy-making levels of government and hence in their influence on decision-making relevant to their field of responsibilities. If the extension service is to be an effective education arm of the ministry, it should be represented in the policy councils of the ministry by a director who is equal with other members of the ministry executive council. He can then be aware of ministerial policies, and have a direct influence on decisions relating to budgets, staffing and programming (Watts, 1984).

It is, however, very rare for one agency alone to be in charge of all extension work in a country; generally different agencies work concurrently. Co-ordination between them is the exception, with resultant overlapping, duplication, and even contradictions in the advice given to farmers. Export crops are frequently the responsibility of separate agencies, usually statutory marketing boards, each for a single commodity, which provide extension services and may even have their own research facilities. Farmers' associations, commercial–industrial interests, co-operatives, etc., may integrate extension activities with their major operations. Rural banks frequently employ their own advisory agents to supervise the use of inputs purchased with the credit supplied by the bank.

697

Even where the extension service is operated by government, responsibility may be assigned to several ministries: agricultural, rural or community development, education, etc. Even in a single ministry, extension work is frequently carried out separately by different departments: crop production, horticulture, animal husbandry, fisheries, forestry, game, community development, water, home economics, etc. In these cases, there is bound to be bureaucratic competition and inflated staffing.

There are therefore many differences among extension systems adopted by different countries. These depend on the size of the country, the important farm products, the existence of effective farmers' associations, the structure of the agricultural branches of government, the size of their budgets, and the types of training in the agricultural schools from which extension workers are recruited. Last, but not least, the political clout of various pressure groups will affect the components of the extension system.

As agricultural technology develops, the number of external advisory agencies increases, and the extension service must adapt to this trend (Barker, 1986).

THE CONVENTIONAL EXTENSION SERVICE

The most prevalent type of extension in the LDCs is the government extension service.

The justification for government assuming responsibility for extension education lies in the fact that the extension programme, though aimed at improving the efficiency of agricultural production, actually contributes to general economic growth and in many cases to export earnings. Without the extension service, public funds invested in development projects would not be used as efficiently as they should be.

Without a service advising farmers on the adoption of innovations, the benefits of agricultural research financed by government may be lost, and potential economic gains foregone. Similarly, investments in rural infrastructure—roads, transport, markets, irrigation schemes, etc.—would not be effectively utilised.

Organisational structure

The conventional extension model is based on a 'downward' diffusion of technical innovations from research institutes to the Department of

Agriculture, through state, district or sub-district layers, to a multipurpose extension agent, to the opinion or formal leaders of a village (or to innovative farmers), and from them diffusion throughout the farming sector (Fig. 21).

Most government extension services have a number of common features (Arnon, 1987):

(1) They are usually organised as territorial hierarchies: headquarters staff in the capital city, subordinate levels at province or region, district and sometimes division or subdistrict, and a broad base of geographically dispersed field workers in the lowest subareas.

(2) They consist of three categories of workers: (a) *extension officers* (EOs), who are responsible for the administration of the service at national headquarters, and at the regional and subregional levels; (b) *extension advisers* (EAs), who work with the farmers at village and farm level; and (c) *subject matter specialists* (SMSs), who provide technical advice and guidance to the extension officers in all matters pertaining to their respective specialities, and in particular, on special programmes.

Basically, the conventional structure of government extension services is as follows.

At headquarters, a director of extension is responsible for planning, co-ordinating and monitoring of the entire service. He must develop extension techniques, define regional targets for national production and realistic work programmes for extension officers; he must monitor the effectiveness of the extension service, develop training courses for extension staff and lecture at these courses. He is also responsible for effective liaison between research and extension, participates in identifying problems for research and their priorities. He is assisted by a number of heads of departments; these are generally based on commodities or groups of commodities.

Heads of departments provide guidance in administrative, technical and financial matters to the regional extension officers, and monitor their activities.

Depending on its size, the country is divided into a number of administrative units and subunits; at each level, an extension officer is responsible for the administration of the extension service in his area, providing leadership for the various subject matter specialists and co-ordinating their activities as well as supervising extension agents.

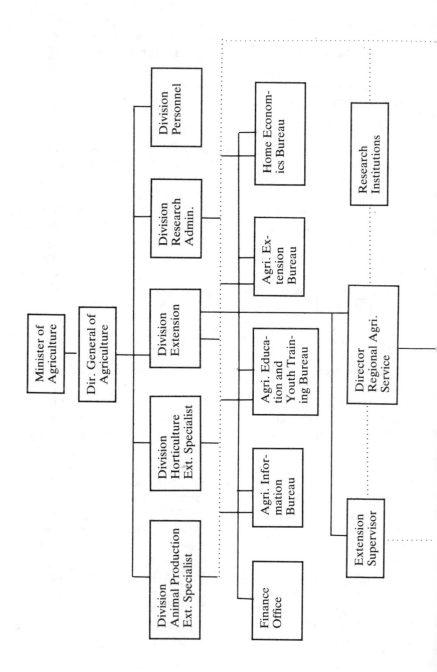

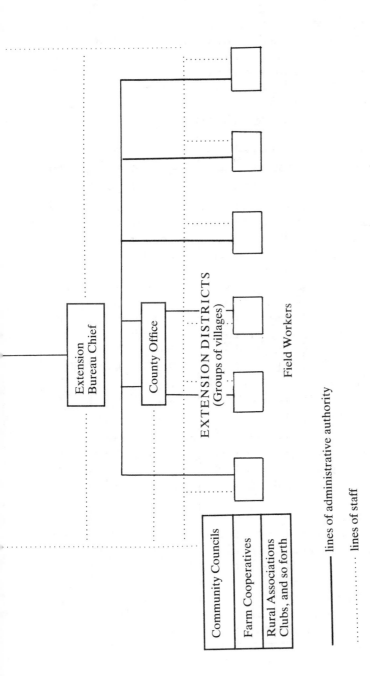

Fig. 21. Organisation chart of a conventional extension service. (From Claar & Bentz (1984), reproduced by permission of FAO, Rome.)

Down to the regional level, the function of the extension officers is basically administrative, and is differentiated mainly by the size of the areas under their jurisdiction. They are generally responsible for the implementation of the rural development programmes in their respective areas of jurisdiction and for supervising the work of the extension officers reporting to them. It is at the two lowest levels, the division and circle, that extension workers assume their distinctive roles.

The divisional extension officer (DEO)

The main functions attributed to the DEO within the divisional boundaries are:

(1) To determine the most appropriate farming systems under the ecological and economic conditions prevailing in his division.
(2) To prepare agricultural development programmes.
(3) To collect information on the agriculture of the division, and social and economic changes affecting farm life.
(4) To ensure the supply of necessary equipment and materials for the effective implementation of agricultural programmes.
(5) To ensure that government schemes for rural development projects within his area are operating satisfactorily.
(6) To determine training needs of his staff.
(7) To monitor the performance of the staff under his control.
(8) To foster and maintain good public relations in the community.
(9) To assist with in-service and farmer training by giving lectures, organizing field days, etc.

From the above it is evident that the duties of the DEO entail a mix of administrative and technical functions. The technical functions are frequently submerged by the administrative responsibilities. This can only be avoided by providing adequate administrative support for the DEO.

The extension adviser (EA)

The EA is the key element in the extension programme, working directly with farmers and thereby translating government policy into reality at field level. Generally, EAs are required to carry out a wide variety of functions. In addition to purely advisory work, they may have to carry out most ministry programmes at local level: 'they may sell and distribute inputs, perform regulatory services, arbitrate disputes, collect agricultural data, and handle credit and subsidy programmes. In fact, they become the local representatives of government rather than full time extension workers' (Swanson & Claar, 1984). The implications of this misconception of the functions of the EA have been discussed in Chapter 18.

Subject matter specialists (SMSs)
Inevitably, EAs will encounter problems with which they cannot cope, and therefore require the backing of more specialised extension agents, subject matter specialists, who serve as consultants to the EAs, drawing on detailed knowledge of a specific commodity or discipline. SMSs are preferably stationed at the appropriate experimental station, so that they can maintain close contact with their counterparts in research.

Their primary role, however, is to support the EAs, who are mainly generalists, by giving them guidance and support: identifying the problems, proposing effective solutions, participating in the preparation of demonstration fields, preparing bulletins and pamphlets; they are also active in the in-service training programmes of the extension staff and co-operate in experimental work with their research colleagues.

The SMSs should increasingly look to farming systems research as a source of appropriate technology to recommend to farmers. They should be trained in the FSR approach, so that they can participate actively in multilocationary testing and technology verification (Denning, 1985).

THE SHORTCOMINGS OF THE CONVENTIONAL EXTENSION SERVICES

There is no disputing the fact that extension efforts in developing countries have not always achieved the anticipated results. 'An agricultural extension service is used by almost every country as a major tool for agricultural development; yet it has proved disappointing in many ways' (Hunter, 1970).

Even in Latin American countries, that have been free from colonial domination for more than a century, 'extension has not been able to materially affect the standard of living of the major portion of the rural population, which continues at subsistence levels. With limited resources and limited numbers of personnel, extension services are not able, under present organizational and limited resources, to reach all small farmers' (FAO, 1971).

In the early 1980s, the situation in most Third World countries had not yet improved markedly, and donors continued to report frustration in trying to improve these systems (Swanson & Claar, 1984). This frustration is exemplified by the statement made by the former chairman of the Board for International Food and Agriculture Development, Wharton (BIFADS sixth birthday: a personal exaugural. *BIFAD Briefs* (1983) 6(4) 1–11): 'If there is an area in which we have been most unsuccessful, it has been in the development of cost-effective and programme-efficient means for the delivery of new scientific and technical knowledge to the millions upon millions of farm producers in the Third World. We know how to harness the creative and innovative forces of science and technology in the war on hunger, but I submit that we still have not been fully successful in technology diffusion.'

An understanding of the reasons underlying this lack of progress is a prerequisite for the efforts required to improve the situation.

Assuming that there is no lack of appropriate, locally tested improved methods, making possible dependable and substantial increases in production, shortcomings of the extension services can be the result of organisational defects and operational weaknesses, ineffective extension personnel, economic constraints, deficient linkages with other institutions involved in development; political, cultural, social and institutional constraints.

It must, however, be stressed that the findings we will present are based on a world-wide review of the extension services in Third World countries. This does not signify that every national extension service necessarily presents all the deficiencies mentioned, and some may even have a few not mentioned here. Some countries may be free of most, if not all of them. The T & V extension system has certainly addressed itself to eliminating some of these deficiencies, in particular the organisational and operational defects (cf. pp. 712–20).

Organisational defects

There are situations in which organisational forms have been imported from other countries without being adapted to accommodate the cultural norms of the community to be served, or without consideration of the available human and capital resources.

Even when this is not the case, the conventional extension service tends to be highly bureaucratic in structure and in its field operations. Moris (1987) lists the major typical organisational deficiencies of such a government service:

- a steep vertical hierarchy;

- a stress on downwards, rather than upwards communication;
- a high rate of transfer within a given level;
- reliance on mainly public funding;
- staff that think of themselves as government officials rather than farm advisers.

Poor management

Agricultural extension in the LDCs has a long history of management failure: perennial problems of poor communication between headquarters and the field staff and vice versa; lack of support and supervision of field workers, and the absence of a detailed realistic plan of work (Adams, 1982). Additional operational weaknesses are imposing roles that are not related to the promotion of agricultural development, inadequate numbers of extension workers and lack of logistic support.

One of the most important weaknesses of an extension service is the lack of a single, direct line of administrative control and technical support. The extension worker is often subordinated to several agencies, and is not fully accountable to any one of them exclusively. Supervision of field staff is often erratic, or may be non-existent. Field visits may be rare, or erratic; the instructions then given based on hasty appraisals. As a result, national policies seldom reach the local extension worker, and feedback rarely reaches the top levels of administration, whose decisions are then made without adequate and up-to-date knowledge of the situation at the operational level (Rogers & Svenning, 1969).

Field workers are often left to decide for themselves which crops to emphasise, what farmers to concentrate on, and how to organise their time. All these factors tend to foster lethargy, and to create feelings of alienation.

This general lack of supervision and purpose contrasts sharply with the clearly defined tasks, strict discipline, and close supervision of field staff that characterise the T & V system of extension.

Poor communications within the service

Communications from the field staff upwards is also frequently defective. Reports are often 'routine, ritual, unusable, unused and unread' (Chambers & Belshaw, 1973). Communication within ministry headquarters is frequently no less defective. At the highest level, the heads of the service are rarely directly involved in development policy making; as a result, the

extension acts as an executive arm, carrying out policies in the formulation of which they had no part. Orders to carry out these decisions move
downwards through the echelons, who have no discretion to adjust the
programmes to the specific socio-economic and ecological conditions in
the areas under their jurisdiction.

Strong headquarters—poor field staff

'There are extension services that have set up strong headquarters with
teams of professional specialists, with university graduates heading
advisory teams operating from well-built offices, with well laid-out plans.
But there it ended: the village worker rarely got training, received a salary
which was a fraction of that received by his superior, and had no means
of transportation. Furthermore, he did not usually know what was
expected of him' (Elkana, 1987).

Elkana concludes that 'unless the extension service focuses its efforts
and means on the support of the village level worker, so that he can
perform his job adequately, there is little chance that it will serve its
purpose.'

Poor planning

Chambers & Belshaw (1973) state that in African countries extension
programmes are often badly designed or inappropriate. There may be a
lack of instructions from headquarters, or an overload of programmes
flowing out from the centre without any systematic appraisal of compatibility in their demands on staff time, 'the later ones burying the earlier'.

The extension worker in some countries is expected to perform numerous roles, amongst these 'advocate, teacher, organizer, enforcer of regulations, planner, catalyst, coordinator, fee collector and communication
specialist' (Safrenko, 1984). To these functions one may add the more
congenial roles of preparing demonstration plots and involvement in on-
farm trials.

In one district in Kenya, Leonard (1977) found that the official programmes for a year, when worked down to the level of location, varied in
their demands on staff time, month by month, from 18 per cent to 474 per
cent of the working days available!

The main weaknesses in planning extension work, are summarised as
follows:

(a) unrealistic targetry, with junior field staff not involved in settling the targets;
(b) no systematic ordering of priorities between the competing demands of different programmes on field staff time; and
(c) no systematic work planning for field staff.

Inadequate number of extension workers and logistic support

Because of financial constraints and shortages of trained personnel, the number of extension workers in many LDCs is wholly inadequate to ensure a satisfactory coverage of the areas for which they are responsible.

This shortcoming is further compounded when the requisite logistic support is not forthcoming: visual aids, equipment, transportation, communication channels, guidance by subject matter specialists, etc.

The opposite situation is in evidence in other LDCs, where 'there are simply more agents than can be supported. In recent years, at least, extension has been an employment agency more than a development agency. And in almost no case is there significant support for research' (McDermott, 1987).

In many African countries, for example, the extension worker who is usually less well trained and has to work under far more difficult conditions than in developed countries, with fewer facilities, is often supposed to work with several thousands of farm families (Nour, 1969).

In a survey of 295 Philippino farmers, it was found that 88 per cent were aware of the extension worker's presence in the village. However, when asked how often they had contact with the extension worker, the replies were as follows: never, 22 per cent; seldom, 45 per cent; often (once or twice a month), 18 per cent; and very often, 15 per cent (Lu, 1968). Insufficient numbers of extension workers have led to the 'palm tree type' of personnel structure. Because it is easier to produce and employ a limited number of graduates at the higher level than the large numbers required at intermediate level, the top échelons of the extension service may have too few mid-level technicians to work with them. As a result, little work is done in the field and contacts between the extension service and farm people are minimal (Chang, 1969).

In view of the frequently diffuse form of rural settlement and the resultant dispersal of the potential clients over large areas, the mobility of the extension workers is of considerable importance in relation to their ability to function effectively. A fairly typical example is the regional office of

Esmeralda in Ecuador, responsible for farming families dispersed in an area of 200×80 km with poor roads and an inefficient public transport system. The efficiency of the extension worker is therefore entirely dependent on having an appropriate vehicle at his disposal (Ampuero, 1981).

Ineffective extension personnel

Many of the earlier leaders of the national extension services were trained in the USA or Europe, and were imbued with extension 'doctrines' that were inappropriate to Third World countries. This is especially true for Latin American countries. Other officials in the higher ranks of the service are frequently lacking in the specialised academic preparation for their roles in extension, similar to that usually provided for other vocations. There are still too few universities which include a chair of agricultural extension in the curricula for agricultural or veterinary science.

The village-level worker has the extremely difficult task of introducing new concepts and techniques to a tradition-bound community; this requires skills and technical knowledge as well as an understanding of the people, their problems and attitudes. To be successful the village-level worker requires 'the teaching ability of the schoolmaster, the skills of the farmer, the persuasive capacity of the politician, the understanding of the social worker and almost infinite patience' (FAO, UNESCO & ILO, 1971). Yet this difficult task is generally allocated to the lowest paid and least educated members of the government service.

Technical inadequacy

Extension workers are frequently technically incompetent in agriculture. The little training they receive is limited to technical subjects, whilst the economic and social aspects of field-level extension work are neglected. There is little teaching in communication techniques, and very little in-service training is provided to keep the workers abreast of technological and other changes (FAO, 1976).

The level of knowledge required by the extension worker depends on the professional level of the farmers he serves. As agriculture progresses, extension workers must keep ahead of farmers. When they fail to do so, they soon run out of content for their extension activities, and farmers quickly recognise that their 'advisers' have little left to offer them.

Poor working conditions

Chambers & Belshaw (1973) describe the problems and disincentives facing field agricultural staff in African countries due to poor terms of service, living conditions and working conditions. Pay and allowances are less than those of their peers working for parastatal or private sector organisations; housing is not provided, or is extremely poor; good work is not seen to be rewarded, and poor work is not seen to be penalised. Working conditions are also hard and physically frustrating. The climate is difficult, the roads are poor and transport inadequate.

Conditions are not too different in many Latin American countries. Rice (1974) mentions 'low wages, no fringe benefits, inadequate gasoline allowances, jeeps with broken axles going without repair, primitive visual aids, little fertilizer, seed, pesticides and other materials to run demonstrations. Most importantly, no in-service training to speak of, and no agents to spare for training if adequate course work could be organised. The salary situation is just as debilitating as the materials problem. Wages are low in the civil service, and good men are attracted away if and when opportunities arise in the private sector or in those autonomous public agencies that can set their own salary scales. There is not much chance of further promotion, since extension is a bottom-heavy service with lots of agents, a few chiefs, and nobody in between.'

Local bodies and the private sector do not contribute to the financing of the extension activities, the former because there is no local governmental base, and the latter is either too poor, or maintains its own advisory service. In Mexico, Brazil and Argentina, the federal governments pay about half the budget, whilst state, provincial or county governments, the private sector and voluntary organisations pay the rest.

Attitudes and behaviour

Many extension workers alienate farmers by showing a sense of superiority, combined with a lack of understanding of the cultural values and social mores of their clients. This may be compounded with a dislike of physical labour and 'getting their hands dirty'.

A good deal of the resistance to the proposals of the extension worker can be ascribed to a lack of confidence in his objectivity. The more unbiased he is, and the less he is suspected of making a personal profit from the adoption of his proposals, the more influential he will be.

Deficient linkages with other bodies involved in development

The linkages of the extension services operating with other bodies involved in agricultural and rural development have generally been poor. In view of the importance and scope of this problem, it will be treated in detail in Chapter 22.

Political, cultural and social constraints

Not all the reasons for the failure of extension efforts can be ascribed to deficiencies in the services. The efforts of the extension service may fail because of political, cultural or social factors. The forces opposing change may be too strong: local notables, politicians and traditional leaders may oppose the new service, because they fear they may lose their traditional control over the farmers. Alternatively, the potential benefits may be judged insufficient to overcome social and cultural constraints.

Lack of an appropriate technology

Efforts to modernise agriculture cannot be successful unless innovations are available that have a high benefit/cost ratio and have been shown to be adapted to a given environment by adequate location-specific testing. This is of course a function of an effective research organisation with an appropriate research programme, etc., generally lacking at the time the extension services were established, and in many cases are still deficient.

Inadequate funding

Many of the shortcomings enumerated derive from the basic constraint of a chronical shortage of funds, making it impossible for a national extension service to operate adequately. The limited resources usually made available, 'could not possibly make sufficient impact, and in fact assured failure from the beginning' (FAO, 1971).

IMPROVING THE EFFICIENCY OF THE EXTENSION SERVICES

The long list of deficiencies in many of the existing extension services does not imply that improvement is impossible because of the numerous constraints encountered, mainly by Third World countries.

Faced with the inadequacies of their extension services, many governments have opted for one of the following solutions:

- To maintain the existing structure, but improve its *modus operandi*; the best known approach to achieve this objective is the Training and Visit System.
- Administrative decentralisation of the extension service: involves the transfer of functions and decision-making to regional, provincial, and district administrative units that are part of the central government structure, and are financed and monitored (not controlled) by a central ministry.
- Delegation of functions to regional or functional development authorities, parastatal organisations, or special project implementation units that operate free of government regulations concerning personnel recruitment, contracting, budgeting, procurement, etc. Their advisory services are integrated with their other developmental activities.
- Transfer of responsibility for advisory work to private or quasi-public organisations, such as universities, farmers' associations, rural co-operatives, agro-business, etc.

Though there are occasions when organisational forms have been imported from other countries without being adapted to accommodate the cultural norms of the community to be served, or without consideration of the human and capital resources available, failure of the extension services can rarely be attributed to the form of organisation adopted, but is mostly due to operational weaknesses.

An extension service requires continuity and stability in its work. It cannot be effective if its official guidelines are changed with each change of ministers. An extension law is necessary that defines the mission of extension, the means of financing, and the overall responsibilities of the service (Watts, 1984).

Whatever the circumstances, prerequisites for an effective extension service include:

- a statutory basis and mission;
- effective linkages with research and with farmers' organisations;
- state financial support;
- an adequate number of well-trained and motivated staff;
- continued in-service training;

- provision of adequate incentives for the staff: reasonable salaries, advancement on merit, professional development, social services (housing, schooling, etc.)
- adequate supporting services: field offices, transportation, communication systems, library services, etc.

It should, however, be clear that structural or operational improvements will not result in a more general and equitable rural development, unless the goals of the extension services clearly reflect an unequivocal commitment towards achieving this goal.

THE TRAINING AND VISIT SYSTEM OF EXTENSION

Among the various efforts to improve the functioning of conventional extension services, the best known is the Training and Visit System (T & V). First introduced in India, Turkey, Burma, Nepal, Sri Lanka and Thailand, the World Bank, since the 1970s, has helped some 40 countries in Asia, Africa, Europe, and Latin America to adopt the T & V extension system. In India alone, 13 major states have adopted the system (cf. pp. 660–1) (Baxter & Thalwitz, 1985).

The system, as it was conceived by Daniel Benor, and subsequently adapted in the course of a decade, is described in detail by Benor & Harrison (1977) and will only be summarised briefly here.

Basic principles

As a management system, the T & V system is a farmer-focused professional extension service which aims to overcome the drawbacks of the conventional government extension systems by insisting on a set of principles (Benor, 1987):

- *Professionalism* of a well-trained staff that maintains close contact with researchers and whose recommendations are based on reliable and relevant research results.
- *A single line of command* within a ministry or department, with all extension staff administratively and professionally responsible to a single authority.
- *Concentration of effort*: extension workers devote all their efforts to extension, to the exclusion of regulatory, social, political or other activities. Each staff member of the service has a clearly defined task; the service initially focuses on a small number of important

crops, expanding gradually its field of activity.
- *Time-bound work* based on regularity of visits, regularity of training, regular research/extension workshops.
- *A field and farmer orientation* that takes as its starting point the farm, with its problems and constraints.

Structure and mode of operation (Fig. 22)

The conventional structure of the extension service, as described in the previous chapter remains basically unchanged: the T & V system is also based on village level workers, supported by subject matter specialists (SMSs) working within a management structure which establishes a clear single line of responsibility.

It is in the basic extension techniques, encompassing concepts of goal definition, work planning and monitoring at each échelon of the service, that the 'training and visit' system differs from conventional extension services. Essentially it is based on a systematic, time-bound programme of visits and training. Under this system 'schedules of work, duties and responsibilities are clearly specified and closely supervised at all levels'.

Whilst the basic conventional hierarchical structure remains unchanged, each level of the service has a span of control narrow enough to permit close personal guidance and monitoring of the level immediately below. Hence the regional extension officer (REO) supervises four to eight divisional extension officers (DEOs). Each DEO is in charge of six to eight extension advisers (EAs) or village extension workers (VEWs). The REOs are supervised either directly by headquarters or by an intermediate échelon, depending on the conditions pertaining in each country.

In the T & V system, the envisaged staffing arrangement consists of about 12 per cent administrative and supervisory staff, about 7 per cent SMSs, and about 81 per cent village extension advisers.

The entire organisation is based on the grassroot unit: the VEW circle, which encompasses a reasonable number of farm families to be assisted by one VEW; this number is flexible, and depends on the physical environment, the density of the rural population, the cropping system, etc.

The village extension adviser (EA)
The EA must live in the area in which he is active. The maximum number of farm families that a EA can cover so as to make a significant impact on the agriculture of an area depends on many factors, such as the potentialities for increased production or diversification, the types of crops

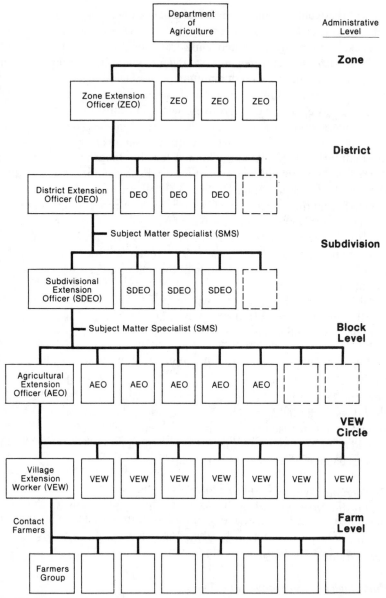

Fig. 22. Organisation pattern of intensive extension service in one of the states in India (From Cernea *et al.* (1985), reproduced with permission of the World Bank, Washington D.C., USA).

grown, the technical and educational level of the farmers and of the extension worker, to name just a few.

These ratios may vary from one in 300 in special cases, up to one in 1200 in areas with a low cropping density and a few dominant crops.

In any case, the EA cannot maintain personal and regular contact with all the farm families in his area. He therefore divides his area into eight groups of farms. From each group he selects, in consultation with the village leaders, about 10 per cent of the farmers who will serve as 'contact farmers', and on whom he will concentrate his main efforts. These should be small farmers, typical in the kind of farming that they are practising, but of good standing in their community, whose opinions are likely to be respected and their example followed. They must indicate their willingness to accept the advice of the EA to try out new practices and to explain them to several neighbours, relatives and friends who are farming in their vicinity. The spread of the improved practices will be further enhanced by group activities. In this way, the EA can make a major impact on relatively few farmers, but the adoption of the improved practices can spread rapidly to other farmers.

Initially, the contact farmers will be encouraged to apply the recommended practices on part of their farm only, and expand as they gain experience and confidence.

The divisional extension officers (DEO)
The DEO spends one day a week on training sessions (alternately: the new one he conducts himself and the one conducted by the SMSs) and four days in the field, monitoring and guiding the work of the EAs. He should be able to visit two EAs and their farmer groups in a day. His visits should be prescheduled and timed so that over a period of several months he sees each of his EAs and their groups.

The regional extension officer (REO)
REOs should spend at least half their working time on field supervision of the extension service and the training programme. They are seconded by a team of SMSs comprising an agronomist, a plant protection specialist and a training specialist.

The subject matter specialist (SMS)
The SMS is the major link between field extension workers and researchers. He remains, however, basically an extension agent, able to diagnose problems in his area of competence, to understand farmers' con-

straints, to tell the researchers what difficulties the farmer is likely to encounter when attempting to adopt new technologies. The SMS therefore needs to be trained, not only in his special field, but also in extension methodology (Verkatesan, 1985).

Priorities

Benor and Harrison (1977) stress that initially the extension effort should be focused on improving agricultural management practices, such as better land preparation, improved seedbed and nursery maintenance, the use of good seed, seed treatment, timely operations, weeding, proper spacing of plants, etc. The increased use of purchased inputs, such as fertilisers, should not be recommended initially. The rationale for this approach is that improved cultural practices produce sure results, involve no risks, require more work (amply available) and require little cash. Further, purchased inputs or costly investments, such as tubewells, cannot give full benefit until practices are first improved.

This approach is in contradiction with the line of thought that stresses the need for a complete package of improved practices. True, fertilisers, high-yielding varieties (HYVs) and tubewells cannot give full benefit unless management practices are improved, but it is equally true that, conversely, the practical effort and time invested in improving management practices will give very meagre results unless the necessary inputs are provided which enable a worthwhile increase in yield to be achieved. The question therefore arises: how to reconcile the excellent results of the system claimed by the World Bank, and the low probability of achieving significant increases in yield from management practices only? The answer to this question is fortunately provided in the annex to the brochure describing the Benor system, in which three case studies of the application of the method are reported in detail. In every one of these cases, HYVs, fertilisers and pesticides were used in conjunction with improved management practices. For example: farmer Shri K. N. Deb Moswami, in the state of Assam, who had installed a tubewell, started growing HYV paddy-rice and wheat, using fertiliser, and found that he was losing money! On the advice of the EA (working within the reorganised training and visit system) 'he adopted line sowing, grew seedlings on raised seedbeds, provided for sufficient spacing and plant population. He also gave top dressings of fertilisers at the rate of 60 kilograms per hectare, supplemented by farmyard manure, and took care to remove weeds. Pest attacks worried him but he has now learnt to use pesticides, can talk about plant diseases in English terms, and mention the cures too.' A clear case of a classic 'package programme' and definitely in contradiction to the stated philosophy of the system. The other case histories similarly show that the excellent results obtained from the 'Training and Visit system' were based on recommendations including purchased inputs and improved management (Arnon, 1987).

Evaluation of the T & V system

The T & V system has been functioning for several years in a number of countries, and it is therefore possible to evaluate the system, and define its problem areas and shortcomings, even though it is generally difficult to quantify the results of extension work (Pickering, 1983).

The T & V system deliberately aims at reaching the small and disadvantaged farmers without, however, neglecting the better-off farmers. The system is under continual pressure by elite groups to monopolise its services; therefore, the extent to which the T & V system effectively reaches the mass of small and marginal farmers is a fair measure of its success (Cernea, 1981).

In a detailed case study of T & V by Feder *et al.* (1985) the main conclusions were as follows: 85 per cent of the contact farmers were being visited regularly, and the majority of the non-contact farmers also had contacts with the village-level workers. There was some bias in favour of the larger farmers, but it was very small. As experience with T & V increased, there was an overall increase in visits to farmers by extension workers: the contact farmers received fewer, and the non-contact farmers more visits. The crop yields on farms that relied on T & V as their main source of information were found to be higher than those relying mainly on other sources of information.

A by-product of the replacement of the existing deficient extension systems by T & V was a stimulation of government investment in agricultural research and extension (Howell, 1984).

T & V and farming systems research
Moris (1987) makes the interesting observation that the T & V system and FSR complement each other. FSR is designed to strengthen upwards feedback from the farm level to the research workers, whilst T & V strengthens downwards communication of proven innovations. There can, however, be points of friction between the two systems. FSR can require a lot of time for the co-operation demanded from the extension field

workers, an ancillary assignment specifically discouraged by the T & V approach.

Problems and disadvantages

The T & V system provides an opportunity to reorganise and make more efficient use of a pool of currently badly managed manpower. However, adopting T & V requires a complete reform of the existing extension services and their integration into a unified service. Considerable investments are required to strengthen the cadres, to redeploy the village-level staff, to provide means of transport, and to finance the regular training of staff. Also essential to the success of T & V, is a strengthening of the agricultural research system, in particular the establishment of experimental stations in ecological regions not covered by the existing system (Pickering, 1983).

All this makes the changeover to T & V a very expensive process. As a result, the externally assisted projects bequeath a high level of recurrent expenditure (Howell, 1982), which many countries are unable to maintain, resulting in a severe curtailment of the programme when financial aid is no longer forthcoming. In LDCs that attempt to establish T & V without outside assistance, shortage of funds may prevent a full implementation of the system, with resultant deficiencies in staff and training. Much of the criticism of T & V relates not to the system itself, but to lack of feasibility, given financial and manpower constraints (Howell, 1984).

The T & V system also implies fundamental changes in the philosophies, organisation and operation of government agencies involved. Its adoption may conflict with powerful vested interests, threatened by loss of control over a large work force. It may have to overcome bureaucratic inertia, and to change the work habits of the extension staff. A reorientation of research objectives may be necessary, in order to solve the problems of the food crops grown by the small farmers. All these requirements are a potential source of conflicts (Pickering, 1983).

The extension staff have to be retrained and new attitudes and motivations instilled, such as the desire to continuously update their professional knowledge, to improve their understanding of the social structure of the rural community in which they operate. They must develop empathy with the small farmers, readiness to understand and evaluate their problems, and to convey relevant information up the chain of command. All this requires a level of involvement by subject matter specialists (SMSs) in the field which is often lacking in practice. In Africa, in particular, the low number of SMSs 'is probably the greatest limiting factor to successful T & V' (Howell, 1984).

To be successful, the T & V system must be adapted to local conditions and the system has sufficient flexibility to allow successful adaptations to be made; however, this flexibility relates to procedures and not to basic principles. Not always is the new system implemented as conceived by T & V. In Thailand, for example, the village contact farmers were not selected by a village committee, but chosen by the headman, and there has been little evidence of farmers communicating with contact farmers. In certain areas the lack of human resources makes it difficult to establish the interlocking system required by T & V. Poor supervision and guidance by senior staff result in the inability of the field staff to deliver an adequate professional service to farmers (Sukaryo, 1983).

An example of how non-adherence to the principles of T & V can bring the method into disrepute is mentioned by Hayward (1987). In Malawi, extension works carefully in accordance with T & V principles, excepting that the farmers' groups meet at block gardens rather than on farmers' fields. 'These gardens are mostly atypical and the lessons apparent in real farming situations are lost to the group, to extension and to research. Farmers become bored with stale messages related to inappropriate crops and their commitment to the programme declined.'

The T & V approach to the single function, whereby all efforts should be exclusively concentrated on professional extension advice, cannot always be maintained in practice because of the conditions prevailing in many LDCs. Howell (1982) writes that in a thorough study of extension, it was found that the only way to make it effective, was for it to be linked to access to input supplies and credit, where these were provided by the public sector.

According to Röling (1987) the T & V system 'is running into heavy waters wherever it is being applied outside monocrop systems such as padi-rice or irrigated wheat'. He concludes that agricultural information systems serving rain-fed areas and complicated farm systems have to find other solutions than the basic T & V model.

Conclusions

Cernea (1981), in his review of the implementation of T & V system, concludes: '. . . the high integration of the various parts of the system, the consistency of its organisational features, its built-in devices for organisational monitoring and maintenance make it an organisation in the sociological meaning of the concept, and not just another administrative service'.

It must be stressed that T & V is based on certain principles which must be adhered to. However, certain details of management, such as the ratio of extension worker to farmers, number of days between visits to contact farmers, intervals between training courses, etc., are not fundamentals of the system, but need to be adjusted to specific geographical and socio-economic conditions, the stage of development of the target population, the professional level of the extensionists, etc.

Moris (1987) gives a few examples of modifications that can be made to suit countries where resource restraints do not permit full implementation according to the original model. Training sessions can be held once a month, when field staff come in for their salaries, rather than every two weeks. The village-level worker can meet groups of farmers, rather than individuals. Contact farmers may be rotated every 3–4 years. Such modifications may dilute the impact of T & V, but are acceptable as long as the system adheres to the principles outlined above.

GROUP EXTENSION METHODS

Tully (1981) observed that the social structure of rural communities in the developed and developing countries is characterised by reference groups. The group is permanent, because interaction between most of its members is lifelong. In addition to interpersonal relationships, group action makes it possible to achieve group goals.

Group norms become established based on held views, beliefs, and attitudes. Group members who do not conform to these norms are considered deviants and lose their influence within the group. Therefore, if one of the group is the only one to adopt a new technique, he may be considered by the other group members as deviating from accepted group norms and he may lose his influence in the community.

According to this concept, choosing contact farmers, progressive farmers, etc., as initiators of new techniques may actually be counterproductive: they may be considered as deviants by their reference groups because of their close association with the extension worker and their receptiveness to new technology. As a result, the innovations may not be adopted by the community.

The failure of the conventional extension methods to reach the majority of the farmers, namely the small, subsistence farmers, has led to the use of group extension methods in communicating with the rural poor (Garforth, 1982).

The objective of achieving a broad-based integrated rural development in which the access to research, extension and inputs is more equitable, and in which farms participate effectively, 'would require extension not to approach people individually, but rather in natural groups, such as communities, capable of deciding on the distribution of productive assets among the members, of setting priorities and deciding on specific action for integrated development, and of forming specialised agricultural development committees for dealing with the extension workers' (Röling, 1982).

An increasing number of countries (such as India, Bangladesh, Indonesia, Mali, Cote d'Or, Thailand) are employing the participatory approach in extension as a way of reaching large numbers of small farmers effectively and involving them in the planning, learning activities, group action and feedback to extension of problems and needs. These groups have the additional advantage of facilitating access to services, credit, inputs, land and marketing (Ghonemy, 1984).

The justification for the group approach is based on three major arguments (Garforth, 1982):

• more people can be reached by extension workers;
• groups provide a more effective learning environment;
• many of the proposals made by extension are more appropriate for group action by the rural poor than for individual action.

Levesque (1961) found that such discussion and action groups can decrease the mutual distrusts among farmers, and, if the extension worker participates, the distrust against him will also decrease.

The introduction of group extension methods has important training implications for extension staff. For most extension workers this approach is outside their general experience and training. More emphasis is required on communication techniques and group dynamics, both in initial training and during in-service training (Garforth, 1982).

Among the programmes based on the concept of group methods are 'animation rurale' in Senegal, the rural reconstruction movement in the Philippines, the co-operative education system in Tanzania and the 'community development programme' of India, and others.

Media forums

A similar concept, adopted in India, is the so-called media forums patterned on the informal discussion groups which are part of most peasant cultures.

Very briefly, the basic elements of these forums are as follows: each village chooses two representatives, called 'conveners', one male, the other female; after a period of training at special centres, each convener receives a transistor radio. Once or twice a week, the village forum members meet, men and women separately, to hear special, regularly scheduled radio programmes; the conveners initiate group discussions and report to the broadcasters on questions raised by the group.

Paraprofessionals ('animateurs', 'conseillers', conveners) are in a better position to communicate their newly acquired technical knowledge, because they are selected by the community in which they work, than a professional extension worker with several years of formal education. They also feel more accountable to the community in which they serve (Garforth, 1982). The accountability to local groups may be of more significance in the successful transfer of knowledge to farmers than the specific extension techniques used or the frequency of extension contacts (Morrs *et al.*, 1978).

Forum strategies have been successful in countries like China and Tanzania, where people are organised at grassroot levels. They have also been successful as pilot projects, carried out by skilled, committed people with all necessary resources at their disposal. Most large-scale applications have failed, due to ineffective management, insufficient resources, and lack of political commitment (Röling, 1982).

Administrative decentralisation of services

Many countries, even those that carry out most of their activities by central direction and control, have chosen to decentralise their rural development programmes and with them, their extension services. The decentralisation consists in the transfer of planning, decision-making and management from the central government to the regional level.

The major reasons for decentralisation have been the disappointing results from central planning and management; the inflexibility and lack of responsiveness of central bureaucracies; the need to reduce the disparities between the sectors of the rural population, and between rural and urban areas. Equitable growth policies need programmes tailored to local conditions and supported by the people they are intended to help. In some countries, decentralisation was adopted as an objective unto itself, as a means of developing self-reliance, participation, and accountability. The decentralisation of the extension service may appear at first sight to parallel the regionalisation of the research system. The resemblance is, however, more apparent than real: the decentralisation of the extension service is according to political, administrative boundaries, whilst the regionalisation of research is in accordance with agro-ecological zones.

DELEGATION OF EXTENSION FUNCTIONS TO REGIONAL OR FUNCTIONAL AUTHORITIES

Statutory commodity boards

Commodity boards have been established in many countries to produce, and especially market, high-value crops efficiently and economically. In Jamaica for example, a number of boards promote the production, processing and marketing of sugar, bananas, coffee, coconuts, and have sole responsibility for marketing their respective commodities. The three biggest boards, for sugar, bananas and coconuts respectively, have established their own research and extension services.

Farmers producing a crop under the aegis of a commodity board, sign individual contracts to produce the crop using practices recommended by the board's extension workers, and the board undertakes to buy the crop at a guaranteed price.

Farmers have little choice but to use the inputs and to follow the technical recommendations given ('contract extension'). Failure to do so may result in a product that will not be purchased by the board, and is therefore unsaleable.

Usually a single adviser is assigned to a zone and handles all aspects of technology transfer, providing both technical advice and the necessary inputs. The extension function in this system is therefore well-integrated with all other aspects of technology transfer.

The focus is on a single commodity; each phase of production and marketing is fully organised. The system can therefore be characterised as a vertically integrated production system (Swanson & Claar, 1984).

Some governments use commodity boards for export crops in order to capture both the foreign exchange and surplus income generated, without much sharing with the farmers, and/or without reinvesting in the system. This leads in the long run to stagnation because of lack of incentives discourages the farmers from participating (Swanson & Claar, 1984).

Experience has shown that whilst the extension services provided by statutory boards can be very effective, they have a number of disadvantages when viewed from the national perspective. A commercial and monocultural attitude towards agriculture leads to a fragmentation of extension education. This not only increases expense and is an inefficient way to use scarce human resources, but their teachings and their promotion of individual enterprises often run counter to the objectives of the farmer whose interest is to develop a combination of crop and livestock enterprises that will yield him maximum net returns, rather than to confine his farm to any one crop to the exclusion of others. They may also run counter to national policies, which may wish to de-emphasise research and extension in certain commodities whilst promoting others found to be more consistent with national requirements (Arnon, 1987).

Regional development boards of societies

In several francophone West African countries, the 'parastatals' or intervention societies, which are organised on a regional basis, are used as major extension agencies.

In Senegal, for example, a number of 'sociétés de développement' operate under the aegis of the Ministry of Rural Development. Each société is responsible for all aspects of development of the region assigned to it, but it also has national responsibility for the dominant crop of the region. Hence, the extension service of each société works both on a regional basis for all the agricultural commodities produced in a specified area, and on a national scale for a specific commodity (Arnon, 1978).

Each of the societies has its own policies and operating procedures. There is no central extension service, capable of co-ordinating extension on a national scale, providing centralised training of the extension workers, giving professional support to the field staff and ensuring liaison with the research service. In addition, the system has all the disadvantages of the extension services provided by statutory boards that have been outlined above.

Land reform authorities

In many Latin American countries special public institutions have been vested with responsibility of implementing land reform and colonisation programmes. Most of these include extension activities in their field work, employing extension field or reform agents. Some co-operate with the regular extension service, others ignore it. In Venezuela, the Instituto Agrario Nacional (IAN) has a small extension directorate, but relies primarily on the extension service of the Ministerio de Agricultura y Cría (MAC); the latter have assigned up to 60 per cent of their extension workers to the service of the beneficiaries of land reform.

By contrast, the Colombian land reform authority, INCORA, has its own field staff of over 500 persons, most of whom devote part of their time to extension work. There appears to be very little collaboration with the extension arm of ICA (Rice, 1974).

Special project implementation units

Defined area programmes

These are frequently donor-assisted projects for a particular area, and are focused on a set of production problems. The objectives are commodity-oriented and aim to increase agricultural output in the project area, thereby demonstrating that agricultural development can be achieved if an integrated approach is adopted. The intended clientele are all the farmers in the project area, though in many cases, the large farmers have been able to capture most of the inputs, credit and marketing services available (Swanson & Claar, 1984).

These projects create their own management structure and technical support systems, in order not to be dependent on the existing bureaucratic institutions. They are able to pay salaries higher than the government institutions, and are therefore able to attract personnel from them. Also, by supplying transportation, inputs, credit etc., they are usually effective in achieving their aims (Swanson & Claar, 1984).

Crop campaigns

The crop campaigns are similar in concept to the defined area campaigns discussed above. The crop campaign as an extension tool was first adopted in Costa Rica in the late 1940s, and has since proliferated in many Latin American countries. In Argentine, the extension service is almost entirely oriented to crop campaigns. Other countries in which this system serves as an important extension tool are Brazil, Peru and Bolivia.

The so-called 'vertical crop campaigns' are based on a complete analysis of the production–distribution system, and the development of a complete set of policies that provide technical, institutional and economic regimes favourable to the producers without upsetting consumer welfare goals (Rice, 1974).

TRANSFER OF RESPONSIBILITY FOR AGRICULTURAL EXTENSION TO NON-GOVERNMENT ENTITIES

In many countries, non-governmental entities have undertaken partial or entire responsibility for agricultural extension. These may be non-profit entities, such as universities (land-grant colleges, agricultural universities) or voluntary organisations, professional associations, such as farmers' associations, rural co-operatives, rural credit institutions, or private sector enterprises. The last mentioned may be very diverse, comprising processors of agricultural commodities, marketing firms for export, agribusiness, etc.

Finally, there are many countries, developed and developing, in which private consultants, generally under government licence, are active.

Under certain circumstances, as we shall see, private sector extension can be extremely effective in agricultural production and rural development. More attention is therefore being given in many countries, to privatisation of the extension services, and/or cost recovery. The idea has merits, but also limits. Private extension is rarely co-ordinated with government extension, and concentrates on particular cash crops, neglecting the food crops, which may be important even to the farmers producing the cash crops. The bias to big farmers is even more pronounced than with government extension workers (Baxter, 1986).

Universities and agricultural colleges

The traditional land-grant college model of the USA is based on combining higher agricultural education, research and extension within the same institution. The adoption of this model in developing countries has often been advocated wherever US aid was active in promoting agricultural development.

Because the American extension system was one of the main contributors to the phenomenal increase in production achieved by American farmers, this success serves as an excellent argument that others should adopt the same system. However, the transfer of responsibility for extension from government ministries to agricultural colleges or universities has generally not given satisfactory results in developing countries (Arnon, 1987).

The United States extension service was evolved to meet the needs of a farm population at a relatively advanced stage of development and not in order to mobilise them for development, as is the primary goal in developing countries. It came in response to pressures exercised by a politically powerful farmers' sector; literacy and exposure to a well-developed system of mass media was widespread; urban jobs were available for farm labour made redundant by modern technology; *per caput* incomes were high (Barraclough, 1973). It is highly improbable that a system developed for these circumstances should be appropriate for the mass of rural people in underdeveloped countries. The family-size but large commercial owner-cultivators who are the mainstay of the American extension effort are a small minority of the farmers in developing countries, the majority consisting of small subsistence farmers unable to adopt new technologies even if they are made aware of them, unless a whole package of prerequisites is assured.

In this context it may be relevant to note that the American extension system has been largely ineffective in rehabilitating the rural population of the backward and depressed areas in the USA, until a special organisation was established for this purpose in the 1930s: the *New Farm Security Administration* (Barraclough, 1973).

An Agency for International Development (AID) report states that in too many cases attempts were made to impose the US form of extension upon a foreign university for which a US university had a contract with AID. Sometimes a new institution was established in competition with an existing one (USAID, 1968).

An example of the problems encountered when establishing the US extension model in a developing country is provided by India.

Large-scale involvement in extension by the universities in India is relatively recent. Where agricultural universities have been established (and the intention is to establish one at least to each state) they have been patterned on the model of the land-grant colleges. Confrontation with the extension services of the ministries of agriculture has already led to a re-evaluation of the function of the universities in the field of extension. The problem was whether the university should aim at direct contact with the farmers, thereby actually taking over the extension service, or should it confine itself to training extension workers, and provide the framework for farmers' courses.

In India, extension has to serve a very large number of small farmers, many of them illiterate, who need—in addition to technical advice—help in obtaining supplies, credit, equipment, etc. For these reasons, the extension staff consists of a large number of simply trained staff. Several agencies are involved, and these require close co-operation; certain programmes (soil conservation, insect control, credit recovery, etc.) may involve elements of control and even compulsion. All these functions require a government controlled co-ordinated service, with statutory functions and an administrative framework for thousands of workers. It is doubtful that a university can, or should, undertake such roles. 'This would immediately turn the university into a bureaucratic organisation indistinguishable from the existing government service.' (Hunter, 1970).

Conflicts have already arisen between the agricultural universities and the state ministries of agriculture. In India and the Philippines these have been resolved by the university undertaking to confine itself to the training of extension specialists and research in extension education whilst co-operating with the existing extension service of the ministry of agriculture and providing support for the field work of the latter.

The reservations outlined above regarding the suitability of the American extension service organisation in no way lessen the importance of the basic principles underlying the US extension system and the findings of extension research in developed countries, provided they are critically examined in the light of the conditions of the individual countries, and then properly adapted to those conditions.

Farmers' associations

In certain countries, government has encouraged the formation of farmers' associations and delegated to them the responsibility, partly or entirely, for extension education of their members.

As these associations are also responsible for the allocation and distribution of inputs, such as water, fertilisers and credit among their members, as well as for the marketing of their products, they are in a position to: (a) decide on the approved practices they wish to recommend for adoption; (b) provide the inputs required at the right time; (c) provide the credit required to purchase the inputs; (d) supply advice and ensure effective supervision of the correct use of the inputs; and (e) guarantee repayment of credit after harvest (Yudelman et al., 1971).

According to Röling (1982), experience has shown consistently that when 'farmers are critical, demanding and have some political clout to in-

fluence extension work' they receive excellent extension. In consequence, the best way to improve the extension service is to organise farmers and make them aware of their rights *vis-à-vis* extension, and of the
fact that their taxes are being used to pay for the service.

A corollary of this concept is that extension services will function best
if they are run directly by the farmers' own organisation. Programmes and
objectives are determined by the members of the groups themselves; the
ways these are implemented are different from association to association.

Advantages

When national grouping of these associations is organised, they can exert
considerable influence on national policy in general, and on the orientation of the research programmes of the national agricultural research system in particular.

Rice (1974) summarises the advantages enjoyed by farmers' associations (and other non-governmental bodies) in promoting agricultural
modernisation in general and extension work in particular, as follows:

- Government agencies generally have difficulties in getting adequate
  funds for an efficient extension service; commodity associations can
  raise taxes for this purpose, generally to supplement government
  subsidies.
- Government agencies are restricted by regulations in the salaries
  they can pay, and have great difficulties in retaining the best extension workers; the associations have no such restrictions.
- The associations and their employees are largely free of the political
  pressures of a government bureaucracy, which are largely responsible for the constant turnover of personnel, low morale, and lack of
  motivation of government extension workers.

The success of many farmers' associations in promoting agricultural
progress of their members, proves that a well-organised and well-equipped extension service, integrated with an efficient credit plan, can be a
major factor in modernising agriculture in the LDCs.

Disadvantages

The major disadvantages of the farmers' associations are:

- they tend to be dominated by the rural elite, and the interests of the
  small farmer may be neglected;
- the associations have been mainly concerned with cash crops, and
  the food crops have frequently been neglected;

- they have led to a proliferation of extension units, and a consequent fragmentation, duplication and weakening of the national extension service.

Handing over the responsibility for extension to independent bodies is only feasible in countries in which farmers have the necessary experience and will to finance (at least partly), operate and control their own extension system.

Credit institutions

The supervised credit institutions are usually autonomous; their salary scales are not subject to government regulations, and wages and fringe benefits are attractive enough to lure away the best extension workers from the government service. The supervised credit agents generally do double duty: they serve as loan officers and as extension agents, a combination that is anathema to conventional extension thinking, but has proved to be highly effective, even if delinquent farmers attempt to avoid the loan officer, and thereby lose out on advice.

In general, the supervised credit operations have a good reputation. Generally, the credit is made conditional on the farmers adopting improved practices, and the lending institution provides the technical advice and monitoring of implementation. The Caja Agraria in Colombia, though not strictly involved in supervised credit, requires small farmers to present a farm plan when applying for credit (Rice, 1974).

Commercial firms

Generally, an innovation involves the promotion and sale of a new product. Many large firms involved in producing inputs for farming, not only engage in research but also hire agents to actively promote their products. These generally use the same techniques of extension as do the extension workers of the public bodies: personal contacts, demonstrations, mass media, etc. They are generally better paid, often more motivated (receiving a percentage of the value of their sales) and more aggressive than their official counterparts. They may therefore play a considerable role in the diffusion of new products. Their main drawbacks are: possible lack of objectivity, with resulting oversell of their products and lower credibility; their tendency to concentrate on the larger farmers and neglect the small farm units which require a larger investment in effort with a smaller expectation of returns.

When a farmer is selected by a company to serve as their agent, these drawbacks may be mitigated, because of the need to maintain credibility with one's neighbours, who regard him as a peer rather than in his role of agent (Rogers & Yost, 1960).

Co-operatives

The adoption of new technologies by farmers can foster special activities of the co-operatives of which they are members, such as the introduction of irrigation, the supply of improved seeds, improved equipment, farm chemicals, etc., and credit. Therefore the co-operatives have a legitimate interest in promoting these innovations.

A further step forward occurs when action is taken to organise co-operative systems of production on the basis of improved methods, specialisation through division of labour, joint control of insects and other pests, etc.

As a prelude, irrigation systems may be developed, land holdings re-allocated and consolidated. The joint adoption of improved technology makes it easier to overcome individual and social constraints; guidance provided by the co-operatives may be mainly directed towards quality control and storage and processing of farm products, whilst advice on routine production procedures is left to the government extension service (Arnon, 1987).

Occasionally, co-operatives are set up for the express purpose of providing extension to their members. For example, in Argentina, the Rural Society for Agricultural Experimentation (SREA) consists of organisations of large farmers who study the results of various farm methods being used by the members. In 1966, there were 73 of these co-operative groups, each consisting of 12 to 16 large farmers, each employing an agronomist or other specialist to analyse farm records and advise members concerning improvement of methods.

In Brazil, two large and effective co-operatives provide service and advice to members, and there are a few similar organisations in Peru, Colombia and some other countries (Hopkins, 1969).

Herzberg (1975) lists the following advantages of extension provided through rural co-operatives:

- It is possible to initiate meetings of the co-operatives' members at regular or *ad hoc* intervals for extension activities, which lightens the work load of the extension workers considerably and increases their efficiency.

- The co-operatives broaden the perspectives of their members, making it possible for them to relate to common problems and to develop common solutions. This in turn facilitates the contribution of the extension service to the solution of these problems.
- Social motivation of the members of the crops is enhanced, generating group dynamism which in turn facilitates the adoption of innovations.

Pickering (1987) stresses the importance of co-operatives growing in response to farmers' felt needs and not as a result of pressure from above. He states that the level of success of the co-operatives is almost inversely proportional to the level of administrative interference by government.

Where the co-operatives are successful, they may fund their own extension workers, who receive training and technical support from the government extension service. Where co-operatives are successful, they should be encouraged to take over extension activities at the local level.

Processors

Processors of agricultural products, such as sugar factories, tobacco companies, fruit and vegetables conserve factories, etc., have a vital interest in obtaining supplies of raw materials according to planned schedules. Many therefore contract with farmers within a reasonable radius around the processing plant to produce the required commodity. These contracts generally include the provision of inputs, credit and guaranteed prices, as well as individual advice. This system is certainly an effective method for promoting the adoption of improved practices.

An interesting example of the advantage of extension provided by a processor over that of the official extension service is described by Rice (1974):

> In two districts of the Boyaca Highlands in Colombia, Duitama and Sogamosa, in certain sections of which wheat and barley are the predominant sources of income. Trends in wheat and barley production had not been demonstrably affected by the official extension service. By contrast, farmers who had contracted to grow barley for the breweries had improved their level of technology rapidly. Because the brewery purchases for taste, all the contracting farmers were growing the same improved variety requested by the brewery. Fertiliser rates were in accordance with the recommended practice. Most of the producers were small farmers.

The sugar companies in Kenya provide their outgrower farmers with various incentives: land preparation, seed cane, weed and pest control, harvesting and transport. The farmers are paid when their cane is delivered at the factory (Onyango, 1987).

Exporters

Companies exporting agricultural commodities frequently provide extension advice to the producers. For example, the United Fruit Banana Company of Ecuador, which is both a producer and a buyer of fruit from small farmers, provides farmers with credit under a ten-year contract, and undertakes to buy the fruit at fixed prices. Improved practices, developed and tested on the company's own farms, are transmitted to the contract-farmers, who are also provided with any required inputs (Rice, 1974). This system is very similar to that of the processors described above.

Contract-farming

Contract-farming, in which farmers agree to provide produce of a certain quality and adhere to strict husbandry instructions in return for an assured market and economic prices, is a system adopted by processors, exporters (including marketing boards, cf. p. 723). This approach is characterised by the close integration of all the functions of the agricultural system, but for a single crop. Extension workers employed by these organisations can give individual attention to a relatively small number of farmers. The recommended practices have generally been carefully verified, and the farmers can rely on all the supporting services required. The farmers have a strong financial incentive to adopt the recommendations of the company's extension worker.

The major advantage this system has over the conventional extension system is that the instructions the farmer receives are mandatory and not advisory only, as in the latter case. 'The processor succeeds with extension not because he can control the farmer, but first, because he can offer the farmer a reliable cash income, and second, because he takes over from the farmer much of the farm and marketing management burden' (Rice, 1974).

The success of contract-farming is usually dependent on the company involved having a monopoly over the supply of seed or planting material, processing and marketing. Therefore contract-farming is only suitable for commodities that have to be marketed (export crops) or processed (milk, sugar) through the company's agency. This is generally not feasible for food crops for local consumption. Producers can easily by-pass the agency and the recovery of credits on farm inputs. The collection of a tax would also be difficult (Adams, 1982).

The success of the contract-farming system is proof of farmers' willingness to comply with strict instructions when they perceive it is in their interest to do so. However, the high degree of control exercised by the processor also has its negative aspects. It limits initiative, reduces the freedom of choice of the farmer and strengthens the existing passivity that characterises small-scale subsistence farmers. It may cause resentment, especially if the farmer is obliged to make investments in inputs that he considers too expensive or risky in the light of expected benefits. However, obliging the farmer to follow recommendations as to the area to be planted, crop rotation, choice of variety, dates of planting, scheduled harvesting, insect and disease control, etc., can be highly effective in moving him from subsistence to modern farming whilst reducing the risks that are generally inherent in this process. It should be borne in mind that the farmer undertakes his contractual obligations of his own free will; he has the choice of not participating in the programme in the future, but will have been exposed in the meanwhile to modern methods, and will have increased his income to a level that will enable him to buy yield-increasing inputs for the crops he may wish to grow as a free agent.

Private consultants

Private consultants, frequently operating under government licence, are increasingly becoming involved in agricultural extension, both in developed and developing countries. Rogers (1987) describes how AID organised a group of about 60 small farmers in Carico (Chile), operating farms of 9–16 ha. The group hired a university-trained agronomist with practical experience, who operated an agricultural consulting firm, to provide technical and management assistance for a fee, based on a percentage of the market value of the crop.

Extension services consisted of individual farm visits every 4–6 weeks, depending on season. Group seminars were also held, at which discussions focused on crop varieties, sources of inputs, bank credit applications, etc.

With few exceptions, net farm incomes increased as a result of this procedure.

SUMMARY

A study by USAID (1983) of agricultural credit, input and marketing services, concluded that public, private and mixed extension systems each have advantages in particular situations:

- public institutions are preferable when benefits are diffuse, public policies need changing, and/or increased equity is a primary goal;
- mixed public and private entities are the best when agricultural services not only require intensive, responsive and flexible management, but also need political influence to achieve programme objectives;
- private sector extension performs best when flexible management and direct and continuing interaction with farmers is needed.

Therefore, private sector extension does have an important role, both in developed and in Third World countries, and can usefully complement government extension systems 'for particular farmers, in particular areas, producing particular crops, and/or using particular inputs.'

Private sector extension cannot completely substitute for public agencies in those circumstances where direct benefits do not accrue to the firms involved, target populations are diffuse or remote, where infrastructure is inadequate, when production consists mainly of basic food grown by subsistence farmers (Roberts, 1987).

Chapter 20

Human Resources in Extension

PROFESSIONAL REQUIREMENTS

In a previous chapter, it has been stressed that one of the major factors responsible for the deficiencies in many extension services is the low professional level of the extension personnel. Hence the importance of a careful selection of future extension workers, and the need for appropriate formation and in-service training.

There are generally four categories of extension personnel: administrative, technical (subject matter specialists (SMSs)), supervisory (extension officers (EOs)), and field or village-level workers (extension advisers (EAs or VEWs)).

Skills required

Cusack (1983) mentions the following skills ideally required from an extension worker: the ability to understand scientific information and to select the research proposals that can be beneficial to the farmers he serves; the ability to interpret and simplify technical information in a language the farmer can understand; skills in group dynamics, human relations and communication.

To these skills, Byrnes & Byrnes (1971) add the following requirements: The extension worker should have the ability to weigh and to recommend alternative strategies, based on the knowledge of market outlooks, agricultural policies, availability of credit, etc.; he should be willing and able to demonstrate new techniques to the farmer, even if this entails physical work and practice; he must be familiar with the customs, values and ways of thinking of the farming population he serves. He must realise that traditional farmers are suspicious of the motives of innovators because of their long historical experience of being socially exploited. He must learn to gain the respect and trust of the villagers, and this he can only do by understanding them.

Level of education required

An important problem is the desirable level of training and education required by the extension agents.

Education has been shown to determine extension skill mainly indirectly, by its effect on the attitudes of the extension worker. Those who had a higher level of education were found to be less bureaucratic and more professional in their work than those who had received only a certificate level education.

This finding does not accord with the general assumption that education tends to increase the gap between the extension worker and the farmer. Low-level workers with little formal education tend to rely on the authority of their position rather than on their knowledge as a source of power, leading to an autocratic and directive approach to farmers (McKillop, 1981). However, these findings do not appear to be applicable under all circumstances.

In the early stages of development, the objective needs of the farming community do not justify aspiring to a high level of education of the extension agents. In a study carried out by the Allahabad Agricultural Institute (1958) in India, it was found that village-level extension workers were more effective in reaching the villagers than were extension agents with high-school or university educations. On the other hand, if the agent is not sufficiently ahead of the farmer in agricultural knowledge and general educational level, he will not achieve the prestige required to influence his 'client'.

In all probability, the former type of extension worker is effective only during an initial period of village development. During this period, able farmers, technicians with secondary school training and agricultural teachers may have to undertake the tasks of village-level workers.

Under these circumstances, staff with tenth-grade educational level frequently serve as paraprofessional extension workers. They may, at best, have a very rudimentary scientific background. In order to be effective, these people need sufficient practical experience, as well as pre-service and in-service training.

Vocational agricultural training at high-school level can provide an adequate basic education for workers beginning a career in extension, at this stage of development, provided they have practical field experience, basic science and a good foundation for eventual higher education.

Paraprofessionals, particularly when selected from within the community in which they work, may actually be better in communicating their new technical knowledge than a professional with several years of formal education (Garforth, 1982). In a review of over 80 projects involving paraprofessionals, Colle *et al.* (1979) found that they are generally highly motivated in their work; that their cultural affinity to their clients, whose socio-economic background they share, facilitates effective communication. Living in the village, they provide more continuous contact, and are always available for advice.

The paraprofessionals also feel more accountable to the community they serve than do most extension workers. This accountability to local groups is of more significance in the successful transfer of knowledge to farmers than the specific extension methods used, or the frequency of extension contacts (Morss *et al.*, 1978).

There are considerable differences in selection procedures, qualifications, training and remuneration of paraprofessionals. They are mainly used to teach local farmers and to serve as liaison between the farmers and professional extension workers (Garforth, 1982).

In research on the effectiveness of the extension service in Kenya, Leonard (1977) found that secondary education had a uniformly detrimental effect on the work performance of agricultural extension workers in the current Kenyan context, one similar to that of many developing countries. Leonard ascribes the negative impact of education on the extension worker 'to the creation of expectations that could not be fulfilled, leading to a low level of job satisfaction and poor commitment to an extension career'. He was, however, able to show that in those fields in which good work by the extension agent led to substantial promotion, the better educated agents performed better than primary school leavers.

Leonard concluded from his study, that improved training of extension personnel will only cause frustration unless proper incentives for good work are provided. The improved capacity of the extension worker can, however, only be used effectively if the other prerequisites for the adoption of improved technology by the majority of the farmers exist or are also provided.

Generalists or specialists?

A difficult balance needs to be kept between the sophistication of the farmer and that of the field workers (Röling, 1982). This problem must be considered at two levels: (a) that of paraprofessional village-level workers in the early stages of development, and (b) university-trained extension workers for commercial farmers.

Regarding the paraprofessional extension worker, it is argued that by limiting his advice to a narrow range of subject matter, or to a specific crop, it is possible to use staff with relatively low qualifications and limited pre-service training; the extension worker can gain greater credibility if he is able to cope very well with strictly limited problems, and that advisory work on key topics leads to quicker results (Garforth, 1982).

This may possibly be an adequate solution in the early stages of transition from subsistence farming to modern farming; as farmers progress the extension worker must keep ahead, and his expertise must not only improve in depth, it must also be widened. This can only be achieved by an appropriate and continuous in-service training programme.

At a certain stage in the development process, paraprofessionals can no longer play a useful role, and the extension service must be manned by university-trained professionals.

From the foregoing, it is clear that the efficiency of the future extension worker will largely depend on how the university responds to the fundamental question as to what it should aim at producing—generalists or specialists? In big countries that are highly developed, there may be justification in training top-rank specialists, but the problem changes when one has to consider low-level and backward agricultures with diversified farming. In such cases, the university should mainly aim at turning out agronomists with wide backgrounds, who will be fairly competent to handle the problems they will encounter in daily life. A more limited number need to be trained as specialists in different branches of production and disciplines to serve as subject matter specialists.

The general situation in most countries is currently one in which the generalist can call on SMSs to advise him and to demonstrate special techniques or to diagnose specific problems.

In countries in which farmers have reached, or are reaching, comparatively advanced levels of know-how, the role of the grassroots extension worker becomes increasingly specialised, 'blending into that of the specialist in a subject matter of particular relevance to the farming system in question' (Pickering, 1987).

Where agriculture is already highly sophisticated and fully mechanised, the 'general adviser becomes an anachronism. Advances in technology are now so rapid, that no adviser can remain wholly effective in a very modern agriculture, unless he specialises in a limited area of competence' (Barker, 1986).

The need to adjust to greater specialisation involves changes in functions, redeployment and retraining of at least part of the staff.

Subject matter specialists (SMSs)

In many countries, the SMS is seen to be the weak link in the two-way continuum: researchers–extension workers–farmers. Efforts to recruit, train, and develop a cadre of SMSs deserves a very high priority. In countries with rudimentary linkages between research, extension and farmers, the SMS is a potential catalyst to bring all parties and their sometimes conflicting interests into a mutually supportive continuum of endeavours' (Pickering, 1983).

The SMSs play a key role in every type of extension organisation: they have to keep abreast of developments in agricultural research in their own country and elsewhere; they have to participate in research projects, in particular in FSR; they have to be involved in the systematic training of the field level works and provide them with on-the-spot advice and backstopping; they have to participate in demonstrations, field days, workshops, seminars, etc.; they have to help in the preparation of bulletins, visual aids and other means of mass communication.

It is the SMS who helps extension staff to lay out on-farm verification trials, collects and collates their results for incorporation into new extension messages or for consideration by research workers in designing additional research programmes.

These roles require that the SMS be trained to an appropriate level of general understanding of the farming systems in his geographic area of responsibility. He should also have specialised training in a relevant area such as agronomy, plant protection or animal husbandry (Pickering, 1983).

The SMS should be expected to work for an advanced degree, and the service should provide the incentives and possibility to do so.

Supporting staff

An effective extension service will require specialists in communications, programme development, training and evaluation (Claar & Bentz, 1984).

Administration and supervision

The *director of extension* is responsible for all the aspects of the organisation including: supervision, personnel management, salary administration, facilities management, fiscal management, programme development; co-ordination, implementation and evaluation. Most of these activities will have to be delegated to others.

The director should have a university degree, and several years of experience at various levels of the organisation (Claar & Bentz, 1984).

Middle management personnel: because of the need for delegation of authority by the director, a team of managers and supervisors is required. Composition of the team, the numbers involved, and the qualifications required will depend on the size of the country, the nature and level of its agriculture, the resources available, etc. The number of managerial levels will depend on the size, complexity and functions of the organisation.

The key function of the *agricultural extension officers* (AEOs) is the effective monitoring of the work of the village-level workers. This function is of critical importance for the effectiveness of the entire service. Monitoring does not simply signify administrative supervision but also guiding, advising and encouraging the VEWs in their work.

The qualifications of middle-level administrators and supervisors should be a university degree with special training in extension education.

The need to train senior and middle-level officers for the extension service is increasingly recognised. In India for example, a national programme (MANAGE) was initiated in 1985 with World Bank assistance, to improve the management competencies of senior and middle-level officers by providing an understanding of the general principles of management as tailored to agricultural extension requirements (Rivera, 1987).

In order to make possible the adoption of modern communication systems in the LDCs, it will be essential for extension staff to be carefully trained, so as to become familiar with the potentials and requirements of the new technologies.

University courses

Many faculties or colleges of agriculture have established departments of agricultural extension. In Taiwan, for example, such a department at the College of Agriculture of the National Taiwan University has a four-year programme in agricultural extension that offers training in five areas: all-university courses; agricultural subject-matter; basic sciences; natural, life and rural sociology; teaching methods, including adult education and psychology; and extension methods, including programming, evaluation, philosophy and comparative extension (Axinn & Thorat, 1972).

In a number of Latin American countries, the faculties of agriculture have signed agreements with the national extension services, whereby groups of students of the last grades spend a training period of about three months as assistants in field extension agencies (FAO, 1971). This arrangement has the additional advantage that it provides the extension service with the opportunity to evaluate prospective candidates for extension work, before their being appointed.

INDUCTION AND IN-SERVICE TRAINING

Whatever the background of the candidate for the extension service, suitable induction and in-service training needs to be provided.

Induction

All people entering the extension service should have a clear conception of the objectives and practices of the organisation in which they intend to serve.

The objective of the induction programme is to reduce the initial stress felt by the new employee related to the performance of his tasks; to familiarise him with the history, structure, policy and procedures of the organisation; the economic, political and social influences that affect the people in the area to which he is assigned; and the professional conduct that is expected of him (Malone, 1984).

All newly appointed staff should first be assigned to work as assistants to experienced and capable extension workers, who then serve as mentors to the new recruits.

In-service training

The objectives of in-service training should be:

(a) to keep up with research: by regular meetings between researchers and extension workers, joint colloquia, etc.;

(b) to impart basic knowledge, not only in the fields directly related to agriculture but also in sociology, economics, psychology, etc.;

(c) to improve extension methods, by constant evaluation of methods, the joint study of research findings and extension methods, the exchange of experience.

A programme of in-service training should include a variety of methods, such as workshops, seminars, symposia, study tours of farming areas, visits to experiment stations, short courses on specific subjects, discussion groups which include extension workers, researchers, and other disciplines, etc. (FAO,1971).

In-service training is a never-ending process. It cannot consist exclusively of what is provided by the services; the people involved must themselves continue training by personal effort—reading, individual learning, etc.

INCENTIVES

Promotion

Promotion in many extension organisations is more easily achieved through an office job, where daily contact with one's superior is possible, than through hard work in the field, in a remote location.

Few extension services have devised effective methods for evaluating the performance of field workers. The most common evaluation procedure is based on the supervisor's confidential report, even though the field staff may be left unvisited and unobserved for a long time. This leads to a major concern of the workers to impress the supervisor, rather than to provide the farmers with quality service (Moris, 1987).

The evaluation of an extension worker's effectiveness should be based on (Baxter & Thalwitz, 1985):

- the extent to which he is familiar with the technological recommendations suited to his farmers' needs and resources;
- the extent to which his farmers have adopted improved technology;
- the extent to which he has given relevant feedback to researchers about farmers' problems and their reaction to proposed innovations.

To these criteria we would add: the degree of his personal involvement in on-farm trials, the quality of his demonstration plots, his active participation in in-service training programmes.

In Zimbabwe, a prerequisite for promotion is the successful completion of an in-service training programme, consisting of 19 courses to be taken by all extension staff over a period of three years (Hayward, 1987).

Status

RAPA/FAO (1985) recommend that the status of agricultural extension workers should be equated to that of agricultural researchers by: (a) professionalising titles and giving the equivalent designation for corresponding ranks/grades; (b) providing appropriate incentives and recognition; and (c) offering equal opportunities for professional advancement.

Some countries are already trying to make status and salaries of extension workers comparable to those of the researchers.

NUMBERS OF EXTENSION WORKERS AND LOGISTIC SUPPORT

A difficult problem is that of determining the minimum number of extension workers required to make a significant impact on the agriculture of a region. The desirable ratio of extension workers to farms will, of course, depend on specific conditions prevailing in each region, such as potentialities for increased production or diversification, the technical and educational level of the farmers, the density of the farming population, the mobility of the extension worker, to name just a few.

A tentative and rule-of-thumb conclusion, established at a conference on extension in East Africa, sponsored by FAO, is that the minimum objective should be a ratio between 1:350 to 1:1000 (FAO, 1962).

The actual ratio varies considerably from country to country: in India and Zambia it is about 800, in Bolivia, over 8000 (FAO, 1975). Actual contact rates may be far lower: they amount to only about 2·5 per cent of all farms in Paraguay and 3·3 per cent in Ecuador (Herzberg & Antuna, 1973).

In many cases in which extension programmes have been most successful (the Gezira in Sudan, Comilla in Bangladesh, and Monkara in Chad) these have been carried out with a massive investment in a limited region, often to the neglect of other parts of the country (FAO, 1976).

In general, it is possible to say that in most LDCs, the number of extension workers is so disproportionate to the number of producers that it prevents extension programmes from reaching or significantly influencing the majority of the farmers.

Ratio of SMSs to VEWs

Another factor to consider is the ratio of SMSs to front-line extension workers. World-wide this ratio is: Europe, 1:12–1:13; USA, 1:20; Africa and Asia, 1:16–1:17. Since the SMSs provide the essential link between researchers and extension workers, this wide ratio may be a serious constraint in the expert advice and training provided to the village-level workers (Claar & Bentz, 1984). These authors further point out that most SMSs in the developing countries hold only a first university degree, and frequently lack experience. In the developed countries they are generally experienced professionals with advanced degrees.

Chapter 21

The Adoption of New Technology

THE DIFFUSION OF NEW TECHNOLOGIES

'For someone to change, he/she must want to know how to, and be able to' (Galjart, 1971).

Adoption of a new technology must be preceded by technology diffusion, e.g. the act of making new technology known to the potential adopters. Diffusion is therefore the link between R & D and adoption. Effective diffusion is an essential but not sufficient condition for adoption. The farmers of a given 'target category' must not only be made aware of an available technology, they must also be convinced that adoption is in their best interests (Pinstrup-Andersen, 1982), and above all, they must *be able* to adopt the proposed technology.

Diffusion research has played a central role in extension theory and practice. The early conclusions drawn from this research and their application to extension practice have been misleading and the results disappointing.

Extension practice still continues to make heavy use of the former conclusions of diffusion theory – 'the old tenets continue to be basic fodder for extension trainees' (Rogers, 1983), even though their originator has changed his views (Rogers, 1962).

Diffusion research has led to the development of new approaches in extension, especially the use of 'target categories' (Röling, 1987).

The Rogers model of diffusion of innovations

Until quite recently, the generally accepted model of diffusion of a new technology within a farming community was that described by Rogers

(1962) (Fig. 23). According to this concept, an innovation is first adopted by a very small group of people – 'the innovators', who are sufficiently educated to realise the potential for profit of the innovation, who are sufficiently rich to invest in the inputs required, and can afford to take the risks involved. After the innovation proves to be a success, the innovators are followed by the 'early adopters', namely those who are less prone to take risks, but can afford the expenditure involved in applying the innovation. The new technology spreads at an accelerated pace, and the bulk of the farmers follow suit—they are called the early and the late majority, respectively. The last to adopt are the 'laggards', the poorest, eldest, most conservative and the most averse to take risks.

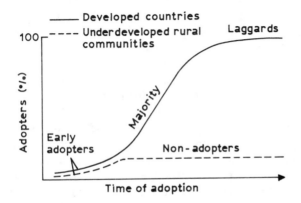

Fig. 23. Contrasting patterns of adoption of agricultural innovations in developed countries and in developing countries.

The progressive farmer strategy

The gradual acceptance of innovations by more and more people has led to the inference that people differ in their willingness to accept new ideas. In other words, it has been inferred that a psychological trait is involved which has been variously called progressiveness, innovativeness, venturesomeness or a willingness to take risks (Fliegel, 1984). It has led to the 'progressive farmer strategy', also called the 'trickle down approach' which was thought to allow one extension worker to cater for hundreds or even thousands of farmers (Röling, 1984).

From this concept two practical conclusions have been drawn for extension strategy: (a) it is useless for the extension worker to devote himself to the laggards, who will either not adopt an innovation or be the last to do so; (b) extension efforts should be concentrated on the innovators, who are most open to new ideas and have the means to do so. Once these have adopted the new practice, the others will follow in any case.

Unfortunately, this 'practical' approach has the greatest appeal in those countries which are poorest in extension resources—human and material. 'Diffusion is generally seen as a God-sent process which assures the trickle down of income and welfare generating ideas and which thereby guarantees their distribution among all members of a population' (Röling *et al.*, 1981).

The progressive farmer strategy gives an appearance of justification and legitimacy to the tendency of extension workers to spend most of their time with the large and prosperous farmers, with whom they feel socially more compatible and with whom it is far easier to achieve visible results and recognition. In several Latin American countries for example, where 1– 2 per cent of the farm families control over two-thirds of the good farm land, they receive an even higher proportion of the extension service's efforts (Barraclough, 1973).

Progressive farmers are eager for extension advice, they *demand* assistance. They complain if they feel they are neglected and are powerful enough to affect the career of the extension worker (Röling, 1987).

The current practice of extension services to provide intensive assistance to a small number of innovative, educated, and generally well-off farmers is strengthened by the following factors (Röling *et al.*, 1981):

- the impact on agricultural production, in the short term, is greater than when the extension effort is diffused over a large number of smaller farmers, because the large farmers are the major potential producers of commercial crops and export earnings;
- the progressive farmers are eager to follow advice, and not much time is required to convince them of the desirability of an innovation;
- they have the means to try out new ideas;
- they have the political clout with which to endanger the career of workers who bypass them;
- they provide an intellectual challenge to the extension worker who also learns from them what to tell others.

The validity of the progressive farmer strategy

Efforts to explain adoption behaviour in the past have focused on innovativeness as an individual trait, and on other personal attributes which tend to coincide with innovativeness (Fliegel, 1984).

Since the early 1970s, the realisation has grown that the focus on individual characteristics is misleading. Galjart (1971), from his studies of smallholder farmers in Brazil, was amongst the first to suggest that in addition to reluctance to adopt innovations, one should consider the inability to adopt new practices. Other extension specialists started questioning the validity of the 'trickle down' approach. Instead of the innovativeness of the early adopters, it was their higher access to resources that was emphasised; at the same time, the so-called 'laggards' were shown to be people with little or no access to resources (Engel, 1976). Blaming people for lack of innovativeness was replaced by blaming the system (Caplan & Nelson, 1973).

Havens & Flinn (1975) showed that smallholder farmers who could not obtain credit, were unable to take advantage of the introduction of an improved coffee variety, because they were unable to wait three years for the plantings to come into production, and were therefore left behind. They were quite willing to adopt the new variety, realising its potential benefits, but were unable to do so.

A Kenyan agricultural extension worker, cited by Röling et al., (1981), explained the problem very aptly: 'My farmers are too poor to follow my advice'.

Besides the inability of subsistence farmers to devote any part of their meagre resources to the purchase of inputs required for innovations, and their inability to take risks, one should add the limited access of the small farmer to information in a highly hierarchical society, dominated by a small clique of influential farmers, such as prevail in the rural areas of many developing countries.

It may even be counterproductive to introduce innovations through the most innovative farmer in a community. In order to maintain the existing social status, as well as the economic advantages accruing to the innovator, the early adopter may very well be motivated 'to withhold, delay, or distort information, if such behaviour is to his advantage' (McAllister, 1981).

Research workers were also becoming aware that a major reason for the lack of impact of research work on the majority of the farmers in the LDCs was due to the exclusive preoccupation of the extension service with the large farmers. Arnon (1987), from his observations as a consultant on agricultural research in a number of LDCs in Asia, Africa, and Latin America stated:

> Unfortunately, it is in the Third World countries that the diffusion of innovations does not, as a rule, follow the pattern implied in the progressive farmers' strategy. In these countries, the village community consists generally of a minority who are able to adopt an innovation, and need only to be made aware of its potential benefits and *a vast majority of small farmers who lack the means needed to make adoption possible*. Even after the progressive and richer farmers in their area have adopted a new practice, they do not follow suit. Instead of a minority of laggards, there is a majority who are *non-adopters*, not by choice, but by force of circumstance.

Röling (1982), in summing up the experience of agricultural extension in LDCs, expressed the same thought in even stronger terms: 'In rural communities that are very differentiated in terms of access, it is nonsensical to expect innovations to diffuse from high-access farmers to low-access ones. In general, innovations will only diffuse within groups of people who are homogeneous in access.'

The strategy of favouring the strongest and most progressive farming sector has had a disruptive influence in developed countries, too. In the USA, the 'extension doctrine' explicitly demanded that extension personnel should work within the power structure of local communities, and focus their efforts on the larger, more progressive farmers in order to achieve rapid progress.

As a consequence of this strategy, it was the larger and more efficient farmers who were the first to adopt new technologies, in particular labour-saving innovations. They were therefore motivated to expand their farming operations and enabled to do so. The smaller farmers could not compete and were forced out of agriculture, a process that accelerated after the Second World War.

In the USA, fortunately, the expanding industrial and service sectors were able to absorb most of the redundant labour from the rural sector.

The need to change extension strategy

Notwithstanding the findings described above of the lack of validity of the 'progressive farmers approach' in the heterogeneous rural societies of the developing countries, the extension services, in nearly every country in the world, continue to limit their efforts to the so-called innovators or progressive farmers (euphemisms for the larger and wealthier farmers).

A major difficulty encountered in attempts to reorient extension efforts to the small farmers, mentioned by Röling (1982), is the fundamental disbelief of the extension workers in the ability of small farmers to innovate. The field workers may pretend to acquiesce in the new policy, but continue to select as contact farmers people 'who could not be called small by any stretch of the imagination'.

This is not exclusively a problem of the LDCs. In the Netherlands, for example, 65 per cent of farm visits by extension workers are to 10 per cent of the farmsteads, and the propensity of farmers to ask for advice rises with farm size (Haverkoort, 1976). Röling (1982) recounts that an effort by an agricultural administrator to get his field workers also to visit a few smaller farmers every week failed completely, as the field workers soon saw their prejudices confirmed and refused to continue the assignment.

In Israel, with its declared policy of following a strategy of equity, the extension service still uses the trickle down approach. In a study on agricultural development of settlements in a formerly desert region in the southern part of Israel, Arnon & Margolies (1984) found that 68 per cent of farmers received no or little advice from the field workers and 41 per cent had not introduced any innovations on their farms since they were settled on the land some 25 years before. When these findings were presented to the Regional Head of Extension, he responded to Arnon's indignation at this situation, by stating that it was the declared policy of the service not to spread their limited resources too thin, and to concentrate their efforts on the progressive farmers, relying on the diffusion process, as 'did every other country in the world'.

Conclusions

Following the new insights described above, it is no longer possible to justify extension strategies which make it possible for the rich farmer to become richer whilst the poor farmer remains poor or even becomes poorer.

The progressive farmer strategy is now seen as a contribution to the rapidly increasing gap between the privileged and the underprivileged sectors of rural society. In fact, the question is being asked whether extension is not contributing to poverty (Röling *et al.*, 1981).

Swanson *et al.* (1984) include research, as well as extension, in this indictment. They write: 'If research and extension continues to focus on increasing agricultural production, particularly by increasing the output of large-scale farmers or units, they are likely to contribute to more acute rural poverty, hunger and/or forced rural/urban migration of small farmers.' It is certain that no national research or extension service desires to be the cause of such negative and antisocial developments; in order to avoid doing so, they must be committed to pursuing a balanced approach that is based on both increased production and social equity.

Developing countries cannot afford to allow the non-adopters to stagnate, both for economic and social reasons. A two-pronged extension strategy must therefore be adopted: conventional methods can be followed for those farmers *able* to adopt innovations, whilst special efforts must be designed to make it possible for the disadvantaged sector to follow suit.

CATEGORISATION

Homogeneous target areas

The most persistent, vocal, and reasoned advocate of a new approach to the problem of diffusion of innovations is Röling (1984, 1987). Following the observation that 'in heterogeneous rural societies horizontal diffusion could easily take place while vertical diffusion would not' (Fig. 24),

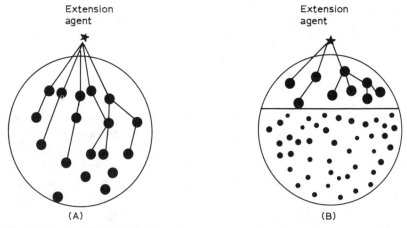

Fig. 24. Two models of diffusion of innovations. A. In a homogeneous population (with equal access to resources). B. In a heterogeneous population (with differential access to resources).

Röling (1984) stated unequivocally that the 'progressive farmer strategy' can be used only in categories of farmers which have been carefully identified as homogeneous, and with innovations which have been developed to suit the characteristics of those homogeneous categories. Hence the need to develop new approaches based on the use of 'target categories' or 'homogeneous social systems'. These have also been called 'recommendation domains' (CIMMYT, 1979) and 'diffusion domains' (Wotowice *et al.*, 1986).

Major categories of farmers

The need to reach the major socio-economic categories of farmers, including small farmers, women, young people, ethnic minority groups, etc., is increasingly recognised in recent years. Swanson *et al.* (1984) wrote that instead of working with selected progressive farmers in a community, and assuming that technology will trickle down, it is important to recognise that there are different socio-economic categories in rural communities, and that 'their circumstances are sufficiently different to mean that appropriate technologies must be developed and adapted by agricultural research, and then targeted, through appropriate extension programmes to reach these major categories'.

Targeting agricultural research and extension on the disadvantaged sectors of the farming population can only be *one* of essential measures. While horizontal diffusion can take place in a relatively homogeneous category, the ability of these sectors to adopt appropriate new technologies will still depend on additional conditions, such as access to resources, market opportunities and adequate infrastructure.

Röling (1982) suggests that instead of grouping farmers as progressive or traditional, large or small, it would be more accurate to group them into 'high-access' and 'low-access' farmers, thereby recognising the fact that farmers have different access to resources such as land, water, credit, inputs, markets, advisory and other services.

Extension strategies favouring the underprivileged rural sectors

There is a serious political will in many developing countries to adopt policies for the advancement of the underprivileged rural sectors.

A redefinition of extension priorities in the light of these policies requires 'a reappraisal of the concepts, philosophy, and effectiveness of conventional extension services' (Lüning, 1982).

Attempts to reorient the extension activities in favour of the under-privileged sectors in the rural areas have taken different forms in different countries. However, whichever method is adopted, it must incorporate the following elements if it is to be successful (Röling, 1982):

- identification of the relevant homogeneous target groups;
- assessment of the requirements and potentials of the various target groups, learning from below;
- participation by local target groups, and helping them to develop into grassroot organisation;
- ability to take decisions quickly, under a unified or co-ordinated command. This requires the decentralisation of powers and functions from higher levels of administration to the field levels, effective co-ordination among agencies at all levels, freer communication among various hierarchical levels and better information and monitoring systems (Mathur, 1983);
- access to a research system that is capable of coping with the problems relevant to the target group;
- ability to reach various hard-to-reach target groups, such as women;
- political commitment and support;
- a staff recruited from among sons and daughters of small farmer families.

Without creating 'constituencies' among the disadvantaged farming sectors, one cannot develop effective service agencies (Sabatier & Mazmanian, 1978). This implies that a research–extension system, to be effective, requires that the 'users' form a system which can exert some countervailing power over the agencies serving them (Röling & De Zeeuw, 1983). In other words, there is need for a formal procedure whereby farmers' representatives can ensure that their interests are duly taken into account by extension and research. Such a procedure is not only to the advantage of the farmers concerned, but also essential to ensure the effectiveness of the extension and research effort.

Leonard (1977), from his studies on the extension service in Kenya, is convinced that the substantial bias of agricultural extension services in favour of the wealthier and more progressive farmers can be lessened if ministry of agriculture decision-makers show the necessary political will. The measures he proposed are: (a) a carefully redefined extension strategy; (b) guidelines for working with the middle and bottom rungs of farmers; and (c) a greater emphasis on group activities.

Extension strategies cannot be divorced from other administrative efforts to improve the economic situation of low-access farmers, such as the articulation of appropriate national economic possibilities, especially on remunerative producer prices, efficient marketing arrangements and realistic exchange rates (Ramakrishnan, 1983). Many other measures need to be taken, which are beyond the scope of this book (see Arnon, 1987).

In brief, the categorisation of the extension clientele into homogeneous extension domains is essential as a means of achieving a political reorientation in favour of the low-access farmers, and also as an essential prerequisite for effective extension work.

THE MAJOR TARGET CATEGORIES

There is major agreement that extension activities must be client-centred, and must therefore be guided by clients' needs; it is also widely appreciated that the extension service, as a public sector agency, has an obligation to serve the needs of all the multiple client-groups found in rural populations, both in developed and developing countries (Swanson & Claar, 1984). Because the rural sector is extremely heterogeneous, extension must concern itself with the problems of each major client-group separately, for the reasons detailed in the previous sections.

High-access farmers

Innovativeness
In relation to innovativeness, high-access farmers can be found on a continuous spectrum, from innovators up to, and including laggards.

As a general rule, almost all high-access farmers will, in due course, adopt new technologies that have proven to be profitable.

A unique category of 'laggards' are very large landowners, who are so rich, and whose income from the land is so great, that they have no motivation to adopt new practices. Typical of this group are many latifundistas in the Latin American countries, who control vast areas of farmland and rangeland.

In the words of a Latin American participant at an international workshop on the adoption of new practices in agriculture: 'Why should our farmers (sic) attempt to increase the yields of wheat?—to change their Cadillac every year instead of every two years?'

The owners of the latifundias tend to transfer the income they extract from the land to urban areas, for industrial or financial investment and for luxury consumption (Yudelman *et al.*, 1971).

Extension strategy

It makes high economic sense for the extension service to devote part of its efforts to high-access farmers: the impact on total agricultural production can be rapid and considerable; high-access farmers are major potential producers of marketable food-crops and of export earnings; they are generally eager to follow advice; by definition, they have the means to try out new ideas and to take a reasonable amount of risk; they have political clout and they provide an intellectual challenge to the extension worker who can also learn from them what to tell others (Röling *et al.*, 1981).

For this category of farmers, the trickle-down approach is completely adequate, and the investment of a relatively small extension effort using modern and sophisticated extension methods, can be extremely productive.

Effect on other rural sectors

Unfortunately, improving the agriculture of the high-access farmers, can entail negative effects on the less-privileged sectors of the rural population, as was amply exemplified in the countries in which the 'green revolution' has occurred.

Though the 'green revolution' has relatively few economies of scale, it has led to differential effects at farm levels, causing an increase in the disparity between different levels in the rural population of each country. Almost everywhere, it has favoured the rural elite for the following reasons (Dasgupta, 1977):

- it increased the profits and assets of the commercial farmers, and therefore their economic power;
- dependence of the small farmer on the large farmer increased, because most means of production are owned by the latter;
- inputs supplied by government are delivered to the villages through the intermediary of the dominant sector, who own most retail shops for fertilisers, seeds and other inputs;
- the extension village workers tend to maintain contact with the larger farmers, to the exclusion of the disadvantaged sector.

Larger landowners generally have the technical knowledge and the incentives for innovation which the small farmers lack, so that the former are in a far better position to grasp the opportunities provided by the 'green revolution'.

The extremely unequal distribution of income also limits access to education, even at primary level, of a large proportion of the rural population, thereby ensuring a continuation of the vicious circle.

In brief, the small farmer is discriminated against in every area. 'Extension agents concentrate on the large farmers, credit agencies favour the low-risk borrowers, fertiliser and pesticide salesmen prefer the large buyers' (Griffin, 1973).

The problems outlined above should not be construed as an indictment of the 'green revolution', without which agriculture in the countries in which it has been a success could not have broken out of its stagnation. They do, however, highlight the need to adopt policies that will improve the ability of the small, impecunious farmer also to benefit from new opportunities provided by agricultural innovations.

Low-access farmers

Governments in the LDCs, United Nations, the World Bank and other organisations are increasingly extending support to small farmer development, both on grounds of equity to redress the imbalance created by earlier efforts at modernising agriculture, and for sheer expediency, to improve agricultural output by a large sector of the rural population (World Bank, 1975). This is of course reflected in the policies of the extension services.

Orienting efforts to serve small farmers is often a cost-effective way to increase production of most commodities and regions. There is considerable empirical evidence, especially in Asia, that smallholdings have the capacity to achieve higher productivity from scarce land resources. The example of Taiwan has been mentioned in a previous section. In the African context, experience has also shown that over a wide range of mixed farming, in lands of medium to high potential, smallholdings have proven more productive than large units. Kenya's rapid agricultural growth, for example, is a proof of the desirability of promoting smallholder agriculture. Orientation of agricultural efforts to small farmers, is therefore not only equitable, but can also be an economically efficient policy (Ramakrishnan, 1983), provided efforts are sharply focused on this category.

Definition
By whatever name low-access farmers are called (small farmers, peasant cultivators, subsistence farmers, poor farmers, disadvantaged rural sector, etc.), they can be defined as a category of farmers 'which seems to have been left behind in the normal process of development and for whom special efforts are needed by governments as an essential element of poverty alleviation strategies' (Ramakrishnan, 1983).

Paradoxically, part-time farmers are frequently in a better position to adopt new technologies than subsistence farmers whose only income is from their farms. Income from off-farm sources can provide the cash required to buy yield-increasing inputs to which subsistence farmers generally have no access.

Extension strategies
It is impossible for field extension staff to conduct a dialogue with millions of individual farmers. They can be reached more easily through their own organisations, and governmental agencies should help them in setting up such self-help organisations. Existence of such participatory organisations will also create external pressures on government agencies to be more responsive to the needs of their clients (Mathur, 1983).

Since the disadvantaged sectors are mostly illiterate and resigned to their marginal position it would be illusory to expect them to take the initiative and organise themselves successfully for group action.

External initiatives and funds are essential to start these groups, in particular to train group organisers and pay their wages until the groups become self-sustaining.

Experience has shown that if a peasant group functions satisfactorily, people in neighbouring villages become interested and ask for help from local leaders from the successful village (Van Heck, 1979).

The major problem encountered in these endeavours is to provide guidance and assistance to these groups whilst at the same time strengthening their ability to decide and act for themselves.

Technical information for the low-access farmers should be based on the results of Farming Systems Research (FSR). The extension personnel should help to identify representative farmers and should be involved in these research activities.

Swanson *et al.* (1984) recommend the formation of technical committees comprising research personnel, SMSs and other senior extension workers, which will determine, on the basis of the findings of the FSR, what technical recommendations should be promoted during the next growing season. Representation of supply agencies and/or input dealers, as well as representatives of the agricultural development bank and/or credit societies should also be invited to attend the meetings of the committee.

Recommendations resulting from the deliberations of the technical committees should be made available to the extension workers, so that they can plan a series of on-farm demonstrations in their respective service areas.

At the conclusion of the growing season, these demonstration plots should be used for farmers' field days, as well as to collect data for the next year's educational campaign.

In addition to working with contact farmers and the demonstration plots, extension workers should also carry out other extension activities, which should be planned as part of an overall campaign that uses an appropriate combination of mass, group, and individual extension methods. The chosen combination will depend on the type of technical recommendation being promoted, the types of farmers being served, and the stage reached by the farmers in the adoption process.

In terms of extension strategy for low-access farmers, the first innovations to be introduced should deal with the staple food crop(s) of the area.

Difficulties

Governments often find that programmes involving the small farmers the most difficult to make effective (World Bank, 1983). Concern for this sector is a relatively recent phenomenon, and there is little previous experience to guide extension services in their operations. Administrative methods appropriate for other general programmes for the rural community as a whole often fail to reach low-access farmers as a separate target group.

The conduct of development operations continues to be hindered by 'such characteristic features of bureaucracy as overcentralization, proliferation of units and staff, absence of effective co-ordination, endless interterdepartmental rivalries and conflicts, lack of accountability, little or no incentives, slow decision-making processes, rigidity of rules, preference for *status quo*, etc.' (McCallum, 1981).

Efforts at improving performance of agencies serving the small farmers are relevant only in the context of a prior political commitment to improve the conditions of the disadvantaged rural sector; without such commitment, 'such efforts are not likely to be effective or sustained over a period of time' (Ramakrishnan, 1983).

Even with a serious political will, the actual ability of the extension staff to serve the mass of the small farmers is often limited by factors beyond their direct control. Appropriate research results may not yet be available, markets far away, prices for food crops may be too low, inputs and seeds may not be available, etc.

WOMEN AS A TARGET CATEGORY

Female labour participation

In most developing countries women are an essential part of the farm labour force. Females as a percentage of the total agricultural labour force were estimated by the International Labour Office as follows (ILO, 1982):

Sub-Saharan Africa	47·2
North Africa, Middle East	25·2
South and Southeast Asia	40·2
Latin America	19·0
Caribbean Region	54·0

These figures indicate that women's participation in agricultural production is greater than generally assumed, and that neglect of their role may jeopardise the realisation of development goals.

The nature of the work done by women is largely determined by physical and economic factors, as well as by the traditions, concepts and social values of different societies.

In Africa south of the Sahara, women are the primary work force on small farms. They do most of the hoeing, weeding, harvesting, transporting, storing, processing and marketing. They share equally with the men in planting and the care of domestic animals. They use few, if any, modern tools or implements. In Asia, women provide much of the work for the staple crop—rice. In the Near East and in Latin America, they are also active on the farm, but to a lesser degree than in Africa and Asia (FAO, 1986b).

It is generally assumed that the husband and father is the head of the family, and responsible for the management of the family farm on which he does most of the work. However, in many rural areas in the developing countries, migration of men to the towns, plantations and mines in search of off-farm work has resulted in the women becoming the actual heads of the family and the mainstays of small-scale farming.

It is estimated that 18 per cent of the rural households in the developing countries are headed by women, and that in a number of countries the figure may be as high as 40 per cent (FAO, 1984).

But even when the men remain on the farm, the women may be carrying an unfair share of the work burden. In Africa, for example, in addition to typical women's work (bearing and raising the children, preparing the meals, carrying water and collecting fuel) women may be carrying up to 70 per cent of the farming work-load. Wherever this is possible, the men produce cash crops, disposing as they wish of the cash income, whilst the responsibility for growing the food for the family rests entirely on the woman.

The role of the women in farming has generally not been recognised and adequately rewarded. The extension workers, with the exception of those concerned with home economics, rarely discuss farming problems with the women, and only exceptionally do the women receive training in crop production. Where women are the heads of the household, 'they encounter more difficulties than men in getting access to land, credit, technical services and commercial market outlets' (FAO, 1984).

Women's workloads further increase, as a rule, with the intensification of agriculture if the tasks involved are manual; men's roles increase if the tasks are mechanised (FAO, 1984).

Categories of rural women

Rural women, as such, cannot be treated as a single category. Ahmed (1985) identifies four broad groups: women from landless households, from small cultivators' households, large farms, and female-headed households.

Women in all four categories have in common the roles of reproduction, child rearing and household maintenance, but the need for, and the impacts of, technological change vary between the groups, so that different policies for each are required. The most disadvantaged, requiring the most urgent attention, are the female-headed households. The proportion of rural households with female heads was found to range from 25–35 per cent in several countries like Sudan, Ghana, and Malawi (Youssef & Hetler, 1984).

Women's influence on farm decisions

In certain societies, the wife has an important influence on the farmer's decision-making. The wife is influential in the allocation of resources for agricultural production: she often controls certain inputs, manages certain production processes and may control the disposition of certain outputs (Rockefeller Foundation and ISNAR, 1985).

Constraints faced by women in the adoption of new technologies

Many of the factors that limit the ability of subsistence farmers to adopt improved technologies also affect rural women, only more so.

Carr (1985) ascribes the limited impact of new technologies on the welfare of rural women to the following factors†:

Neglect by the extension service
Women as a target group are generally neglected by the extension services, with the possible exception of advice on home economics. 'After 30 years of freedom, not only have over 95 per cent of the women in a backward part of India remained illiterate, but the extension structures of nearly every other development department have hardly touched them, excepting one—to "execute" a sterilisation programme' (Jesudasan *et al.*, 1980).

Information is rarely designed for, or aimed at women, on the assumption that men are the decision-makers and income earners in the family, and that therefore any advice communicated to them will somehow get through to their wives. This is true even for farming systems in which women contribute more labour than men, and also for specific activities carried out by female labour. This bias can lead to many absurdities, such as the following:

In Liberia an attempt was made to introduce simple, low-cost hand weeders. For effective use, these require that the rice be planted in straight rows. This principle was explained to the men, but it is the women who do the planting, and this is traditionally *not* done in straight rows. The innovation, which could have been most beneficial, was therefore never adopted (Carr, 1985).

†Though Carr's paper refers specifically to rural women in Africa, most of her generalisations are applicable to rural women in most Third World countries.

In a study of maize storage in Nyanza district, Kenya, it was found that chemical insecticides for improved storage were made available to men, who were also instructed on their proper use. However, women are responsible for grain storage, so that only 4 per cent of the women knew how to apply correctly the insecticide, with the resultant high risk of its application by the majority of users (Kumar, 1985).

Technological innovations of importance to women, which could lighten their work load considerably, such as improved water pumps and home crop processing equipment, become unusable after a time because the women are not trained in the proper operation and maintenance of the equipment. This leads to frequent breakdowns and a high incidence of equipment that remains unrepaired. The women have no cash to pay for the repairs and the men have no incentive (Carr, 1985).

In Gujerat State (India), in the 1960s, the Amal Dairy Co-operative found that when men attended their buffalo management courses, the results in terms of improvement on the farm were minimal. In contrast to the previous examples, the men in this case were persuaded to let the women come to specially devised courses for women only, and the results were most positive. Not only was there an improvement in animal health and productivity, but the women were exposed to ideas and knowledge which led to an improvement in the quality of their lives and that of their families (Cherry, 1987).

Rural women have very little opportunity to learn about crop production, animal husbandry or co-operatives.

A study conducted in Africa found that farms with an adult male present were four times more likely to have a member trained at a farmers' training centre, than farms without an adult male, and 14 times more likely to have received information about agricultural credit (FAO, 1984).

Access to credit

Provision of credit is almost exclusively made to men, thereby ignoring the independent roles of women in dual (husband–wife) households, and the high proportion of female-headed farmsteads (Doyle *et al.*, 1985).

The major reasons why credit is less available to women are: (a) they have no land-titles as collateral (land-titles are generally held in the men's name), and (b) the credit is frequently made available through co-operatives of which membership is mostly male. Lack of credit prevents women from investing in equipment and inputs that could alleviate the drudgery of their daily tasks, improve their productivity, and/or provide an additional source of income with which to improve the welfare of the family (Carr, 1985).

Time constraints

The many work responsibilities of women, which are extremely time-consuming, may be very important factors in determining the acceptability of a new technology. This consideration is especially relevant for Africa, where women may be primarily responsible for food crop production. However, in Asia and Latin America the work load of the women may also have a strong influence of the type of technologies for small farms that are chosen (Rockefeller Foundation & ISNAR, 1985).

Exclusion from the development process

Agricultural development projects are frequently planned with little or no consideration of the role of women or the possible negative effects of such projects on the welfare of the women and their children. The projects are designed as if the husband/father is the breadwinner and the wife only has domestic responsibilities, completely ignoring that in many developing countries women till their own fields and/or have their own sources of income. This assumption has led, for example, to irrigation projects, in which the men are provided with land and water, to be used mainly for cash crops, whilst the women are deprived of their fields and kitchen gardens which are crucial for their subsistence production (Dey, 1981).

Numerous examples of such 'blindness on the part of male and urban and foreign planners toward the important role and activities of rural African women' have been reported (Stevens, 1984).

Lack of authority

Another constraint faced by women who head the homestead, is that while they are left with the responsibility for managing the farm, they are generally without authority for changing the farm's operations.

If, perchance, a wife learns of an innovation she would like to try, she may be unable to do so without her husband's permission, especially if a cash expenditure is involved (Youman & Holland, 1983).

Inappropriate technology

The choice of a technology that is inappropriate for women has often resulted in a worsening of their conditions. For example, the introduction of large-scale palm-oil mills in Nigeria took the processing of the palm fruit out of the hands of the women. At first sight, this is a positive development, as it lightens the work load of the women. Actually, it deprived them of a source of income they used to derive from the home processing of the fruit, which adversely affected the nutritional standards of the family.

The need to eliminate the neglect of rural women

Though the neglect of rural women is largely the result of deeply rooted prejudices regarding the role of women in traditional societies, these attitudes should not be allowed to perpetuate a situation that is an obstacle to progress. The need to eliminate discrimination against rural women is being increasingly recognised since the 1970s. The work of Boserup (1970), the first to treat the subject in depth, was followed by a flood of papers and articles on the subject (Ahmad, 1981).

The neglect of rural women in development policies in general, and extension activities in particular, has resulted in a very limited impact of potentially beneficial technologies and the development of technologies that are neither acceptable nor useful to rural women, or respond to a high-priority need (Carr, 1985).

Safrenko (1984) gives an example from Gambia of the positive effect that can result from providing rural women with training and essential inputs. The Ministry of Agriculture, in response to requests by women to be involved in cash crop production, an area traditionally assigned to men, set up a special project to train 30 women in the production of onions for export. The ministry, together with the Gambia Co-operative Union, provided fertiliser and seed and undertook to purchase the crop at an agreed price. Within two years, the project grew to involve 900 women, providing them with incomes and, more important still, the possibility of joining co-operatives.

Swanson *et al.* (1984) summarised the situation in regards to rural women as follows:

> . . . it is essential that the technology and related needs of this important group of food producers be addressed by those institutions serving the agricultural sector (especially research and extension) so that women farmers should not be further disenfranchised from the development process. Women farmers should be considered as an equally important target for extension services as other small farmers, young farmers, and other significant categories of farmers.

Changes in policy essential

The Lagos Plan of Action for Africa calls upon the governments to recognise that women are a vital element for solving the food crisis and to 'upgrade their skills and lessen their labours by appropriate measures'. Policies have to be adopted to improve women's access to land, credit, inputs and supporting services; to increase their participation in rural organisations, and to provide off-farm employment (Ghonemy, 1984). Of particular importance is the need to ensure equal access for women to cooperatives and credit facilities.

Research policies

Research workers should be encouraged to consult rural women before designing new technologies, in order to ensure that the research and development efforts are beneficial for women too, and not detrimental to their interests (Carr, 1985).

The national research systems, as well as the international centres should develop a long-term strategy that also involves consideration of women's interests (Rockefeller Foundation & ISNAR, 1985).

Extension policies

The introduction of new agricultural practices, technical assistance, income-generating projects (such as cash-cropping) have often widened the gap between men and women (Table 6).

TABLE 6
Areas of access to non-formal education by gender

Area of activity	Percentage of participation	
	Men	*Women*
Crop production	85	15
Animal husbandry	80	20
Co-operatives	90	10
Arts and crafts	50	50
Nutrition	10	90
Home economics	0	100

Source: UNECA/FAO (1974)

Special efforts are needed in agricultural education, extension and training programmes for rural women. Traditional home economics programmes should be broadened to include agricultural and income-earning skills.

Women advisers in the extension services are few, nor is much work aimed directly at women, in spite of their important role in agriculture. Extension services should make a greater effort to attract female workers. This in turn would motivate young women to enrol in agricultural schools, colleges and faculties of agriculture, to prepare themselves for extension careers (Swanson & Claar, 1984).

In some countries a certain amount of progress is being made. In the Philippines, for example, where women are often the leaders in the farm household, extension agents work with them directly, and in certain areas up to one-third of the extension workers are women (Elkana, 1987). Other countries, such as Thailand, Lesotho and Egypt, are training female extension workers.

Other alternatives are being adopted: training female paraprofessionals (Mali), training women to carry out their own demonstration plots (Zaire), and channelling extension activities through women's groups (Bangladesh) (FAO, 1984).

The following recommendations were made at the CGIAR Seminar on Women and Agricultural Technology in 1985 regarding extension service activities specifically aimed at women (Rockefeller/ISNAR, 1985):

- upgrade the skills of extension workers in improved technologies for women;
- wherever possible, train male and female extension staff to work with both male and female farmers;
- set up village-based training courses with men and women farmers, involving practical field work;
- train women in accounting, bookkeeping and management skills;
- develop a two-way dialogue between extension workers and men and women farmers.

An original proposal by Carr (1985) is to include a nutrition/health component in the extension advice aimed at *men*, to help them understand the importance of these aspects so as to be more supportive of women's activities in these fields, as well as understanding the importance of projects aimed at reducing the work loads of women, or increasing their income.

Development planning and implementation

The role and needs of women in designing development programmes and institutional structures are frequently neglected.

The involvement of women in all development efforts will help to: (a) influence thinking at policy-making levels; (b) avoid a negative impact of the projects on women; and (c) increase women's self-confidence and desire to play an active role in the development process.

EXTENSION WORK WITH YOUTH

Rural young people are another client group that have received too little attention from the extension services in the developing countries.

Swanson *et al.* (1984) in reviewing the historical development of agricultural extension in Europe and North America, conclude that the application of science to agricultural production was closely paralleled with formal and non-formal agricultural education opportunities for young people. They state that 'it appears very short-sighted to expect today's youth in the Third World to continue progress towards developing a strong agricultural sector and to develop positive attitudes towards agriculture and improving life in the rural areas if they have little or no access to educational programmes that will increase their agricultural skills and knowledge'.

In the rural areas of many LDCs, most young people are out of school and engage either in unpaid labour around the house or farm, or are periodically employed in low-paying menial jobs.

Extension could play a major role in mobilising and motivating young people to participate in the rural development of their respective countries.

The potentialities of youth movements and organisations as an educational factor can be exploited for extension purposes.

Experience in many countries has shown that rural youth is very receptive to new ideas and can make a considerable contribution in influencing the attitudes of the older generation.

Youth movements can give rural youth a sense of mission, dedication and values which can give their lives new content. This is the one great advantage that developing countries have over developed countries, namely, they have a plenitude of challenges, objectives and undertakings to offer their youth; the developed countries are relatively poor in this respect, and for this reason are burdened with the problems of a disillusioned youth with all its attendant evils (Arnon, 1987).

Community-based rural youth and young farmers' organisations need to be developed and practical training in agriculture, home economics, group leadership and progressive living need to be provided by appropriate extension education programmes.

In some countries, rural youth programmes have been organised. These programmes include 4-H type rural youth clubs, young farmer programmes that are more occupationally oriented, and a variety of national youth service organisations and youth settlement schemes (Swanson *et al.*, 1984).

Youth clubs

A major problem encountered in organising rural youth clubs is that they generally concentrate on the youth of high-access farmers who already have an educational base to build on, hence their impact is limited and the majority of rural youth are left out of these programmes.

Young farmer programmes

The objectives of these programmes tend to be similar: leadership development, citizenship, personal development, and career or occupational development (Swanson *et al.*, 1984).

In *Japan*, an important feature of the extension service is the farm youth training programme carried out on special farms, of which over 50 have been established (Axinn and Thorat, 1972). First conceived during the depression period of the 1930s, when the need was felt to counteract the effects of the depression on farm youth, the programme has since been expanded and has proven its value in normal periods. Training is given to farm youths with at least nine years of education; the courses are of one to two years' duration, during which the youths live, learn and work together. The programme emphasises the improvement of both agriculture and home life. The number of participants is about 5000 boys and 1000 girls every year.

In *Burkina Faso* the Centres de Formation de Jeunes Agriculteurs have an important role in agricultural and rural development. There are almost as many 'centres' as there are extension agents. They often function as

substitute primary schools rather than as centres for training older teen-
agers in agricultural skills, as the name implies, and as was the initial intention (Benor, 1987).

Youth service organisation

Most developing countries are faced with the problem of the growing
number of young people leaving villages and migrating to towns, with
little or no prospects of employment. Youth services can provide a short-
term, partial, but still important contribution to the solution of this problem.

A solution that has been adopted in many countries is to organise unemployed youth that has congregated in the cities into semi-military
organisations, such as civil conservation corps, workers brigades, development service corps, etc. Besides engaging in the type of public works
mentioned above, the development service corps may be used to open up
new areas and to lay out new farms in which they can be settled.

The young people in the youth services are engaged in programmes
aimed at: (a) training in agriculture, with eventual settlement on the land;
(b) carrying out works relating to rural infrastructure, housing, public
work projects, etc.; and (c) providing services to the community, such as
education and health. The first two programmes aim at providing employment opportunities to undereducated youth, and the third is mainly based
on semi-skilled, skilled and educated youth (Rossillion, 1967).

The youth programmes are generally organised as a branch of the
army; the ranks, disciplinary methods and uniform are more or less the
same and they also include a certain amount of military training. Initially,
they generally aim at creating potential community leaders rather than
large-scale enrolment of unemployed youth, and therefore the services
usually rely on volunteers from among those conscripted for the normal
military service. Some of these services include both men and women,
whilst in most countries there are separate branches for women.

Youth settlement schemes

A number of countries in Africa and Latin America have established
special units in their armed forces to provide vocational training in agriculture with the objective of leading to their resettlement.

The 'Service Civique' of the Ivory Coast is fairly typical. The service
has established a number of training farms, in which soldiers, after com-

pleting their first year of compulsory military service, receive an additional year of agricultural training with a military framework. It was hoped that military conscription and discipline would help overcome the negative attitudes towards manual work prevalent among many young men in African societies (Shabtai, 1975). Based on conscription, the first attempts were not a success; most of the conscripts, after completing their military service, did not return to their villages, but drifted to the towns.

Where facilities for farming are favourable, as in the Sudan Gezira scheme, the results have been far better, with over 70 per cent of the youth trained for farming returning to the farms (Nour, 1969).

THE GAP BETWEEN AVAILABLE KNOWLEDGE AND PREVAILING PRACTICES

In most countries, there is a considerable gap between the yields that can be achieved by applying available knowledge and the actual yields achieved by the farmers. Naturally, the gap is greater in the case of traditional farmers, as compared to farmers using modern inputs and practices.

Examples of the gap between potential and actual yields are presented in the following data.

The results of a comparison between average state yields in the Rio Gran de Sul (Brazil) and yields obtained following the use of a recommended package of practices are presented in Table 7.

TABLE 7
Average state yields compared to yields from improved practices

| Crop | Yields (kg/ha) | | Per cent increase |
	Average	From improved practices	
Rice	2600	7000–9800	169–277
Maize	1100	6000–7400	445–573
Wheat	900	2800–3900	211–333
Soybeans	1200	3000–3500	150–192
Beans	800	1500	88
Potatoes	4600	30000	552
Pasture (meat)	40–90	350	289–775

Source: Malavolta & Rocha, 1981

The differences in yield levels between experiment station, a farm demonstration plot and an average farm in Pakistan are shown in Table 8.

TABLE 8
Approximate yields (units/ha) at a research station, demonstration plot and average farm in Pakistan

Crop	Research station	Demonstration plot	Average farm
Rice (paddy)	125	100	38
Cotton	88	70	20
Maize	250	200	28
Wheat	150	100	23

Source: FAO (1969)

Yields of cassava in Colombia averaged 6 tons per ha on farms, varying from 0·3 to 34 tons per ha between farms, and on experiment stations averaged 50 tons per ha (Pinstrup-Anderson, 1982).

The gaps in yields between developed and developing countries for three major cereal crops is shown in Table 9.

TABLE 9
Gaps in yields of major cereal crops

Crop	(a) Developed countries	(b) Developing countries	
	kg/ha	kg/ha	in % of (a)
Rice	5840	1981	34
Wheat	2196	1348	61
Maize	4681	1366	29

Source: FAO (1977)

The lower yield levels of farmers in general are obviously due to the fact that they do not operate under the same optimal conditions as do the researchers at the experiment stations. A second reason is that farmers are not interested in obtaining maximum yields, which give a lower economic return than optimal yields. Furthermore, the yields obtained by farmers in the LDCs are usually below the economic ceiling, because of their desire to avoid risks, because prices are unpredictable, or because inputs are unavailable. These are legitimate considerations, but still do not explain entirely the enormous gap in yields between modern and traditional farmers.

Reasons for the technology gap as perceived by farmers

The reasons for the technology gap, as perceived by small-scale farmers themselves, and by the agricultural technicians who served them, have been studied by Bradfield (1981) in different localities in the lowland humid tropics of Latin America (Table 10).

TABLE 10
Rankings on production limiting factors, as given by small farmers

Limiting factor	Peru	Brazil	Colombia
Low prices for products	1	5	2
Lack of:			
credit	2	—	5
seeds, fertilisers	3	—	3
feed for animals	—	—	—
water for irrigation	—	—	—
machinery	4	—	4
land	—	—	—
technical help	3	3	—
transportation	—	4	1
Weeds	5	—	—
Poor health	—	1	—
Exploitation by other people	—	2	—

Adapted from Bradfield (1978)

Clearly these small-scale farmers are aware of the need for yield-increasing inputs and for credit to purchase them. Lack of these inputs is perceived as the major constraint experienced by these farmers. Of interest is the fact that lack of physical resources (land, water) is so rarely mentioned as a constraint.

Bradfield found that the technicians who work with these farmers and know their conditions gave an entirely different priority listing of the problems and constraints faced by the farmers, and tend to see physical and technical factors (lack of transportation, lack of technical help) as the most limiting.

The examples presented above, of the considerable gap between potential and actual yields, are not intended to imply that the obtention of maximum yields is desirable or economically justified. However, the size of the gap is an indicator of the economic benefits that can be expected if extension is successful in narrowing the gap substantially.

The time-lag in the adoption of new practices

After a new technique has been developed by agricultural research, its widespread adoption by farmers may be delayed for a more or less lengthy period.

This time-lag is affected by a number of factors:

* the time required by research and extension before they are prepared to recommend the adoption of a new practice;
* the time interval between awareness of the individual farmer of the new techniques and its adoption;
* the length of the diffusion process in the farming community;
* the type of innovation and the interactions between the new inputs;
* environmental factors;
* social, economic and institutional factors.

When should an innovation be recommended?

A dilemma with which researchers are frequently faced is when to recommend the adoption of a new practice. The natural tendency is, of course, to play safe by testing a new variety or new practice for a number of seasons before recommending its adoption. However, delaying the adoption of a new variety or technique capable of increasing productivity in agriculture can be the cause of potential economic loss as great as that incurred in taking the risks of premature adoption. The responsibility for recommending adoption or delay lies squarely with the researcher and cannot be avoided.

However, even if a research worker is prepared to recommend a new practice, the extension worker may be an obstacle.

Even in a highly developed country, such as the USA, a considerable time-lag may occur between 'awareness' by the extension agent of an innovation and his readiness to urge adoption. The reason may be the same as that which prompts the research worker to delay recommending the practice—the wish to be very sure that the farmer is not exposed to risk and, it goes without saying, fear of risking one's professional reputation. An example of such a delay is given by Rogers and Yost (1960):

extension agents in Ohio, after having been made aware of the effects of stilboestrol, a growth-promoting sex hormone feed adjunct for beef cattle, required about two years before recommending the new practice to the farmers. If such hesitations are typical in a modern commercialised agriculture, they will occur still more frequently when traditional farming is concerned.

The question arises whether a procedure can be devised which shortens the time-lag in large-scale adoption whilst minimising risk due to insufficient testing under local conditions.

The following can serve as general guidelines for such a procedure:

(a) If a practice or new variety has proved to be highly successful in one or more countries with similar ecological conditions there is a *prima facie* justification for a shortened procedure of testing in the country contemplating adoption.

(b) The tests should be carried out for one year under as wide as possible a variety of conditions, in experiment stations and on farmers' farms.

(c) If the results are favourable, a decision should be taken on the extent of adoption for the following season: area, number of farmers, region or regions, etc. It should be stressed that whilst a recommendation based on the results of these preliminary tests is the responsibility of research, decisions in this respect should be taken jointly by research, extension and representatives of the farmers.

(d) Wherever possible, government should insure farmers participating in the adoption programme against excessive risks; this can be done by guarantee (1) against loss of costs of production; (2) against lower yields as compared to the standard practice or variety; or (3) of a minimum level of return per unit area.

At the same time as the programme of adoption is being executed, systematic testing and research on the new practice or variety must be continued, aimed at refining the practice or improving the variety, according to local needs and conditions.

An example of taking a calculated risk in the adoption of a new variety, that showed unusual potential, is that followed by the International Rice Research Institute (Los Banos, Philippines) for the now famous IR8 strain of rice. Normally, a selection is made from 3 to 4 seasons of preliminary yield tests, 2 to 3 seasons of general yield trials and 2 to 3 seasons of regional adaptability tests in different parts of the country (Castillo, 1975).

The IR8 strain began to look very promising by late 1965. In spring 1966, the few kilograms of seed available was space-planted and heavily fertilised, producing an about eight-hundred-fold increase. This nucleus seed was distributed to a number of seed-producers in the Philippines and other countries in the summer of 1966. Its performance was sufficiently impressive to justify its official release in the autumn of 1966.

In 1967, over half of the rice grown under controlled irrigation in Central Luzon (Philippines) were planted to IR8. In India, 500 kg of seed was imported in summer 1966; the same year 20 tons were obtained. By 1967, farmers had planted over 60 000 hectares (Cummings, 1971).

THE COMPLEMENTARY NATURE OF TECHNOLOGICAL FACTORS

'Agricultural innovations can be described as separate discrete entities, but their potential typically lies in their appropriate utilisation in combination with a range of other items of technology, conventional and/or newly introduced. Innovations achieve their potential in packages' (Fliegel, 1984).

Agricultural progress cannot, as a rule, be piecemeal. There is little point in introducing varieties if they are unable to develop their potentials owing to lack of nutrients: there is no justification in adopting practices aimed at producing what might become a bumper crop, if disease prevention and pest control are not carried out and a large part of the crop is thereby lost. Living standards will hardly rise if weed control is ineffective because it has to be carried out by back-breaking manual labour.

For these reasons, single-practice programmes, such as introducing irrigation, applying fertilisers, adopting high-yielding varieties, using good seeds, controlling pests, etc., usually give poor results.

Another aspect that adds to the argument against single practices, even when effective, is that at best they provide only small increases in yield over traditional farming and therefore have very little impact. Increases of 10 or even 20 per cent over an average grain yield of 500–600 kg/ha that is usual for traditional farming, are smaller than the normal fluctuations in yield due to climate, and will therefore not even be attributed by the farmer to the improved practice. It is also doubtful whether the increase is sufficient to justify the cost of the input required. In the early stages of development, it is almost essential that spectacular increases of at least double or triple the normal yield be obtained. This is usually not possible with a single practice, but may be reasonably expected with an appropriate 'package' of practices (Arnon, 1987).

Perrin & Winkelman (1976), in a review of the adoption of new wheat and maize varieties in Africa, Latin America, India and Turkey, concluded: (a) differences in adoption rate between small and large farms could only be partly explained by differences in access to information, availability of inputs, marketing facilities, farm size and aversion to risk; (b) the most persuasive factor was the differences in expected yields: Adoption accelerated only when significant increases in yield could be made in the agroclimatic environment of those farmers presently not adopting.

'Package programmes'

Under the so-called 'package programme' approach, the extension service aims at providing an integrated technological proposal, whilst taking care of possible bottlenecks to adoption.

The basic concept underlying the programme is that agricultural progress will be more rapid, and the adoption of improved practices will be more effective if (Malone, 1965):

- A 'package' of complementary, improved production practices, locally adjusted to climate, soil and irrigation conditions, is established by specialists.
- The package deal is applied across an entire farming community, by helping whole groups of farmers in each village to make a break with tradition by adopting the package of improved practices.
- The technical supplies (clean seed of improved varieties, equipment for improving seed-bed formation, fertilisers, and plant protection chemicals) needed for the execution of the programme, and the necessary credits to finance the plans, are made available to the villages in time and in sufficient quantity, together with a number of supporting services, that are required for this purpose. These include suitable transport and marketing arrangements, adequate storage facilities, seed and soil testing laboratories, workshops and credit institutions.
- A general educational programme clarifying the benefits to be derived from the package plan, including demonstration plots, is executed. Each demonstration is carried out on two plots in a farmer's field, on one of these plots the package of improved practices is applied, whereas the other is farmed in the traditional way.

- A simple farm plan is worked out with each farmer participant, indicating the crops he will grow and the supplies he will require.

Sadikin (1983) describes an integrated scheme known as the BIMAS programme which was developed and applied in Indonesia for rice production. This consisted of a package of inputs, among which were extension, credit, good seed, fertilisers, and insecticides. The technology package also included land preparation, and timely application of the various inputs; it was first tried by students of the Bogor Agricultural University on a 50-ha verification and demonstration plot on farmers' fields. The yield increases obtained encouraged the farmers to experiment further. Using the experience gained, the provincial agricultural services, through their extension arm, launched a large-scale introduction of this technology package. This was first directed to well-irrigated areas with adequate infrastructure and good farmers who were known innovators. Inputs were made available at the farm gate and on time; irrigation water and plant protection were assured; marketable surplus found ready buyers; and the farmers organised themselves so as to solve the day-to-day problems co-operatively.

The campaign achieved yields well above the national average. The Government supported the BIMAS programme with research, training, extension, rural credit, inputs and a floor-price policy. BIMAS management boards were established at national, provincial, and district levels. Millions of farm families all over the country were involved. The introduction of new, high-yielding rice varieties further encouraged the spread of the programme.

Over the years, the BIMAS programme has undergone continuous change. Improvements have resulted from reviews, reorganisation, adjustments and the introduction of insect-resistant varieties.

The latest organisational development is a special intensified production programme, called INSUS, which involves a group approach to extension, relying on the active participation of farmers in decision-making on inputs to be purchased, fertiliser application rates, plant protection schedules, and water management. There is a guaranteed price for the rice produced.

Sadikin (1983) concludes: 'the BIMAS and the INSUS programmes are excellent demonstrations of how an attractive and profitable technology package can be adapted and adjusted by farmers themselves to suit their own needs'.

Based on the positive experience with the rice programme, Indonesia is now adapting the BIMAS and INSUS approaches to production programmes for other commodities, including maize, grain legumes, vegetables and poultry.

Formulation of 'technology packages' for a single crop

CIMMYT has adopted an intermediate procedure, whereby teams of agronomists and economists concentrate on formulating technology practices for a single crop, instead of engaging in full-scale farming systems research, prior to proposing complex package programmes.

These teams do not attempt to identify 'optimum' packages of practices, but 'to forge good approximations—technologies which promise more incomes with acceptable risks to representative farmers, in the expectation that each farmer will, after adoption, adjust the recommended practices to fit his own particular circumstances' (Winkelmann & Moscardi, 1979).

In their pre-extension research, three types of on-farm trials are carried out by the CIMMYT research teams:

(a) *Yes–no trials* designed to study the major effects and interactions of the factors thought to be the most critical in limiting production. These are mainly in form of factorial designs, in which two levels of inputs or practices are examined—one at current farmers' levels and the other at a significantly higher level.

(b) *'How much' trials* aimed at identifying levels at which income-seeking, risk-averting farmers might be prepared to adopt practices identified as limiting in yes–no trials.

(c) *Verification trials*—after researchers and farmers are convinced that an appropriate strategy is available, that is consistent with the farmer's circumstances and promises a significant improvement in income at an acceptable level of risk, the strategy is verified on a number of representative sites. Recommendations are made after confirmation by the verification trials.

Shortcomings of package programmes

The 'package' concept has come into some disrepute, not because of the principle underlying this concept, namely the synergistic effect obtained when several components of the package are applied together, but because of the tendency to offer the farmer a package that is beyond his means, or constitutes too risky an investment (Gomez, 1985). On the other hand, when attempts are made to compromise, and leave out certain elements of the package in order to make it more affordable, these can result in failure of the project.

As is often the case in LDCs, it has proven to be difficult to translate a promising theory into practice. Moris (1987) mentions three essential conditions for putting together a technological package:

(a) to have an adequate base of field-tested research;
(b) the economic testing of the package, so that the cost of the package, including inputs, labour and field operations, etc., in relation to prices and expected yield increases can be estimated;
(c) the components of the package are weighed from the farmer's perspective, taking into account different resource availabilities and the potential returns from these resources if used in a different way.

An example of a package that did not conform with these requirements is described by Moris (1987). In Kenya, data on mean labour input required by five alternative options show that the officially recommended package by the Ministry of Agriculture would require 325 man-days per ha in contrast to the traditional farmers' system requiring 142 man-days per ha. If one compares the unit of return at planting time, the ministry's package yields 5 shillings/hour, whereas the farmers' own practices yield nearly 19 shillings/hour.

Chapter 22

The Agricultural Information System (AIS)†

Effective technology development and transfer depends on an interactive, holistic system, that Röling (1987) calls *the agricultural information system*. The system includes: a research subsystem; a dissemination subsystem; a user subsystem.

The system must perform six basic functions in order to ensure the initiation and the continuation of the information flow process (Nagel, 1979): (a) identification of problems at the producer level; (b) generation of innovations; (c) validation under farmers' conditions; (d) dissemination; (e) utilisation; and (f) evaluation.

The basic model which combines the subsystems, the six functions, and the areas of responsibility of the subsystems is given in Fig. 25.

RESPONSIBILITY FOR THE BASIC FUNCTIONS OF THE AIS

A clear definition of responsibility for each of the functions of the AIS, and an equally clear definition of the areas in which joint action by two or more subsystems is required, are essential for the efficient functioning of the entire system.

That the research subsystem should be responsible for the generation of innovations and that the extension service should be responsible for the dissemination process appears *a priori* an unassailable statement.

---

†Also called Agricultural Technology Management System (ATMS) or Agricultural Knowledge System (AKS) (Kaimowitz, 1987)

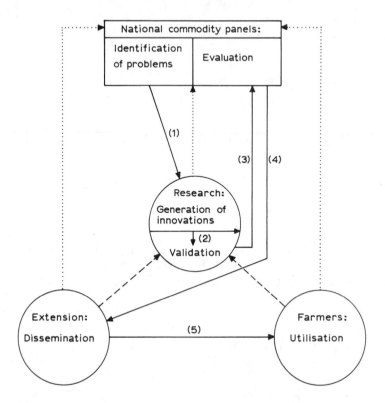

Fig. 25. The Agricultural Information System: showing the three subsystems (circles) and their respective areas of responsibility for the six basic functions of the system; a joint decision-making framework (the National Commodity Panels); the flow process of innovation (1)–(5); flow of information to the National Commodity Panels (- - - -►).

However, these simple definitions are not generally accepted by the subsystems concerned, as shown by the following statements:

'Researchers' role is to produce research results seeking alternatives for the future; advisers' role is to arrive at technologies practically and profitably utilisable by the industry now, when needed' (Elkana, 1986). In other words, generation of innovations is the responsibility of both the research and extension subsystems: the former should engage in 'research for the future', the latter in research of immediate applicability. Such a distinction is not only untenable, but must lead to duplication of efforts and resultant inefficiency.

Barker (1986) adds the following comment: 'ideally an extension service needs to have a research and development capability, both to provide information for advisory work and to establish the creative status of the organisation'. One could argue with equal logic, that 'ideally a research organisation should have an advisory and development capability, in order to ensure that the findings of research are brought to the farmers rapidly, efficiently, and without distortion'. This is exactly what the Colombian Agricultural Institute attempted to perform, and the result was an unmitigated disaster for research, extension and development (cf. p. 171).

A different concept is that the research sub-system should concern itself exclusively with basic research, and the extension service with adaptive research. This concept was adopted in Great Britain, on the premise that research should be insulated against political interference. The result has been a costly dual system of research, in which the research carried out by the Agricultural Research Council is indistinguishable from that carried out by the extension arm of the Ministry of Agriculture & Fisheries (cf. pp. 21–2).

We have argued (cf. p. 00) that the premise that the role of agricultural research is concluded when a solution to a given problem has been achieved in the laboratories and the fields of the experiment station and that adaptation of these results is the responsibility of the extension service, inevitably leads to the following consequences: (a) the alienation of the research worker from the real needs and potentials of the majority of the farming community; the research sub-system actually disengages itself thereby from up to 90 per cent of the national research effort, which can, and should be carried out in fields away from the central or regional experiment stations; and (b) the extension service has no alternative but to involve itself in agricultural research, a function for which it is not organised, and does not have the professional capacity to fulfil, unless it establishes a complete research system, as exemplified by the British model.

However, even in those research organisations which accept responsibility for research right up to the stage at which a proposal can be adopted by the farmer after rigorous on-farm testing, there are countries in which the dissemination subsystem still feels impelled to carry out its own research, often under different designations. The legitimisation of this approach is that 'most advisers feel that unless they test and try out technology applications locally in their own farmers' settings, they cannot honestly and convincingly make sound recommendations' (Elkana,

1986). As a result, the extension workers spend up to 50 per cent of their time on experimentation, duplicating what has already been done and farm-tested, and this in situations in which the number of extension workers is not adequate for the tasks they need to undertake.

Recapitulation

There are situations in which the dissemination subsystem has genuine grievances against the research subsystem: when the latter is either exclusively concerned with basic research or does not provide appropriate and adequately farm-tested technologies. In both cases there are two alternative pathways open to the two sub-systems:

For the research sub-system: (a) to accept full responsibility for research aimed at solving the major problems of the farming sector, and providing appropriate and on-farm tested technologies; or (b) to retire to its ivory tower, and accept a situation in which the dissemination sub-system assumes this responsibility.

For the dissemination sub-system: (a) to accept a clear definition of the roles of research and extension respectively, whereby research is responsible for the generation of innovations in all its stages, and extension is responsible for their dissemination; or (b) to engage in both functions of generation and dissemination of technology.

One would assume that elementary logic and best national interests would dictate that both sub-systems should adopt their respective (a) alternative, and make all efforts for full and fruitful co-operation. Unfortunately, reality in this matter is not generally in accord with the dictates of logic or what is in the best national interest. In many countries, the research system confines itself to its ivory tower, and the extension agency tries to be self-contained. 'It is not uncommon for agricultural extension agencies trying to do research, while agricultural research agencies are trying to do extension. The net result is poor research performance by extension and few farmers covered by research' (RAPA/FAO, 1985).

From the foregoing, it should not be assumed that there is anything wrong with extension workers, as individuals, being involved in experimentation, or research workers with extension activities, a subject with which we will deal in detail in a subsequent section of this chapter.

THE INTERFACE BETWEEN RESEARCHERS, EXTENSION WORKERS, AND THE FARMER

'Research systems that cannot transmit findings to the extension service and to farmers make little practical contribution. Extension work that is not sustained by results obtained through research has little value, and may even be detrimental' (ISNAR, 1982).

Liaison between research and extension workers

The stereotype concept of the channels of communication between researcher, extensionist and the farmer is that the extension worker communicates research findings to the farmer and transmits the latter's problems to the researcher, who then incorporates these problems into his work programme: Researcher ⇄ extension worker ⇄ farmer.

The simplicity and logic of this model is misleading and it rarely functions as described. It does, however, ensure that the research worker is effectively insulated from direct contact with farmers.

It is essential that the generation and diffusion of new technologies be effectively linked, whatever the institutional setting in which the two sub-systems operate. In the absence of effective linkages, the process of dissemination of innovations is disrupted, the effectiveness of the research system and its relevance to farmers is reduced. 'Agricultural researchers and extension workers find in each other a convenient alibi if the farmers ignore their message' (Elz, 1984). In brief, whilst the need for close co-operation between research and extension appears to be axiomatic, its achievement in most countries is the exception rather than the rule. Extension work and research are usually organised in different services, but this is not necessarily the reason for the general tendency towards separation of interests and even alienation between the two services.

ISNAR (1982) teams have found in many countries 'that extension workers without close links to research pass on improvised technical messages; and some research is carried out in isolation, building up technical solutions, some of which are unused or unusable'.

Kaimowitz (1987) confirms these findings:

Repeatedly, around the world, the linkages between research and extension have been identified as one of the weakest areas of agricultural technology systems. Moreover, linkage problems are not unique to developing countries, agriculture, or the public sector. We have encountered numerous examples in the literature of similar problems in developed capitalist and socialist countries, in industry, and in the social sciences. In fact, problems in this area appear to be universal and may, indeed to some extent, be inevitable. What makes the linkage problems particularly severe in public-sector institutions charged with promoting agricultural development in developing nations is the general weakness of these institutions, the great cultural and educational differences which exist between researchers, technology transfer workers, and farmers in these countries, and the urgency of their need to increase agricultural production, particularly among small producers.

In a sampling survey carried out by Fernandez (1977) among staff from six research institutes and eight extension services in Latin America, the author found 'that 82 per cent of the researchers felt that the extension personnel made too little effort to learn about and to transfer the technologies generated, and 75 per cent of the extensionists felt that the new technology was not acceptable to smallholders'.

Despite their intrinsic interdependence, various forces promote the separation between extension workers and researchers: different professional orientations, and means of accountability; different methodologies and the advantages of separate organisational frameworks for the two services (Baxter & Thalwitz, 1985).

Rice (1974) describing the extension services in the Latin American countries, mentions that already in the initial stages of US aid in setting up these services: 'it sometimes required a strong chief of the US agricultural contingent to force the research specialists to leave their experiment stations and visit the field extension centre'. Conversely, 'one could visit a Centre where the extension technicians were not on speaking terms with the research technicians, and where the American leadership was not concerned with forcing the two to co-operate'.

Reasons for the alienation between the two services

Some of the reasons for the lack of co-operation between researchers and extension workers have been briefly mentioned in the preceding section.

Lecomte (1964) lists the following complaints about research workers made by extension workers:

1. they shut themselves up in their laboratories or experimental fields without bothering about working conditions at all, or only very little;
2. they do not pay enough attention to economic factors or to the extension workers' points of view and actual requirements;
3. they write up the results of their experiments in language which is at best involved, and sometimes virtually a secret code; and
4. they often refuse to commit themselves to a firm opinion and make too many refinements and reservations about their conclusions.

Wyckoff (1965) adds another complaint, 'that too much research is carried out which has no application, research which cannot be interpreted, or research which finds the "wrong" conclusions (e.g. does not agree with the preconceived ideas of the extension specialist)'.

Tension between research and extension is in evidence even when optimum administrative relationships exist. Hildreth (1965) attempts to explain these tensions by proposing three hypotheses:

1. Tension is caused by a difference in orientation to the variables of time and depth of analysis. Research considers time as the variable and depth of analysis as the standard to be met; as depth of analysis is a function of time (to a considerable extent), research will not readily agree to severe time restrictions. Extension, on the other hand, is under constant pressure to provide answers, and time is therefore the standard they have to meet. As they are dependent, to a large extent, on research for these answers, tension ensues.
2. On the other hand, research workers complain that extension workers: (a) do not really trust or accept their findings; (b) do not ask research workers for information when they need it; and (c) do not make it clear to research workers just how important or far-reaching are the problems which they meet out in the field.
3. In addition, researchers tend to be irritated by constant demands by the extension workers to provide them with answers to urgent problems. They feel that these demands, even if justified, disrupt their long-range research. They tend to overlook the advantages to be derived from these contacts: an awareness of the farmers' problems and a source of ideas for new research objectives.

In short, the opposing claims are that research workers are not adequately informed about actual practical questions and that they do not properly communicate their findings to those who would take advantage of them, and vice versa.

These attitudes easily lead to a situation in which the extension worker, instead of serving as a link between research and the farmer, becomes, on the contrary, an obstacle between them. Lack of contacts between research and extension easily lead to conflicting conceptions and opposing instructions to farmers, to the detriment and confusion of the latter.

Practically everybody is agreed that such a situation is incompatible with the needs of the agricultural community and must be avoided, nowhere more so than in developing countries. Lip-service to this idea is, however, not sufficient, nor can the problems be solved only by contacts at the individual level.

IMPROVING THE LINKAGES BETWEEN RESEARCH AND EXTENSION

In principle, there are two pathways for promoting effective two-way communication between the research and the dissemination sub-systems: (a) institutional frameworks, and (b) functional relationships between researchers and extension workers.

The first approach consists of adopting a common administrative framework for the research and extension functions; the second is based on collaboration between two separate systems, each with clearly defined responsibilities and resources, for research and extension respectively, and with institutionalised functional links between researchers and extension workers.

Common institutional frameworks

A total fusion of research and extension into a single institution has been adopted in a number of countries. In Argentine, for example, the National Institute for Agricultural Technology (INTA) has equal responsibility, on a national scale, for research and extension activities.

A more common approach, is for research and extension to form separate departments within the Ministry of Agriculture, or to belong together in the various commodity departments of the ministry.

The major advantages claimed for combining research and extension into a common administrative framework are (Herzberg & Antuna, 1973):

- improved availability of information on innovations for the extension workers;

- the research workers are kept constantly aware of the technical, social, and economic problems encountered by the extension agents in their contacts with farmers;
- the time-lag between research findings and their adoption is reduced;
- improved understanding between research and extension workers, working jointly towards a common goal;
- involvement of the researchers in the extension work makes it possible to reduce the number of extension specialists;
- the research centres can undertake the in-service training of the extension workers.

At first sight, such a highly integrated approach, in which the same organisation is responsible for the creation of new technology and its transfer to farmers has the considerable advantages outlined above, of which the most important is probably an improved flow of communication between research and extension.

This is the solution that has been adopted by INTA in Argentina, and has been functioning since its inception in 1958.

The extension division of INTA has achieved an excellent reputation (Rice, 1974), a very rare case in regards to national extension services. This success has been attributed to a number of factors:

- the close links with research personnel, who are in charge of the whole research–extension process;
- the emphasis on training, which includes periodic year-long sessions for every extension worker at the Graduate School in Agricultural Sciences at Castelar (which offers a Masters degree in extension);
- the prestige of the director and his deputy, who have been with programme since the establishment of INTA;
- INTA is well financed (cf. p. 160) and able to pay attractive salaries and to equip the service adequately;
- its major clientele are the larger, wealthier and more sophisticated farmers.

Does the INTA example prove that combining research and extension under the umbrella of a single Institute really improve the relationship between them?

First, we must point out that there is considerable experience of having the two services under a common umbrella, when both research and extension were units of ministries of agriculture, and the relations between them were, in most cases, notoriously bad.

Second, the INTA example does not prove that the improved relationship between the two services is due to the common administrative framework.

Howell (1984) writes that in some countries that he visited, in which agricultural research had been separated from the Ministry of Agriculture, home to the extension service, it was claimed that this separation impaired effective extension support. But in other countries, where the two services are located within the Ministry of Agriculture, similar complaints were made.

Howell's findings confirm the general experience that the inclusion of research and extension in a single administrative framework does not, in itself, ensure good collaboration between the two functions. In another Latin American country, Colombia, the National Institute (ICA) had been charged with integrating the research and diffusion functions precisely in order to improve the transfer of research findings to the farmers. The result: 'the extension division in ICA continued to have no direct contact with the research division and promulgated indiscriminately the application of foreign methods inappropriate to the realities of the country' (Trigo *et al.*, 1982).

Even where research and extension are merged into a single service, there is often a professional distance between extension workers and scientists which inhibits common discussion and analysis of farm-level constraints to adoption (Farrington & Howell, 1987). In one country, for example, the allocation of resources to research was disproportionally higher than allocation to extension. This developed a high superiority complex among the research staff and an inferiority complex among the extension staff functioning under one agency (RAPA/FAO, 1985).

It does not appear to be necessary or even desirable to force into a common administrative framework two disparate services. Even though they have a common goal, they require entirely different administrative, technical and other supporting services. Differences in objectives, working methods, professional qualifications, criteria for professional advancement, are bound to cause friction when both services are in the same administrative organisation.

Actually, a good case can be made against integrating research and extension in a single national institute. Their functions are so basically different, that they require distinct administrative frameworks if they are to operate efficiently.

A strong linkage with the authorities and bodies responsible for development is absolutely vital for both services, but the nature of the linkage required by each of them is entirely different.

The duty of research is to solve problems and provide new technology which can be promoted by the development authority. The linkage with the latter is exclusively concerned with policy and the consequent programming of research. Such a linkage does not require a common organisational framework; we have seen that it can actually be inimical to research.

Extension, by contrast, is an *operational* arm of development; its proper place is therefore *within* a body directly responsible for agricultural development, with which it can, and should, form institutional bonds.

It is quite possible to achieve an improved working relationship between the two services even if they function in two separate, though closely linked, administrative frameworks. The measures required to achieve this aim will be discussed in the next section.

Functional linkages between separate research and extension services

It is generally accepted that close integration of research and extension activities is a critical necessity. In the previous section, we have shown that institutional and administrative integration do not, as a rule, provide the answer; what can and should be achieved is *functional* integration, based on close working relationships in practice, between researchers and extensionists.

Ultimately, linkage activities are performed by individuals, whose behaviour is influenced by their training, experience and incentives (Kaimowitz, 1987). The institutions involved, if willing, can provide the training, the incentives, and the formal frameworks to initiate and foster these linkages, even when the two services operate in separate administrative units.

A common policy 'umbrella'

The two administrative units, for research and extension respectively, can both function under the aegis of the ministry of agriculture. Though the research institute may have administrative and functional autonomy, it should remain linked to the ministry, which has authority over policy and programming of research (cf. p. 258). The extension service, by contrast, is usually a division within the ministry. The objectives and programmes of both systems are therefore based on the same overall development policy, in the achievement of which they are heavily dependent on each other.

The planning of research and extension work

Research–extension–farmer co-operation should begin at the planning stage. Apart from the commodity committees that function at national level (cf. pp. 391–2), each regional station should have an advisory committee in which research, extension, local government and farmers are represented and which makes proposals defining general goals and priorities and the allocation of the resources available for regional research to the various problem areas. A number of subcommittees need to be established for each of the main commodities or groups of com-

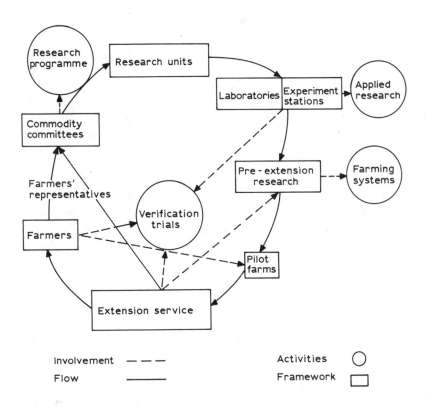

Fig. 26. The agricultural research process: involvement of extension workers and farmers in planning and implementation.

modities of the region. These subcommittees consist of the research and extension specialists in the area of competence of the subcommittee, and one or more representative farmers. The subcommittees evaluate the results of current research, decide on what research findings are ripe for adoption, discuss the extension programme for the transfer of their findings to farmers, and indicate priorities for future research.

They also decide on the adaptive research trials needed, which should be conducted on farmers' fields to test the validity of research findings from the regional experiment stations and to ascertain farmers' acceptance of these findings.

These institutionalised contacts between research, extension and farmers (Fig. 26) ensure that research is not divorced from farming realities and that innovations are adopted by consensus. The farmer is assured that his problems are brought to the attention of the research worker and that the necessary attempts are being made to provide a solution according to priorities determined in consultation with his own representatives. The extension worker feels that he is a full partner in this planning process, and has a say in on what innovations should be adopted, and on when and how this should be done.

The psychological importance of this form of contact and reciprocal influence of research worker, farmer, extension worker and local government cannot be overstressed.

However, farmers' representation in the planning process involves certain problems in most developing countries, which will be discussed in more detail.

Participation of farmers does not signify that proposals for research themes must come from them only, though their initiatives in this respect should be welcomed. The research workers have a primary role in the formulation of the research programme; but their proposals must be communicated to their 'clients' so that there is opportunity for dialogue and the possibility of influencing the final outcome.

Discussion and decision-making on the planning of research and extension work at two different levels—national and regional respectively, may appear to be cumbersome procedures and a possible source of conflicting interests. It is indeed more difficult to implement than decision-making from the top. However, recognition of the need to accommodate location-specific problems, requirements and potentials within the framework of national priorities and available resources should be the overriding considerations, and this requires grassroots participation, if it is to be effective (Arnon, 1987).

Joint service units

Certain units can be established that provide services for research and extension, such as libraries, central and regional laboratories for the diagnosis of soil fertility problems, fertiliser recommendations, water requirements, identification of diseases and pests, etc.

Joint publication unit

A central unit for the production and dissemination of documentation for use in the extension service, and collaboration in the preparation of regular radio and television programmes, newspaper and magazine articles, bulletins and reports, could also promote effective two-way communication between the two services.

Joint locations for both services

An ideal solution is to have the regional extension service located on the same site as the regional experiment station, where they could have their own buildings and facilities. At least the SMSs, who should be the major link between extension and research, should have their offices at the regional stations.

Liaison units

Some countries have set up a liaison unit between research and extension services in order to improve communication between the two bodies.

Nigeria is trying this approach in several of its research institutions. Sudan has a 'Contact Unit' in the Agricultural Research Corporation. Similar set-ups are found in Kenya and Zambia.

In the Netherlands, liaison officers are stationed in the research institutes, and are responsible for supplying information to the SMSs in extension districts, and for informing research on field problems encountered by the SMSs. Because of the small size of the Netherlands, the liaison officers are able to cover the whole country (Zuurbier, 1983).

Experience has shown that liaison units can result in the new bodies becoming an additional barrier between research and extension. This can happen when the new body 'becomes another specialised group with its own language, objectives and procedures' (Rubenstein, 1957) and attempts to serve as a go-between, instead of promoting direct contacts between research and extension workers, such as field days, symposia, workshops, etc.

PERSONAL LINKAGES BETWEEN RESEARCH AND EXTENSION WORKERS

Ultimately the linkages are performed by individuals; it is therefore at the personal level that the opportunities for fruitful co-operation between the two services have the greatest potential. It is, however, also at this level that problems are most apt to arise because of personal incompatibilities or narrow ambitions. Individuals may lack motivation or incentives for carrying out joint functions. They may feel antagonistic to their professional counterparts, and have little respect for them. These constraints can be overcome by creating frameworks for joint activities which ultimately prove that co-operation can be more rewarding to the individuals concerned, than separate and antagonistic action.

On-farm experimentation

The major area in which the activities of research and extension workers converge in the field is on-farm experimentation (Fig. 27). Innovations resulting from agricultural research must be tested for their adaptability

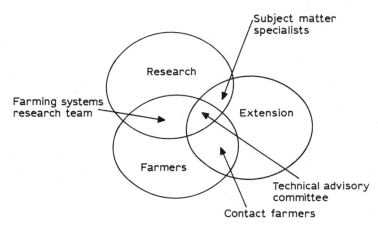

Fig. 27. An alternative conception of a technology development, transfer and utilisation system. (From Swanson *et al.* (1984), reproduced by permission of FAO, Rome.)

and relevance to the farmer's environment. Experiments under actual farming conditions should be the final and essential stage of testing a new technology. Typically, new varieties, recommended fertiliser rates, plant production methods, etc., should be validated by on-farm testing, either as single practices (verification trials, testing or adaptive research) or by their incorporation into appropriate farming systems (pre-extension research, farming systems research). These are all an integral part of the responsibilities of the research subsystem. However, the research organisation does not have the field staff spread throughout the country required for the systematic testing, under many environments, of proposed innovations. Nor does it need to have such a staff, because, ideally, extension staff should be involved in these activities. For the extension service, collaborating with the research institution, in the field, in a country-wide network of on-farm trials, can be a major factor for improving the morale, motivation and professional competence of its staff. The best extension workers are not satisfied with the exclusive role of 'postmen' between researchers and farmers. The majority feel a need to participate in the research process, by testing on the farm of one of their farmers, before recommending a new practice. This is a legitimate desire, and should be encouraged, but not as an individual, quasi-private undertaking. The trials in which the extension workers should participate should form part of a country-wide validation programme, under the aegis of the research organisation, and should be implemented in co-operation between the researcher whose proposals are being validated and the extension workers.

Collaboration in on-farm testing is the single most effective way to forge strong links between the workers of the two services. The extension agents can make important contributions to farming systems research and other forms of on-farm testing. Incidentally, these activities are also a meeting ground for personnel of both services with the farmers, a subject which will be treated later in more detail.

Conversely, when research activities, under whatever name, are carried out as a separate function by the extension service, this is the single most effective way to erect barriers between the two services, and to weaken the effectiveness of both research and extension. The result is competition, with conflicting institutional and personal interests, instead of the collaboration we have described.

Research and extension are separate and different professions, and as such require personnel with different qualifications and attributes. For the research organisation, the on-farm trials are not individual tests, but form part of a regional effort to provide improved and tested technologies to different categories of farmers, operating under different ecological and socio-economic environments. For the individual extension worker, *his* trial stands on its own, intuitively he draws conclusions from what he has observed, and 'sells' the information to his clients. Lacking professional qualifications for research, he is unaware that his conclusions may be erroneous and hence misleading. Interseason variability within a location may be, and frequently is, far greater than inter-location variability in a given season. Only multilocational trials, that have been properly planned to enable appropriate statistical analysis, and carried out for several seasons can make it possible to draw valid conclusions which can be safely recommended to farmers.

When the extension workers undertake field trials as an individual, independent activity, not only is the effectiveness of the validation process undermined, but the extension effort is also weakened. In Israel, for example, the original intention was for extension workers to devote 10–15 per cent of their time for on-farm trials in co-operation with the research workers; gradually two processes took place: (a) a breaking away from co-operation in a regional framework, to individual experimentation, and (b) an increase in the time devoted to field experimentation, which is now on an average 40–50 per cent of the extension agent's work load (Elkana, 1987). The contribution of extension to new technologies has diminished, because of the fragmented, individualistic approach, and the effectiveness of the extension service impaired, because half of the time of its workers is taken up with a function that is not extension. And last, but not least, links with research are thereby weakened, and common interests replaced by competition for funds and conflicting interests.

Recapitulating

On-farm experimentation in its various forms provides a comprehensive, logical and efficient framework for researcher–extensionist–farmer linkages, whereby the researcher contributes his specific competence in the field of research and a national approach to problem-solving; the extension worker contributes his knowledge of local conditions and familiarity with the farmers involved; and the farmer contributes his practical farming experience and insight in the problems with which he and his peers are faced.

The functional division of work between the three partners in on-farm experimentation can be summarized as follows:

The researcher: (a) planning the on-farm trials on a regional basis within a national research project; (b) collating and analysing the data, and drawing conclusions of nationwide applicability; and (c) regular monitoring of progress in the field.

The extension worker: (a) helping the researcher select suitable locations for the trials and identifying the farmers participating; (b) layout of the trials; (c) supervision of the implementation of field operations in the experiment by the co-operating farmer; (d) regular observations: recording phenological data, incidence of diseases, pests etc.; (e) notes on farmers' practices, etc.; (f) supervising harvesting and recording plot yields.

The farmer: carrying out all the field operations involved in the trial.

All three together: regular visits to the trial, visual evaluation of the results; implications for future work and for adoption of innovations by the farmer.

Dividing *institutional responsibility* for the activities involved in the research process, between the research organisation (experiment station and laboratory research) and the extension service (on-farm experimentation)—as is frequently proposed—inevitably introduces a divisive element in what can potentially be an ideal framework for collaboration between the two institutions.

Joint projects

An example of a joint project undertaken by research and extension workers is the National Agricultural Research Project which has been initiated in India on a pilot basis. A number of zonal research stations have instigated status reports on adoption patterns and production constraints of all major crops and crop recommendations. These reports are compiled by researchers, SMSs and extension supervisors with the collaboration of extension agents and farmers selected by stratified sampling. Eventually, it is anticipated that data on some 2000 different farming situations will be obtained and regularly revised.

In parts of India, therefore, researcher–extension worker collaboration in a recognisably common enterprise has been strong enough to override the structural gap between agricultural university-employed scientists and department of agriculture-employed extensionists (Farrington & Howell, 1987).

Involvement of senior research workers in in-service training of extension workers

Field-days and seminars should be held regularly in which the results of research should be communicated to, and discussed with, the extension personnel.

A very useful recommendation made by Howell (1984) is to hold a one-month residential training course every year, before the main crop season. The courses should be held at regional experiment stations, and would forge links between research and extension staff that could be further strengthened in monthly training days for different extension circles.

Personnel movement between the two services

Personnel movement between the two systems should be made possible, and on occasion even encouraged. This would apply mainly to SMSs who could be taken from the ranks of the research workers; conversely, extension specialists, after a number of years of extension work, could change places with colleagues in research.

Experienced extension workers could make excellent farm managers for the regional experiment stations, supervisors of on-farm experiments, etc.

These changeovers would generally involve a period of induction training.

FARMERS' PARTICIPATION

The active involvement of farmers in the planning and implementation of extension programmes 'is one of the major determinants of success in programmes designed for small farmers' (Morrs *et al.*, 1975).

Farmers' participation is important at all stages, from identifying research needs and determining priorities (cf. p. 388–9), field testing results, conducting demonstrations, evaluating acceptability, up to identifying social, economic and institutional constraints (Cernea *et al.*, 1985).

Traditionally, the research workers have determined their programmes without consulting the farmers, and extension workers have decided what technical advice to transmit to farmers in order to improve their productivity and living standards. The image of an effective information system in which research produces the information, extension passes it on to farmers and farmers use it is a false one. If the information system is to be effective, farmers should be active participants in providing information and in asking for the kinds of information they consider useful to them (Tendler, 1982).

Until recently, the role of power exerted by organised farmers has been overlooked, even though it is known to have been 'an essential ingredient of agricultural development in industrial nations and in the development of effective services for progressive farmers in agrarian countries' (Röling, 1987).

Rogers & Yost (1960) consider 'user control' as one of the important factors for the success of the Land-Grant College model. Through this control, farmers have been able to make research as well as extension accountable to them, and thereby have ensured relevant contributions from the two services.

What is important, is to create formal procedures whereby farmers *participate* in the decision-making process, instead of having to exercise political pressures as adversaries in order to achieve a positive response to their requirements (see Chapter 7).

The concept of farmers' participation is gaining ground. Committees have served as a forum for the determination of local priorities and implementing district-level plans at the village level in Botswana, Kenya, Sri Lanka and elsewhere (Garforth, 1982).

Farmers' representation

The most satisfactory way of choosing farmers' representatives is to leave the farmers to select them themselves. The elected farmers will then feel under an obligation to their sponsors to represent their interests and to pass on what they have learned.

However, the major problem is the lack of more or less formal local organisations of rural people that would enable them to make their needs known to the authorities in general, and to research and extension in particular, as well as to choose their representatives on decision-making bodies.

In some countries, government has been successful in encouraging the formation of farmers' associations which play a significant role in agricultural development. The participation of their representatives in the work of the commodity committees can have a significant beneficial impact on the orientation of research and the adoption of research findings.

However, the number of developing countries in which such farmers' associations are active is still restricted. In the majority of cases, rural life is still dominated either by rich landlords, or by a gerontocracy, or some other elite group.

This brings us to the question of how should the representative farmers to serve on the commodity committees be chosen in the absence of formal organisations. It is frequently stated that attempts to replace traditional leaders, or approach farmers directly rather than through their traditional leaders, can be counterproductive and should be made very carefully.

There is no doubt that the readiness to adopt new ideas and practices is influenced by the nature of the leadership and the degree of control it exercises over the community. Reliance on the 'traditional leaders' may be justified as a short-term, tactical approach—but normally a strengthening of their traditional stranglehold (economic, moral and social) on their people will be self-defeating in that it will reinforce the conservative, anti-change tendencies which are very strong in any case and will stifle the initiative of more progressively minded members of the clan and village. Bose (1961) has a case when he states 'that it is necessary to change the entire cultural pattern of an underdeveloped society before we can hope for permanent technological changes'. Dumont (1962) also stresses the evil influence of the gerontocracy of the villages. By making the strongest economic power the preserve of the chiefs of large families, many primitive societies entrust the levers of progress to the oldest, who are often the least receptive to modern techniques.

The absence of formal organisations that would be able to nominate their representatives for the commodity committees should not discourage attempts at ensuring the farmers participation in research and extension planning.

The only possibility is to co-opt informal leaders, or those who, with proper encouragement, can develop into such leaders. In every rural community can be found farmers who have commonsense and native intelligence, a modicum of initiative and a desire to better their situation, and who are otherwise fairly representative of the majority of the farmers in the region. It is from among these that members of the regional commodity committees should be chosen. Incidently, the same guidelines can be used for the choice of farmer co-operators for managing pilot farms and for on-farm experimentation.

Possibly, the involvement of small farmers in decision-making bodies at grass-root level, of relevance to the improvement of their lot, will help to increase the awareness of their peers of the advantages of organising themselves into local farmers' associations.

Farmers participation in experimentation

Farmers' participation is of course not confined to the planning of agricultural research and extension, a subject that has been treated in Chapter 7. Farmers can and should be involved in all on-farm trials. The concept of farmers' participatory research (FPR) has arisen in response to pressures for greater involvement of farmers in the process of technology development.

FPR is seen as the management of on-farm trials by farmers and researchers jointly. In practice, it has a number of additional aspects, including: consultation with farmers of the problems to be investigated; evaluation by the farmers of the results of the trials, and the dissemination of the results by farmers to farmers.

In a review of recent field experience with FPR in several countries, it was found that in institutional terms, most FPR has been conducted hitherto by the international centres, universities, and voluntary organisations, with few examples from the national research services (Farrington & Martin, 1987).

Difficulties in fostering farmers' participation

In the 1950s, the 'community development' approach was adopted on a large scale in the LDCs, and by 1960, over 60 nations had undertaken national or regional development programmes. By 1965, most of these programmes had been drastically reduced or terminated.

In rural development programmes, farmers' participation was a major goal, as well as a major tool for achieving the objectives of the programmes, and experience of many years has shown 'that it is a most difficult and elusive goal to attain' (Holdcroft, 1982). Participation by nearly all segments of rural society was rarely accomplished in any of the community development programmes. Though these programmes were more concerned with social problems than with increasing agricultural productivity, the difficulties experienced in obtaining farmers' participation and the lessons learnt, are fully relevant to extension efforts to promote technological innovations and to farmers' participation in the formulating of agricultural research and extension programmes.

Holdcroft (1982) ascribes the failure of the community development approach to the fact that these programmes did not attack the basic structural barriers to equity in the rural communities. The village-level workers aligned themselves with the rural elites, ignored the poor, and tended to direct villagers, rather than develop a local leadership.

Little attention was given to ensuring that the rural poor should also share in the benefits from the programme. As a result, poverty and food scarcity were not reduced, and disparities in wealth between the large farmers and the small farmers became even greater, and more widespread.

The major implications of the failure of the community development approach for farmer participation in extension and research were:

- participation, a major goal in community development strategy, proved to be difficult to attain, because 'the structural barriers to greater equity were not addressed';
- instead of promoting participatory democracy, self-reliance and local initiative, the village-level workers were paternalistic and 'directed the programmes';
- the village-level workers were generally secondary school graduates, and were directed to work with established leaders, thereby strengthening the existing power structure;
- the rapid expansion of the community development programme made it necessary to recruit poorly trained village workers, to whom too many responsibilities in too many villages were assigned.

Since the self-help extension methods are based on the same principles, and use the same methods as the community development programme they should draw on the latter's experience and avoid the pitfalls outlined above.

RELATIONS BETWEEN RESEARCH WORKERS AND FARMERS

The importance of feedback from grass-root levels to research, regarding the needs of farmers and the constraints they face, has always been stressed by researchers, extension workers and farmers alike. In the past, this feedback was considered to be exclusively a function of the extension worker. We have already seen how weak and ineffective the traditional linear model of researcher–extensionist–farmer has generally been. This stereotype linear model should be replaced by a triangular model, which allows direct contact between researcher and farmer and *vice versa*:

Meeting the farmers

Hildebrand (1985) stresses that researcher–farmer contacts should not be limited to the research workers who are members of FSR teams, or carry out other forms of on-farm trials. 'Even national commodity team scientists should maintain researcher–farmer linkage. This linkage not only provides the researchers with a better understanding of the farmers' situation, it also creates confidence and a sense of accomplishment that is too often lacking among research personnel.'

Unfortunately, direct contacts between researchers and farmers are generally infrequent, and take place mostly on field days in which research results are presented to audiences of extension workers and selected farmers. However, 'in these cases communication is also limited to a one-way flow of information: the scientist comes to teach and demonstrate, not to listen. Comments or criticisms by farmers are usually treated in a defensive manner, rather than as a basis for dialogue on the latter's problems' (Doorman & Cueves Pérez, 1987).

The existence of joint commodity committees and of pre-extension research units, and the two-way communication channel of the extension service does not absolve the individual research worker from the need to acquaint himself personally and directly with the problems of the farming community he is supposed to serve. Journeys through the area served by the experiment station and talks to farmers and their advisers will help him to orient his research work to the real needs of farming. The greater the success of the research and extension efforts, the more dynamic will agriculture become, and the greater will be the need for the research worker to remain aware of the changes that are occurring in farming practice.

Conversely, visits by the farmers and extension personnel to the experiment station should be encouraged by open-days, field days and organised visits, lectures, seminars, etc.

We have stressed the importance of on-farm research in its various forms for forging links between research and extension workers. On-farm research also provides an opportunity for creating an effective researcher–farmer linkage. Research conducted on farms, in partnership with farmers, and on problems of direct and immediate concern to the latter, provides farmers with confidence that the researchers are really interested in improving their lot.

Learning from farmers

A related principle is that the effectiveness of agricultural research and extension workers will be greatly increased if the attempt to learn from the traditional farmers at the same time as they try to teach them. Many failures in introducing innovations into traditional agriculture occurred because research and extension workers did not understand the rationale of the traditional farmers' methods.

Therefore, the research worker and the extension worker must learn to know local conditions, and the habits and traditions of the people they are to serve. A practice that may appear at first sight to be irrational may actually be a necessary adaptation to local conditions.

The classic example is the former attitude of agronomists trained in the European tradition to the long fallows which are characteristic of shifting cultivation in the humid tropics. The traditional methods were considered a terrible waste of resources; however, the many efforts made to replace the long fallow by more modern methods—before learning the reasons that had led to the adoption of the traditional practice— were costly failures.

Farmers as a political support constituency

A completely different aspects of research–farmer interactions to those described above, is the need for a farmers' constituency that is able and willing to exert influence on the legislative and executive budgetary process in favour of agricultural research.

This political support of the research system forms a critical element in the capacity of the research institutions to attract public funding and other forms of administrative support essential for the development and effective functioning of the institutions (Ruttan, 1982).

For developed countries, the farming constituency, which has shrunken until it no longer accounts for more than 3–5 per cent of the active population, has lost much of its political clout, a fact reflected in the gradual decline, both in real and relative terms, of financial support for agricultural research.

In the LDCs, the support for agricultural research forthcoming from finance and planning ministries is likely to fluctuate with changes in foreign exchange demands and periodical food shortages; public pressure from a strong support constituency can have a strong influence in attenuating this tendency.

LINKAGES BETWEEN THE AGRICULTURAL RESEARCH ORGANISATION AND POLICY-MAKERS

Generally, policy-makers do not have a clear or accurate perception of the potential contribution that agricultural research has made, and can make, to agricultural development. 'Policy makers frequently consider the work of the researchers as irrelevant to their problems and expressed in an incomprehensible jargon' (Vinyu, 1979).

Unless the agricultural research leaders can convince policy-makers (and these include, besides the politicians—the planners, the financial authorities of the public sector, the rural constituency, etc.) that agricultural research is not a luxury, but an activity that no country can afford to neglect, 'it may be extremely difficult to develop a strong research system with secure funding, rational priorities, and the ability to attract and retain high-quality staff' (ISNAR, 1985).

Agricultural research leaders, in many countries, do not even attempt to achieve an active exchange of information with policy-makers. This neglect is compounded when the research institution does not integrate its work with the national development objectives.

To convince political leaders of the need to support agricultural research, which by its very nature is a long-term undertaking, is far from easy. 'The ultimate results of their support may not be evident until their own terms of office are long over and there is likely to be little short-term advantage for them, even if there may ultimately be substantial benefits for the community' (ISNAR, 1985).

Qasem (1985) concludes:

There is no one identifiable process through which issues like this are resolved in developing countries. In some cases pressure comes from the foreign agencies which provide loans, technical assistance, and/or objective advice. In others, support from the politicians comes when the country faces a crisis caused by technical problems which are expected to be solved by competent research institutions. In all cases, resolution of such questions requires patience, endurance and a power of persuasion on the part of research leaders and practitioners.

In summary, as long as policy-makers remain sceptical about the value of agricultural research, it will be difficult to convince them to allocate sufficient funds. Agricultural research will therefore remain ineffective, thereby confirming the scepticism of the policy-makers.

It therefore follows that agricultural research leaders cannot remain passive in this situation but must act to change it. The need to convince policy-makers of the potentials inherent in applying the results of agricultural research on a national scale is as important as publishing papers. Doing good research is not enough, it must be perceived as an effective tool in promoting agricultural development.

First and foremost, every effort should be made to show that the research programmes are congruent with the national agricultural development plans, and that the realisation of development targets is largely dependent on the effective implementation of the results of this research.

The policy-makers should be enticed to visit the research stations, see on-farm trials, and pilot farms. They should be invited to attend research seminars, to present their problems, and to learn what is being done to solve them. They should also be involved in the research planning (cf. pp. 396–403).

A research organisation that shows initiative and drive in these matters, instead of the usual passivity and going hat in hand to request budgetary allocations barely sufficient to maintain the status quo, can achieve very much for improving its ability to do good and important work.

RESEARCH ORGANISATION–UNIVERSITY RELATIONSHIPS

In most developed countries, linkages between public-funded agricultural research and university research are weak or non-existent, excepting at the personal level. The kinds of research undertaken by the two institutions are frequently indistinguishable, the major difference being in the orientation of the research: development oriented in the case of the public institutions; determined by the interest or inclination of the individual faculty member in the second case (or by the interests of a funding agency).

This dual system of research has its *raison d'être* in the developed countries; it is a luxury that the LDCs can ill-afford.

The major functional linkages between public-funded research and the universities that are desirable and possible are:

- participation of the universities in the research planning panels at national and regional levels;
- contract research by faculty staff on behalf of the national research institute;
- teaching appointments for the research institution staff at the faculties of agriculture;
- joint university–research institute graduate and post-graduate training programmes.

All these issues have been treated in detail in previous chapters.

In some countries, institutional linkages between the national agricultural research system and the universities have been formalised. In India, for example, linkages between the research institutes and the state agricultural colleges are being developed by specially designed forums:

- eight regional committees of the Indian Council for Agricultural Research (ICAR) meet annually, where interaction occurs between ICAR and university scientists and development officers from the state departments of agriculture;
- national commodity workshops of ICAR, to which university staff and extension directors are invited to participate;
- local management committees of the Transfer of Technology Project;
- The National Level Advisory Committee of the Ministry of Agriculture and Rural Development, to which ICAR scientists are invited to participate.

The land-grant college model and the agricultural universities are special cases of research–teaching linkages within the same institution, a subject that has been treated in detail in Chapter 3.

LINKAGES OF RESEARCH WITH THE GENERAL PUBLIC AND THE MEDIA

Because of the diminishing influence of the farming community (cf. p. 193) on which agricultural research traditionally depended for support, it has become vital 'to make society and decision-makers realize that increasing agricultural productivity is important for consumers as well as for producers; it will stop rising food prices and will increase their country's competitive position in foreign markets; it is also essential for the success of the overall development process. To change the traditional attitude of indifference or even opposition to research, to one of full-hearted support, requires time and effort' (Elz, 1984).

It is therefore essential to keep the public aware of the problems that research is endeavouring to overcome, and informed of the successes achieved. 'A good relationship with the media is indispensable for communication between research workers, consumers, government, and the private sector. The mass media coverage should be aimed at national and local levels. It should be handled in a professional way, using specially trained personnel. To build up a good public image and favourable public opinion is a task that requires a major effort and cannot be neglected' (Alves, 1984*b*).

Television and radio are effective media for this purpose, though the press should not be neglected. Open days at the laboratories of the research institution, exhibitions, field days at the experiment stations also play an important role. Even the individual visitor should be made welcome, and not treated as an intruder, as unfortunately is often the case.

Some of the topics which need to be brought to the attention of the public are enumerated by Elz (1984):

- research as a means of reducing or stabilising food prices and improving nutrition;
- efforts of research to improve the quality of food, and to reduce environmental pollution;
- research as an instrument to decrease reliance on imports and increase export earnings;
- research as a means to stabilise and increase rural income and as a factor for social stability.

EXTENSION AGENT–FARMER INTERFACE

Doorman & Cuevas Pérez (1987) describe the relationship between extension agent and farmer in the Dominican Republic. Their description is applicable to the situation in many LDCs.

In the relationship between the extension agent and the farmer, communication is mainly a one-way process. Extension agents transmit the recommendations of the research institutions to farmers, who may, or may not adopt them. These recommendations are supposed to apply to the whole country, without taking into account the considerable differences between the regions. Also, the objectives of the farmers may be entirely different to those of the originators and transmitters of the technology packet. The former strive for a maximum income with a minimum input, while the latter aim at maximum production levels. The extension workers may recognise that part of the lack of interest of the farmers can be attributed to a limited applicability of the recommendations, but they also consider most farmers as too uneducated and too traditional, in the sense of resistance to changing their practices, to adopt new technology.

The authors conclude: 'leaving aside the validity of the latter assumption, it does seem clear that the extension agents are in a difficult position. They find themselves in a situation in which they have to transmit a message with limited applicability, the contents of which they are unable to influence, and in which their clients are not interested'.

In the generally heavily centralised and authoritarian extension services, decisions about policy and strategy are concentrated at those levels of the organisation which are the most remote from the grassroot level.

LINKAGES OF THE EXTENSION SERVICE WITH POLICY-MAKERS

The political environment in any country is as critically important to the success or failure of an extension effort as it is for research. Effective extension therefore requires a linkage to governmental planning, strategy and policy, all of which have an impact upon the support accorded to extension (Watts, 1984).

Government officials and politicians are generally unaware of the complex social, cultural, economic and institutional problems involved in the adoption of new methods by the majority of the farmers. They expect quick results, which are not forthcoming, nor can they be forthcoming. The ensuing disappointment is reflected in the lack of support for agricultural extension. This in turn further contributes to the difficulties with which the extension service is faced, creating a vicious circle from which it is extremely difficult for the service to extricate itself.

One of the objectives of extension is to encourage the disadvantaged rural sectors to organise themselves into self-help groups (cf. pp. 811–2). A strong and effective extension service can therefore be perceived as a threat to the power elite, causing further problems for the service.

In those countries in which the extension service projects a poor image, this will not be changed unless a concerted effort is made (a) to improve the image, and (b) to make the policy-makers aware of the potential importance of an effective extension service.

OTHER INSTITUTIONAL LINKAGES OF EXTENSION

Extension services in many LDCs have not developed strong links with other institutions. Besides the linkages with research, already discussed in detail (cf. pp. 788–99), functional linkages with all other bodies concerned with rural development—statutory bodies responsible for regional or national commodity programmes, credit agencies, suppliers of farm inputs, educational bodies, professional organisations, etc.—are essential.

In various Latin American countries there has been a tendency to separate development actions from extension activities, and in certain cases they are even considered incompatible. The consequences of this approach have been discussed in the overview of extension activities in Latin America.

Linkages between extension services and agricultural education

Agricultural extension personnel need basic extension training which should be provided by the agricultural colleges and universities. Conversely, the agricultural colleges and universities need feedback from the extension services on the problems and needs of farmers in order to improve their teaching and research and to make these relevant to the situation in their respective countries. Linkages between extension and agricultural education are therefore mutually beneficial (RAPA/ FAO, 1985).

Some countries have institutionalised these linkages. In the *Philippines*, for example, the regional agricultural colleges and the universities have developed their own links with the extension service. The University of the Philippines correlates its extension programmes with those of the government agencies and also makes contributions to their extension activities. The university conducts seminars, workshops and training courses for extension staff of the Ministry of Agriculture and Forestry.

In the regional agricultural colleges, there are also joint efforts in the production of instructional and informational materials, and sharing of facilities and equipment.

In *Bangladesh*, the agricultural university and its graduate training institute provide in-service training for the ministry's extension personnel. In the *Republic of Korea*, an institutional co-operation committee is responsible for institutional co-operation between agricultural extension and education. College professors are appointed as joint staff members of the rural development administration for active participation in the research and extension work (RAPA/FAO, 1985).

Summing up their review of extension–agri-education linkages, RAPA/FAO (1985) conclude: it should be clear that the function of the universities is *extension education* and not extension service. As such, extension activities of agricultural colleges and universities should be: non-degree training, specialist support, information packaging and dissemination, and action-research.

This, of course, does not preclude individual members of faculty staff of being involved in the activities of the extension service as in the examples given above.

References to Part Six

Adams, M.E. (1982). *Agricultural Extension in Developing Countries.* Longman, Harlow.

Ahmad, M.M. (1981). Resource allocation to agricultural research in Pakistan. In *Resources Allocation to Agricultural Research,* ed. D. Daniels & B. Nestel, Proceedings of Workshop, Singapore. International Development Research Centre, Ottawa, pp. 55–60.

Ahmed, I. (ed.) (1985) *Technology and Rural Women: Conceptual and Empirical Issues.* George Allen & Unwin, London.

Alarcon, E., Artega, M.A. & Restrepo, L.F. (1984). *Sintesis del Plan Nacional de Transferencia de Tecnologia Agropecuaria.* PLANTRA–ICA, Bogotá.

Alcantara, Cynthia H. de (1976). *Modernizing Mexican Agriculture: Socio-Economic Implications of Technological Change 1940–1970.* UN Research Institute for Social Development, Geneva.

Allahabad Agricultural Institute (1958). *Extension Evaluation.* The Leader Press, Allahabad.

Alves, E.R. de A. (1983). Desafios de la Investigación Agricola en el Brazil. In *Seminario Internacional Sobre la Generación de Información y Cambio Tecnologico en la Agricultura.* Montevideo, pp. 15–37.

Alves, E.R. de A. (1984*a*). Notes on dissemination of new technology. In *Brazilian Agriculture and Agricultural Research,* ed. L. Yeganiantz, EMBRAPA, Dept. of Diffusion of Technology. Brasilia, DF, pp. 221–46.

Alves, E.R. de A. (1984*b*). Major issues in research allocation. In *Planning and Management of Agricultural Research,* ed. D. Elz. World Bank, Washington, DC, pp. 27–39, 122.

Ampuero, E. (1981). *Organización de la Investigación Agrícola para Beneficiar a los Pequenos Agricultores en America Latina.* Cornell International Agriculture, Cornell University, Ithaca, NY.

Andersen, J. (1964) The liaison between research and extension. Paper presented to FAO Ad-hoc Meeting on the Relationships between Agricultural Research and Agricultural Practice, Berne.

813

Ansorena, L. (1972). La extension rural en el desarrollo. *Desarrollo Rural en las Americas*, **IV**(1), 80–96.

Arnon, I. (1978). Problemas que enfrenta la investigación para transferir sus resultados a las practicas agricolas. In *Organización Institucional y Administración de la Investigación Agricola*, ed. A. Marzocca. IICA, OEA Zona Sur. Montevideo, pp. 25–35.

Arnon, I. (1987). *Modernization of Agriculture in Developing Countries*. Wiley, Chichester.

Arnon, I & Margolies, Julia (1984). (*Regional Planning and its Results*) (in Hebrew). Settlement Study Centre, Rehovot.

ARS (1984). *Agricultural Research Service Technology Transfer Plan*. USDA, Washington, DC.

Axinn, G.H. & Thorat, S.T. (1972). *Modernizing Agriculture—A Comparative Study of Agricultural Extension Education Systems*. Praeger, New York.

Baharsjah, S. (1985). Improving research–extension linkages: the Indonesian experience. In *Research–Extension–Farmer*, ed. M.M. Cernea, J.K. Coulter and F. Russell. A World Bank & UNDP Symposium. World Bank, Washington, DC, pp. 28–33.

Barker, M. (1986). Perspectives for advisory work in the context of the future evolution of agriculture. Paper presented at the Seminar on the Future of the Advisory Services, Louvain. Commission of the European Communities.

Barraclough, S.L. (1973). *Agrarian Structure in Latin America*. Lexington Books, Lexington.

Baxter, M. (1986). New developments in agricultural extension. Paper presented at the Sixth Intern. Farm Management Congress, Minneapolis. World Bank, Washington, DC.

Baxter, M. & Thalwitz, W. (1985). National policies and institutional constraints to linking research with extension in Asia. In *Research–Extension–Farmers*, ed. M.M. Cernea, J.K. Coulter and F. Russel. A World Bank & UNDP Symposium. World Bank, Washington, DC, pp. 42–48.

Benor, D. (1987). Training and Visit extension: back to basics. In *Agricultural Extension Worldwide. Issues, Practices and Emerging Priorities*, ed. W.H. Rivera and Susan, G. Schramm, Croom Helm, London, New York, Sydney, pp. 137–48.

Benor, D. & Harrison, J.O. (1977). *Agricultural Extension—The Training and Visit System*. World Bank, Washington, DC.

Blair, B.D. (1986). Dissemination of pest management information in the Midwest, USA. In *Advisory Work in Crop Pest and Disease Management*, ed. J. Palti and R. Auscher. Springer-Verlag, Berlin, pp. 231–33.

Blum, A. (1987) *The Development of Agricultural Extension in the Arab Sector of Israel and the Gaza Strip*. Faculty of Agriculture. Hebrew University, Rehovot.

Bose, S.P. (1961). Characteristics of farmers who adopt agricultural practices in Indian villages. *Rural Soc.*, **26**, 138–46.

Boserup, E. (1970). *Women's Role in Economic Development*. St. Martins Press, New York.

Bradfield, S. (1981). Appropriate methodology for appropriate technology. In

Transferring Technology for Small-scale Farmers, ed. N.R. Usherwood. ASA Spec. Publ. 41. American Society of Agronomy Madison, Wisc., pp. 23–33.

Busch, L. & Lacy, W.E. (1983). *Science, Agriculture and the Politics of Research*. Westview Press, Boulder, Co.

Byrnes, F.C. & Byrnes, K.J. (1971). Agricultural extension and education in developing countries. In *Rural Development in a Changing World*, ed. R. Weitz. MIT Press, Cambridge, Mass, pp. 326–51.

Caplan, N. & Nelson, S.D. (1973). On being useful: the nature and consequences of psychological research on social problems. *Am. Psychologist*, **28**, 199–211.

Carr, Marylen (1985). Technologies for rural women: impact and dissemination. In *Technology and Rural Women: Conceptual and Empirical Issues*. ed. I. Ahmed. Allen & Unwin, London, pp. 115–53.

Casas, J., Carrière, P. & Lacombe, P. (1979). *La Recherche Agronomique et la Diffusion du Progrès Technique en Union Soviétique*. Ecole Nat. Sup. Agronomique, Montpellier.

Castillo, Geilia, T. (1975). *All in a Grain of Rice*. Southeast Regional Centre for Graduate Study and Research in Agriculture. College, Laguna, the Philippines.

Cernea, M.M. (1981). Sociological dimensions of extension organisation: the introduction of the T & V system in India. In *Extension Education and Rural Development*, Vol. 2, ed. B.R. Crouch and S. Chamala. John Wiley & Sons, Chichester, New York, Brisbane, Toronto, pp. 221–35.

Cernea, M.M., Coulter, J.K. & Russell, F.A. (1985). *Research–Extension–Farmer*. A World Bank and UNDP Symposium. World Bank, Washington, DC.

Chambers, R. & Belshaw, D. (1973). *Managing Rural Development*. Institute of Development Studies, University of Sussex, Brighton.

Chang, C.W. (1969). The situation, problems and trends in agricultural education and training in Asia and the Far East Region. Regional paper, *World Conference on Agricultural Education and Training*. FAO, UNESCO and ILO, Copenhagen.

Chen, C. (1981). Effective extension approaches applied in Taiwan. In *Extension Education and Rural Development*, Vol. 2, ed. B.R. Crouch and S. Chamala. John Wiley & Sons, Chichester, New York, Brisbane, Toronto, pp. 33–40.

Cherry, Mary (1987). Communication in agriculture. In *Communication in Agriculture*. Symposium, Royal Agricultural College, Cirencester, pp. 1–7.

CIMMYT (1979). *Demonstration of an Interdisciplinary Approach to Planning Adaptive Agricultural Research Programmes*. Eastern Africa Economics Programme, Nairobi.

Claar, R.D. & Bentz, R.P. (1984). Organisational design and extension administration. In *Agricultural Extension: A Reference Manual*, ed. B.E. Swanson, FAO, Rome, pp. 161–85.

Colle, R.D., Milton, J.E., Taylor, Ellen & Berman, P. (1979). *Paraprofessionals in Rural Development*. Concept Paper. Centre for International Studies, Cornell University, Ithaca, NY.

Committee on African Development Strategies (1985). *Compact for African Development*. *INTERPAKS*, **4**(1).

Cummings, R.W. (1971). Agricultural research and technology. In *Behavioural Change in Agriculture,* ed. J.P. Leagans and C.P. Loomis, Cornell University Press, Ithaca, NY, pp. 79–83.

Cusack, D.F. (ed.) (1983). *Agroclimate Information for Development: Reviving the Green Revolution.* Westview Press, Boulder, Co.

Darling, H.S. (1970). The place and function of the university faculty in the national research programme. In *The Organisation and Methods of Agricultural Research,* ed. J.B.D. Robinson, Ministry of Overseas Development, London, pp. 22–28.

Dasgupta, B. (1977). *Agrarian Change and New Technology in India.* UNRISD, United Nations, Geneva.

Denning, G.L. (1985). Integrating agricultural extension programmes with farming systems research. In *Research–Extension–Farmer,* ed. M.M. Cernea, J.K. Goulter and F. Russell. A World Bank and UNDP Symposium. World Bank, Washington, DC, pp. 113–35.

Dey, J. (1981). Gambian women: unequal partners in rice development projects. *J. of Dev. Studies,* **17**(3), 109–22.

Dirección de Planeación (1983). *Transferencia de Tecnologia Agropecuaria en Colombia Con Enfasis en el ICA.* ICA, Bogotá.

Doorman, F. & Cuevas Pérez, F. (1987). Relationships between research, extension and farmers in the Dominican Republic. In International Workshops on *Agricultural Research Management.* ISNAR, The Hague, pp. 83–92.

Doyle, J.J., Gandin, B.E., & Paling (1985). Women and agricultural technology: understanding the needs of both the direct users and the ultimate beneficiaries of ILRAD's research. In *Women and Agricultural Technology: Relevance for Research.* Report from the CGIAR Inter-Centre Seminar on Women and Agricultural Technology. Rockefeller Foundation and ISNAR. The Hague, pp. 115–22.

Dumont, R. (1962). Accelerating African agricultural development. *Impact of Science on Society,* **12,** 231–53.

Elkana, Y. (1986). What will advisory services be like in the context of the future evolution of agriculture? The Israeli viewpoint. Paper presented at the Seminar on the Future of the Advisory Services, Louvain, Commission of the European Communities.

Elkana, Y, (1987). The extension worker and the farmer. In *Communication in Agriculture.* Internal Symposium Royal Agricultural College, Cirencester, pp. 17–27.

Elkana, Y. & Sagiv, Y. (1981). Agricultural extension service in Israel. *Kidma,* **22**(2), 7–14.

Elz, D. (ed.) (1984). *The Planning and Management of Agricultural Research.* World Bank, Washington, DC.

Engel, P. (1976). *Agrarische Struktuur en Verspreiding van Vernieuwingen Scriptie Voorlichtingskunde.* Agricultural University, Wageningen.

Evenson, R.E. (1985). The economics of extension. Paper presented at the Reading Conference. Economic Growth Centre, Yale University.

FAO (1962). *Development Centre for East, Central and Southern Africa.* ETAP Report No. 1566. ETAP, Rome.

FAO (1971). *Rural Extension in Latin America and the Caribbean*. Rome.

FAO (1975). *Agricultural Extension and Training*. Rome.

FAO (1976). *The State of Food and Agriculture, 1975*. Rome.

FAO (1977). *Monthly Bulletin of Agricultural Economics and Statistics*, **26,** No. 9.

FAO (1984). Women in developing agriculture. In *The State of Food and Agriculture*. Rome, pp. 107–55.

FAO (1986*a*). Proceedings of the *Research Extension Linkage* Workshop, Hanoi.

FAO (1986*b*). *Population and the Labour Force in Rural Economies*. Economic and Social Development. Paper 59. FAO, Rome.

Farrington, J. & Howell, J. (1987). *The Organisation and Management of Agricultural Research. Current Research Issues*. ODI Discussion Paper 20, London.

Farrington, J. & Martin, Adrienne (1987). *Farmer Participatory Research: A Review of Concepts and Practices*. ODI, Discussion Paper 19. London.

Feder, G., Slade, R.H. & Sundaram, A.K. (1985). *The Training and Visit Extension System: An Analysis of Operation and Effects*. World Bank, Washington, DC.

Fernandez, F. (1977). Objectives and content of training at the international centers of agricultural research. Paper presented at the CGIAR Forum on Training. Washington, DC.

Fletcher, J.T. (1986). Crop protection: the role of the agricultural development and advisory service in England and Wales. In *Advisory Work in Crop Disease and Pest Management,* ed. J. Palti and R. Auscher. Springer-Verlag, Berlin, Heidelberg, New York, Tokyo, pp. 177–93.

Fliegel, F.C. (1984). Extension in communication and the adoption process. In *Agricultural Extension: A Reference Manual,* ed. B.E. Swanson, FAO, Rome, pp. 77–88.

Fraser, C. (1987). Communication for agriculture in Italy. A story of needs and opportunities. In *Communication in Agriculture* International Symposium, Royal Agricultural College, Cirencester, pp. 55–67.

Gabriel, T. (1987). Agricultural advice and extension. In *Communication in Agriculture*. International Symposium Royal Agricultural College, Cirencester, pp. 9–16.

Galjart, B.F. (1971). Agricultural development and social concepts: a critique. *Rural Sociology*, **36**(1), 31–42.

Garforth, C. (1982). Reaching the rural poor: a review of extension strategies and methods. In *Progress in Rural Extension and Community Development,* Vol. 1, ed. G.E. Jones & M. Rolls. John Wiley & Sons, Chichester, New York, Brisbane, Toronto, pp. 43–70.

Ghonemy, El. M.R. (1984). *Development Strategies for the Rural Poor*. FAO Economic and Social Paper 44. FAO, Rome.

Gomez, A. (1985). A farming system's approach to identifying farmers' production problems. In *Research–Extension–Farmer*, ed. M.M. Cernea, J.K. Coulter and F. Russell. A World Bank & UNDP Symposium. The World Bank, Washington, DC, pp. 63–70.

Griffin, K. (1973) Policy options for rural development. In *Rural Development and Employment,* ed. E.D. Gotch, Ford Foundation Seminar, Ibadan, pp. 18–79.

Havens, A.E. & Flinn, W.L. (1975). Green revolution technology and community development: the limits of the programs. *Econ. Dev. and Cultural Change,* **23,** 469–81.

Haverkoort, A.W. (1976). Met peilen van de voorlichtings behoefte. *Bedrÿfsontwickeling,* **7**(11), 802–12.

Hayward, J.A. (1987). *Commitment to Agricultural Extension.* Address presented at the 35th International Course on Rural Extension, Wageningen.

Herzberg, J. (1975). *Staatliche Landliche Beratungsdienste als Instrument der Entwicklungsförderung in Südamerika.* D.L.G. Verlag Frankfurt (Main).

Herzberg, J. & Antuna, S. (1973). *Analytical Study of the Extension Services for Ecuador and Paraguay.* FAO, Rome.

Hildebrand, P.E. (1985). Researcher–farmer linkage for technology and agricultural development. In *Agricultural Research Policy and Organisation in Small Countries:* Agricultural University Wageningen, pp. 70–75.

Hildreth, R.J. (1965). Tensions between research and extension workers—three hypotheses. *J. Farm Econ.,* **47,** 838–40.

Holdcroft, L.E. (1982). The rise and fall of community development in developing countries, 1950–1965. A critical analysis and implications. In *Progress in Rural Extension and Community Development.* Vol. 1, ed. G.E. Jones and M. Rolls. John Wiley & Sons, Chichester, New York, Brisbane, Toronto, pp. 207–31.

Hong, P. (1981). The joint farming operation in Taiwan: a new approach to effective co-operative food production. In *Extension Education and Rural Development,* Vol. 2, ed. B.R. Crouch and S. Chamala. John Wiley & Sons, Chichester, New York, Brisbane, Toronto, pp. 87–107.

Hopkins, J. (1969). *The Latin American Farmer.* USDA, Washington, DC.

Howell, J. (1982). *Managing Agricultural Extension: The T & V system in Practice.* Discussion Paper 8. Overseas Development Institute, London.

Howell, J. (1984). *Conditions for the Design and Management of Agricultural Extension.* Agricultural Administrative Unit, ODI, London.

Hsieh, S.C. & Lee, T.H. (1966). *Agricultural Development and its Contribution to Economic Growth in Taiwan.* Joint Commission on Rural Reconstruction, Taipei.

Hunter, G. (1970). *The Administration of Agricultural Development: Lessons from India.* Oxford University Press, London.

ILO (1982). *Yearbook of Labour Statistics.* Geneva.

INTERPAKS (1985). Ambivalent US lessons in pre-extension history. *INTERPAKS Interchange,* **2**(3), 1–3.

ISNAR (1981*a*). *Kenya's National Agricultural Research System.* The Hague.

ISNAR (1981*b*). *The Agency for Agricultural Research and Development in Indonesia.* The Hague.

ISNAR (1982). *Annual Report.* The Hague.

ISNAR (1985), *Serving National Agricultural Research Systems.* Lessons from Country Experiences. 1980–1984. The Hague.

Jain, T.C. (1985). Constraints on research-extension linkages in India. In *Research–Extension–Farmer.* A World Bank and UNDP Symposium. World Bank, Washington, DC.

Jesudasan, V., Roy, P. & Koshy, T.A. (1980). *Non-formal Education for Rural Women*, Allied Publishers, New Delhi.

Jones, G.E. (1982). The Clarendon letter. In *Progress in Rural Extension and Community Development*, Vol. 1, ed. G.E. Jones and M.J. Rolls. John Wiley & Sons, Chichester, New York, Brisbane, Toronto, pp. 11–19.

Kaimowitz, D. (1987). Research—technology transfer linkages. In *International Workshop on Agricultural Research Management* ISNAR, The Hague, pp. 109–13.

Kumar, S.K. (1985). Women in agriculture in Sub-Saharan Africa. In *Women in Agricultural Technology*. Rockefeller Foundation and ISNAR, The Hague, pp. 169–89.

Lacy, W.B., Pigg, K., & Busch, L. (1980). Clients, colleagues, and colleges. Perceived influences on extension agents. *Rural Sociology*, **45**, 469–82.

Lecomte, R. (1964). *Relations entre la Recherche Agronomique et la Vulgarisation. Situation en Belgique*. Report FAO *ad hoc* Meeting on Relationships between Agricultural Research and Agricultural Practice. Berne.

Leonard, D.K. (1977). *Reaching the Peasant Farmer, Organisation Theory and Practice in Kenya*. University of Chicago Press, Chicago and London.

Levesque, A. (1961). *Etude sur la Marche de la Vulgarisation*. Synergic-Roc, Paris.

Lionberger, H.F. & Chang, H.C. (1981). Development and delivery of scientific farm information: the Taiwan system as an organisational alternative to Land-grant Universities—US style. In *Extension Education and Rural Development*, Vol. 1, ed. B.R. Crouch and S. Chamala. John Wiley & Sons, Chichester, New York, Brisbane, Toronto, pp. 155–83.

Lu, H. (1968). Some socio-economic factors affecting the implementation at farm level of a rice production programme in the Philippines. PhD Thesis, University of the Philippines.

Lüning, H.A. (1982). The impact of technological change on income distribution in low-income agriculture. In *Progress in Rural Extension and Community Development*. Vol.1, ed. G.E. Jones and M.J. Rolls. John Wiley & Sons, Chichester, New York, Brisbane, Toronto, pp. 21–42.

McAllister, J. (1981). Rural innovators: a struggle for power. In *Extension Education and Rural Development*. Vol.1, ed. B.R. Crouch and S. Chamala. John Wiley & Sons, Chichester, New York, Brisbane, Toronto, pp. 135–45.

McCallum, A. (1981). *Why Improve the Organisation and Administration of Agricultural Development in the Near East?* Report of a Regional Expert Consultation held in Nicosia. FAO, Rome.

McDermott, J.K. (1987). Making extension effective: the role of extension/research linkages. In *Agricultural Extension Worldwide. Issues, Practices and Emerging Priorities*, ed. W.M. Rivera and Susan G. Schramm, Croom Helm, London, New York, Sydney, pp. 89–99.

McKillop, B. (1981). Role of the change agent in Papua New Guinea. In *Extension Education and Rural Development*, Vol. 2, ed. B.R. Crouch and S. Chamala. John Wiley & Sons, Chichester, New York, Brisbane, Toronto, pp. 123–29.

Malavolta, E. & Rocha, M. (1981). Recent Brazilian experience on farmer reaction and crop response to fertilizer use. In *Transferring Technology for Small-Scale Farmers*, ed. N.R. Usherwood. Spec. Publ. 41. Am. Soc. Agron. Madison, Wisc., pp. 101–13.

Malone, C.C. (1965). Some responses of rice farmers to the package program in Tanjore District, India. *J. Farm Econ*, **47**, 256–68.

Malone, V.M. (1984). In-service training and staff development. In *Agricultural Extension—A Reference Manual*, ed. B.E. Swanson. FAO, Rome, pp. 206–17.

Mathur, H.M. (1983). Improving the performance of agricultural development administration. Paper presented at Expert Consultation on Organisation and Management Structures for Rural Development. FAO, Rome.

Moris, J.R. (1987). Incentives for effective agricultural extension at the farmer/ agency interface. In *Agricultural Extension Worldwide. Issues, Practices, and Emerging Priorities*, ed. W.H. Rivera and Susan G. Schramm, Croom Helm. London, New York, Sydney, pp. 199–224.

Morrs, E.R., Hatch, J.K., Mickelwait, D.R. & Sweet, C.F. (1978). *Strategy for Small Farmer Development: An Empirical Study of Rural Development*. Agency for International Development, Washington, DC.

Nagel, U.J. (1979). Knowledge flows in agriculture: linking research, extension and the farmer. *Zeitschrift für Auslandische Landwirtschaft*, **18**(2), 135–50.

Nour, M.A. (1969). The situation, problems and trends in agricultural education and training in the African region. Regional Paper, World Conference on *Agricultural Education and Training*. FAO, UNESCO & ILO, Copenhagen.

OECD (1981). *Agricultural Advisory Services in OECD Member Countries*. Paris.

Onyango, C.A. (1987). Making extension effective in Kenya: the district focus for rural development. In *Agricultural Extension Worldwide. Issues, Practices and Emerging Priorities*, ed. W.H. Rivera and S.G. Schramm. Croom Helm, London, New York, Sydney, pp. 149–61.

Padmanagara, S. (1985). The joint formulation of extension messages by research and extension staff in Indonesia. In *Research–Extension–Farmer*. A World Bank and UNDP Symp. World Bank, Washington, DC, pp. 136–43.

Perrin, R. & Winkelman, D. (1976). Impediments to technical progress on small versus large farms. *Am. J. Agric. Econ*, **58**, 888–94.

Pickering, D.C. (1983). Agricultural extension: a tool for rural development. In *Agricultural Extension by Training and Visit*, ed. M.M. Cernea, J.K. Coulter and F. Russell. World Bank, Washington, DC, pp. 3–13.

Pickering, D.C. (1987). An overview of agricultural extension and its linkages with agricultural research: the World Bank experience. In *Agricultural Extension Worldwide. Issues, Practices and Emerging Priorities*, ed. W.H. Rivera and Susan G. Schramm, Croom Helm. London, New York, Sydney, pp. 66–74.

Piñeiro, M., Fiorentina, R., Trigo, E., Balcazar, A., & Martinez, Astrid (1982). *Articulación Social y Cambio Tecnico en la Produción de Azucar en Colombia*. Instituto Interamericano de Cooperación Agricola. San José.

Pinstrup-Andersen, P. (1982). *Agricultural Research and Technology in Economic Development*. Longman Inc. New York.

Qasem, S. (1985). Research policy linkages. In *Agricultural Research Policy and Organisation in Small Countries.* Agricultural University, Wageningen, pp. 58–63.

Ramakrishnan, S. (1983). *The Importance of Orienting Rural Development Organisations to Service Small Farmers.* Expert Consultation on Organisation and Management Structures for Rural Development. FAO, Rome.

RAPA/FAO (1985). Report of the Expert Consultation of *Linkages of Agricultural Extension with Research and Agricultural Education.* FAO Regional Office for Asia and the Pacific, Bangkok.

Rice, E.B. (1974). *Extension in the Andes.* The MIT Press, Cambridge, Mass.

Rivaldo, O.F. (1987). Strategies for strengthening the Brazilian agricultural research system. In *The Impact of Agricultural Research on National Agricultural Development.* ISNAR, The Hague, pp. 161–81.

Rivera, W.M. (1987). India's agricultural extension development and the move toward top-level management training. In *Agricultural Extension Worldwide. Issues, Practices and Emerging Priorities,* ed. W.H. Rivera and Susan G. Schramm. Croom Helm, London, New York, Sydney, pp. 225–50.

Roberts, N. (1987). Successful agricultural extension: its dependence upon other aspects of agricultural development. The case of public sector extension in North-East Africa. In *Agricultural Extension Worldwide. Issues, Practices, and Emerging Priorities,* ed. W.H. Rivera and Susan G. Schramm. Croom Helm, London, New York, Sydney, pp. 75–88.

Rockefeller Foundation and ISNAR (1985). *Women and Agricultural Technology: Relevance for Research.* Report from the CGIAR Inter-Centre Seminar on Women and Agricultural Technology. The Hague.

Rogers, E.M. (1962). *Diffusion of Innovations.* The Free Press, Glencoe, Ill.

Rogers, E.M. (1983). *Diffusion of Innovations,* 3rd Edition. The Free Press, Glencoe, Ill.

Rogers, E.M. & Svenning, L. (1969). *Modernization among Peasants.* The Free Press Glencoe, Ill.

Rogers, E.M. & Yost, M.D. (1960). *Communication Behaviour of County Extension Agents.* Ohio Experiment Station, Bulletin. 850. Wooster.

Rogers, W.L. (1987). The private sector: its extension systems and public –private coordination. In *Agricultural Extension Worldwide,* ed. W.M. Rivera & Susan G. Schramm, Croom Helm, London, New York, Sydney, pp. 13–21.

Röling, N. (1982). Alternative approaches in extension. In *Progress in Rural Extension and Community Development.* Vol. 1, ed. G.E. Jones and M.J. Rolls. John Wiley & Sons, Chichester, New York, Brisbane, Toronto, pp. 87–115.

Röling, N. (1984). Agricultural knowledge: its development, transformation, promotion and utilisation. In *Proceedings of 6th European Seminar on Extension Education,* ed. R. Volpi and F.M. Santucci. Centro Studi Agricoli, Borogo a Mazzano, Italy, pp. 124–205.

Röling, N. (1986). The structure of advisory services in the context of the future evolution of agriculture: Conceptual aspects. Paper presented at the Seminar on the Future of the Advisory Services, Louvain. Commission of the European Communities.

Röling, N. (1987). *Extension Science—Information Systems in Agricultural Development.* Agricultural University, Wageningen.

Röling, N., Ascroft, J. & Chege, F.W. (1981). The diffusion of innovations and the issue of equity in rural development. In *Extension Education and Rural Development*, Vol. 1, ed. B.R. Crouch and S. Chamala, John Wiley and Sons, Chichester, New York, Brisbane, Toronto, pp. 87–115.

Röling, N. & De Zeeuw, H. (1983). *Improving the Quality of Rural Poverty Alleviation*. Final Report of the Small Farmer and Development Cooperative. IAC, Wageningen.

Rossillion, C. (1967). Youth services for economic and social development: a general review. *Int. Lab. Rev,* **95**(4), 1–12.

Rubenstein, A.H. (1957). Liaison relations in research and development. In *Human Relations in Industrial Research Management,* ed. R. Livingston and S.H. Milberg, Columbia Univ., New York, pp. 222–40.

Ruttan, V.W. (1982). *Agricultural Research Policy*. University of Minnesota Press, St. Paul, Mn.

Sabatier, P. & Mazmanian, A. (1978). *The Conditions of Effective Implementation: A Price to Accomplishing Policy Objectives*. University of California, Davis.

Sadikin, S.W. (1983). Overcoming technology gaps. In *Agricultural Research for Development: Potentials and Challenges for Asia,* ed. B. Nestel. ISNAR, The Hague, pp. 28–33.

Safrenko, A.J. (1984). Introducing technological change: the social setting. In *Agricultural Extension–A Reference Manual,* ed. B.E. Swanson. FAO, Rome, pp. 56–76.

Sakson, N. (1986). Chief characteristics of extension in Poland. *INTERPAKS Exchange,* **3**(3), 1–2.

Shabtai, S.H. (1975). Army and economy in tropical Africa. *Econ. Dev. and Cultural Change,* **23**, 687–701.

Stavis, B. (1978). Agricultural research and extension services in China. *World Development,* **6**, 631–45.

Stevens, Yvette (1984). *Technologies for Rural Women's Activities—Problems and Prospects in Sierra Leone*. ILO, Geneva.

Stevenson, R. (1984). Agricultural extension at a crossroads? In *Proceedings, 6th European Seminar on Extension Education,* ed. R. Volpi and F.M. Santucci. Borogo a Mazzano, Italy, pp. 36–39.

Streeter, C.P. (1972). *Colombia. Cambios en la Agricultura, el Hombre y los Métodos*. The Rockefeller Foundation, New York.

Sukaryo, D.G. (1983). Farmer participation in the Training and Visit system and the role of the village extension worker: experience in Indonesia. In *Agricultural Extension by Training and Visit. The Asian Experience,* ed. M.M, Cernea, J.K. Coulter & J.F.A. Russell, World Bank, Washington, DC, pp. 18–25.

Swanson, B.E. (1984). ed. *Agricultural Extension—A Reference Manual*. FAO, Rome.

Swanson, B.E. & Claar, J.B. (1984). The history and development of agricultural extension. In *Agricultural Extension—A Reference Manual,* B.E. Swanson, FAO, Rome, pp. 1–19.

Swanson, B.E., Röling, N. & Jiggins, Janice (1984). Extension strategies for technology utilization. In *Agricultural Extension—A Reference Manual,* ed. B.E. Swanson. FAO, Rome, pp. 89–107.

Tendler, J. (1982). *Rural Projects through Urban Eyes: An Interpretation of the World Bank's New Style Rural Development Projects.* Staff Working Paper 532. World Bank, Washington, DC.

Thomason, I.J. & Toscano, N.C. (1986). Plant protection advisory work in the University of California Cooperative Extension Service. In *Advisory Work in Crop Pest and Disease Management,* ed. J. Palti and R. Auscher, Springer-Verlag, Berlin, Heidelberg, New York, Toronto, pp. 220–30.

Trigo, E. Piñeiro, M. & Ardila, L.J. (1982). *Organización de la Investigación Agropecuaria en America Latina.* IICA, San José.

True, A.C. (1928). *A History of Agricultural Extension Work in the United States 1785–1923.* USDA Misc. Publ. 36. Washington, DC.

Tully, Joan, (1981). Changing practices: a case study. In *Extension Education and Rural Development.* Vol. 2, ed. B.R. Crouch and S. Chamala. John Wiley & Sons, Chichester, New York, Brisbane, Toronto, pp. 79–86.

UNECA/FAO (1974). *The Role of Women in Population Dynamics Related to Food and Agriculture and Rural Development in Africa.* UNECA, Addis Ababa.

USAID (1983). *Strengthening the Agricultural Research Capacity of the Less Developed Countries: Lessons from Aid Experience.* Washington, DC.

Van Heck, B. (1979). *Participation of the Poor in Rural Organisation.* FAO, Rome.

Verkatesan, V. (1985). Policy and institutional issues in improving research–extension linkages in India. In *Research–Extension–Farmer.* World Bank and UNDP, Washington, DC, pp. 13–27.

Vinyu, V.V. (1979). The role of research in solving problems of the developing countries: a Third World view. In *Give Us the Tools,* ed. D. Spurgeon. Intern. Development Research Centre, Ottawa, pp. 177–88.

Volpi, R. (1982). Borgo a Mazzano and the problems of agricultural extension. In *Progress in Rural Extension and Community Development,* Vol. 1, ed. G.E. Jones & M. Rolls, John Wiley & Sons, Chichester, New York, Brisbane, Toronto, pp. 189–206.

Watts, L.H. (1984). The organisational setting for agricultural extension. In *Agricultural Extension—A Reference Manual,* ed. B.E. Swanson. FAO, Rome, pp. 20–39.

Wharton, C.R. (1965). Education and agricultural growth: the role of education in early stage agriculture. In *Education and Economic Development,* ed. C.A. Anderson and M.J. Bowman. Aldine Press, Chicago.

Winkelmann, D. & Moscardi, E. (1979). *Aiming Agricultural Research at the Needs of Farmers.* CIMMYT, Londres, Mexico.

World Bank (1975). *Rural Development: Sector Paper.* Washington, DC.

World Bank (1983). *Strengthening Agricultural Research and Extension: The World Bank Experience.* Washington, DC.

Wotowice, P. *et al.* (1986). *Research Recommendation and Diffusion Domains: A Farming System Approach to Targetting.* Conference on Gender Issues in Farming Systems Research and Extension. University of Florida Gainesville.

Wyckoff, J.B. (1965). Closer cooperation between research and extension. *J. Farm Econ,* **47**, 3834–37.

Yen, S.S.T. (1985). *General Aspects of Agricultural Extension Programmes in China.* Proc. on Agriculture in China, Inst. Intern. Develop. and Education in Agriculture and Life Sciences.

Youman, D. & Holland, D. (1983). Extending FSR Results in Lesotho. *FSSP Newsletter* **1**(3), 11–12.

Youssef, N. & Hetler, Carol, B. (1984). *Rural Households Headed by Women. a Priority Concern for Development.* World Employment Programme Research Paper. ILO, Geneva.

Yudelman, H., Butler, G. & Banerji, R. (1971) *Technological Change in Agriculture and Employment in Developing Countries.* Development Centre OECD, Paris.

Zhou, F. (1987). Organisation and structure of the national agricultural research system in China. In International Workshop on *Agricultural Research Management,* ISNAR, The Hague, pp. 219–23.

Zuurbier, P.J.P. (1983). *De Relatie Tussen het Landbouwkundig Onderzoek, de Voorlichting en de Boer in Nederland.* Minis. Landbouw en Visserÿ/BSA, The Hague.

Index